Wastewater Treatment Systems

Wastewater Treatment Systems

Modelling, Diagnosis and Control

Gustaf Olsson
Lund University
Sweden

Bob Newell
The University of Queensland
Australia

British Library Cataloguing in Publication Data

A CIP catalogue record for this book is available from the British Library
ISBN 1 900222 15 9

The paper used in this publication meets the requirements of ANSI/NISO Z39.48-1992 (Permanence of Paper)

Typeset by the authors using the LATEX typesetting system
Printed in the UK by The Book Company, Ipswich

First edition published 1999
10 9 8 7 6 5 4 3 2 1

Published by IWA Publishing,
Alliance House, 12 Caxton Street, London SW1H 0QS, UK

Preface

Writing technical textbooks is a masochistic thing to attempt. We think we may have survived - though only time will tell. There are of course numerous people to thank:

- Since they must continue to live with us, we first thank our long-suffering families - Kirsti in Sweden (and our now grown-up children), Joyce and Janet in Australia (and also Craig who escaped to the United States).
- Then there are our colleagues with whom we share coffee breaks and long discussions - Ulf Jeppsson, Christian Rosén, Sven-Göran Bergh, Morten Hemmingsson, Gunnar Lindstedt in Industrial Electrical Engineering and Automation (IEA) at Lund University, Bengt Carlsson at Uppsala University and Ian Cameron and Christine Smith in the Computer Aided Process Engineering (CAPE) Centre at The University of Queensland.
- Tova Lidbeck made a special contribution by suffering through a tour of the Ryaverket treatment plant in Göteborg, Sweden and then producing some excellent water-colors for the book cover and to start each part of the book.
- Gustaf specially thanks Professor Emeritus John F. Andrews, USA, for sharing so much of his experience and knowledge of wastewater.
- Thanks to Sydney Water (Australia), Loganholme Wastewater Treatment Plant, Queensland (Australia), SYVAB - Himmerfjärdsverket, (Sweden), GRYABB - Ryaverket, Göteborg (Sweden), Käppalaverket, Lidingö (Sweden), Malmö Sewage Works Sjölunda, Malmö (Sweden), Ronneby Wastewater Treatment Plant (Sweden), The Swedish Water and Waste Water Works Association, and Ann Arbor Wastewater Treatment Plant, Ann Arbor, Mich. (USA) for data used in the examples and exercises of the book.
- Last but by no means least are the many students who suffered through courses in several countries attempting to fathom draft chapters.

Then there is the equipment (it is now many computers later) and the software (a few different operating systems, several flavours of LaTeXand numerous graphics programs).

Thanks are also due to a couple of hideaways in the Swedish forest and in the Australian bush which in no small way made such an undertaking possible.

Gustaf Olsson
Bob Newell
July, 1999

Contents

Chapter 1

Introduction

What is an introduction? We can introduce the book itself to the reader. We can introduce the topic of the book. We can introduce the parts of the book. We can make suggestions on how the book might be used. And of course we have many types of readers to consider. Well we will try to do all these things for all you readers.

1.1 Introducing the Book

This is a book about the control of biological wastewater treatment plants. So what is control?

Figure 1.1 presents a conceptual view showing the four components of any control system: the process, the measurement, the decision-making and the implementation.

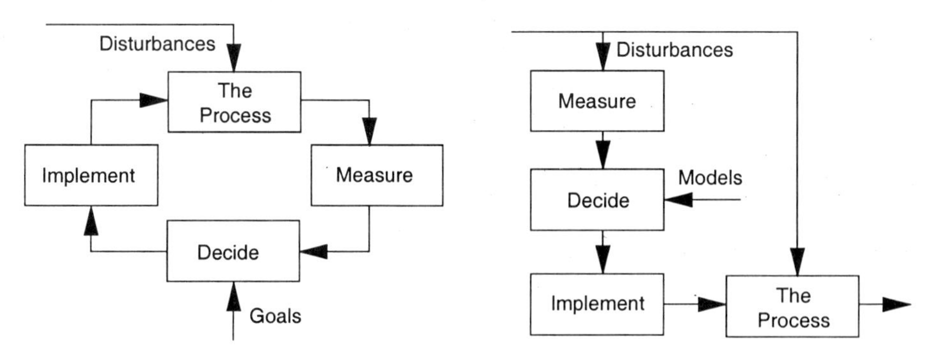

Figure 1.1: Feedback and feedforward control systems.

This conceptual view is applicable to any process and any form of control. At one extreme, it may be the manager measuring by "gut feel", deciding heuristically and implementing by memo. At the other extreme, it may be a computer measuring by sophisticated chemical analyser, deciding by a model-based optimisation algorithm and implementing via automatic process instrumentation, untouched by human hand. In both cases they are guiding a process to achieve specified goals despite disturbances from the surrounding environment. This classical view on the left in Figure 1.1 is called "feedback", since we feed information on the status of the process back to the inputs to steer it where we wish to go.

We also utilise the other basic concept, that of "feedforward" compensation, shown on the right in Figure 1.1. This is where we measure the disturbances (when we can) and take anticipatory actions to compensate or cancel out the effects of the disturbance. We need models of the effects of the disturbance and the manipulated variable to do this. Feedback control in the shower is feeling the temperature and adjusting the tap, while feedforward is hearing the toilet flush and adjusting the tap in anticipation. It is better to fix the problem in this way before it occurs, but we need a model to tell us how much compensation is required. We usually combine feedforward for fast anticipatory response, where possible, and feedback to take care of model inaccuracies in our compensation, and the disturbances which we do not measure.

The book will explore both of these concepts, implemented by simple controllers, sophisticated controllers and by human supervision.

The key component of control is the process. Our process is the biological wastewater treatment plant. The emphasis is on the activated sludge process for municipal wastewater, also often called the biological nutrient removal (BNR) process. There is some discussion of processes using fixed biological films and of anaerobic digestion. The principles are equally applicable to industrial wastewaters, though these are not covered specifically.

We cannot control what we do not understand. That is why the first part of the book - Modelling - is about the process. The second part - Diagnosis - is about so-called open loop control, doing it with a human in the loop. It includes *detection* to find out what is going on in the process. The *diagnosis* is the procedure to find the cause of a disturbance in the process. Once this is done, a human operator may correct for the disturbance. The third part - Control - is about so-called closed loop control, doing it without a human in the loop. It has also been said that we cannot control what we cannot measure, hence the fourth part of the book - Instrumentation - which discusses sensor technology and measurement principles. The last part - The Future - is the authors putting their heads on the chopping block, predicting the unpredictable.

It is instructive to examine the proceedings of a workshop held at Clemson University in the USA in 1974 entitled "Research Needs for Automation of Wastewater Treatment Systems" (Buhr *et al.* (1975)). It was a gathering of 104 US wastewater industry people, contractors and academics with a few key overseas visitors. Wastewater plants had only recently considered sensors, computers, automation and control and the workshop reviewed what had been done and what was needed to progress. It is depressing reading in that most of the needs identified are still needs twenty five years later. Why progress has been painfully slow is discussed below and in more detail in the book. The needs identified at the workshop included:

- communication between researchers and the industry,
- training in a process systems approach,
- development of simple effective dynamic models,
- techniques for model validation, and
- development of control strategies.

Though some progress has been made, these needs are still largely there today. That is why we think it was worth the pain of writing this book. We exaggerate: sometimes it was fun. If we could remember, we would tell you when that was. Because the meeting of wastewater plants and a systems approach to modelling and control is relatively recent, there will be four classes of readers:

1. Those expert in control and wastewater treatment - a few of you, but you won't buy the book anyway since you know it, but you never know.

2. Those expert in control but not wastewater - you should read the first part on the process entitled Modelling and browse the examples.

3. Those new to control but expert in wastewater - you can read the parts called Diagnostics and Control and maybe even Sensors.

4. Those new to both control and wastewater - read it all, and good luck. If you are a student there are exercises too.

You can all read the last part - The Future - for a laugh. And if you read it again in a few years time, it will be even more amusing.

For the computer literate, there is also a collection of MATLAB programs and functions that are mentioned throughout the book. They will run on both the professional and student versions of MATLAB. They will enhance your learning and be a bit of fun too. You can do things to the process you would not dare to at the plant, or even in the laboratory. They might even be useful as aids to process operation, especially the Toolbox of general functions, but also the simulations, but beware. Read the book carefully first, especially the chapters on "Model Fitting", before applying the tools - tools must be used properly. Section 1.3 below introduces the five major parts of the book in a little more detail.

1.2 Introducing the Topic

1.2.1 Why Control?

"Why control a wastewater plant? Mine has been running fine!" Perhaps! Water has been going in and coming out the other end, most of it, and it looks pretty good when it comes out. At least, it looks better than when it went in. But how good is it? Will it meet the regulations and the scrutiny of tomorrow? Can you trim your operating costs? Will your plant handle the inevitable load increases? How big a load will it take before more volume is needed?

Effluent standards will get tighter. There are even indications in some countries that tomorrow's regulations must be met on the basis of spot checks, not monthly averages! Or maybe a group of your plants has a combined annual discharge limit, and you must decide the best way to tune your plants! Failure to comply will cost dearly, especially as wastewater treatment and government regulators become increasingly independent! Even recurrent budgets are getting tighter, and will get even tighter. It is not so easy to get capital now either! And when you have to fight harder for resources, perhaps even from a bank, a good public image can help.

It is all about the bottom line. Advanced control is not the answer to all your prayers, but it can help. It has helped lots of others. Real results have shown a reduction of two and more in quality variations. A series of diverse case studies showed an average of 6 per cent saving in operating costs and remarkably short payback times (Marlin *et al.* (1987a)).

"That advanced control might work in an oil refinery, but my process is different!" A familiar refrain to a control engineer. Strangely, you hear it from all sorts of industries, even some oil refineries who point the finger somewhere else.

Is the activated sludge process really different? Like almost any process, the answer is NO and YES. No, all processes can be controlled to perform better. It is the process knowledge, the sensor technology and the way the plants have been designed and built that may limit what can be achieved today. But, where there is a will, there is a way. Yes, wastewater processes do have some unique features:

- the daily volume of wastewater treated can be huge,
- the disturbances in the influent are enormous compared to most industries,
- the influent must be accepted and treated, there is no returning it to the supplier,
- the concentrations of nutrients (pollutants) are very small, even challenging sensors,
- the process depends on microorganisms, which have a definite mind of their own,
- the reliable separation of the effluent from the biomass is challenging, and
- the dollar value of the product in the marketplace is remarkably low.

Of course other processes have their unique features too. Crude oil has a zillion components, some crude oils are like smelly water, and some have to be jack-hammered out if they cool. The components include carcinogens and worse, and are highly flammable or explosive. Some of you know what **they** have in their wastewater.

What are really different are the attitudes and incentives of the different industries. Of course the attitudes often depend on the incentives. Wastewater, food, and minerals all claim to be different. Really they just have not had the incentives until today to put in the groundwork that the oil industry started in the 1970s. But now they are starting, and they will progress much faster than the oil industry did, thanks in part to the oil industry. In ten years wastewater treatment plants will have as sophisticated control as the oil industry, perhaps even more sophisticated because of those differences we listed above.

1.2.2 Incentives

An IAWQ workshop on incentives for and constraints on Instrumentation, Control and Automation (ICA) in Wastewater Treatment (Olsson (1993)) listed the following incentives:

Effluent quality standards. An increasing public awareness will be an efficient driving force. The formation of national water quality management strategies will support the development of better wastewater treatment. In many countries there are increasingly severe penalties for pollution offences. The effluent quality and a good performance has to be guaranteed consistently, which will have a great impact on instrumentation, control and automation. The legislators often look at the best available technology. The effluent quality standards, however, have to be technically and economically feasible, where the need for the receiving water should be the driving force.

Economy. This may be the primary driving force. In most countries the standards are formulated as permits. Then it is a driving force to change the emphasis from compliance to compliance at least cost. In some countries the authorities charge according to the effluent pollution, where different components are given their weighting factors. Whatever the rules are, there is a driving force to minimise the costs and get more out of existing systems. In particular one has to look at savings in energy, chemical, and personnel costs.

An improved operation by ICA may also lead to deferred capital expenditure. This is likely the most significant economic factor.

To extend a plant to larger capacity by traditional design methods will be too expensive in the long run. Instead, process dynamics and operational methods ought to be included in the design procedure in order to make the best use of the investments. As a consequence the safety and capacity margins will decrease which requires a reliable operation. It is natural to look for the performance limits of current designs. The process has to be well understood to ensure a good control, particularly the mechanisms that govern nitrification and denitrification in a cold climate.

Recycling of process water will become an issue in the future, making wastewater treatment part of an overall production process. This means, that wastewater treatment can not be seen as non-profit. Instead this gives further incentives for source control, since raw material and products can be saved in the production industry.

Plant complexity. Plant complexity may be one of the most important driving forces. Since nutrient removal will become more common the plant complexity will grow. The need for consistent operation further emphasises the importance of ICA.

It will become more and more important to understand the impact of load variations.

The increasing quantities of sludge and decreasing opportunities for landfill makes it necessary to develop better control methods for sludge treatment and disposal processes.

Improved tools. There will be an increasing need to measure dynamic changes, not only of the influent wastewater (such as toxic warnings) but of variations in the plant itself. There has to be a proper awareness of the required sampling frequency. This will further emphasise the need for better on-line sensors. The constant improvements of sensors (like respirometers) is in itself an incentive to promote the control applications.

Most simulation models of today, describing the structured behaviour of activated sludge plants, are far too complex to be used for operation. They can not be verified by on-line plant measurements, and are consequently inadequate for predicting the plant behaviour. For advanced control it is important to obtain models with appropriate complexity. They should be updated by on-line measurements and be able to predict the plant behaviour for an appropriate time. The process simulators are becoming available, but still the operational models have to be further developed in order to make the simulators valuable for operation.

1.2.3 Constraints

The same workshop (Olsson (1993)) also listed the constraints on faster development in the use of ICA in wasterwater:

Legislation. In many countries there is a lack of a tight regulatory standard, which makes it unnecessary to tighten the control on plants. The enforcement of the effluent regulations is often quite inadequate. Many regulatory standards are not adapted to the need of the receiving water. Instead they are static and do not consider dynamic variations.

Education, training and understanding. In most countries the public awareness has been small until recently. This has caused delayed investments in better treatment facilities. Operators are not always adequately educated to deal with on-line instrumentation and control. Many of them do not know where and how to take proper samples and how to condition them. Most environmental engineers (including operating engineers as well as designing engineers) would need more basic education in dynamics in order to appreciate the need and potential for ICA.

Manufacturers and suppliers are often not aware of the need for better flexibility and control. It is too late to start thinking about the control system when the design is already fixed.

It has to be realised that the management is part of the overall control of a plant. Many managers are not technically trained, which may lead to misunderstandings and inconsistent operation of a plant.

Economy. Since the wastewater treatment industry is basically a non-profit industry, there have been quite unclear goals as to what ICA could do. Automation has been considered costly and has not been part of the initial design. Instead, the design has been completed first, and then ICA is introduced. This means that a proper weighting between construction and operation costs is often not achieved. An adequate use of ICA may in fact be a crucial factor that can make a plant run consistently and economically.

Measuring devices. There is a lack of reliable on-line sensors today. As an example the reliability has to be significantly increased for suspended solids meters, ammonia sensors, phosphorus sensors and chemical oxygen demand (COD) monitors. Redox meters have a considerable drift. There is a lack of fatty acids monitors for the control of anaerobic processes. In many cases operators are not sufficiently qualified to use and maintain instrumentation systems. Most sensors today do not allow long periods without extensive maintenance.

Plant constraints. There is still a lot of uncertainty about the efficiency of ICA technology. Since there is a lack of good plants demonstrating the success of ICA, many designers are still conservative. Plants are constructed with large safety margins. One looks for tried methods. However, it also means that too little flexibility and controllability is built into most plants. In fact, today there are disincentives for innovative design!

Software. Today there is a lack of software standardisation. Most software systems are proprietary and are often quite unfriendly for the user. It is consequently an educational problem to make the users competent in using the software. However, there is also a need to make software more user-friendly and flexible, so that the specific needs of a plant operation can be met.

1.3 Introducing the Parts

Here we will introduce the major parts of the book in more detail and discuss their interrelationships.

1.3.1 Modelling

This part contains eight chapters: one on wastewater treatment processes, two on dynamic modelling of these processes, one on empirical models, three on aspects of fitting models to plant data, and one on wastewater simulators. The goal of this part is to improve your process understanding, your understanding of models, and particularly how to use them correctly.

Modelling and simulation is the basis of improved control of any process. This is clear if you examine Table 1.1, taken from "Current uses for dynamic simulators" (Hudson (1991)), which lists some of the many uses being made of the results from dynamic process simulators and the models therein within the general process industries.

Models and simulation will be of interest to you if you are a practitioner or researcher dealing with the design, study or operation of dynamic systems. Wastewater treatment plants are very dynamic processes and most of the uses of models listed in Table 1.1 are potentially valuable in the design, study or operation of wastewater treatment processes.

Chapter 2 begins with a qualitative introduction to wastewater unit operations and how they are typically assembled into treatment processes, about the basic mechanisms involved (hydraulics, nutrient reactions, chemical precipitation, mass transfer, biomass growth, settling and clarification) and a brief discussion of the dynamics of these mechanisms.

Chapter 3 introduces you to basic dynamic modelling of activated sludge processes. In an easy to understand incremental fashion, dynamic models are built up from simple tanks to biological nutrient removal. The importance of model complexity and the assumptions made are discussed. The chapter is important even if you never write a model yourself.

Chapter 4 is more advanced dynamic modelling of nutrient removal processes, hydraulics, settling and clarification and batch processing. From these two chapters you can gain process understanding and an appreciation of the internals of simulators that you will come across. If nothing else you will be able to "scare the hell out of" simulator salespeople.

Chapter 5 summarises the common forms of empirical and black box models. Included are functional models (equations), network models (using neurons, radial basis functions and wavelets), and qualitative models (digraphs, rules and fuzzy relations).

- Safety and loss prevention
  - Detecting unexpected explosive mixtures during transient conditions
  - Minimising cold and hot emissions during startup and shutdown
  - Detecting hydrate-forming conditions
  - Designing and testing emergency protection systems
  - Blowdown and relief header design
- Conceptual process design
  - Feasibility analysis of novel process designs
  - Technical, economic and ergonomic evaluation of design alternatives
- Control system design
  - New or revamped plant control system design, checks on novel designs, and improvements to existing systems
  - Impact of change-over from trays to packing on the controllability of columns
  - Generating, before commissioning, optimal values for setpoints and tuning parameters
  - Generating algorithms for model-based control
  - Impact of instrument range selection and noise
  - Control valve specification and sizing
- Operational optimisation
  - Startup and shutdown scheduling
  - Early indication of the operating window and methods for expanding it
  - Minimising trips, scheduling for batch processes and pipeline distribution networks
- Troubleshooting
  - Maintaining optimal plant performance during enforced operational changes, for example, long-term equipment failures
  - Accident or failure enquiries; deciding what actually happened and investigating preventive measures
  - Removing unwanted operating characteristics
- Training and education
  - General plant operational know-how
  - Knowledge of specific plant
  - How to use digital control systems
  - Instruction in process and control system design, especially the importance of system interactions
  - Demonstrating the advantages of advanced control systems
- Control room aids
  - Equipment condition monitoring and plantwide performance optimisation
  - Testing of planned actions before actual use
  - Readily accessible self-teaching package for use during spare periods

Table 1.1: Current uses for dynamic simulators

Chapter 6 discusses the planning processes necessary to obtain good experimental data for fitting model parameters. Whether you develop your own models or use someone elses, fitting them to your plant and validating them is crucial to their success. If you don't get this right, the fanciest control strategy or diagnosis technique will come to nought. For software, if not for waste treatment, garbage in is garbage out.

Chapter 7 introduces regression techniques for fitting process models for both steady state data gathered at a number of different operating points and time series data gathered at one operating point. In particular, it stresses the techniques to evaluate whether the fit you obtain is a good one.

Chapter 8 discusses model fitting strategies and issues with particular reference to biological treatment processes and how the model fitting can be approached incrementally.

Chapter 9 discusses simulators for wastewater treatment. This is the tool that utilises the models. Some general principles are presented to help you to understand simulators and information is given on what software packages are available at present and how you might evaluate them for your own needs.

1.3.2 Diagnosis

Monitoring of the process and the early diagnosis of process problems is the subject of this part of the book. Here there is a broader perspective and emphasis on advanced tools for decision support. The strengths of the computer and of the human mind must work together for effective problem diagnosis. There are four chapters.

Chapter 10 introduces the diagnosis problem, the components of a diagnostic system and some general strategies. They are described as data screening, that is to remove irrelevant data, detection of abnormal features and effects, and diagnosis, to analyse the "effects" and find possible causes.

Chapter 11 introduces Statistical Process Control (SPC), which is an extremely powerful framework for the detection and diagnosis of process problems. The monitoring and detection tools are introduced and the human organisation required to exploit these tools is discussed. Traditional SPC tools, advanced multivariate techniques and cluster analysis are introduced.

Chapter 12 presents examples of the use of on-line parameter estimation for problem detection and diagnosis. The examples include the estimation of settling characteristics and the exploitation of respirometry.

Chapter 13 discusses Knowledge Based Systems (KBS) which can be a very effective way of encapsulating human expertise and experience. Such systems have been especially applied to the diagnosis of sludge bulking problems.

1.3.3 Control

This part of the book presents some basics of control and an overall control strategy for biological nutrient removal (BNR) processes. As we have said - you cannot control what you cannot understand - but after the part on Modelling you are ready to Control. There are eight chapters.

Chapter 14 introduces the goals and objectives of operating wastewater treatment processes more specifically than our previous discussion of incentives. This forms the basis for the decision making in controlling your plant, whether implemented by humans or by computers. A hierarchical strategy is proposed for BNR plants based on the inherent dynamic characteristics of the processes involved.

Chapter 15 discusses disturbance characteristics, modelling disturbances and disturbance rejection.

Chapter 16 discusses manipulated variables. What can be adjusted to control the process? How sensitive are these variables and what are the constraints? We will revisit the comments above on the flexibility of plant designs.

Chapter 17 introduces feedback control - a single process variable controlled by a single manipulated variable. Topics such as on/off and proportional- integral-derivative (PID) control algorithms, stability and controller tuning will be discussed. To illustrate the principles, we will look in detail at dissolved oxygen control.

Chapter 18 will tackle model based control. It will start gently with feedforward control and progress through to nonlinear multivariable controllers. We will only talk of some of the numerous control techniques - those that have been used successfully in industry and that are applicable to BNR processes. Soft sensors and state estimators will also be discussed since they help the hard sensors feed the information to the controllers. Here the attention is on the bioreactors and the control of the nutrient removal processes.

Chapter 19 introduces batch plant control including equipment, representation and design of sequences and an introduction to the sequencing of the sequenced batch reactor (SBR).

Chapter 20 discusses plant wide control. Here our perspective is broader, covering two or more of the process units and even as broad as considering the utilisation of the sewer system. Applications include hydraulics, especially the damping of disturbances, and sludge inventory control. The principal technique is optimisation, with discussion of objective functions, constraints and solution techniques.

Chapter 21 presents a systematic approach to advanced control cost benefit studies. How can you justify applying advanced control and how can you show that it really worked? An example is presented from the set of case studies mentioned above.

1.3.4 Instrumentation

We cannot control what we cannot measure. Advanced control must have measurements, so this part of the book addresses measurement. Three chapters make up this part.

Chapter 22 presents more generic measurement principles. The characterisation and evaluation of sensor performance is discussed. Some of the dangers of poor sample preparation are presented. Where should sensors be located?

Chapter 23 discusses recent information on the more advanced sensors. Where do they stand from the viewpoints of accuracy and reliability? What mechanisms perform the basic sensing? Here the emphasis is inevitably on sensors such as nutrient analysers and respirometers.

Chapter 24 gives a brief introduction to basic instrumentation systems. This includes actuators, sensors, control equipment of various types, and the design of operator displays and reports.

1.4 Concluding the Introduction

We hope that you have been well introduced. We have tried to tell you what the book is about and why you should be interested in the book and the topic - the control of biological wastewater treatment plants. We have given you an outline of the parts - Modelling, Diagnosis, Control and Instrumentation - and an outline of the chapters in each part. You should be able to zero in on your interests and on the information you need. Now we can do no more than to hope you enjoy reading more, and that you learn something to help you achieve your goals more effectively.

1.5 Further Reading

For information on the history of ICA in wastewater treatment:

1. Andrews (1969), Andrews (1974), Busby & Andrews (1975), and Andrews (1992).
2. IAWPRC - IAWQ Conferences on ICA, IAWPR (1977), IAWPR (1981), IAWPRC (1985), IAWPRC (1990), Jank (1993), Lynggaard & Harremoes (1996), Briggs (1997).

3. the Clemson conference in Buhr *et al.* (1975).

For information on ICA in other industries:

1. Kane (1987) in petrochemicals.
2. McFarlane (1995) and Mittal (1997) in food processing.

1.6 Exercises

1. List examples of feedback control in your industry, in your plant, in your home and in your everyday life.
2. List examples of disturbances in the control loops listed.
3. Formulate the criterion for the control in your examples.
4. Repeat the previous exercises for examples of feedforward compensation.
5. List and rank the incentives and constraints on improved operation of your business, of your process or of your town.
6. What is the reason for controlling a wastewater treatment system?
7. List examples of disturbances in a wastewater treatment plant.

Part A

Modelling

Part A

Modelling

Chapter 2

Plant Dynamics

Why model processes? In Chapter 1 we discussed why we need to control processes in general and biological wastewater processes in particular. We also said that mathematical models form the basis of advanced control as you will see when you read chapters in Parts B, Diagnosis, and Part C, Control. Indeed Table 1.1 presented the many uses of simulations based on dynamic process models. Before we look in detail at dynamic modelling as we will in Chapter 3, there are a few preliminaries.

Process engineers will ask immediately "why dynamic models?". A major flaw in process engineering courses is the emphasis on steady-state modelling, and this spills over into the design of processes with insufficient flexibility for control purposes, which we also saw mentioned in Chapter 1.

Section 2.1 discusses why dynamics are necessary. The corollary of "you cannot control what you cannot understand" is of course that "You cannot model what you cannot understand". Therefore we will look firstly at the processes we wish to model. Sections 2.2 and 2.3 will discuss wastewater processes and the basic mechanisms which drive these processes. Section 2.4 discusses the dynamics of wastewater processes in qualitative terms.

So, we decide that we want a dynamic model for control or diagnosis or another of the many uses listed in Table 1.1. But what sort of model do we need? This question is addressed in Sections 2.5 and 2.6. Having learnt how your process works and decided what sort of model is appropriate to your purpose, it is then time to get down to specifics. The next chapter details how to write a specification for your model and then develop the model.

2.1 Disturbances

One important reason why we are interested in dynamics when we deal with wastewater treatment is the nature of the influent disturbances. Very few wastewater treatment plants have the luxury of an influent at a constant flowrate and of a constant composition. Typically influent characteristics, flowrates and nutrient loadings and in some parts of the world temperature, vary by factors of two to ten. Figure 2.1 shows the daily variations so typical of raw wastewater.

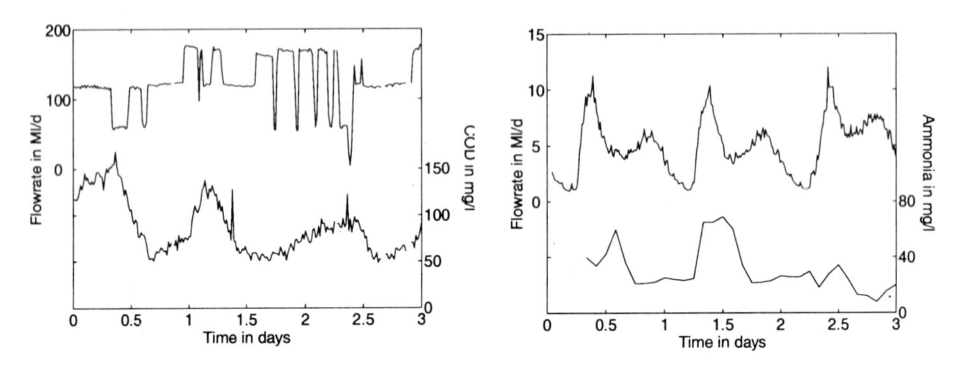

Figure 2.1: Large and small plant disturbances.

On top of the almost regular daily variations are weekly and annual cyclic variations. In many plants there are also self-inflicted disturbances, such as filter wash water returned to the primary settler (Figure 2.2). These are often not buffered and can cause major disturbances to biological processes (introducing oxygen into the anaerobic zone for example) and to secondary clarification (by hydraulic shocks). Hydraulic disturbances are the most common and also the most difficult to control. Not only influent flowrate variations but also poor pumping operations contribute to operational disturbances. Clarifier upsets are often caused by inadequate pump control. Some of the internally generated disturbances (that should not occur!) are quite fast. A sudden flow increase caused by the start of a pump will influence the flow propagation along the plant within a fraction of an hour.

Then there are the unexpected (Figure 2.2): rain storms, snow melts, industrial discharges, toxic releases, etc. A wastewater treatment plant is NEVER at steady state.

While the primary goal of a treatment plant is to achieve an average reduction in nutrient levels, the secondary goal is disturbance rejection, to achieve good effluent quality in spite of the many disturbances. Modelling and simulation are key tools in the achievment of these two goals.

2.2 Treatment Plant Processes

The treatment of municipal and some industrial wastewaters generally uses a combination of primary, secondary and tertiary treatment. Primary treatment simply screens and settles large particles and skims off floating greases and oils. Secondary

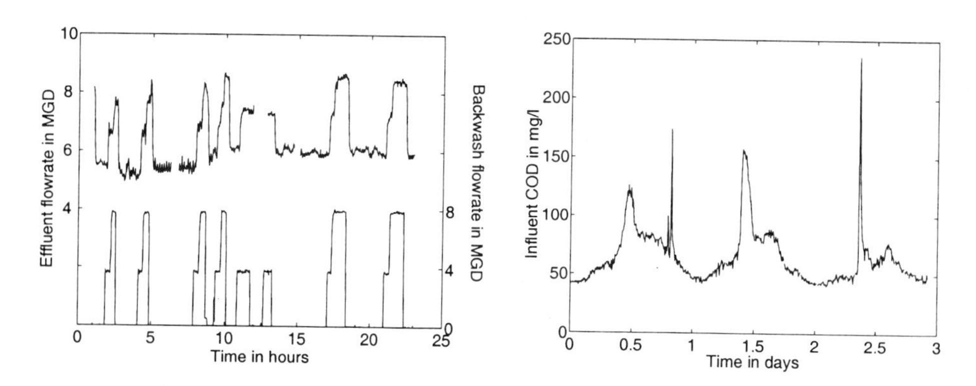

Figure 2.2: Filter-wash and unexpected disturbances.

treatment biologically removes organic carbon and in newer plants soluble nitrogen and/or phosphates. Tertiary treatment attempts to limit the microorganisms and other pathogens in the treated water by membrane filtration or deep-bed filters and some form of disinfection using chlorine, ozone or ultraviolet light. Chemical precipitation of phosphates is common during primary and/or tertiary treatment.

The secondary or biological treatment of wastewater utilises a number of key processes, unit operations in the process engineering terminology. These include primarily processes for biological reactions and processes for settling and clarification (Figure 2.3). Chemical precipitation to remove phosphorus is sometimes utilised just before the settling and clarification.

Figure 2.3: Aerial photo of a treatment plant.

The biological reactors involve several reaction types:

- Anaerobic fermentation
- Anaerobic activated sludge
- Anoxic activated sludge
- Aerobic activated sludge (Figure 2.4)

and an increasing number of physical configurations:

- Mixed and partially mixed tanks
- Tanks with loose or fixed packings
- Fixed-film processes such as trickling filters
- Sequenced batch reactors

With increasing pressures to reduce volumes, the variety of fixed-film configurations increases daily.

After the reaction processes, the other important class of processes is separation - separation of water from biomass and/or other solids. Traditionally this has involved simple settling tanks and clarifiers but again the pressure on volume reduction is increasing the interest in filtration - both using conventional filtration equipment and using membranes.

Figure 2.4: Activated sludge aeration tank.

There are many flowsheet configurations, often of a proprietary nature, which attempt to obtain the optimal combination of anaerobic, anoxic and aerobic conditions for the microrganisms. Chemical treatment often replaces or complements the biological reactions. Commonly used flowsheets are shown in Figure 2.5.

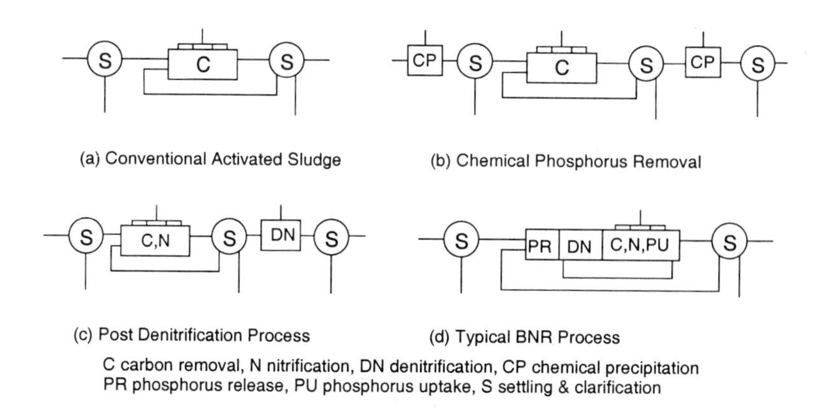

Figure 2.5: Typical WWT process flowsheets.

The flowsheets used in a particular instance depend on:

- the influent characteristics,
- the relative importance of carbon, nitrogen and phosphorus removal,
- the experience of local operators, and
- the expertise of local consultants.

2.3 Mechanisms in the BNR Process

While there is quite a range of processes and equipment involved, there are only a limited number of basic mechanisms involved - hydraulics, nutrient reactions, chemical precipitation, nutrient mass transfer, biomass growth, settling and filtration.

2.3.1 Hydraulics

Wastewater flows through a treatment plant predominantly by the force of gravity, and there are many elegant civil engineering design features to minimise head losses. The volume of wastewater is so large that the amount of pumping must be minimised. There is often only a single set of wastewater pumps to lift it from the sewer to the plant inlet. Otherwise, most pumping involves the recycles of mixed liquor and sludge and the small sludge wastage streams. Vessels are almost exclusively concrete tanks of various shapes and sizes with levels controlled by overflow weirs. For many purposes they can be assumed to be of constant volume.

2.3.2 Nutrient Reactions

The reactions involved in the removal of nutrients from wastewater biologically can be conveniently classified into the removal of organic carbon, nitrogen and phosphorus. Phosphorus is often also removed chemically so this will also be discussed.

The removal of soluble organic carbon occurs primarily by the aerobic or anoxic growth of a class of organisms called heterotrophs. The organic carbon becomes either additional biomass or carbon dioxide in soluble and gaseous form. Aerobic growth occurs where oxygen is available and anoxic growth occurs where there is no oxygen but there are nitrates which can be utilised by the organisms as an oxygen source. Insoluble organic carbon must first be converted to soluble organic carbon by hydrolysis, an enzymatic reaction catalysed by enzymes secreted by the biomass. This is illustrated in Figure 2.6.

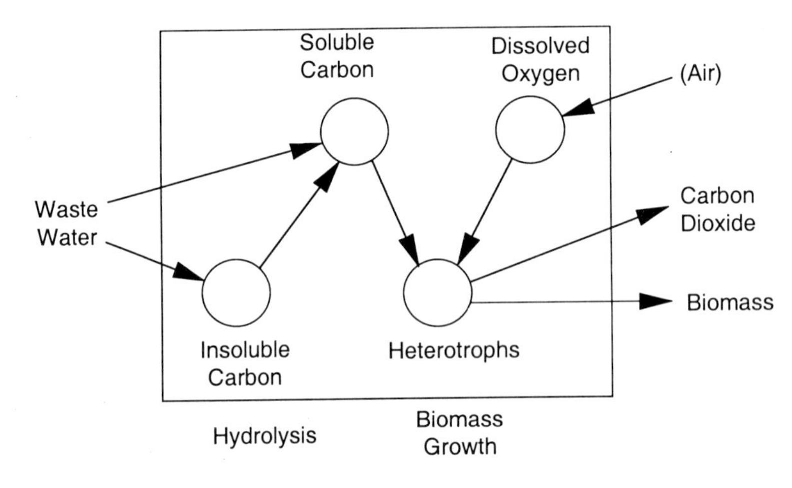

Figure 2.6: Principal reactions in organic carbon removal.

Influent nitrogen is primarily in the form of ammonium compounds. It is removed in a two-step process, nitrification and denitrification, illustrated in Figure 2.7.

Nitrification occurs by the aerobic growth of a class of organisms called autotrophs which converts ammonium compounds to nitrites/nitrates. The growth rate of autotrophs is significantly lower than that of heterotrophs and is also more sensitive to their environmental conditions. In cold climates the growth rate can fall to the extent that they are effectively washed out of the system halting the nitrification process. Denitrification occurs by the anoxic growth of the heterotrophs converting nitrates/nitrites to gaseous nitrogen and consuming carbon. There is substantial evidence for so-called simultaneous nitrification and denitrification (SND) in aerobic conditions. This is thought to occur by anoxic conditions occurring on the inside of biomass flocs.

Biological phosphorus removal is an extremely complex procedure (Figure 2.8). Influent phosphorus occurs primarily as soluble ortho-phosphates and appears to be finally removed as insoluble poly-phosphate nodules inside a class of organisms

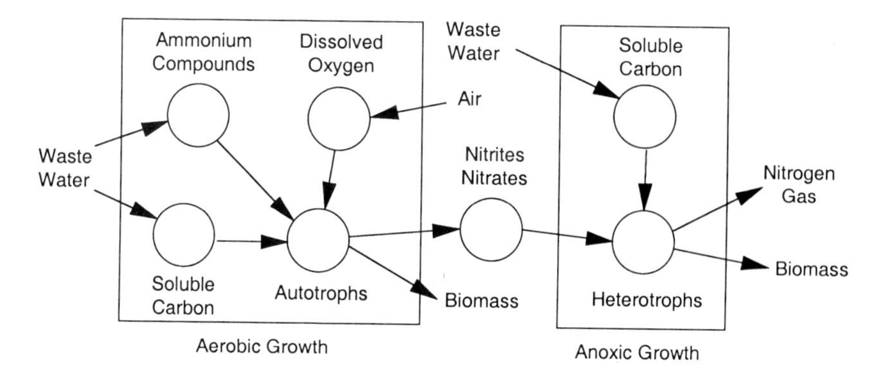

Figure 2.7: Principal reactions in nitrogen removal.

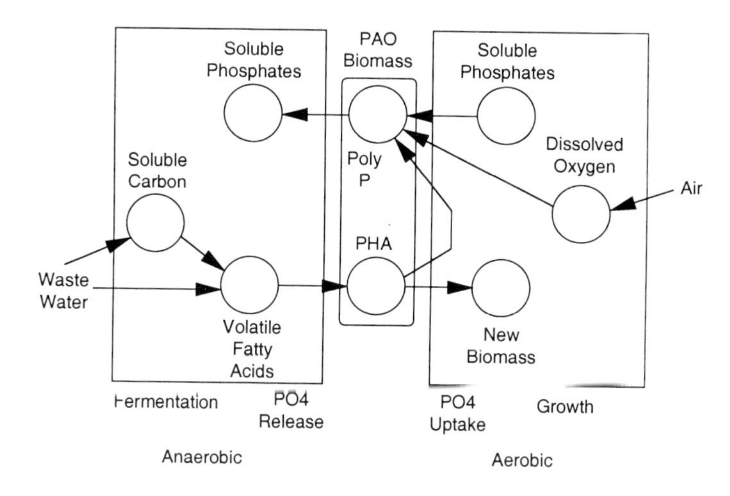

Figure 2.8: Principal reactions in phosphorus removal.

called simply bio-P organisms or phosphorus accumulating organisms. It must be said that there is still much to be learnt about the mechanisms of bio-P removal. The mechanisms as currently understood are summarised by the IAWQ Activated Sludge (AS) Model No 2 (Gujer *et al.* (1994)). The mechanism is a balance between anaerobic release and aerobic uptake of soluble phosphates. During the release the bio-P organisms consume volatile fatty acids (VFA) and internally store poly-hydroxyl-alkanoates (PHA) while the poly-phosphate nodules are released as soluble phosphates. During the uptake the PHA is consumed and soluble phosphates converted to poly-phosphate nodules. The existence and quality of the VFA available in the anaerobic zone appear to have much to do with the success or otherwise of a net phosphorus removal. The net removal probably also relies on the growth of the bio-P organisms which occurs in aerobic conditions when there is sufficient stored PHA. There is also some evidence of growth and phosphorus uptake in anoxic conditions.

As a result of the quest for phosphorus removal, anaerobic fermentation of soluble organic carbon into volatile fatty acids (VFA) has become an important feature of activated sludge processes. The fermentation must occur in conditions unfavourable to methanogens which would break down the fatty acids to methane. Sometimes the configuration of the sewer system is such that sufficient fermentation occurs in the sewers before the treatment plant. Otherwise prefermenters are constructed, which generally convert primary settled sludge to a VFA-rich stream for the anaerobic reactions.

2.3.3 Chemical Precipitation

In many situations it is more economically favourable to chemically precipitate the phosphorus by the addition of aluminium or iron compounds such as alum, pickle liquor, ferric hydroxide or ferric chloride. The chemicals are generally added just before the primary settling and/or the secondary clarification. Although precipitation reactions are complex and only partially understood, the primary action appears to be the combining of orthophosphate (see Chapter 23 for basic description of phosphorus) with the metal cation. Polyphosphates and organic phosphorus compounds are probably removed by being entrapped, or adsorbed, in the floc particles. Aluminium ions combine with phosphate ions as follows:

$$\begin{array}{l} Al_2(SO_4)_3 \cdot 14.3\ H_2O + 2\ PO_4^{---} \rightarrow \\ \quad 2\ AlPO_4 \downarrow +3\ SO_4^{--} + 14.3\ H_2O \end{array} \tag{2.1}$$

The precipitate is indicated by the arrow pointing down on the right side of the reaction. The molar ratio for Al to P is 1 to 1 and the weight ratio of commercial alum to P is 9.7 to 1. Usually the required dosage is substantially greater than the stoichiometric quantity. One of the competing reactions is with natural alkalinity (see Chapter 23) as follows:

$$\begin{array}{l} Al_2(SO_4)_3 \cdot 14.3\ H_2O + 6\ HCO_3^{-} \rightarrow \\ \quad 2\ Al(OH)_3 \downarrow +3\ SO_4^{--} + 6\ CO_2 + 14.3\ H_2O \end{array} \tag{2.2}$$

Iron reagents precipitate orthophosphate by combining with the ferric ion:

$$FeCl_3 \cdot 6\ H_2O + PO_4^{---} \rightarrow FePO_4 \downarrow +3\ Cl^- + 6\ H_2O \tag{2.3}$$

The molar ratio is 1 to 1. As with aluminium, a greater amount of iron is required in actual precipitation than this chemical reaction predicts. One of the competing reactions with natural alkalinity is:

$$\begin{array}{l} FeCl_3 \cdot 6\ H_2O + 3\ HCO_3^{-} \rightarrow \\ \quad Fe(OH)_3 \downarrow +3\ CO_2 + 3\ Cl^- + 6\ H_2O \end{array} \tag{2.4}$$

Ferrous sulfate also forms a phosphate precipitate with a Fe to P molar ratio of 1 to 1. Chemical-biological treatment combines chemical precipitation of phosphorus with biological removal of organic matter and nitrogen and sometimes biological phosphorus removal.

2.3.4 Nutrient Mass Transfer

If biological reactions are to occur, the reactants must be transported to and the products transported away from the microorganisms. A lack of reactants or excess of products slows the biological reactions.

Perhaps the best known mass transfer mechanism is the transfer of oxygen from gaseous air (occasionally enriched with extra oxygen) into the water and to the organisms. Dissolution is enhanced by a larger gas-liquid interfacial area (more and smaller bubbles) and a larger concentration difference between the surface and the bulk liquid (by oxygen enrichment). Transport within the liquid is promoted by stirring, but this should not be vigorous enough to physically disturb the organisms.

Mass transfer within the free flocs of organisms in stirred vessels is not well understood and its significance to biological reaction rates is a matter of speculation. Undoubtedly substrates (products) must be transferred into (out of) firstly the floc and then the membrane of the microorganism. Whether concentration profiles within the flocs are significant enough to create a different environment (aerobic outside to anaerobic inside) is unknown, but certainly postulated as one mechanism for simultaneous nitrification and denitrification (SND).

On the other hand, mass transfer in fixed films of organisms is better understood and is known to be crucial to the performance of reactors with immobilised organisms.

2.3.5 Biomass Growth

The maintenance of a viable sludge inventory with the right proportions of different species of organisms in an active state is very important to the operation of biological nutrient removal. This is so not only for the actual nutrient removal but also for the formation of good settling flocs. It is an aspect of activated sludge plant operation that requires much study and there are many myths.

The principal difficulties are the lack of unambiguous measurements, the relatively slow and disparate growth rates of the microorganisms, competing environmental requirements and a lack of knowledge about what conditions are favourable and unfavourable for the different classes of organisms. This is further complicated by the lag phases experienced with microorganisms, delays before they become inactive in unfavourable conditions or before they become active in favourable conditions. They also have a memory, so that their behaviour depends on prior treatment. This applies not only to the classes of organisms mentioned above, but also to classes of organisms that cause problems when they emerge - foaming and bulking (sludge floating rather than settling) are the best known examples.

2.3.6 Settling

The primary mechanism involved in the separation of insoluble particulates, including biomass, from the water and soluble substances is settling. Settling of dense particles which approximate spheres in shape is well understood. The mechanisms involved in the settling of biomass, large irregular flocs with a density very close to that of water, is much less, if at all, understood. The primary mechanisms are shown in Figure 2.9.

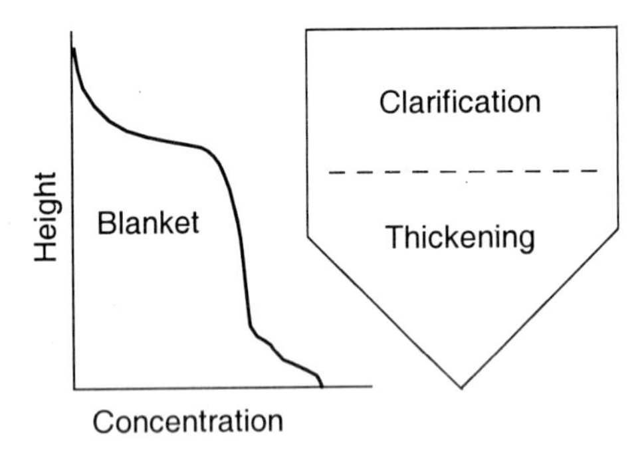

Figure 2.9: Primary mechanisms in clarifiers.

Clarification is the mechanism involved in the concentration gradient of very fine particles in the "clear" water above the sludge blanket level. Such "unhindered settling" is well understood but, for low density particles, the effects of turbulence and flow irregularities can be substantial. Changes in inflow or sludge withdrawal can cause hydraulic shocks as well as velocity changes.

Thickening is the mechanism involved in the concentration gradient of settled particles below the sludge blanket level. Such "hindered settling" and "compaction" is less well understood, particularly when complicated by the sludge withdrawal rakes and conical shapes of typical vessels. Vessel geometries and flow patterns vary considerably as shown in Figure 2.10.

Flocculation is another mechanism whose importance is just beginning to be widely recognised. Flocculation occurs in the region of the feed, when the velocity drops to a sufficiently low level. Inflow effects may be at least partly due to disruptions to the flocculation, which can drastically change the floc size distribution.

2.3.7 Filtration

Filtration is not a common process for primary biomass separation at present, although tertiary filters are often employed for polishing or final reduction of suspended solids. The mechanisms are well understood, if somewhat empirically.

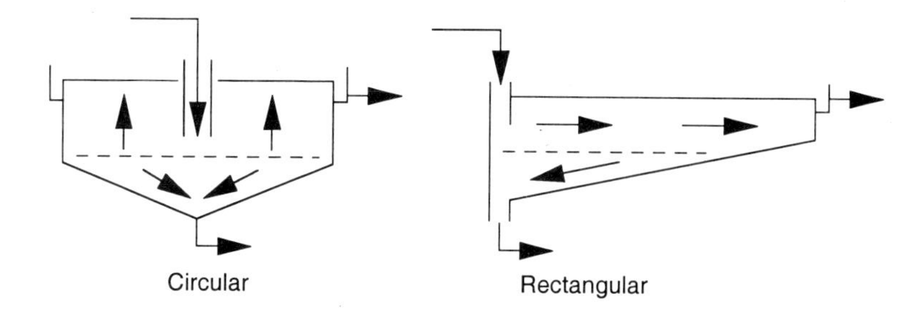

Figure 2.10: Principal clarifier configurations.

They primarily involve measurement or prediction of medium and cake resistances to flow. Again membranes are not regularly used except in sampling systems for nutrient analysers. Large scale use is being trialled for tertiary treatment, where irreversible fouling is a much slower process.

2.4 Dynamics of the BNR Process

The dynamic behaviour of most real process systems can be characterised as some combination of two basic responses.

1. The dead time response (Figure 2.11) for intensive variables (concentrations and temperatures) results from flow through pipes and plug-flow in vessels. The example shows a perfect dead time response, in reality some dispersion (or spreading) occurs due to a degree of turbulent mixing which will round off the corners of the response.

2. The exponential response (Figure 2.12 response (a)), commonly known as an exponential lag or first order lag, typically results from a well-mixed volume. Higher order exponential responses (Figure 2.12 response (b)) are common and result from well-mixed volumes in series.

The exponential response is characterised by the time constant, shown in Figure 2.13 (the time to 63 per cent of the final change for a step response). High order responses are characterised by a number of time constants, the principal rise being characterised by the largest or dominant time constant.

For a simple tank the time constant is equal to the residence time, the tank volume divided by the volumetric flowrate through the tank, a term better known to designers. The time constant and residence time are scale independent in that doubling both the flow and the volume will result in the same times and the same dynamic response.

All processes involve a wide range of speeds of the dynamic responses of different parts of the process. Wastewater treatment processes are no exception. We can classify these dynamics by speed into very fast, fast, medium and slow. We

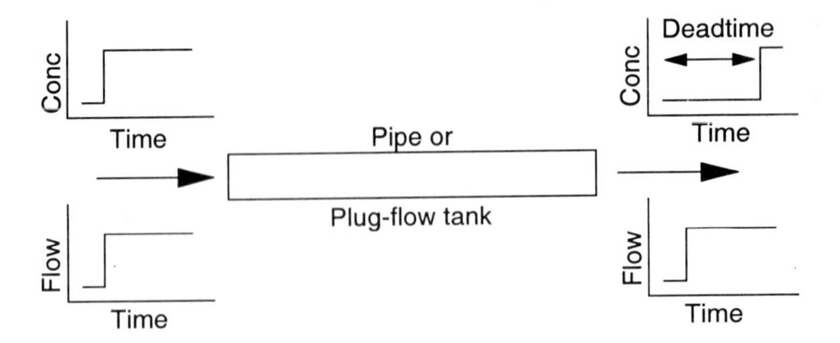

Figure 2.11: dead time responses for intensive variables.

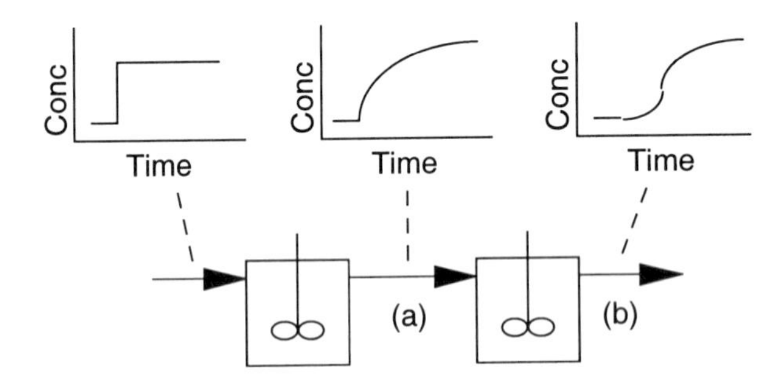

Figure 2.12: Exponential responses in tanks.

will ignore the class of very fast dynamics (very small time constants) involving pumps, compressors, valves, and sensors (excluding nutrient analysers which often have significant dead time responses). We will examine the source of the other dynamics before we attempt to classifiy them.

The basic dynamics of BNR processes are governed by the basic mechanisms discussed in the previous section. We will discuss these in turn.

2.4.1 Hydraulics

Flow of liquids through vessels introduces different dynamics for extensive variables, flowrate, and for intensive variables, nutrient concentrations. The flowrate out of vessels reacts to inlet flowrate changes essentially instantaneously if it is a closed vessel or with a time lag if it has an outlet weir. The size of the lag depends on the weir geometry.

Intensive variable, concentration or temperature, dynamics for flow through vessels are characterised by the mean residence time, vessel volume divided by flowrate. This parameter is commonly known as the hydraulic residence time or

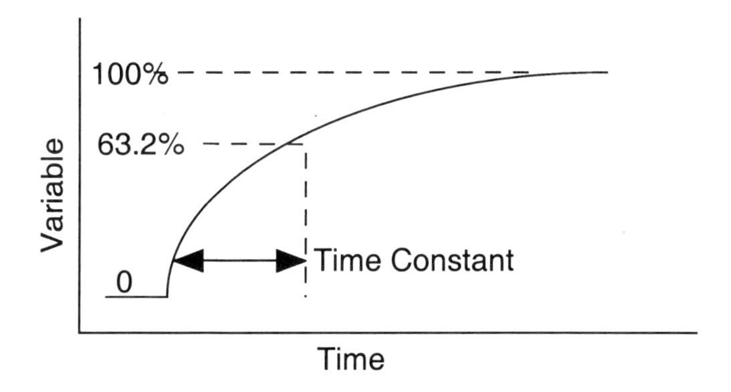

Figure 2.13: Time constant of exponential lag.

HRT and makes simple scaling of designs possible. The spread of the residence time depends on the flow patterns within the vessel which will fall somewhere between plug flow, no spread giving a dead time response, and perfectly mixed, a first-order lag with time constant equal to the residence time. The time constant is typically of the order of minutes to hours.

Flow down pipes causes no delays to the extensive variable flowrate, but causes a transport delay for intensive variables such as nutrient concentrations. The dead time is equal to pipe length divided by liquid velocity. These are unlikely to be significant with the possible exception of return activated sludge which could experience dead times of a few minutes, and poorly designed analyser sampling systems which have been known to have dead times of hours.

Recycle flows slow down the dynamics of intensive variables. For a well-mixed vessel, the time constant of the dynamic response is $(R + 1)$ times the mean residence time of the vessel, where R is the recycle ratio (recycle flowrate divided by inlet flowrate).

2.4.2 Nutrient Reactions

Nutrient removal by the biomass in a well-mixed reactor effectively speeds up the dynamics for that nutrient concentration compared to the residence time of the tank. If the effluent concentration is one tenth the influent concentration, the time constant is one tenth the reactor residence time. It follows that time constants for nutrient concentrations will be within an order of magnitude of the reactor residence times, and generally smaller, depending on nutrient removal ratios and recycle ratios. Thus time constants of the order of 1 to 10 hours could be expected. Recirculation of nitrate in a pre-denitrification system is a fast process, since the flowrate is so high. Consequently the nitrate concentration can be changed within minutes.

2.4.3 Chemical Precipitation

The most important aspect of chemical precipitation is to achieve rapid and complete mixing. Once the chemicals are mixed with the wastewater, the dynamics are very fast, responding in a matter of seconds.

2.4.4 Mass Transfer

The transfer of oxygen from a gaseous form to a dissolved form takes place within a time scale of 15-30 minutes. A change in air flowrates therefore does not immediately affect the dissolved oxygen concentration in the aerator. The respiration rate may change within minutes due to changes in substrate loading or toxic inputs. This will result in DO changes that take place in the time scale determined by the DO dynamics.

2.4.5 Biomass Growth

If biomass growth and sludge wastage are in balance, the sludge dynamics time constant is equal to the solids residence time (SRT). The SRT is the solids holdup divided by the solids wastage, where the suspended solids in the effluent should be neglected with great care. This is typically of the order of several days for the activated-sludge process. Significant changes in the proportions of different species within the biomass could take longer, maybe several weeks for slow growing species.

2.4.6 Clarification

The flow patterns in a clarifier will approximate plug flow upwards for clarified effluent and plug flow downwards for return activated sludge. This would typically involve a dead time of 1-4 hours in each stream.

2.4.7 Response Times

It follows from the previous discussion that the dynamics of the typical BNR process can be classified as fast, medium, and slow as shown in Table 2.1.

Speed	WWT mechanism
Fast	Flow dynamics Dissolved oxygen
Medium	Concentration dynamics Nutrient removal
Slow	Biomass growth

Table 2.1: BNR process dynamic response times.

As we will see later, this classification influences the type of model we develop and also the design of control strategies.

2.5 The Purpose of the Model

A single model would never be sufficient for all the different purposes for which we mathematically describe an activated sludge process. Instead one has to look at a spectrum of models. The purpose of the model will significantly influence the structure and the complexity of it (Figure 2.14).

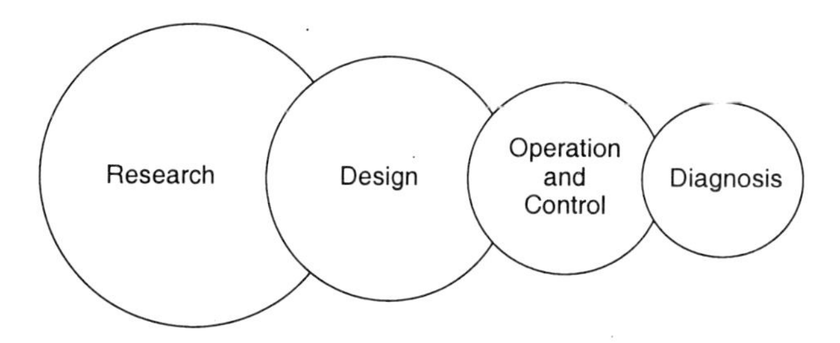

Figure 2.14: Complexity of models.

One may distinguish between models for:

- research - to explore the consequences of new knowledge and to adequately explain significant phenomena within the process;
- design - to explore the influence of different input disturbances and parameter changes in order to meet the required goals;
- operation and control - to make adequate use of on-line instrumentation and laboratory measurements to construct control actions that will bring the process towards the desired performance, despite external and internal disturbances;
- diagnosis - to interpret patterns in plant data, detect abnormalities and suggest causes.

A key feature of a model will be its ability to predict the plant behaviour. Again, the issue of time scale comes in. The prediction horizon can be quite different for different purposes. If the prediction horizon is of the order weeks, then the biological phenomena become essential. However, if it is only of the order of hours, then organism growth can generally be neglected. For operation, the prediction horizon is connected to the control action or the manipulated variable. To control sludge mass balances in the system, there has to be an adequate prediction of the sludge distribution for the next few hours. The forming of bulking sludge or foaming caused by filamentous growth takes many days or weeks. To control it - if possible at all - may take a shorter time if the organisms are killed (by chlorination). To change the growth conditions will take weeks. Consequently the prediction horizon has to be of that order as well.

2.5.1 Models for Design

A model for design can of course not be fully validated until the plant is built. Such a model is aimed at describing all important mass balances of the system in such a way that the volumes and the flowrates of the system can be adequately designed, taking disturbances into consideration. In order to faithfully describe an activated sludge process a large number of phenomena have to be taken into consideration, such as

- characterisation of the influent;
- hydraulics of each tank;
- hydrolysis of different substrates of the influent;
- carbonaceous, nitrogenous and phosphorus removal mechanisms;
- organism growth and decay;
- sludge clarification and thickening mechanisms.

The influent wastewater is characterised by its composition, not only in organic carbon, nitrogen and phosphorus fractions, but also in soluble, particulate, biodegradable and unbiodegradable fractions. The biological part of the activated sludge process has been subject to a lot of research. The organisms are represented by at least one class of heterotrophs and one or two species of nitrogen consuming autotrophic organisms. The main reactions are added for each class of organisms and each class of available substrate.

The best known model is probably the IAWQ (former IAWPRC) AS Model No 1 (Henze *et al.* (1987a)), describing the reactions for organic carbon and nitrogen removal. The main emphasis of the model is the biological reactor, while the settler dynamics is treated comparatively superficially. Typically the number of state variables varies between ten and fifteen for each completely mixed reactor for models described in the literature. The AS Model No 1 contains 13 states, describing carbonaceous and nitrogenous removal, each state with several reaction rate and stoichiometric parameters to be determined. The more recent IAWQ AS Model No 2 (Gujer *et al.* (1994)), which adds the dynamics of phosphorus removal, contains 19 states for each completely mixed reactor.

In most models available the clarifier has been treated as a pure concentrator, sometimes with a time lag or time delay. More structured models that incorporate both the clarification and the thickening phenomena have been presented. Still the dependence of the settling parameters on the biological conditions of the sludge is not straight-forward. Usually it is assumed that there is no biological activity outside the bioreactor. There are, however, indications that some biodegradation takes place in the settler.

Most of the growth models are generally described by Monod type expressions, as are various substrate limitations and inhibitions (here they are called switching functions). However this is simply a convenient empirical expression which seems to fit the data well, but there are also other less popular expressions.

2.5.2 Models for Control

In a model used for control purposes there are other demands. Its complexity is determined by two major factors,

- the effect of the control actions, and
- verifiability.

A model based control problem can be considered as consisting of two parts, prediction and control. In the first a model is used to predict the measured or estimated variable at the next sampling time. The controller is then given a value such that the predicted value will approach the target value as closely as possible. This indicates that the time scale of the model has to be related to the time scale in which the manipulated variable can influence the process. Furthermore, one has to ensure that the manipulated variable has enough "authority" to change the output variable. Chapter 16 discusses this in detail.

Since a real plant has time varying parameters, it has to be possible to adequately update the model parameters. This problem of identifiability and also verifiability has been addressed recently. Basically a complex design model has to be reduced so that its parameters can be identified, given available plant data. Therefore appropriate complexity becomes crucial.

2.5.3 Models for Diagnosis

Diagnosis means "the conclusion after a critical study". It is a Greek word meaning "to distinguish" and "to know". The idea of diagnosis in process control is to find indicators that become a simple reflection of reality in order to get a rational basis for decision making. Observations and measurements are the basis for diagnosis.

The wastewater treatment system is defined by a number of state variables, typically concentrations of various substrate and organism types. Usually they cannot be directly measured. Instead, the process is observed by measurements that are related to the state variables. For the diagnosis the complete plant with all its equipment has to be considered.

Monitoring the machinery and the plant equipment forms the basic level of diagnosis. Simple indicators will warn the operator if a motor is not running or a pressure is getting to high or too low. Alarms of equipment behaviour or basic physical parameters - like flowrates, pressures, or levels - are essential pieces of information. At the other extreme, biological parameters will change slowly. Floc settling properties depend among other things on the species of organisms. It is important to detect early signs of flow settling changes. Once they are apparent we may have bulking sludge, and it is too late to make simple corrections. Therefore early warning systems become critical in biological wastewater treatment.

Process knowledge is an essential ingredient of diagnosis. A primary observation may lead to a whole chain of examinations of the problem, exactly as in medical diagnosis. For example, a high turbidity of the effluent may be the primary warning. The reason may be a too high influent flowrate, poor sludge settleability, or a high sludge blanket. By back-chaining and asking questions (automatically or manually), the diagnosis will hopefully find the initial cause for the problem.

In some cases one can make a sophisticated calculation based on measurements and a process model in order to back-calculate non-measureable variables. Such a procedure is called estimation, and will be further discussed in Chapter 12. Unfortunately, even sophisticated models like the IAWQ AS Models No 1 and 2, cannot predict sludge bulking or foaming. Furthermore, the coupling between the biology and sludge settleability is still unknown to a large extent. Therefore, a good diagnosis has to rely on both mathematical models (that reflect what is quantitatively known today) and empirical experience.

How does one know what model structure to try? Well, one doesn't. For the real process there is never any "true model" anyway. One has to try several different approaches.

2.6 The Complexity Issue

Needless to say, the activated sludge process is a complex process. However, considering measurement data from a full scale plant, it may be considered "data rich" but "information poor". This is related to the question of complexity. It is important to distinguish between two fundamentally different types of description, data description and process (model) description. A data description may be all the outputs from a computer generated simulation, while the process description would be the differential equations that generate the data sequence. The amount of information is quite different in the two cases. In general the amount of information required for the process description is much less than that for the data description. An explanatory model in essence is replacing data by process.

A good model can substitute large amounts of data information with small amounts of process information. When constructing dynamic models for predictive purposes, it is easy to get misled into thinking that complexity implies effectiveness. A too high order of dynamic model may arise from some of the degrees of freedom in the model being used to represent the noise behaviour of the data. The predictive power of a model can be tested against new data.

In wastewater treatment systems, some data may be too noisy to contain much useful information. For example, effluent water concentrations of organic matter or suspended solids from a well functioning plant are quite small, of the same order of magnitude as the measurement inaccuracies. It was pointed out earlier that models of wastewater treatment systems based on only influent-effluent data were quite unreliable.

Engineers use a multiplicity of models in their work. For studying a wastewater treatment we have already indicated some different model purposes. Models for design, research and operation are usually used independently and for specific purposes.

Different design models are mostly used for their explanatory purposes in design. In developing these design models theory and experiment go hand in hand, and there is a continuing struggle to replace data by process. Of course such models also have some predictive role, but this is only verified under rigorously controlled experimental conditions, which isolate the reduced set of circumstances under which the particular model is appropriate. Hence, their predictive roles are seldom used or required in the way that atmospheric models are used for predictive purposes in weather forecasting.

It is true in many engineering applications that the uses of models for explanatory and for predictive purposes can require different amounts of complexity. A collection of models of quite modest complexity (each one associated with one aspect of the plant behaviour) can be used to predict in a limited sense that the plant will work in the way that the designer intended. An altogether different magnitude of complexity is required in order to make specific predictions about the detailed future behaviour of any specific plant at a specific time.

Typically, we may reproduce the main qualitative features of the interaction between organism kinetics and sludge settling properties. However, to model the plant to the extent that we can accurately predict its future requires much more information.

Operation and control can reduce the externally perceived complexity of a plant. This may be illustrated by a simple example. Without closed loop control the dissolved oxygen (DO) concentration may vary significantly due to load variations. This in turn will influence the formation of different organisms. If the DO concentration is kept constant despite load changes, then the organism formations may be more predictable. Furthermore the DO dynamic equation may simply be eliminated from the rest of the system model, since the direct coupling of DO dynamics to the biology has been vastly reduced.

In process engineering, it is common to use plant models having hundreds or even thousands of state variables, while controllers for the same plant have only tens of state variables. The designers need to develop models in order to understand the plant. They will devise a point of departure for a related model building with reduced complexity, that would reflect only those aspects of plant behaviour whose complexity would be successfully matched by a feasible controller design. The control people need to find models that effectively interact with the real plant. This means, that the model should be verifiable from plant data.

This illustrates again that the plant data may be "information poor", in other words, it is difficult to squeeze out process knowledge from plant data. The large models of a plant are usually theoretically constructed and aggregated in a piecemeal fashion, and not generated from real experimentally obtained full scale plant data. The complexity of a useful controller for a plant will normally match the process complexity that can be derived from plant data. We try to squeeze the maximum information of useful predictive machinery from plant data, and we can only control something to the extent that we can predict its future. Any attempt to apply more control to the plant than the process knowledge would allow will result in poor control, process failure or instability.

If a control system cannot perform well enough by a direct feedback of measured variables, then one may incorporate a model of the system which is being controlled. Such a model can be thought of as a trade-off between data (measurement) information and process information.

In control engineering the problem of controlling a system is a mixture of being able to predict its behaviour and to observe (measure) its behaviour, that is a mixture of process and data. As remarked in the previous section, a sophisticated controller uses a model of the plant. Such a model is, however, simulating rather than explanatory. The key weapon for control is feedback, using measurements. Any model has a limit to its predictive capability, and feedback is the only way to reduce the uncertainty.

2.7 Summary

We have introduced in a qualitative manner the biological wastewater treatment process and the basic mechanisms involved. We have also introduced dynamics in a qualitative manner and the concept that different mechanisms respond at different speeds, a very fundamental concept in designing control systems and operating treatment plants.

The need for models was then discussed, together with the other important concept you should remember, different tasks require models of different types and different complexities.

In the following two chapters we will look at the different processes and mechanisms again in a quantitative manner as we address the model development process.

2.8 Further Reading

- On operational disturbances: Olsson (1977), Olsson (1989).
- A recent review of nutrient removal mechanisms can be found in Mino *et al.* (1998).
- IAWQ AS Model No 1 and 2 kinetic model reports by Dold *et al.* (1980), Dold & Marais (1986), Henze *et al.* (1987a) and Gujer *et al.* (1994).
- Variations and extensions of the IAWQ AS Model No 2 have also been proposed by Griffiths (1994) and Maurer & Gujer (1998).

- Good overviews of model development in activated sludge systems: Beck (1977), Beck (1984), Jeppsson (1993) and Jeppsson (1996).
- The discussion on the complexity issue has been inspired by MacFarlane (1993). See also Rasmussen & Lind (1981).
- for information on WWT processes and flowsheets:
    1. Eckenfelder Jnr. (1980)
    2. Henze *et al.* (1995)
- for information on WWT process equipment:
    1. Metcalf & Eddy Inc. (1972)
    2. Hammer (1986)
- for information on WWT process design:
    1. Randall *et al.* (1992)
    2. Joint Committee of ASCE and WEF (1998)
    3. Qasim (1999)
- for information on nutrient removal processes:
    1. Grady & Lim (1980)
    2. Halling-Sørensen & Jørgensen (1993)
    3. Henze *et al.* (1995)
- for information on process dynamics:
    1. Andrews (1992)
    2. Lee *et al.* (1998)
    3. Olsson & Piani (1992)
    4. Seborg *et al.* (1989)

2.9 Exercises

1. List all the disturbances that may affect the influent to the biological treatment section of your treatment plant, their sources, their likely frequency and magnitude.
2. Sketch a flowsheet of your treatment plant and identify the basic mechanisms occurring in each major piece of process equipment.
3. For each of the principal nutrient removal processes in the activated sludge process (carbon removal, nitrification, denitrification, phosphorus release, phosphorus uptake) list the possible consequences of too much or too little of the following nutrients: VFAs, organic carbon, dissolved oxygen, ammonia, nitrates, and phosphates.

4. List the factors that might affect the mass transfer of oxygen and whether the effect is favourable or unfavourable.

5. List the changes in process conditions that might affect the clarification and thickening process.

6. The MATLAB program `Exercise_6` simulates the return activated sludge pipe on a treatment plant. Investigate the effects of step changes in valve position and inlet suspended solids concentration on the outlet flowrate and suspended solids concentration. Explain the effects you observe.

7. The MATLAB program `Exercise_7` simulates COD and dissolved oxygen dynamics in an aeration basin with three well-mixed sections each with its own DO probe. Investigate the effects of step changes in influent wastewater flowrate, in influent COD and in the three air flowrates.

8. Considering just the aeration zone of an activated sludge plant for carbon and nitrogen removal, list the variables and mechanisms you might include in models for research, design, operation and diagnosis.

Chapter 3

Basic Modelling

This chapter will introduce basic dynamic modelling in an incremental fashion from the perspective of activated sludge processes. But first we will introduce some process systems terminology, just so you may be able to understand some of the book, and our prejudices about the process of model development.

3.1 Process Systems Terminology

There are a few concepts of process systems engineering introduced now that are needed to better understand the chapter. If your mathematics is a little rusty you might like to browse through Appendix D on Matrices, Vectors and Scalars and Appendix E on Differential Equations. These appendices are simply very short refreshers on the concepts and terminology.

3.1.1 Process Variables

Any system is conventionally described in terms of:

- **input variables**, which may be further classified into manipulated variables u_i if we have control over them or disturbance variables d_i if we do not.

- **state variables** x_i are the set of independent variables which uniquely determines the state of the process. They are the holdups of component masses, energy and momentum within the process. We can seldom measure (observe) all the state variables.

- **output variables** y_i, also commonly called measurements, which are the variables we can observe and which are related in some way to the state variables.

The key to defining these variables is to clearly *specify the system boundary.* Two mechanisms in the activated sludge bioreactor are oxygen dissolution from an air stream and substrate oxidation using the dissolved oxygen. If oxygen dissolution is the system of interest, air flowrate is an input variable and dissolved oxygen concentration both a state variable and an output. If substrate oxidation is the system of interest, dissolved oxygen concentration is an input variable. In other words, inputs and outputs are relative terms.

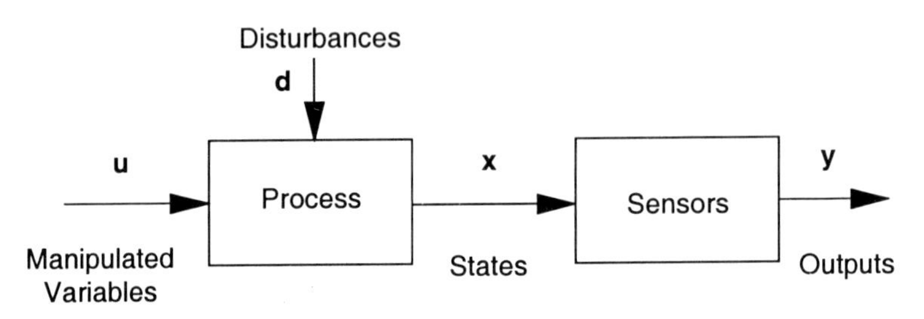

Figure 3.1: Process variable definitions.

The inter-relationship of these variables can be seen in Figure 3.1 and can be expressed mathematically in terms of the general state and output equations:

$$\frac{dx}{dt} = f(x, u, d, p) \tag{3.1}$$

$$y = g(x, u, d, p) \tag{3.2}$$

where p is a vector of system parameters.

This vector notation is interpreted as:

$$\begin{bmatrix} \frac{dx_1}{dt} \\ \frac{dx_2}{dt} \\ \vdots \\ \frac{dx_n}{dt} \end{bmatrix} = \begin{bmatrix} f_1(x_1, \cdots, x_n, u_1, \cdots, u_r, d_1, \cdots, d_s, p_1, \cdots, p_t) \\ f_2(x_1, \cdots, x_n, u_1, \cdots, u_r, d_1, \cdots, d_s, p_1, \cdots, p_t) \\ \vdots \\ f_n(x_1, \cdots, x_n, u_1, \cdots, u_r, d_1, \cdots, d_s, p_1, \cdots, p_t) \end{bmatrix} \tag{3.3}$$

$$\begin{bmatrix} y_1 \\ y_2 \\ \vdots \\ y_m \end{bmatrix} = \begin{bmatrix} g_1(x_1, \cdots, x_n, u_1, \cdots, u_r, d_1, \cdots, d_s, p_1, \cdots, p_t) \\ g_2(x_1, \cdots, x_n, u_1, \cdots, u_r, d_1, \cdots, d_s, p_1, \cdots, p_t) \\ \vdots \\ g_m(x_1, \cdots, x_n, u_1, \cdots, u_r, d_1, \cdots, d_s, p_1, \cdots, p_t) \end{bmatrix} \tag{3.4}$$

Table 3.1 shows a typical variable classification for a simple COD removal activated sludge plant with one well mixed bioreactor and one secondary clarifier assuming constant volumes.

Some unfortunate but very common features are illustrated by this example:

- the number of outputs (measurements) is much smaller than the number of states which means the system may not be observable (we cannot determine the system states from the measurements)

Class	Bioreactor	Clarifier
Manipulated variables	air flowrate	sludge recycle sludge wastage
Disturbance variables	influent flowrate influent COD	
State variables	bioreactor COD bioreactor DO bioreactor SS	effluent COD effluent DO effluent SS sludge COD sludge DO sludge SS
Output variables	bioreactor DO	effluent COD effluent SS sludge SS
Parameters	bioreactor volume oxygen transfer parameters stoichiometric and kinetic parameters	clarifier volume and area

Table 3.1: Typical variable classification.

- the number of manipulated variables is smaller than the number of outputs which is smaller still than the number of states which means the system may not be controllable (we cannot drive the system states to arbitrary desired values with the manipulated variables)

These problems are not unique to the activated sludge process.

3.1.2 Operating and State Spaces

In process systems terminology the *operating space* of a process is defined in terms of the manipulated variables and the feasible operating space is delineated by a set of operating constraints. Constraints may be process related (the biomass needs oxygen), equipment related (maximum pumping rates) or safety related (effluent requirements).

Figure 3.2 shows a possible operating space for our simple example. We usually hope that designers will give us a nice big feasible operating space, although plants have been built which ended up to have no feasible operating space. For processes with more manipulated variables it becomes more difficult to visualise the operating space.

The *state space* is defined in terms of the state variables and is usually a higher dimensional space which is difficult to visualise. Even for our simple example with

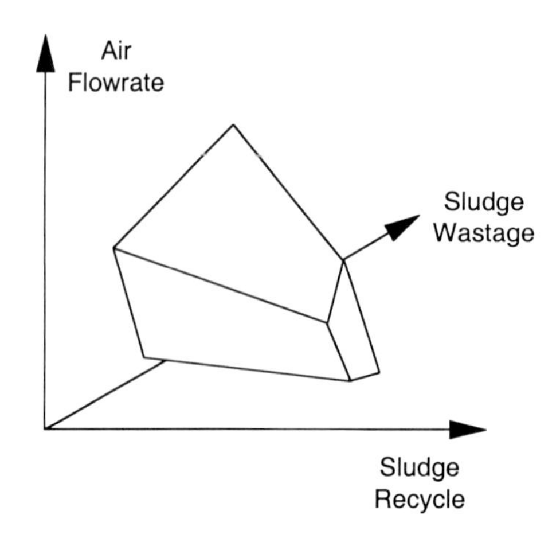

Figure 3.2: Feasible operating space.

a simplified view of the clarifier, it is a nine-dimensional space. Let us know if you can plot that clearly on two-dimensional paper, and we will take out a patent.

For *controllable* processes, there is a mapping between the feasible operating space and an accessible state space. In a controllable system, it is possible to drive the system to any point in the state space in a finite time. Usually there are two major conditions to be satisfied. The control variables have to have a sufficiently strong coupling to the various states. Then, the amplitude of the control variable has to be sufficiently large, so that the process can be properly influenced. With a qualitative term we sometimes call this *control authority* (control authority is discussed in more detail in Chapter 16).

Observability is related to the measurements. In an observable system, it should be possible to indirectly calculate all the states of the system, given the information from the sensors. The information from the various sensors has to be sufficiently independent to provide the full information. Two sensors, telling almost the same thing, may not tell more than only one of them. In other words, if we can determine where we are in the state space from our measurements, then the process is observable.

3.1.3 Steady States

For completeness we should discuss *steady states*, though many would say that the only steady state for an activated sludge process occurs when the plant is shut down. This is generally true for any real process. However, we can also (more correctly, perhaps) consider "steady state" in terms of average values rather than constant values (only some process designers believe that constant conditions exist, bless their souls!).

For a well behaved stable controllable process (no chaos influences or bifurcations), each point in the feasible operating space will have a corresponding "steady state" point in the state space. If the process is unstable, there is no steady state point. If the process is not controllable, then there are many possible steady states for a given point in the operating space. We may pass through other points in the state space during transients or dynamic responses, but we will settle down to operate at the steady state.

3.2 Model Development

Model development involves a number of tasks to be performed which are summarised in Figure 3.3. We will address the first three tasks in this chapter and the remainder in Chapters 6 and 8.

The problem specification is the most important task, yet also the most neglected. It sets the requirements which guide the model development and defines the criteria for model evaluation. The development process and the making of appropriate assumptions will generate the set of equations which we normally call a model. Model verification is the first testing of the model against our requirements.

3.2.1 Problem Specification

Any good project engineer prepares detailed specification documents for constructing a process. There are many good reasons for preparing a project specification for the modeller. In the previous chapter we discussed the two related concepts of "the purpose of the model" and the "model complexity". We will see below, when we begin to develop models, that different modellers can make different but quite reasonable assumptions and come up with radically different models. The difference in the models is both in terms of complexity and also in terms of how faithfully they represent the real world.

As the end user of the model, what do you want? If you don't specify what you want, don't be surprised if you don't get it. With a specification, we may even get what we want in a reasonable time and at a finite cost. The accuracy specifications define a stopping point. The many horror stories that cause many to avoid modelling projects result from poor specifications and poor project management.

3.2.2 Model Development

Sections 3.4 to 3.7 will cover model development in some detail. There are four basic mechanisms involved in activated sludge processes.

Hydraulics - wastewater flow through vessels
Mass Transfer - oxygen dissolution
Reactions - biological nutrient removal and fermentation
Separation - settling and clarification

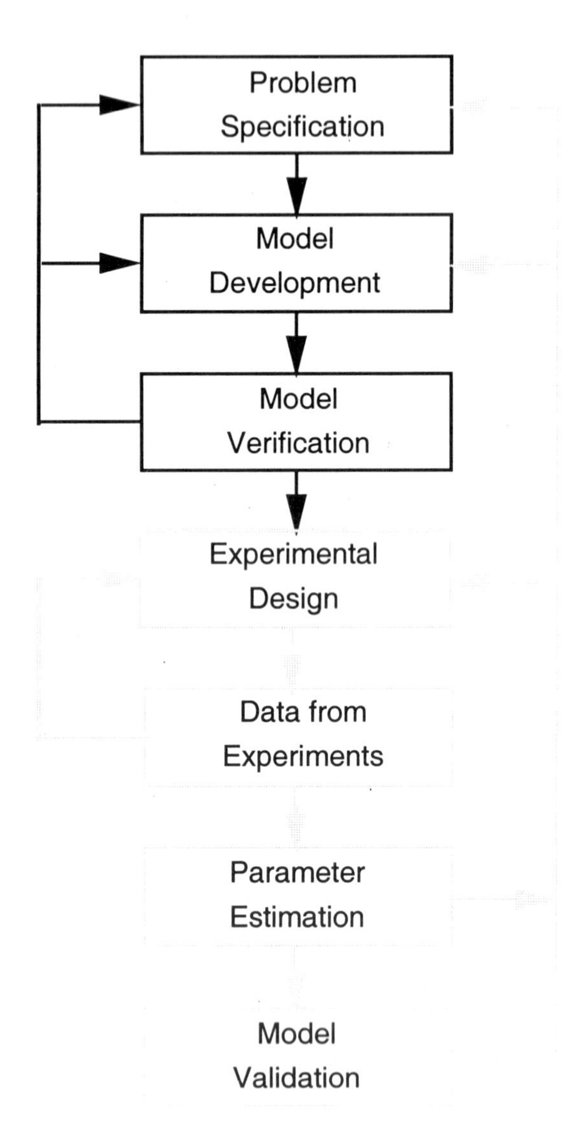

Figure 3.3: The model development process.

The objective of this chapter is to enable you to write and simulate simple models for the first three of these mechanisms. This will then enable you to understand and even write more complex models, such as those in Chapter 4. Understanding the models developed in these two chapters will enable you to explore and better understand the basic mechanisms involved in the biological treatment of wastewater and the commercial simulators that you may use.

3.2.3 Model Verification

Section 3.8 of this chapter will discuss the verification process and some common techniques used to determine whether the model is appropriate to the requirements. It is just the first part of the model evaluation process. The cost of making experiments, estimating parameters and validating the model is considerable. Therefore the more testing of the model before this stage the shorter and less costly will be the complete model development process.

3.3 Problem Specification

The important issues in problem specification are defining the operating region of interest for your model and the accuracy required within that region. For dynamic models, the region must include a time dimension as well as the important input and operating variables.

The quantitative definition of the region and the accuracy is naturally related to the intended use of the model, such as plant design, process optimisation, model-based control, or diagnosis. In all these instances the values specified will depend upon the accuracy required of model predictions. A preliminary design may only require accuracies of 20-30 per cent, but detailed design may require 5-10 per cent. In the case of dynamic models, the time horizon over which model predictions will be made and how the accuracy required varies with the elapsed time must be specified. Because of the feedback involved, model-based control of nutrient removal for disturbance rejection purposes may require accuracies of only 30-50 per cent over 10-20 minutes. An open-loop process such as optimisation of biomass inventory may require accuracies of 5-10 per cent over 10-20 days. Often the accuracy required may be less as time progresses. The specification may also be different for the different variables whose values are to be predicted.

It is the users responsibility not to over-specify the model, to ask for predictions he does not need at accuracies he does not need. It is the modeller's responsibility to produce the simplest (least complex) parsimonious (fewest parameters) model that will satisfy the specified accuracies in the specified regions. An example specification is shown in Table 3.2.

Use of Model:
prediction of transients in effluent C and N concentrations
prediction of transients in aerobic tank DO levels
Users of Model:
process engineer (university graduate, training course B)
shift supervisor (high school leaver, training courses A and B)
Region Definition:
air flowrate: 0 to $15 m^3/hr$
sludge recycle: 50 to $75 Ml/d$
sludge wastage: 0 to $10 Ml/d$
nitrate recycle: 100 to $200 Ml/d$
influent C: 400 to $800 mg/l$
influent N: 50 to $150 mg/l$
Required Accuracy:
C concentrations: within $10 mg/l$ when influent C 600 to $800 mg/l$
C concentrations: within $3 mg/l$ when influent C 400 to $600 mg/l$
N concentrations: within $2 mg/l$ throughout region
DO levels: within $0.3 mg/l$ throughout region

Table 3.2: Example modellers specification.

3.4 Basic Dynamic Modelling

This section will introduce dynamic modelling incrementally to show the simple methodology that can be applied to modelling any process.

3.4.1 Conservation of Mass and Energy

The derivation of dynamic process models is based upon the simple premise that what we put into a vessel either stays in there or it comes out somewhere else. This applies to any physical component (water, nutrient, microorganisms) or to energy.

This simple premise is called the conservation of mass or energy, and the corresponding mathematical equations we write are called mass or energy balances. In this chapter we will not consider energy balances, assuming the activated sludge process operates everywhere at a fixed temperature, though the actual temperature may vary from one simulation to the next. The effect of energy dissipation from pumps is not considered.

The general conservation or balance equation can be expressed as follows:

(the rate of change of vessel contents) = (rate of inflows) - (rate of outflows) + (rate generated) - (rate consumed)

where "the rate of change of vessel contents" can be expressed as the time derivative of the contents, "d/dt(vessel contents)". We write one of these balances for each of the components that we wish to consider in our model (for example water, classes of nutrients, oxygen, classes of microorganisms). This will result in a set of ordinary differential equations, which are then integrated over time to tell us how the amounts of these components vary given certain initial and input conditions. We will use this general balance equation in all the simple examples in the following sections.

3.4.2 Balance Volumes

Before writing down a mass or energy balance, we must define the boundaries for the volume over which the balance will be written. This boundary definition should be explicitly stated in our modelling documentation. Balance volumes are generally defined by phase boundaries, in wastewater treatment this generally means the liquid phase. Phase boundaries are in turn frequently defined physically by the container. We will assume in our modelling that the contents of a balance volume are well mixed, that is that concentrations are the same throughout the volume. This is sometimes called the *perfect mixing assumption.*

In wastewater treatment, tanks are frequently long with liquid generally flowing down the length of the tank. In this case concentrations will change down the length of the tank, so that we have a *distributed parameter system* with concentrations depending on the position down the tank as well as with time. Such systems can generally only be solved numerically by dividing the physical tank into a series of balance volumes down the length of the tank. We then assume each volume is well mixed approximating the smooth change of concentrations down the length of the tank by a series of steps (Figure 3.4). This is then called a *lumped parameter system* which is much simpler to model and simulate.

The size of the balance volumes, which need not be equal, is generally decided by experience, or preferably on the basis of a residence time distribution study (Chapter 8).

3.4.3 Modelling Assumptions

This sounds all very simple, and it is. However, it is possible for five groups of people to model the same simple process and come up with five different models, or even more than five.

This happens because we make assumptions about how the process operates. We do this either because we are not sure how it operates or because we wish to keep the model simple. As you would expect, different groups of people will make different assumptions. Which assumptions are made should relate to the intended use and the intended user of the model, the region over which the model needs

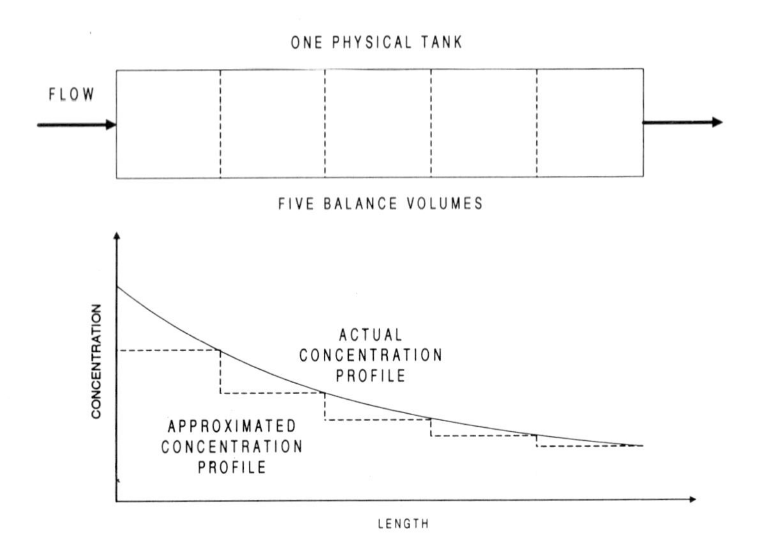

Figure 3.4: Approximating long tanks by a series of well mixed balance volumes.

to represent the process, and the required accuracy. All of these should be in the model specification discussed in the previous section.

It is very important in modelling to carefully write down the assumptions that you make. You or someone else might want to come back and change them in the future. Often they will need to be modified or removed in order to satisfy the specifications. Few modellers will select all the appropriate assumptions the first time. Of course there is unlikely to be a uniquely correct set of assumptions.

3.5 Modelling Hydraulics

Let us consider the simple example of a tank with one inflow, one outflow, and a single component (water) shown in Figure 3.5. There are no reactions involved (this is the first assumption), so water is not generated or consumed in the tank.

Our general mass balance equation becomes:

d/dt(water in tank) = (inflow rate) - (outflow rate)

We might introduce some mathematical symbols and get:

$$\frac{dM}{dt} = m_{in} - m_{out} \tag{3.5}$$

where M is kg of water, m is water flowrate in kg/s, and t is time in seconds. Each term in our equation must have the same units, in this case kg/s. This is the **dimensionality check** which you should apply every time you write or change

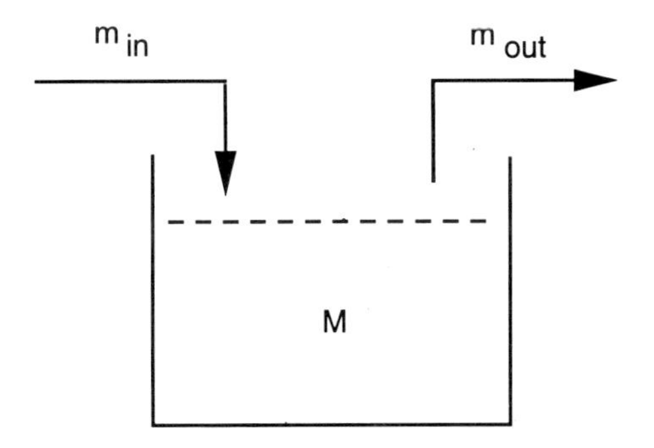

Figure 3.5: Simple tank example.

an equation. All our modelling groups should get this result. But this is where diversity begins.

Example 3.1 (Constant volume tank) *Group 1 decides that the contents of the tank are effectively constant (this is the second assumption). This means that* $dM/dt = 0$ *(that is, M does not change) and they rewrite the model as:*

$$m_{out} = m_{in} \tag{3.6}$$

Figure 3.6 shows the response of the outflow rate (bottom plot) to a step change in the inflow rate (top plot), simulated by the MATLAB program `Example_3_1.m`.

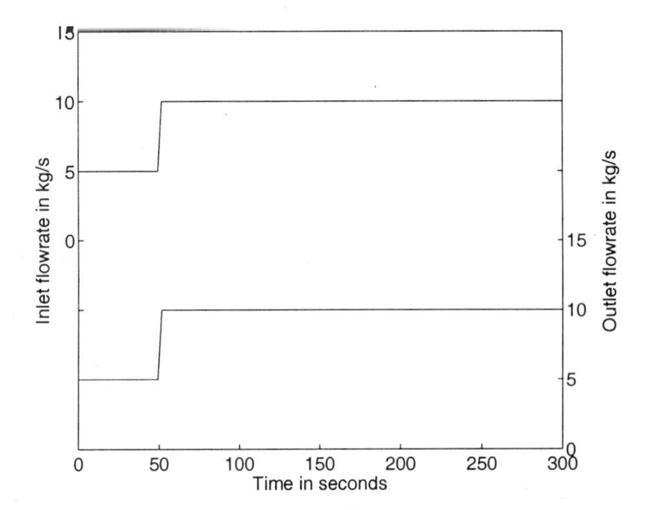

Figure 3.6: Response of constant volume tank.

Example 3.2 (Tank with overflow weir) *Group 2 decides that the outflow is determined by an overflow weir (this is a different second assumption). They consult a book to find the Francis weir formula which gives:*

$$m_{out} = 1.015 \rho n L h^{1.5} \tag{3.7}$$

where ρ is the density, n is the number of weirs, L is the weir length in metres, and h is the height of the water above the weir. The exponent of h will vary depending upon the weir shape. The value of h can be expressed as:

$$h = \frac{M}{\rho A} - H \tag{3.8}$$

where A is the tank area and H is the weir height. As well as these two algebraic equations, we also have our original mass balance:

$$\frac{dM}{dt} = m_{in} - m_{out} \tag{3.9}$$

You can see that two groups can come up with quite different models. Figure 3.7 shows the response of the outflow rate (bottom plot) for this model to a step change in the inflow rate (top plot), simulated by the MATLAB program `Example_3_2.m`.

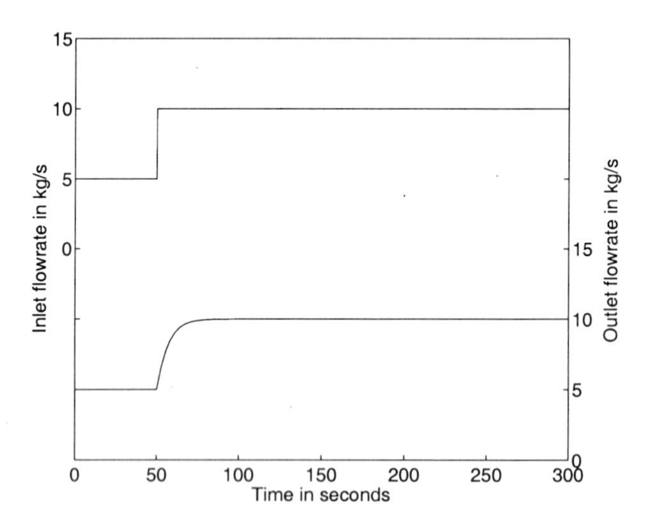

Figure 3.7: Response of tank with overflow weir.

Which model is the right one? It depends! The accuracy of the different model predictions will depend upon how well their assumptions match reality on the plant. As we have discussed under specifications above, you must know what accuracy you need for the purpose you wish to make of the model. Systems with more than one component are modelled in the same way. We can write a mass balance for each component in turn, or an overall mass balance and a mass balance for all but one component. A mass balance for a component is sometimes referred to simply as a *component balance.*

Example 3.3 (Simple buffer tank) *Consider a tank of volume V megalitres with one inflow and one outflow (Figure 3.8). The influent wastewater has two components, water and 200mg/L of biodegradable COD. Assume no water or nutrient generated or consumed in the tank (no biomass present).*

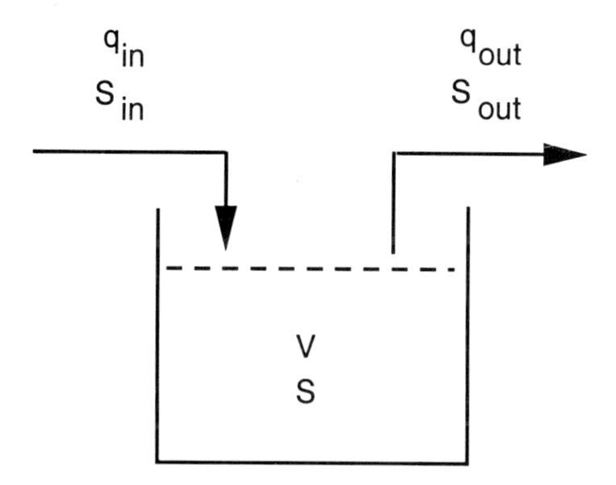

Figure 3.8: Simple buffer tank.

An overall mass balance can be written as:

$$\frac{d(\rho V)}{dt} = \rho_{in} q_{in} - \rho_{out} q_{out} \tag{3.10}$$

where ρ is the density, V is the volume and q is the volumetric flowrate. A component balance for the soluble COD nutrient is:

$$\frac{d(VS_S)}{dt} = q_{in} S_{S,in} - q_{out} S_{S,out} \tag{3.11}$$

where S_S is the concentration (mg/L). A component balance for the pure water is:

$$\frac{d(VS_W)}{dt} = q_{in} S_{W,in} - q_{out} S_{W,out} \tag{3.12}$$

It should be clear that the overall mass balance (Equation 3.10) is simply the sum of the two component balances (Equations 3.11 and 3.12). Therefore, only two of these three equations are independent. The golden rule *is, if you have C components, then you only use C mass balances.*

In wastewater treatment, it is conventional to use the overall balance and a balance for each nutrient, leaving out the pure water balance. It would also be normal to assume equal and constant densities (nutrient concentrations are typically very small), so that the overall mass balance (Equation 3.10) simplifies to:

$$\frac{dV}{dt} = q_{in} - q_{out} \tag{3.13}$$

Equations 3.13 and 3.11 make up our model's mass balances. There are two commonly used assumptions relating the outflow conditions to conditions inside the tank.

1. Assuming constant volume implies that $dV/dt = 0$ so that $q_{out} = q_{in}$.

2. Assuming a well-mixed tank implies that $S_{S,out} = S_S$.

Our model therefore becomes simply:

$$\frac{dS_S}{dt} = \frac{q_{in}}{V}\left(S_{S,in} - S_S\right) \tag{3.14}$$

The MATLAB program `Example_3_3.m` can be used to generate Figure 3.9 which shows a typical outflow rate and concentration responses to step changes in the inflow rate q_{in} and the inflow concentration $S_{S,in}$ respectively.

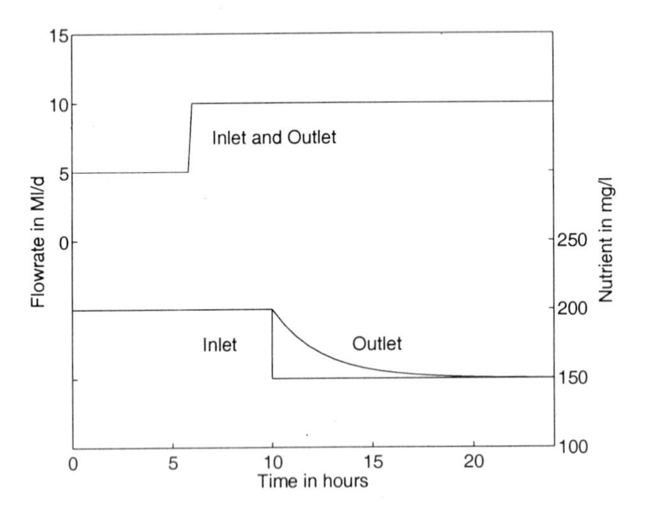

Figure 3.9: Buffer tank flow and concentration responses.

The outlet flowrate will respond instantly to changes in the inflow rate due to the constant volume assumption, but outflow concentrations will always respond slowly to inflow changes as the tank contents buffer or damp the disturbances. The larger is the residence time or time constant, the ratio V/q_{in}, the slower will be the responses.

3.6 Modelling Oxygen Transfer

Mass transfer is the movement of a component between two phases. The notable example in aerated activated sludge vessels is the transfer of oxygen from the air into the water so that it can be utilised by the biomass.

In principle, the oxygen must move from the bulk air to the water surface, dissolve in the water, then move from the surface into the bulk liquid. In waste treatment we typically assume that the extreme turbulence of the air and air bubbles results in the water at the surface of the bubbles being saturated with oxygen.

We can then represent the mass transfer by the equation:

> (rate of oxygen transfer) = (mass transfer coefficient) * (air-water surface area) * (concentration difference)

or in symbols:

$$r_a = K_L a(S_{O,sat} - S_O) \tag{3.15}$$

where S_O and $S_{O,sat}$ are the dissolved oxygen concentration and the saturated dissolved oxygen concentration respectively. The first two terms on the right, K_L and a, are typically combined and related to the air flowrate, often linearly, to give:

$$r_a = K_a q_a(S_{O,sat} - S_O) \tag{3.16}$$

where K_a is a constant which must be determined for a particular installation at a particular operating point. The $K_L a$ is a lumped parameter of several phenomena. The K_L part can be seen as an absorption coefficient and the a part as an area/volume ratio. The area and volume for an open tank without aeration devices refer to the contact area between water and air and to the tank volume respectively. In an aerator, the area is related to the bubble size, which is a function of the aeration equipment and the air flowrate. The contact between the air bubbles and the water is important. Small bubbles rise more slowly than large bubbles and have longer contact time. They also give a larger contact area than large bubbles. The oxygen transfer rate is different for wastewater than for clean water. The proportionality factor between clean water (CW) and wastewater is called α:

$$K_L a = \alpha \ (K_L a)_{CW} \tag{3.17}$$

Similarly the saturation concentration for dissolved oxygen is different in clean water and wastewater:

$$S_{O,sat} = \beta \ (S_{O,sat})_{CW} \tag{3.18}$$

Example 3.4 (Simple aerated tank) *Let us consider dissolved oxygen as a third component in the tank of Example 3.3. The additional component balance for oxygen will be:*

$$\frac{d(VS_O)}{dt} = q_{in}S_{O,in} - q_{out}S_O + K_a q_a(S_{O,sat} - S_O)V \tag{3.19}$$

Making the usual assumptions of constant volume and a well-mixed tank, Equation 3.19 simplifies to:

$$\frac{dS_O}{dt} = \frac{q_{in}}{V}(S_{O,in} - S_O) + K_a q_a (S_{O,sat} - S_O) \tag{3.20}$$

Figure 3.10 shows how the dissolved oxygen (DO) concentration reacts to air flowrate changes. The effect of the air flowrate times DO concentration nonlinearity is clearly seen. This figure was produced by the MATLAB program `Example_3_4.m`.

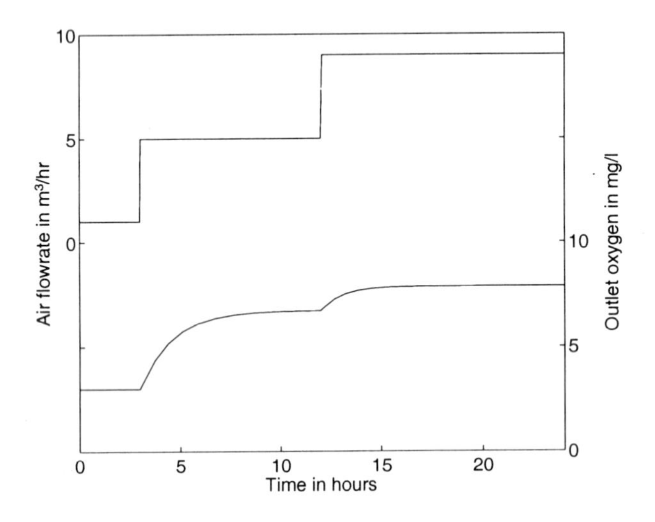

Figure 3.10: Dissolved oxygen responses.

Example 3.5 (Simple respirometer test) *An elementary respirometer test can be described as follows. Sludge is aerated for a while in a bottle. After the aeration has been done, the bottle is closed and the dissolved oxygen (DO) concentration is measured. The DO concentration will decrease until it reaches a very low value.*

There is no influent flow rate and no additional DO, so the model of the DO concentration is simple:

$$\frac{dS_O}{dt} = r \tag{3.21}$$

The DO as a function of time looks like Figure 3.11.

Note, that the slope is linear in the beginning, indicating that r is constant and negative for large values of S_O *. As the DO concentration becomes small the slope gets smaller, which means that r goes to zero as* S_O *goes to zero.*

We will model r as a function:

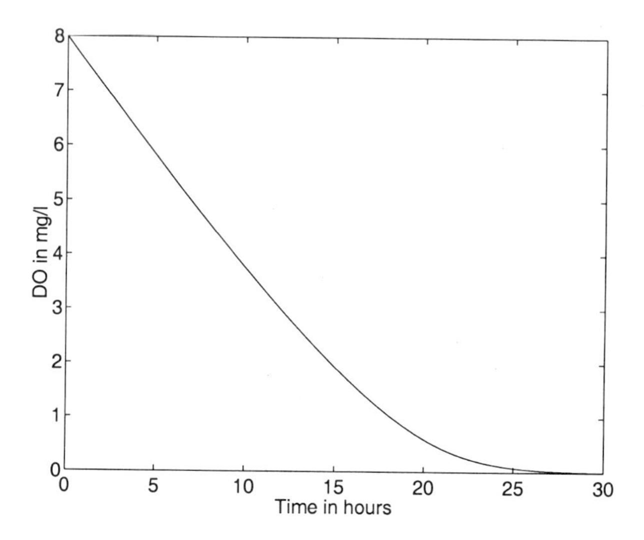

Figure 3.11: A simple respirometry curve.

$$r = -r_{max} \frac{S_O}{K_O + S_O} \tag{3.22}$$

The function makes the respiration rate zero for zero DO level and saturates to a constant value for large DO levels. It looks like Figure 3.12 and is usually referred to as the Monod function. The respiration rate is actually a function not only of the DO level but also of the available substrate. This will be shown in the next section.

Dissolved oxygen (DO) concentration in aerated tanks is typically controlled to a desired constant value by automatic adjustment of the air flowrate. While the appropriateness of this strategy can be debated, we will examine the strategy as a vehicle to illustrate the modelling of an automatic controller.

A DO sensor in the aeration tank generates a measurement signal y that is compared to a desired value y_d to generate an error e:

$$e = y_d - y \tag{3.23}$$

The air flowrate (preferable) or air header pressure is then adjusted according to some control algorithm. The most common continuous algorithm is the so-called Proportional plus Integral (PI) algorithm. The adjustment has two components, one proportional to the error, and the other proportional to the integral of the error (or sum of the errors) over time.

$$u = u_o + K_c \left(e + \frac{1}{\tau_I} \int edt \right) \tag{3.24}$$

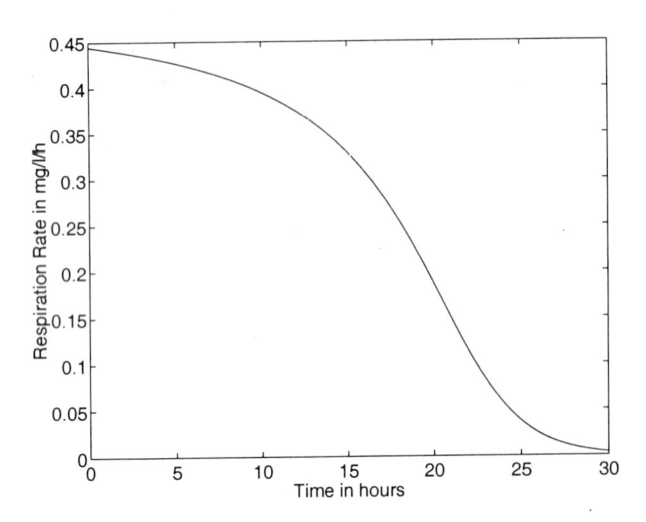

Figure 3.12: Illustration of the Monod function.

where u_o is the actuator signal at steady state. The tunable controller constants are the **gain** K_c and the **integral time** τ_I. Chapter 17 discusses feedback controller tuning.

Example 3.6 (DO controller) *In a simulator which solves differential and algebraic equations, the integral term in the controller is implemented by the following two equations in place of Equation 3.24:*

$$\frac{dz}{dt} = e \tag{3.25}$$

$$u = u_o + K_c \left(e + \frac{z}{\tau_I} \right) \tag{3.26}$$

The first equation integrates the error, calculated by Equation 3.23, and the second calculates the actuator signal. The MATLAB program `Example_3_6.m` *assumes the simplest of models for the DO sensor ($y = S_O$) and air flowrate actuator ($q_a = u$). Figure 3.13 shows the DO response (centre curve) to changes in influent DO (a disturbance) and in the controller setpoint y_d. The top curve in Figure 3.13 shows the air flowrate, corresponding to the control action taken.*

3.7 Modelling Biological Nutrient Removal

Modelling of systems involving chemical or biochemical reactions is only a little more complex. We simply write dynamic mass balances as we have above, only this time the generation and consumption terms are involved.

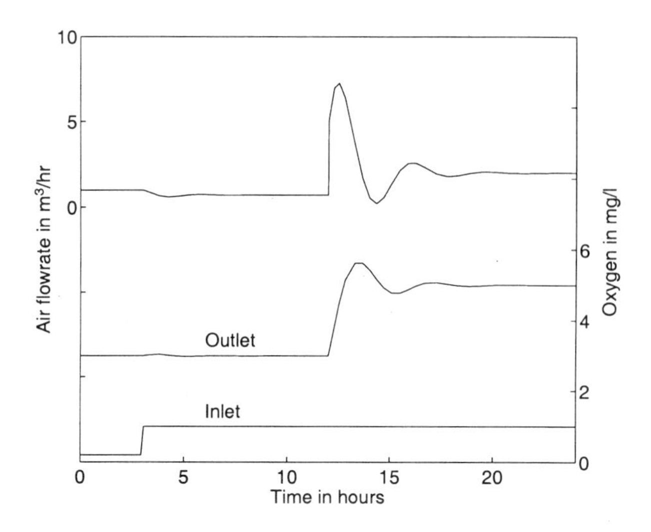

Figure 3.13: Controlled DO responses.

3.7.1 Modelling a Single Nutrient

The simplest example for the biological treatment of wastewater would involve three components, water, nutrient and biomass. For concentrations, we use the IAWQ convention of S for components dissolved in the wastewater and X for solid components (Figure 3.14).

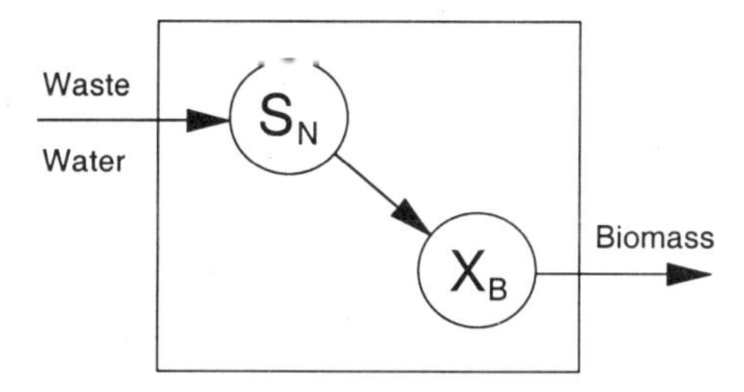

Figure 3.14: Simple nutrient removal.

The microorganisms absorb the nutrient from the wastewater using it to grow and to produce more biomass. If we assume a constant volume tank and constant density, the wastewater mass balance reduces to:

$$q_{out} = q_{in} \tag{3.27}$$

where q_{in} and q_{out} are the flowrates, typically with units of megalitres per day. The biomass mass balance will be:

$$\frac{d}{dt}(VX_B) = q_{in}X_{B,in} - q_{out}X_{B,out} + r_BV \tag{3.28}$$

where V is the tank volume, X_B is biomass concentration (typically mg/L) and r_B is the reaction rate ($mg/L/day$) for biomass growth. The last term on the right-hand side is the only change we have made to the form of our previous hydraulic models. The nutrient mass balance will be:

$$\frac{d}{dt}(VS_N) = q_{in}S_{N,in} - q_{out}S_{N,out} + r_N V \tag{3.29}$$

where S_N is nutrient concentration (mg/L) and r_N is the nutrient reaction rate ($mg/L/day$). We can assume a well-mixed tank so that:

$$X_{B,out} = X_B \quad and \quad S_{N,out} = S_N \tag{3.30}$$

If we can define our reaction or growth rates then the model is complete. Activated sludge stoichiometry and kinetics are conventionally presented in tabular form as in Table 3.3.

Process	Components		Kinetics
	Nutrient N	Biomass B	
Aerobic heterotrophic growth	$-\frac{1}{Y_B}$	1	$\hat{\mu}\left(\frac{S_N}{K_N+S_N}\right)X_B$

Table 3.3: Simple biological kinetics.

The expression for the reaction or growth rate for a particular component is obtained by summing all the terms in the appropriate column after multiplying each by the kinetic expression in the last column. For example,

$$r_B = \hat{\mu}\left(\frac{S_N}{K_N + S_N}\right)X_B \tag{3.31}$$

and

$$r_N = -\frac{1}{Y_B}\hat{\mu}\left(\frac{S_N}{K_N + S_N}\right)X_B \tag{3.32}$$

Note that the expressions 3.31 and 3.32 are Monod type expressions like Equation 3.22.

Example 3.7 (Simple biological reactor) *The biomass component and this simple reaction have been added to the buffer tank of Example 3.3. The results shown in Figure 3.15 were produced by the MATLAB program* `Example_3_7.m`.

You can see the typical fast response of the nutrient and the slow response of the biomass to a sudden halving of the influent substrate concentration.

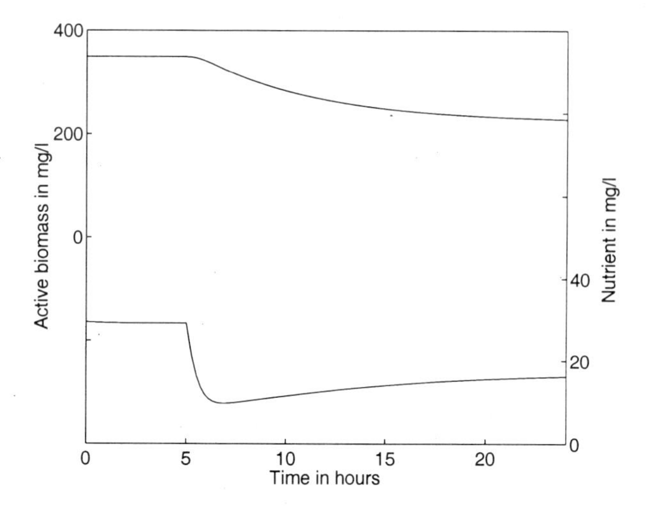

Figure 3.15: Simple bioreactor responses.

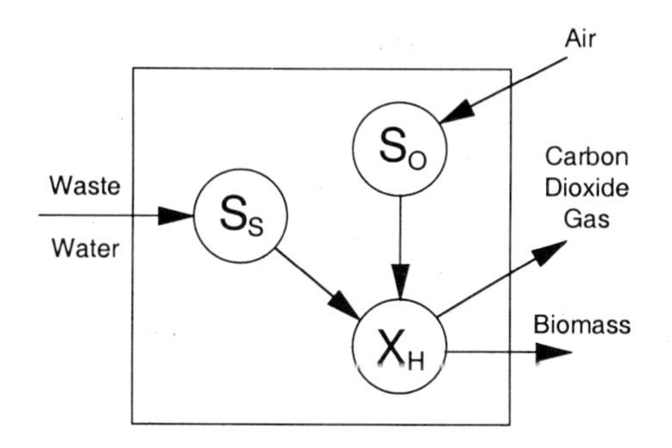

Figure 3.16: Soluble carbon removal.

3.7.2 Modelling Carbon Removal

The simplest example of the biological treatment of wastewater would involve four components; water, soluble carbon, oxygen and heterotrophic biomass. The carbonaceous nutrient and dissolved oxygen are absorbed by the microorganisms which grow to produce more biomass and which release carbon dioxide (Figure 3.16). The heterotrophic biomass mass balance will be:

$$\frac{d}{dt}(VX_H) = q_{in}X_{H,in} - q_{out}X_{H,out} + r_H V \tag{3.33}$$

where X_H is the biomass concentration (mg/L) and r_H is the reaction rate for biomass growth ($mg/L/day$). The carbon mass balance will be:

$$\frac{d}{dt}(VS_S) = q_{in}S_{S,in} - q_{out}S_{S,out} + r_S V \tag{3.34}$$

where S_S is the soluble carbon nutrient concentration (mg/L), expressed as soluble COD or TOC or some equivalent, and r_S is the nutrient reaction rate $(mg/L/day)$. The oxygen mass balance will be:

$$\frac{d}{dt}(VS_O) = q_{in}S_{O,in} - q_{out}S_{O,out} + r_OV + K_La(S_{O,sat} - S_O)V \tag{3.35}$$

where S_O is the oxygen concentration (mg/L), r_O is the oxygen consumption rate $(mg/L/day)$ and the K_La term is the mass transfer from the aeration. We have assumed a well-mixed tank so that:

$$X_{H,out} = X_H \quad and \quad S_{S,out} = S_S \quad and \quad S_{O,out} = S_O \tag{3.36}$$

If we can define our reaction or growth rates then the model is complete.

Process	Components			Kinetics
	Nutrient	Oxygen	Biomass	
Aerobic heterotrophic growth	$-\frac{1}{Y_H}$	$\frac{Y_H-1}{Y_H}$	1	$\hat{\mu_H}\left(\frac{S_S}{K_S+S_S}\right)\left(\frac{S_O}{K_{OH}+S_O}\right)X_H$
Decay of heterotrophs	$1-f_P$		-1	b_HX_H

Table 3.4: Kinetics for carbon removal.

Again activated sludge stoichiometry and kinetics are conventionally presented in tabular form as shown in Table 3.4 from which we can determine that:

$$r_H = \hat{\mu_H}\left(\frac{S_S}{K_S+S_S}\right)\left(\frac{S_O}{K_{OH}+S_O}\right)X_H - b_HX_H \tag{3.37}$$

and expressions for r_S and r_O can be found in a similiar manner. This completes our model for the aerobic removal of soluble carbon nutrient.

We now note that the reaction rate r_H is a function of both the substrate and dissolved oxygen concentrations. Each one of the terms looks like a Monod function. Also note that the yield coefficient Y_H is smaller than 1, so that the term $\frac{Y_H-1}{Y_H}$ is negative.

Example 3.8 (Aerated biological reactor) *The aerated carbon removal reactor described by Equations 3.33 to 3.36 is simulated by the MATLAB program* `Example_3_8.m`*. Some example responses are shown in Figure 3.17 to a four-fold increase in air flowrate at five hours and a 25 per cent decrease in influent COD at ten hours. The range of speeds of response of the DO, COD and biomass is evident.*

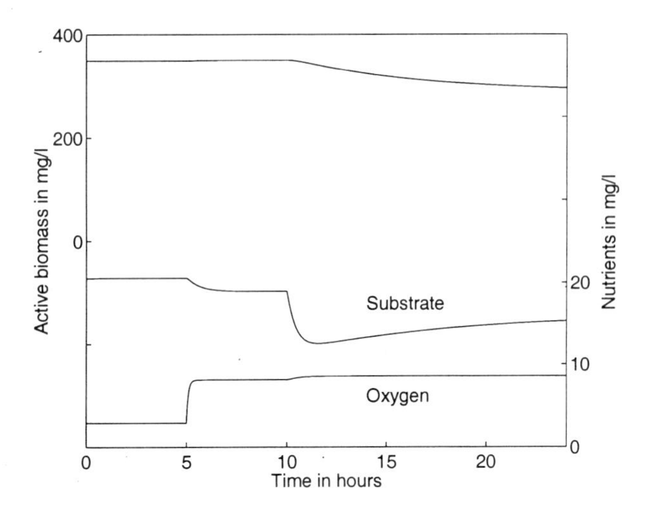

Figure 3.17: Aerated carbon removal responses.

Example 3.9 (DO and OUR) *As indicated in Equation 3.35 there is a close relationship between the DO concentration and the oxygen uptake rate r_O. With the simple complete mix reactor, we will show that the air flowrate can be directly related to the oxygen uptake rate, provided the DO is kept constant by a controller. The oxygen transfer rate can be written as a function of the air flowrate and is usually a function that saturates for larger air flowrates. In simplified analysis it is written as:*

$$K_L a = K_a \, q_a \tag{3.38}$$

where q_a is the air flowrate. For increasing air flowrate then the parameter K_a decreases gradually towards zero. For practical purposes the first and second term on the right hand side of Equation 3.35 can be neglected compared to the other terms. Furthermore, since the DO is assumed to be perfectly controlled, the DO concentration is constant at $(\bar{S}_O)$, so we have:

$$0 = r_O + K_a \, q_a \, (S_{O,sat} - \bar{S}_O) \tag{3.39}$$

and consequently,

$$r_O = -K_a \, q_a \, (S_{O,sat} - \bar{S}_O) = k \, q_a \tag{3.40}$$

where k is a constant. Thus the oxygen uptake rate can be approximately detected by the air flowrate. Note, that we have made a few approximations, so the proportionality does not hold for all load levels.

3.8 Model Verification

Once you have a model, whether you derived it yourself or constructed it from a library of models, the last thing you should do is believe in it. At least, not until it has been verified and validated, and even then be skeptical. This is so for all processes, but particularly so for biological processes where microbial populations change and adapt to varying conditions and our basic understanding remains rudimentary.

Model verification usually involves setting up the model with any reasonable set of parameters and then systematically examining its behaviour to changes in input conditions, system states, and parameters and its behaviour in limiting and extreme situations. The goal is to delineate the operating region in which the model behaves sensibly, and to determine if this region encompasses the region in which you will use the model. If the model does not pass this qualitative testing, then we must go back to the modelling and examine our assumptions or find other models.

Example 3.10 (Carbon removal process) *As an example, consider the model for a simple carbon removal process shown in Figure 3.18 and described in detail in Appendix B. It assumes a single well mixed aeration tank and a perfect clarifier.*

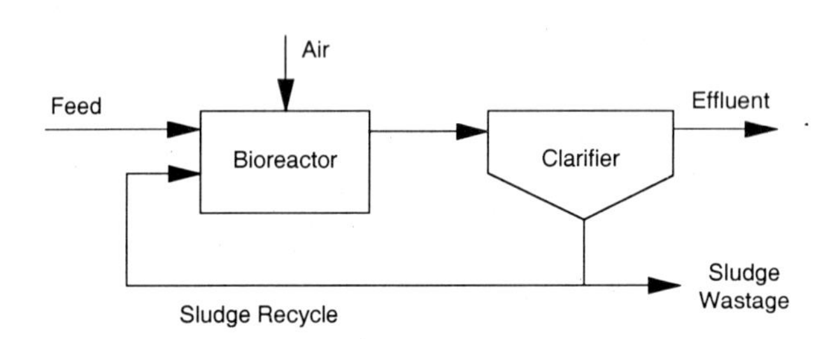

Figure 3.18: Example carbon removal process.

The model consists of three equations (mass balances for substrate, oxygen and heterotrophs), with the numbers of classified variables shown in Table 3.5.

Variable Class	Number
disturbances	2
manipulated variables	3
states	3
outputs	2
known parameters	2
unknown parameters	8

Table 3.5: Variable classification.

The operating space is described by the three manipulated variables: sludge recycle flowrate, sludge wastage flowrate and air flowrate. A systematic examination of the extremes for these variables showed the model to be very robust, except where the oxygen went close to zero. In this case the results were still reasonable, but the equations became ill-conditioned and took a very long time to solve.

Another useful outcome of a systematic study arises if you do a sensitivity study of the model inputs and parameters. This gives you an answer to the question "which parameters to adjust" when it comes to parameter estimation and model validation. Many of the more complex BNR models are not identifiable since there are many more parameters than there are feasible measurements. Assuming reasonable values for insensitive parameters can alleviate this problem.

One way of defining the sensitivity of a dynamic system of equations:

$$\frac{dx}{dt} = f(x, u, d, p) \tag{3.41}$$

is to examine the change in the function $f()$ for changes in the system parameters p.

This was done for our simple carbon removal process using MATLAB program `Example_B_2.m`. Detailed results are shown in Appendix B with a summary ranking shown in Table 3.6 below for the eight parameters. It is clear from this that four of the parameters could be set to literature values, leaving four parameters that could be usefully used for model fitting: $\hat{\mu}_H$, K_S, Y_H and b_H.

Parameter	$\frac{dS_S}{dt}$	$\frac{dS_O}{dt}$	$\frac{dX_H}{dt}$
$\hat{\mu}_H$	High	Medium	High
K_S	High	Medium	Medium
K_{OH}	Low	Low	Low
Y_H	High	High	Medium
b_H	Medium	None	None
f_P	Low	None	None
a	None	Medium	None
b	None	Medium	None

Table 3.6: Dynamic sensitivity results.

The steady state sensitivity of the same system is obtained by solving:

$$f(x, u, d, p) = 0 \tag{3.42}$$

to examine the change in the system states x for changes in the system parameters p. This was done for our simple carbon removal process example with the same program. Detailed results are shown in Appendix B with a summary ranking shown in Table 3.7 below for the eight parameters. It is clear from this that the

Parameter	S_S	S_O	X_H
$\hat{\mu}_H$	High	Very Low	Very Low
K_S	High	Very Low	Very Low
K_{OH}	Very Low	Very Low	Very Low
Y_H	Very Low	High	High
b_H	Medium	Low	Low
f_P	Very Low	Very Low	Very Low
a	Low	High	Very Low
b	Low	High	Very Low

Table 3.7: Steady state sensitivity results.

parameters a and b should be added to the four parameters chosen above to ensure good fitting of steady state values.

This does not answer the question of whether these six parameters are indeed identifiable given the available measurements. Identifiability raises many issues including:

- can all the states be excited by the manipulated variables (is the system controllable)?
- can all the states be determined from the measurements (is the system observable)?
- are the parameters independent (uncorrelated)?

Consider the three equations for our example process (see Appendix B):

$$\begin{aligned}\frac{dS_S}{dt} &= \frac{q_F}{V}(S_{SF} - S_S) \\ &\quad -\frac{\hat{\mu}_H}{Y_H}\left(\frac{S_S}{K_S+S_S}\right)\left(\frac{S_O}{K_{OH}+S_O}\right)X_H + (1-f_P)b_H X_H\end{aligned} \tag{3.43}$$

$$\begin{aligned}\frac{dS_O}{dt} &= \frac{q_F}{V}S_{OF} - \frac{q_F+q_R}{V}S_O + \frac{Y_H-1}{Y_H}\hat{\mu}_H\left(\frac{S_S}{K_S+S_S}\right)\left(\frac{S_O}{K_{OH}+S_O}\right)X_H \\ &\quad +a\left(1-e^{-\frac{q_A}{b}}\right)(S_{O,sat} - S_O)\end{aligned} \tag{3.44}$$

$$\begin{aligned}\frac{dX_H}{dt} &= \frac{q_F}{V}X_{HF} - \frac{q_W}{V}\left(\frac{q_F+q_R}{q_W+q_R}\right)X_H \\ &\quad -\frac{\hat{\mu}_H}{Y_H}\left(\frac{S_S}{K_S+S_S}\right)\left(\frac{S_O}{K_{OH}+S_O}\right)X_H + (1-f_P)b_H X_H\end{aligned} \tag{3.45}$$

If we examine these equations, we find that the two manipulated variables q_R and q_W only appear together in the same term in the third equation.

Therefore they are not independent as far as our model is concerned. Despite this the following digraph (Figure 3.19) shows that all states can still be excited though not independently? Of course there needs to be sufficient excitation, which means that q_A and q_R (or q_W) should be varied throughout the specified operating region.

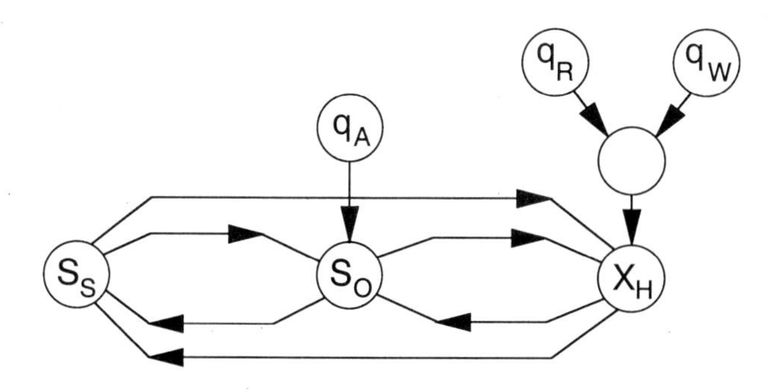

Figure 3.19: Manipulated variable - state connectivity.

In our example, the two measurements, S_S and S_O, are two of the states and Figure B2.10 shows that the third state X_H is well connected with the other two. Only a quantitative observability test can determine for certain whether X_H can be estimated from the other two. These tests are limited in that they apply only to linear systems so linearisation at many points is required.

The correlation between parameters is difficult to calculate a priori. Most model fitting programs will give this data to enable you to evaluate the fit. However, by inspection it is often possible to see likely correlations, for example that between $\hat{\mu}_H$ and K_S (because S_S is normally much less than K_S).

The conclusions of the verification study of our example could be that we have seven variables to estimate, six parameters and one state, and perhaps some doubt on whether we can reasonably estimate both $\hat{\mu}_H$ and K_S, unless we make sure we perform experiments with a wide range of values of S_S.

3.9 Summary

We have presented a technique for building simple dynamic models. These simple dynamic models can be readily extended with other classes of microorganisms and additional nutrients, provided the mechanisms involved are known. The next chapter, Chapter 4 Advanced Modelling, will summarize extensions for nitrogen and phosphorus removal. The IAWQ AS Models are also such extensions.

Once tank and clarifier models are available it is simple to configure quite complex processes. The advanced modelling chapter also discusses the modelling of settling and clarification.

We have also discussed model specification and verification using simple examples. Perhaps the more difficult task is obtaining reasonable estimates of the parameters, particularly those describing reaction rates and settling characteristics. These are wastewater dependent as well as dependent on the particular population of biomass. These are also time dependent. Chapters 6 to 8 on model fitting discuss the problem of parameter determination and subsequent model validation.

However, with some care in the model formulation and parameter estimation and with a reasonable dynamic simulator, dynamic simulations can become extremely useful design and operational tools.

3.10 Further Reading

- for information on modelling processes:

  1. Cameron & Hangos (1999)
  2. Johansson (1993)
  3. Lee *et al.* (1998)
  4. Seborg *et al.* (1989)

- for information on nutrient removal processes:

  1. Henze *et al.* (1995)
  2. the IAWQ AS Models in Henze *et al.* (1987a), Henze *et al.* (1987b) and Gujer *et al.* (1994).

- For distributed parameter modelling of activated sludge tanks see Lee *et al.* (1999).

- more complex BNR models are not identifiable since there are many more parameters than there are feasible measurements as discussed in Beck (1986), Olsson (1989) and Jeppsson & Olsson (1993).

- a quantitative observability test Ayesa *et al.* (1991).

- a detailed discussion of the oxygen transfer rate are found in United States Environmental Protection Agency (1989).

3.11 Exercises

1. Choose a process unit at your treatment plant and write two specifications, for a design model and for a model for operator training.

2. Copy the MATLAB program `Example_3_3.m` and use it to investigate the effect of residence time on the concentration responses in a simple buffer tank.

3. Copy the MATLAB program `Example_3_3.m`, modify the simulation to incorporate an overflow weir rather than making the constant volume assumption, and investigate the effect of this on the responses. Hint: look at the MATLAB program `Example_3_2.m`.

4. Copy the MATLAB program `Example_3_6.m` and use it to investigate the effect of different controller parameters on the DO responses. Can you tune it better?

5. Copy the MATLAB program `Example_3_8.m` and modify the simulation by adding an insoluble COD component which hydrolyses to soluble COD according to the kinetics in Table 3.8 below. Write down the component balance and any assumptions involved. Use the modified simulation to investigate the effect of the new component on the speeds of the concentration responses in the tank. Use values of 3.0 and 0.1 for the parameters K_h and K_X respectively.

Process	Components		Kinetics
	X_S	S_S	
Aerobic hydrolysis	-1	1	$K_h \left(\frac{X_S/X_H}{K_X+X_S/X_H}\right) X_H$

Table 3.8: Simple hydrolysis kinetics.

Chapter 4

Advanced Modelling

This chapter takes a look at a selection of advanced modelling topics relevant to biological wastewater treatment. The goal is to introduce you to the modelling of typical wastewater treatment unit operations and unit processes so that, when you need to, you can read and understand the more detailed modelling literature from manufacturers and researchers. We will introduce nitrogen and phosphorus removal in activated sludge, prefermentation and anaerobic digestion, a brief look at immobilised films, more on hydraulic modelling, settling and clarification and finally a brief discussion on modelling batch processes.

4.1 Modelling Biological Nutrient Removal

This section takes a simplified look at modelling nitrogen and phosphorus removal.

4.1.1 Modelling Nitrogen Removal

The simplest example of the biological removal of carbon and nitrogen from wastewater would involve seven components; water, soluble carbon, oxygen and heterotrophic biomass as before plus ammonia (NH), nitrate (NO) and autotrophic biomass as shown previously in Figure 2.7.

Nitrogen removal involves two sequential steps (Figure 4.1). The aerobic growth of autotrophs consumes soluble carbon, ammonia and dissolved oxygen to produce extra biomass and nitrates in solution. Occasionally this step is further divided into two, one step producing nitrites and the second further oxidising the nitrites to nitrates. The second major step is the anoxic growth of heterotrophs which use the nitrates as a source of oxygen and produce extra biomass and nitrogen gas. The mass balances for heterotrophic biomass, soluble carbon, and oxygen are identical to the previous equations.

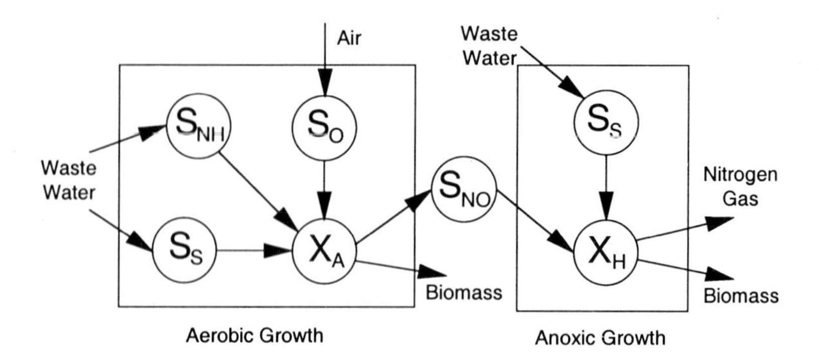

Figure 4.1: Nitrogen removal.

The autotrophic biomass mass balance will be:

$$\frac{d}{dt}(VX_A) = q_{in}X_{A,in} - q_{out}X_{A,out} + r_AV \tag{4.1}$$

The ammonia nutrient mass balance will be:

$$\frac{d}{dt}(VS_{NH}) = q_{in}S_{NH,in} - q_{out}S_{NH,out} + r_{NH}V \tag{4.2}$$

The nitrate mass balance will be:

$$\frac{d}{dt}(VS_{NO}) = q_{in}S_{NO,in} - q_{out}S_{NO,out} + r_{NO}V \tag{4.3}$$

We have assumed a well-mixed tank so that:

$$X_A = X_{A,out} \quad and \quad S_{NH} = S_{NH,out} \quad and \quad S_{NO} = S_{NO,out} \tag{4.4}$$

Tables 4.1 and 4.2 define our reaction or growth rates for carbon plus nitrogen removal from which we can determine the rates r_H, r_A, r_S, r_{NH} and r_{NO} in the usual way.

This completes our model for the removal of soluble carbon and nitrogen nutrients. The full IAWQ AS Model No 1 also includes alkalinity balances and alkalinity effects on kinetics.

Example 4.1 (Carbon and nitrogen removal) *The MATLAB program called* `Example_4_1.m` *simulates the model in Tables 4.1 and 4.2 in a batch reactor (zero inflow and outflow). The results of the aerobic and anoxic phases are shown in Figure 4.2.*

Three phases are clear in the left plot in Figure 4.2, growth of heterotrophs until the carbon substrate is consumed, growth of autotrophs until the ammonium is consumed, and a slow decay of the biomass. The air flowrate was constant so

Process	Components				
	Carbon	Oxygen	Ammonium	Nitrate	Heterotr Biomass
Aerobic heterotrophic growth	$-\frac{1}{Y_H}$	$\frac{Y_H-1}{Y_H}$	$-i_{XB}$		1
Anoxic heterotrophic growth	$-\frac{1}{Y_H}$		$-i_{XB}$	$\frac{Y_H-1}{2.86Y_H}$	1
Aerobic autotrophic growth		$\frac{Y_A-4.57}{Y_A}$	$-i_{XB}-\frac{1}{Y_A}$	$\frac{1}{Y_A}$	
Decay of heterotrophs	$1-f_P$		$i_{XB}-f_P i_{XP}$		-1
Decay of autotrophs	$1-f_P$		$i_{XB}-f_P i_{XP}$		

Table 4.1: Kinetics for carbon and nitrogen removal.

Process	Components	Kinetics
	Autotr Biomass	
Aerobic heterotrophic growth		$\hat{\mu}_H \left(\frac{S_S}{K_S+S_S}\right)\left(\frac{S_O}{K_{OH}+S_O}\right) X_H$
Anoxic heterotrophic growth		$\hat{\mu}_H \left(\frac{S_S}{K_S+S_S}\right)\left(\frac{K_{OH}}{K_{OH}+S_O}\right)\left(\frac{S_{NO}}{K_{NO}+S_{NO}}\right) \eta_g X_H$
Aerobic autotrophic growth	1	$\hat{\mu}_A \left(\frac{S_{NH}}{K_{NH}+S_{NH}}\right)\left(\frac{S_O}{K_{OA}+S_O}\right) X_A$
Decay of heterotrophs		$b_H X_H$
Decay of autotrophs	-1	$b_A X_A$

Table 4.2: Kinetics for carbon and nitrogen removal (contd).

that the DO value increased as the oxygen uptake rate decreased. Again, the right plot in Figure 4.2 shows three phases, an initial decrease of DO to zero with a little nitrification, denitrification and consumption of carbon substrate, and finally microorganism decay resulting in an increase in COD.

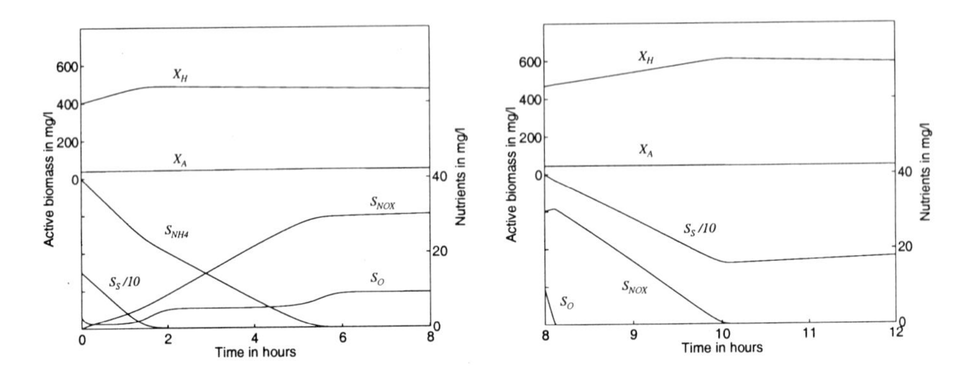

Figure 4.2: Nitrification and denitrification.

4.1.2 Modelling Phosphorus Removal

Phosphorus removal is a much more complex process and is probably best considered diagrammatically in the beginning (Figure 4.3). It was also described previously (see Figure 2.8).

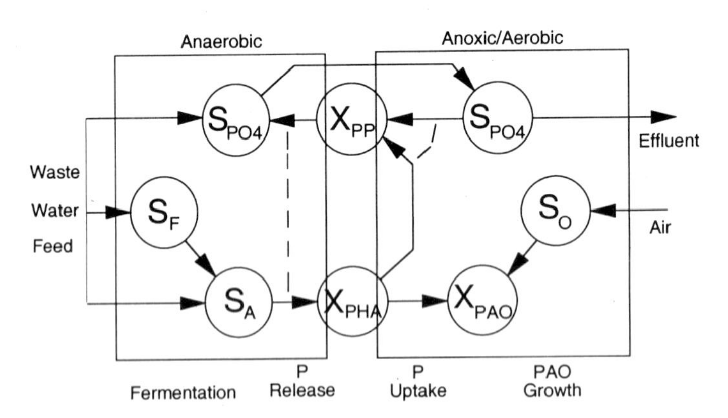

Figure 4.3: Phosphorus removal.

The three X components are all parts of the phosphorus accumulating organisms (PAO). The two dashed lines show linked transformations. Four basic mechanisms are indicated:

- Fermentation of fermentable COD, S_F, to volatile fatty acids (VFA), S_A, which can be utilised by the PAO microorganisms to store carbon as poly-hydroxyl-alkanoates (PHA), X_{PHA}.
- Phosphorus release from poly-phosphate (PP), X_{PP}, into solution at the same time as the VFA is converted to PHA.
- Phosphorus uptake from solution to PP utilising the PHA and dissolved oxygen, S_O.
- Growth of the PAO biomass, X_{PAO}, also utilising the PHA and dissolved oxygen.

One of the complicating factors is that the heterotrophs and the denitrification process also compete for the VFA. The *art* of P removal is to maintain conditions resulting in PAO growth and in more P uptake than P release. It remains an *art* since the correct conditions remain a subject of some debate and may well be dependent on the wastewater and other environmental conditions (such as alkalinity, redox potential, and temperature).

This model of the P removal process introduces five more mass balances and states compared to the carbon and nitrogen removal model - S_A, S_{PO4}, X_{PHA}, X_{PP}, X_{PAO}. Tables 4.3 and 4.4 represent the basic stoichiometry and kinetics.

Process	Components						
	S_F	S_A	S_{O2}	S_{PO4}	X_{PHA}	X_{PP}	X_{PAO}
Fermentation	-1	1					
P Release		-1		Y_{PO4}	1	$-Y_{PO4}$	
P Uptake			$-Y_{PHA}$	-1	$-Y_{PHA}$	1	
PAO Growth			$1-\frac{1}{Y_H}$	$-i_{PBM}$	$-\frac{1}{Y_H}$		1
PHA Breakdown		1			-1		
PP Breakdown				1		-1	
PAO Breakdown				ν_P			-1

Table 4.3: Kinetics for phosphorus removal.

The aerobic and anoxic growth of heterotrophs on fermentable substrate S_F and volatile fatty acids S_A have identical stoichiometry, but have different factors which multiply the normal kinetic rate expressions (those in the last column of Table 4.2). These additional multiplicative factors are for S_F:

$$\frac{S_F}{S_F + S_A} \frac{S_{PO4}}{K_P + S_{PO4}} \tag{4.5}$$

and for S_A:

$$\frac{S_A}{S_F + S_A} \frac{S_{PO4}}{K_P + S_{PO4}} \tag{4.6}$$

Process	Kinetics
Fermentation	$q_{FE}\left(\frac{K_{O2}}{K_{O2}+S_{O2}}\right)\left(\frac{K_{NO3}}{K_{NO3}+S_{NO3}}\right)\left(\frac{S_F}{K_{FE}+S_F}\right)X_H$
P Release	$q_{PHA}\left(\frac{S_A}{K_A+S_A}\right)\left(\frac{X_{PP}/X_{PAO}}{K_{PP}+X_{PP}/X_{PAO}}\right)X_{PAO}$
P Uptake	$q_{PP}\left(\frac{S_{O2}}{K_{O2}+S_{O2}}\right)\left(\frac{S_{PO4}}{K_P+S_{PO4}}\right)\left(\frac{X_{PHA}/X_{PAO}}{K_{PHA}+X_{PHA}/X_{PAO}}\right)\left(\frac{K_{MAX}-X_{PP}/X_{PAO}}{K_{IPP}+K_{MAX}-X_{PP}/X_{PAO}}\right)X_{PAO}$
PAO Growth	$\hat{\mu}_{PAO}\left(\frac{S_{O2}}{K_{O2}+S_{O2}}\right)\left(\frac{S_{PO4}}{K_P+S_{PO4}}\right)\left(\frac{S_{NH}}{K_{NH}+S_{NH}}\right)\left(\frac{X_{PHA}/X_{PAO}}{K_{PHA}+X_{PHA}/X_{PAO}}\right)X_{PAO}$
PHA Breakdown	$b_{PHA}X_{PHA}$
PP Breakdown	$b_{PP}X_{PP}$
PAO Breakdown	$b_{PAO}X_{PAO}$

Table 4.4: Kinetics for phosphorus removal (contd).

The additional mass balances can be integrated with the nitrogen removal model discussed above by replacing the single component balance for S_S by two component balances for S_F and S_A, and modifying the kinetic expressions in Table 4.2 by multiplying them by the expressions in Equations 4.5 and 4.6 respectively. The full IAWQ AS Model No 2 also includes alkalinity balances and alkalinity effects on the kinetics.

Example 4.2 (phosphorus removal) *The MATLAB program called* `Example_4_2.m` *simulates the model in Tables 4.3 and 4.4 in a batch reactor (zero inflow and outflow). The results of the anaerobic and aerobic phases are shown in Figure 4.4 respectively. The left plot in Figure 4.4 shows phosphorus release, consisting of the companion processes of conversion of acetate to PHA and of insoluble polyphosphate to soluble phosphates. The right plot in Figure 4.4 shows four phases, microorganism growth on acetate, microorganism growth on PHA with phosphorus uptake, continued growth on PHA without release, and finally some phosphorus release due to solids decay.*

4.2 Modelling Prefermentation

Fermentation is the anaerobic breakdown of organic carbon into volatile fatty acids (VFA), which promote biological phosphorus uptake as we discussed in the previous section. A simple single-step fermentation model was presented in the previous section in Tables 4.3 and 4.4. There are two other mechanisms possibly involved. If insoluble organic carbon is used as a feed, as is often the case, this must first be hydrolysed to the soluble form. A less desirable mechanism is the breakdown of VFA into methane.

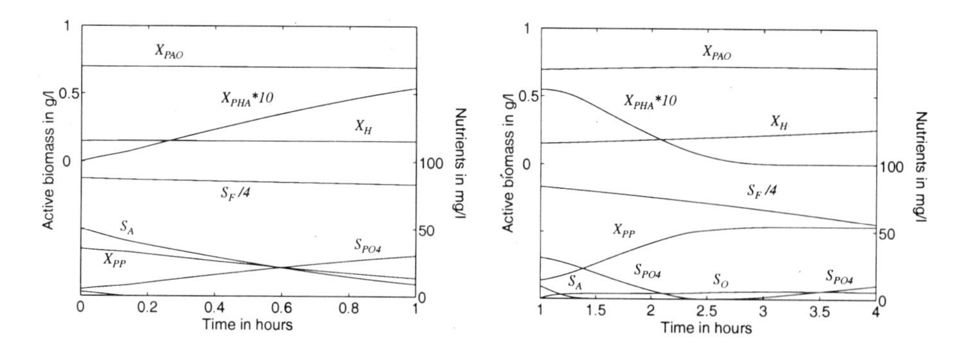

Figure 4.4: Phosphorus release and uptake.

The two classes of organisms involved are the acidogens, which produce the VFA, and the methanogens, which break down the VFA. Hydrolysis is an enzymatic reaction where organisms are indirectly involved by secreting the enzymes. The IAWQ AS Model No 2 model represents fermentation very simply as two sequential reactions as shown in Figure 4.5.

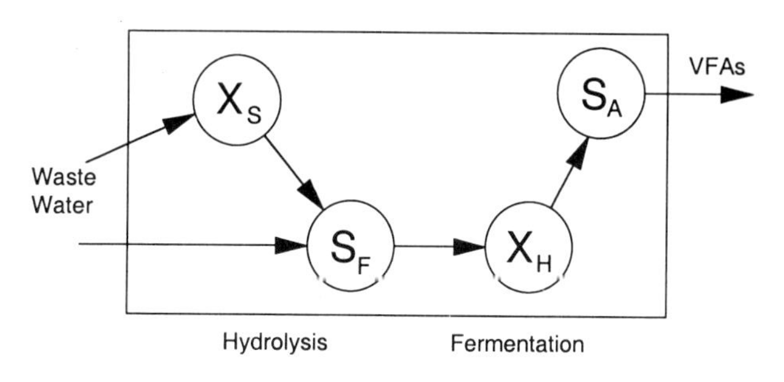

Figure 4.5: Simple fermentation model.

In the first step insoluble organic carbon, X_S, is hydrolysed to soluble fermentable substrate, S_F, by enzyme catalysed extracellular reactions. In some circumstances this step may be mass transfer limited, for example if the solid particles are large so that they have a small surface area. The second step is fermentation, where the heterotrophic organisms convert the fermentable organic carbon to volatile (short chain) fatty acids, S_A. The kinetic relationships are given in Table 4.5.

A more complex model being developed at The University of Queensland involves the hydrolysis of soluble, S_S, and insoluble organic carbon, X_S, to fermentable substrate, S_F, used by the heterotrophic (acidogenic) organisms, X_H, for growth and to produce VFA, S_A. The hydrolysis is catalysed by the enzymes X_E. It also includes a simple model for the growth of the methanogens, X_M, and the production of methane gas (Figure 4.6).

Process	Component			Kinetics
	X_S	S_F	S_A	
Hydrolysis	-1	$1-f_{si}$		$K_h \frac{(X_S/X_H)}{K_X+(X_S/X_H)} X_H$
Fermentation		-1	1	$q_{FE} \frac{S_F}{K_{FE}+S_F} \frac{S_{ALK}}{K_{ALK}+S_{ALK}} X_H$

Table 4.5: Kinetics for simple fermentation model.

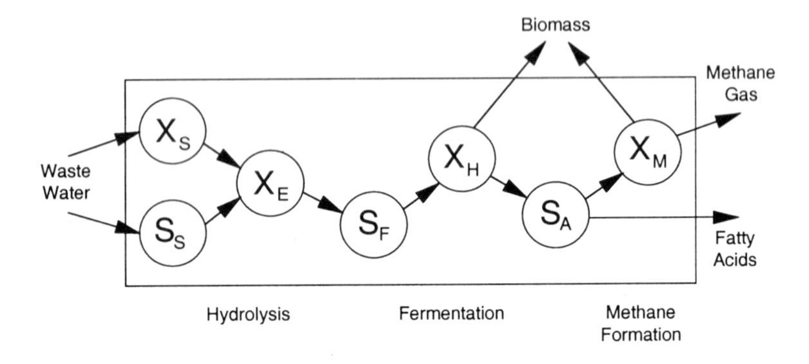

Figure 4.6: Fermentation model.

The kinetic relationships are given in Tables 4.6 and 4.7. The mass balances can be simply formulated and the rates derived from the tables.

Example 4.3 (Fermentation) *The MATLAB programs* `Example_4_3A.m` *and* `Example_4_3B.m` *simulate the fermentation models in Tables 4.5 to 4.7 respectively in batch reactors (zero inflow and outflow). The results are shown in Figure 4.7. The left plot shows clearly the difference in reaction rate between fermentation and the much slower rate limiting hydrolysis. The more complex model results in the right plot show similiar but more complex behaviour.*

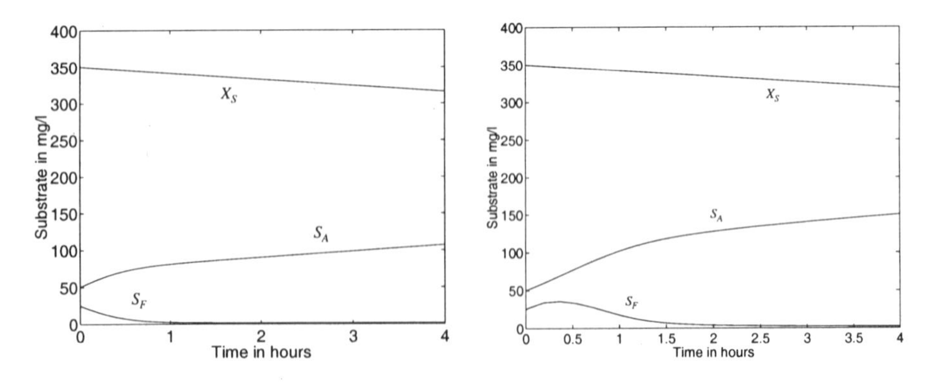

Figure 4.7: Simple and complex fermentations.

Process	Component					
	X_S	S_S	X_{ON}	X_E	S_F	S_{NH}
Hydrolysis of Insolubles	-1			Y_E	$1-Y_E$	
Hydrolysis of Solubles		-1		Y_E	$1-Y_E$	
Hydrolysis of Organic Nitrogen			-1			1
Growth of Acidogens					-1	$-Y_{nx}Y_H$
Growth of Methanogens						$-Y_{nx}Y_M$
Decay of Enzymes		1		-1		
Decay of Acidogens		1				Y_{nx}
Decay of Methanogens		1				Y_{nx}

Table 4.6: Kinetics of fermentation model.

Process	Component				Kinetics
	S_A	CH_4	X_H	X_M	
Hydrolysis of Insolubles					$k_{H,XS}X_SX_E$
Hydrolysis of Solubles					$k_{H,S}S_SX_E$
Hydrolysis of Organic Nitrogen					$k_{H,N}S_{ON}X_E$
Growth of Acidogens	$1-Y_H$		Y_H		$\hat{\mu}_F \frac{S_F}{K_F+S_F} X_H$
Growth of Methanogens	-1	$1-Y_M$		Y_M	$\hat{\mu}_M \frac{S_A}{K_M+S_A} X_M$
Decay of Enzymes					d_EX_E
Decay of Acidogens			-1		d_HX_H
Decay of Methanogens				-1	d_MX_M

Table 4.7: Kinetics of fermentation model (contd).

4.3 Modelling Anaerobic Digestion

The anaerobic digestion process is simply an extension of the fermentation process that was discussed in the previous section. In anaerobic digestion, the methane gas production step is encouraged to reduce the organic carbon content of the solid and/or liquid effluents. The methane gas is either flared or the energy content is partially recovered in gas engines or boilers.

There are a number of reactor configurations. The simple well-mixed reactor is generally used for activated sludge digestion. Two-stage reactors with various schemes to encourage attached growth of the biomass are common for the treatment of high-strength wastes from canneries, breweries, etc. These are becoming very popular to pretreat such waste streams before discharge to the sewer.

The principal operational problem in anaerobic digestors is the high growth rate of acidogens compared to the growth rate of the methanogens. Furthermore, the methanogens are inhibited by low pH, so that the process is potentially unstable and subject to catastrophic failure. A complete washout of methanogens is a too common occurrence. Two-stage reactor configurations (Figure 4.8) attempt to separate the acidogens and methanogens, controlling the pH of the intermediate stream with recycles and the addition of caustic soda. Caustic soda is a major component of the operating cost.

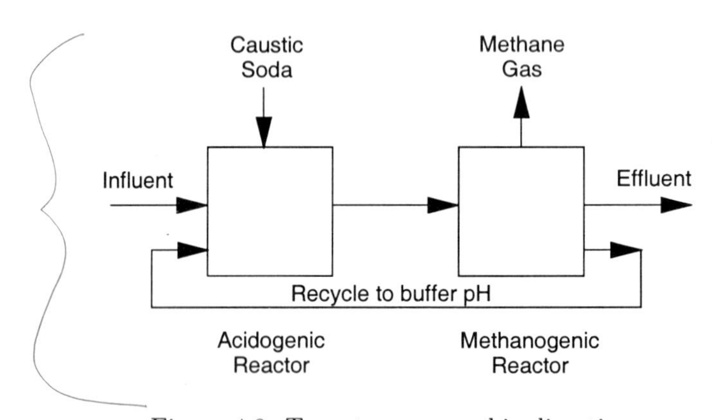

Figure 4.8: Two-stage anaerobic digestion.

Mechanistic or design models have been developed which subdivide the acidogenic biomass into classes producing different VFA components. They also divide the methanogenic biomass into two classes to introduce hydrogen as an intermediate in the production of methane. Hydrogen in the reactor gas is often used as an early indicator of methanogenic inhibition. The models also include ionic balances in order to predict the consumption of caustic soda. Recent modelling work has incorporated sulphate digestion. Hydrogen sulphide often reduces the value of the methane gas, producing corrosive acids when burnt.

Simple models, such as the one in Tables 4.6 and 4.7 in the previous section, can be used for sludge digestion. The prediction of reactor stability, and caustic consumption in two-stage reactors, requires more complex models.

4.4 Modelling Immobilised Film Processes

Immobilised film processes are gaining in popularity as the pressure increases for process intensification. The concept is much simpler than the implementation. If the biomass can be immobilised, then large solids residence times can be achieved in smaller volumes without or with much reduced recycling. There are however a number of problems with the concept:

- the packing must have a surface to which the live microorganisms will attach firmly.
- the packing and flow conditions should be conducive to sloughing off of dead biomass.
- the packing, and hence the biofilm, should have large surface area per unit volume.
- the packing and flow conditions should ensure good mixing of the liquid.
- the packing and flow conditions should ensure good wetting.
- the liquid pressure drop should be low.

A large number of packing types have been tried, both fixed and free floating. The most common fixed film processes are trickling filters for various parts of the activated sludge process and various packed vessels for anaerobic digestion of high strength wastes.

The same biological reactions occur in fixed film processes. The added mechanism of importance is mass transfer, getting reactants to the microorganisms in the biofilm and getting reaction products away. If the liquid phase wets the biomass fully and is well mixed, then diffusion through the biofilm is the primary concern. Since potentially reactants are being used and products formed throughout the biomass there will be concentration profiles within the film as shown in Figure 4.9.

Because substrate concentrations are lower and product concentrations higher than in the liquid phase, reaction rates in the biomass will be slower. If extreme changes in concentrations occur the types of reaction may change, for example if dissolved oxygen drops to very low levels anaerobic reactions will take over from aerobic reactions. More extreme changes will also affect organism growth rates and layering of different organism classes will occur which can affect the mechanical integrity of the film. These factors mean that high surface areas and thin films are ideal.

Substrate and product concentrations within the biofilm can be modelled by considering a thin slice of biofilm parallel with the surface and writing the usual

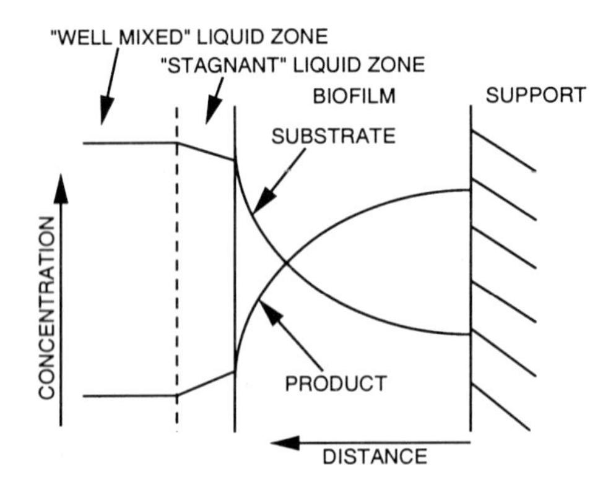

Figure 4.9: Biofilm concentration profiles.

dynamic component balances. This results in a partial differential equation of the form:

$$\frac{\partial S}{\partial t} = D\frac{\partial S}{\partial x} + r \tag{4.7}$$

where S is the concentration, t is time, x is distance into the biofilm, D is the diffusion coefficient and r is the reaction rate. The rates can be obtained from the tables of stoichiometry and kinetics in the usual way. This form of equation increases the complexity of the modelling as we now have two independent variables, time and distance into the biofilm. Generally the stagnant film is considered small and concentrations at distance zero are considered to be bulk liquid concentrations. A more dramatic assumption generally made is that the biomass is homogeneous with a given fixed distribution of types of organisms.

The liquid phase can also be a complex modelling exercise. If the support is free and the liquid well mixed, then the liquid phase can be modelled as we usually model well mixed reactors. Generally reactions are assumed not to occur in the liquid phase. If the support is fixed, there will be concentration gradients in the direction of flow. Component balances can be written over a thin slice perpendicular to the flow direction which will give a similar partial differential equation involving a third independent variable, distance along the liquid flow path.

Mechanistically modelling biomass growth and decay within the biofilm is much more complex and few attempts have been made. Changes in substrate concentrations induce different growth rates and even different microorganisms which causes zones and heterogeneity.

Solving such systems of equations is possible but extremely numerically intensive, especially if the bioreactor geometry and liquid flow patterns are complex. But, if you have more time than your authors, and larger computers, then everything is possible. Solution faster than real time could be a challenge. Appendix E on Differential Equations discusses some techniques. As you might expect, there have been various simplified models developed with more or less drastic assumptions. As you should be aware by now, the more drastic the assumptions the more limited is the usefulness of the models.

4.5 Modelling Hydraulics Again

In Chapter 3 we derived a simple model for the hydraulic propagation in a tank. Here we will now extend the discussion to other types of outlets in tanks. The purpose is to describe how the flowrate in a wastewater treatment system is not the same in all units of the system. Assume that the influent flowrate is suddenly increased by an extra pump being started. Then the effluent flowrate will increase gradually until it reaches the same value as the input. This will take typically a good portion of an hour in a large plant.

Let us consider the mass balance in a tank once more (Figure 4.10).

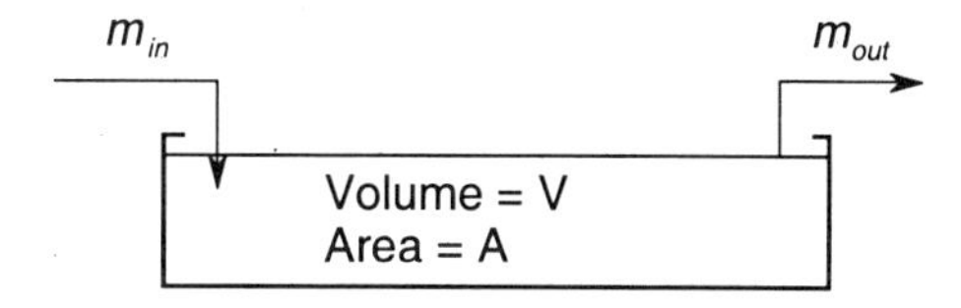

Figure 4.10: Simple tank hydraulics.

We repeat the total mass balance of the tank, Equation 4.8,

$$\frac{dM}{dt} = m_{in} - m_{out} \tag{4.8}$$

where M is the total liquid mass. Now we assume that the influent flowrate is known, while the effluent flowrate m_{out} is determined by the overflow weir type. We have assumed that there are no waves on the surface, and the effluent flowrate will be determined only by the liquid level. Both sides of Equation 4.8 can be divided by the liquid density, so the mass balance will be transformed to a volumetric balance over the tank,

$$\frac{dV}{dt} = q_{in} - q_{out} \tag{4.9}$$

where q are the flowrates (here we express them in $m^3/\min$). Let us assume that the area of the tank is constant (A m^2), and the total liquid height is $H + h$ (m) (see Figure 4.11). The weir height H is constant, so we get

$$\frac{dV}{dt} = A\frac{dh}{dt} = q_{in} - q_{out} \tag{4.10}$$

The total flowrate over all the N weirs can be calculated by considering Figure 4.12, showing one weir of arbitrary shape. The liquid level is h above the weir and the liquid velocity at level z is $v_z(z)$.

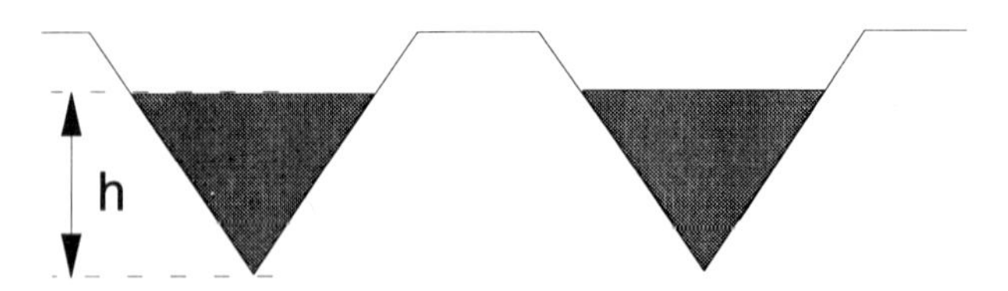

Figure 4.11: Definition of liquid height.

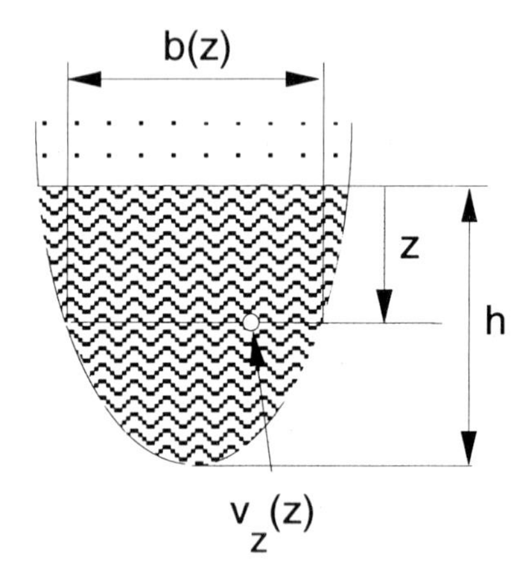

Figure 4.12: The weir geometry.

The total flowrate over all the weirs is obtained by integration:

$$q_{out} = N \int_0^h b(z) \cdot v_z(z) \cdot dz \tag{4.11}$$

where the variables are defined in Figure 4.12. The liquid velocity can be calculated from the Bernoulli equation,

$$p_{stat} = p_0 + z\rho g - \frac{1}{2}\rho v_z^2 \tag{4.12}$$

where ρ is the liquid density and g the gravity acceleration (to be consistent g has to be expressed in $m/\min^2$!). The pressures are assumed to be equal, $p_{stat} = p_0$, so we obtain,

$$v_z(z) = \sqrt{2gz} \tag{4.13}$$

This velocity expression is inserted into Equation 4.11, so the total flowrate becomes,

$$q_{out} = N \cdot \sqrt{2g} \int_0^h b(z) \cdot \sqrt{z} \cdot dz \tag{4.14}$$

Obviously, the flowrate depends crucially on the shape of $b(z)$. For a rectangular shaped weir $b(z)$ is constant, and the integral can be easily evaluated,

$$q_{out} = constant \cdot N \cdot b \cdot h^{3/2} \tag{4.15}$$

which is the Francis formula shown in Chapter 3. For a V-shaped weir $b(z)$ varies linearly with z, and is a function of the angle α of the V-shape. It is straightforward to derive the integral (Equation 4.14) that becomes,

$$q_{out} = constant(\alpha) \cdot N \cdot h^{5/2} \tag{4.16}$$

The flowrate can be made directly proportional to h for a suitably chosen cross section of the weir. If we design $b(z)$ = constant $/\sqrt{z}$ (z>0), then the flowrate becomes

$$q_{out} = k_0 + k_1 \cdot N \cdot h \tag{4.17}$$

where the k parameters are constant. Such a weir is called a Sutro weir, Figure 4.13, and has been used widely to get a device that easily translates flowrate into a linear scale. The flowrate is proportional to h, provided that $h > 0$.

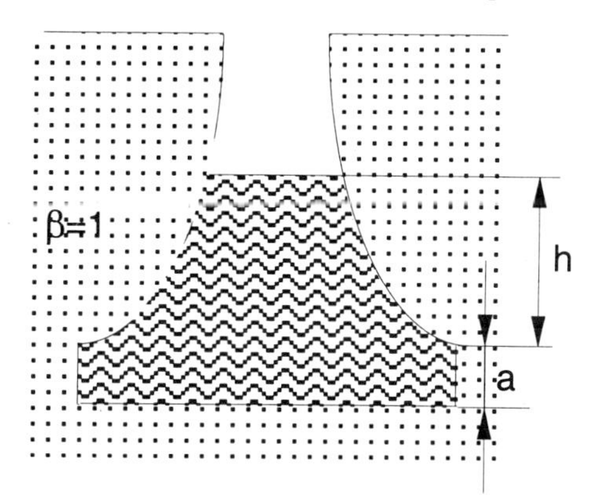

Figure 4.13: A Sutro weir.

We observe that all the three common types of weirs have a flowrate function of the structure

$$q_{out} = k + \alpha \cdot h^{\beta} \tag{4.18}$$

where k, α, and β are constants that depend on the type of weir and the number of weirs. We can now write the balance equation for a tank with outlet weirs as:

$$A \cdot \frac{dh}{dt} = q_{in} - k - \alpha \cdot h^{\beta} \tag{4.19}$$

Note, that the tank area A and not the volume determines the dynamics.

Example 4.4 (Different weir types) *We will now compare the three types of weirs with respect to their dynamic properties. Suppose the influent flowrate has been doubled from 1 to 2* $m^3/\min$ *(corresponding to an increase from 1.44 to 2.88 Ml/day). Before the step all the three liquid levels are the same, 0.05 m above the weir. The tank area is 100* m^2.

The MATLAB program `Example_4_4.m` *produced the simulation results in Figure 4.14. We note that the effluent flowrate with the V-shape weir increases much faster to its new steady state, that is it has a low hydraulic damping. After the step the liquid level has increased slightly. For the Sutro weir, it increases from* 0.05 *to* 0.10 *m, that is* 50 *mm. For the rectangular weir, it increases only* 30 *mm, and for the V-shape only* 16 *mm.*

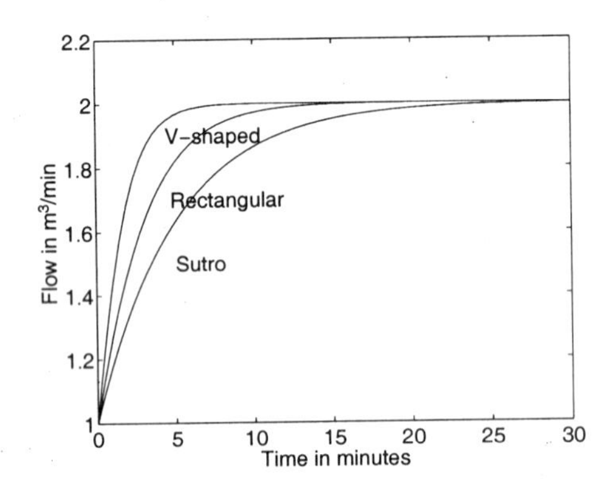

Figure 4.14: Comparison of three weir types.

Such a small volume change, however, has a significant influence on the time it takes for the effluent flowrate to build up. Expressed in another way, we can see that the *time constant* of the Sutro weir is about 50% larger than that of the V-shape weir.

A weir design with low damping does not allow the liquid level to change very much. If a high damping is desired, then the volume of the tank should be allowed to vary. Let us compare two rectangular weir designs with the same weir width b but different number of weirs. For the same effluent flowrate they are at equilibrium at different heights h (see Equation 4.20),

$$q_{out} = constant \cdot N_1 \cdot b \cdot h_1^{3/2} = constant \cdot N_2 \cdot b \cdot h_2^{3/2} \tag{4.20}$$

or

$$\left(\frac{h_1}{h_2}\right)^{3/2} = \frac{N_2}{N_1} \tag{4.21}$$

This is intuitively clear. If the number of weirs is increased, then the total width of the effluent flow increases, so the liquid height becomes smaller. Now, let us see what that means from a dynamic point of view. We will derive a linearised version of Equation 4.19 and examine how quickly the liquid height may change for an influent flow change. Let us assume that the steady state influent flowrate is $q_{in,0}$ and the corresponding liquid height h_0.

In Chapter 5 we demonstrate that a nonlinear equation can be linearised by taking the partial derivatives. In this case we obtain:

$$A\frac{d}{dt}(\Delta h) = \Delta q_{in} - \alpha\beta h_0^{\beta-1} \cdot \Delta h \tag{4.22}$$

where $\Delta h = h - h_0$ and $\Delta q_{in} = q_{in} - q_{in,0}$. Equation 4.22 can be written in the form:

$$T\frac{d}{dt}(\Delta h) = K \cdot \Delta q_{in} - \Delta h \tag{4.23}$$

where T is the time constant and K a gain parameter,

$$\begin{aligned} T &= \frac{A}{\alpha\beta h_0^{\beta-1}} \\ K &= \frac{1}{\alpha\beta h_0^{\beta-1}} \end{aligned} \tag{4.24}$$

Let us compare the time constants for two rectangular weirs with the same weir width but a different number of weirs, N_1 and N_2. Then the ratio between the respective time constants (Equation 4.21) is:

$$\frac{T_1}{T_2} = \frac{A}{\alpha_1\beta h_{1,0}^{\beta-1}} \frac{\alpha_2\beta h_{2,0}^{\beta-1}}{A} = \frac{N_2}{N_1}\frac{h_{2,0}^{\beta-1}}{h_{1,0}^{\beta-1}} = \frac{h_{1,0}}{h_{2,0}} = \left(\frac{N_2}{N_1}\right)^{2/3} \tag{4.25}$$

In other words, if we decrease the number of weirs to half the value, then the time constant would increase by a factor of $2^{2/3} \approx 1.6$, that is about 60% longer.

Let us compare the time constants of a rectangular weir and a V-shaped weir. For the same steady state influent flowrate, the V-shaped weir has a flowrate $\alpha_V h_V^{5/2}$, while the rectangular has the flowrate $\alpha_{Rec} h_{Rec}^{3/2}$. The flowrates are equal:

$$\alpha_{Rec} h_{Rec}^{3/2} = \alpha_V h_V^{5/2} \tag{4.26}$$

The ratio between the time constants is:

$$\frac{T_{Rec}}{T_V} = \frac{A}{\alpha_{Rec}\left(\frac{3}{2}\right) h_{Rec}^{1/2}} \frac{\alpha_V\left(\frac{5}{2}\right) h_V^{3/2}}{A} = \frac{5}{3}\frac{h_{Rec}}{h_V} \tag{4.27}$$

If the weir designs have been made such that the steady state volume of the tanks are the same, then the rectangular weir will have a 60% longer time constant than the V-shape weir. This was actually demonstrated in Figure 4.14.

Why would we wish a larger damping? There is an obvious reason for this. The bottleneck of the operation of an activated sludge process is the settling. The clarifier process is sensitive to flowrate changes. Still there are many treatment plants having primary pumping without any variable speed drives. Consequently, when fixed speed pumps are turned on, there is a hydraulic propagation along the plant, and the clarifier will see a flow increase somewhat later. If the weir designs are such that the flow increase is dampened, then the detrimental effect of the flowrate change is reduced in the clarifier.

Example 4.5 (Flowrate damping) *We will look at a system of a primary sedimentation tank and an aerator, Figure 4.15, simulated by the MATLAB program* `Example_4_5.m`. *The primary influent flow in steady state is 1* m^3/min *(=1.44 Ml/day). We will calculate how a stepwise flowrate change of 50% in the influent will propagate along the plant. We assume that the areas are identical, 100* m^2. *Assume that the weirs in both the tanks are rectangular and the total width of them is 2 metres.*

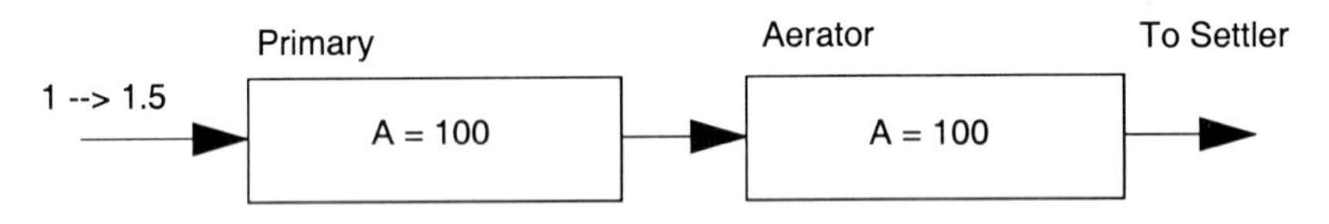

Figure 4.15: Hydraulic simulation example.

At equilibrium the aerator level is 19.8 mm above the weir. For the stepwise 50% increase of the influent, we can see how the flowrates propagate (Figure 4.16). It will take 4.5 minutes before the settler inflow has reached 90% of the influent flow change. The level of the aerator will increase by 6.1 mm due to the flowrate change.

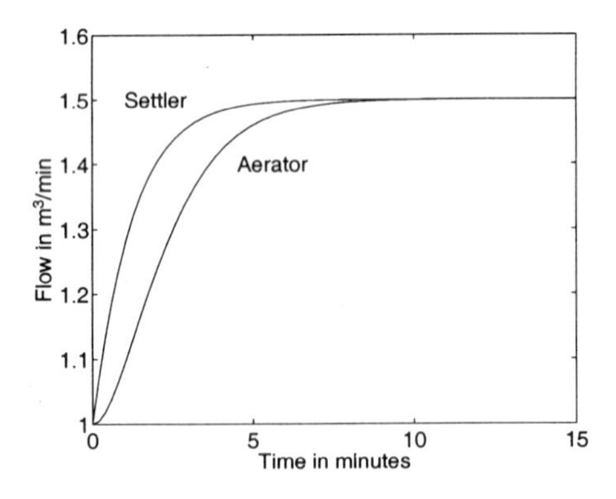

Figure 4.16: Flowrate changes after the primary settler and aerator.

Let us now decrease the total weir width to half the value. This will dampen the hydraulic change at the outlet of the aerator, and now it will take 8 minutes before the aerator effluent flowrate has reached 90% of the total change, Figure 4.17. Better damping is obtained, since the volume is allowed to vary more. The volume variation, however, is quite modest. In this case the liquid level increased 3.6 mm more than in the first case.

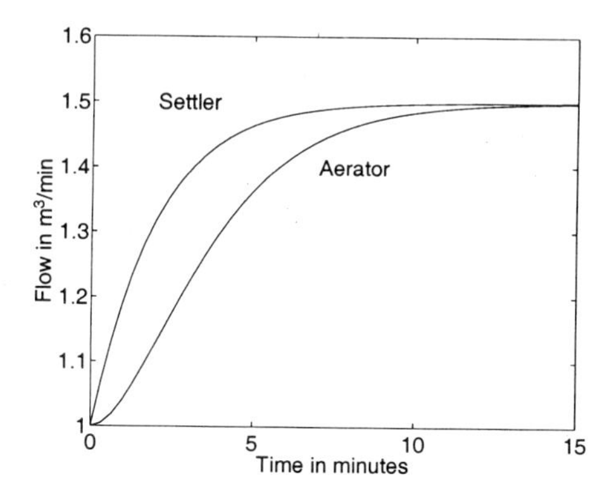

Figure 4.17: Flowrate changes with shorter weir.

Designers of treatment plants traditionally have desired to obtain as small volume changes as possible. Then the pressure variations for the aeration system are minimised, and the aeration would not be very much disturbed. The price for a small volume variability is, however, a large sensitivity for hydraulic shocks to the clarifier.

With a control system, the aeration process can easily be compensated for pressure disturbances. They are indeed very small. In Figure 4.17 the liquid level change was 9.7 mm. For a typical aerator depth of about 4 m, this is only 0.25 per cent variation of the pressure. In other words, there are no reasons to make constant volume more important than a good hydraulic damping. The weir design is a typical example where plant design can be improved by using control system design.

4.6 Modelling Settling and Clarification

The remaining mechanism of primary importance in the activated sludge process is that of separation. Primary sedimentation settles grit and floats grease, but will not be considered here. Typically, in the modelling of primary settling it is usually assumed that the particles are separate and do not interact. The velocity of the particles in the liquid can then be modelled by Stoke's law. Secondary clarification separates the biomass from the treated wastewater and is a key mechanism in determining the quality of the effluent. Biomass in the effluent affects both the clarity and the oxygen demand, which are both key quality determinants.

The mixed liquor flow entering the clarifier splits in two, one flowing upwards and over an overflow weir, and the other flowing downwards and being withdrawn at the bottom (see Figure 4.18). We normally assume no vertical mixing (so-called plug flow). Superimposed on these flows is the settling of the biomass under the influence of gravity. This results in the reduced concentration in the overflow and the increased concentration in the underflow. The variation of biomass concentration with height as well as with time makes this a *distributed parameter system* which makes modelling more involved. Another key feature of clarifier modelling is the *flux model*, the mathematical description of the rate of settling as a function of biomass concentration.

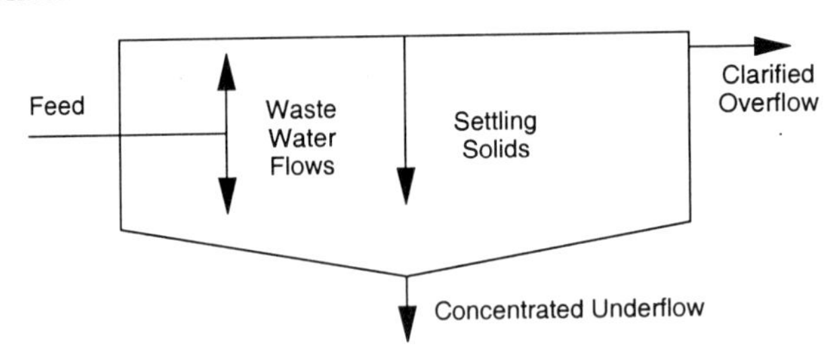

Figure 4.18: Clarifier flows.

Compared to primary settling, the concentration of the particles is so high that there is a significant interaction between the individual floc particles. Thus it is no longer possible to apply Stoke's law. Instead it is assumed that the velocity of the sludge relative to the liquid is a function of the local concentration. This is called zone settling. It was found empirically that, for concentrations typical for settlers, the relative velocity could be described by such a function, illustrated by Figure 4.22a below. The settling behaviour is then described as the flux model. It is worthwhile reminding ourselves that one of the problems with secondary settling has to do with the small differences in density between the flocs and the water. A typical floc may have a density of 0.5% more than the surrounding water. It follows that the settling process becomes extremely sensitive to environmental changes of the settler. A small temperature change or a wind may disturb the settler sufficiently to significantly change the solids distribution pattern.

When the concentration has increased to a large value, a new phenomenon takes place. Then the interaction between flocs is so large that the floc geometry will be influenced and we have reached the so called compaction stage. This happens in the bottom part of the settler. The knowledge of the phenomenon in the compaction zone is still not well known in terms of models. Furthermore, when the concentration is so large the flocs become a significant proportion of the volume. For the dense flocs to settle, some water has to be displaced upwards. Therefore the total mechanism in the compaction stage is a complex combination of sludge moving downwards under compaction and water moving upwards to make room for the sludge.

In the clarifying part of the settler the floc particles behave as individual objects. Since they are not coupled together as in the thickening part, it is not obvious how to predict their behaviour. The flocs are certainly not homogeneous. Instead some of the flocs may appear like round flocs that would settle readily, while others may be filamentous. If a filament appears by itself, it may have problems settling and it will be transported with the upstream water into the effluent. Likewise there are often dispersed organisms that will not settle, so they also come out with the effluent water. The phenomenon of flocculation, that occurs near the feed entry where the velocity decreases, changes the distribution of floc sizes, especially for small flocs. It is therefore a very important factor in clarification performance but is also poorly understood.

If the amount of filaments is relatively small, then they may actually connect flocs and serve as bridges between the flocs. As a result the flocs may filter some of the dispersed sludge particles causing a better effluent. Considering the composition of various organisms in the sludge, it is apparent that it becomes even more difficult to predict the clarifier behaviour. Dynamic modelling of clarifiers has received much less attention compared to modelling of the aeration tank. An elegant solution still evades the activated sludge modellers. Biomass concentrations in outflows, sludge blanket level and sludge inventory are the key values an ideal model should predict.

4.6.1 Multi-Layer Models

The most popular clarifier model considers the clarifier vessel as a number, n, of horizontal slices with the feed into slice m (Figure 4.19). Typically such models consider 10 to 100 slices depending on the required accuracy for predicting the sludge blanket level. Each slice has a bulk flow of liquid and solids either upwards (above the feed) or downwards (below the feed). In addition, solids settle into the slice from the slice above and settle out of the slice to the slice below. Each slice is assumed to be well mixed. For slice $i, i = 1, ..., m-1$ the solids mass balance above the feed looks like:

$$h_i A_i \frac{dx_i}{dt} = q_o(x_{i+1} - x_i) + A_i(f_{i-1} - f_i) \tag{4.28}$$

where

h = height of the slice
A = cross-sectional area of the slice
x = solids concentration
q = liquid flowrate ("o" is overflow, "u" is underflow)
f = settling flux (concentration times velocity)
t = time

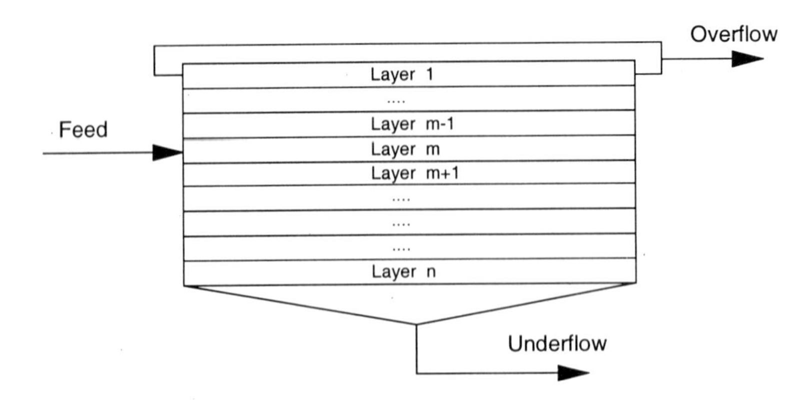

Figure 4.19: Multi-layer model.

For slice $j, j = m+1, ..., n$ the solids mass balance below the feed looks like:

$$h_j A_j \frac{dx_j}{dt} = q_u(x_{j-1} - x_j) + A_j(f_{j-1} - f_j) \tag{4.29}$$

The feed slice m solids mass balance looks like:

$$h_m A_m \frac{dx_m}{dt} = q_f x_f - (q_o + q_u)x_m + A_m(f_{m-1} - f_m) \tag{4.30}$$

It is usually assumed that layers 1 through $n-1$ are zone settling and that the compaction of the sludge takes place inside slice n. This is not modelled explicitly. Instead, the concentration in the slice is adjusted to fit a boundary condition: at steady state the mass flow out of the slice has to be the same as the mass flow in the underflow, out of slice n. The overall mass balance assuming constant volume is simply:

$$q_f = q_o + q_u \tag{4.31}$$

The corresponding model for the soluble nutrients can be either n corresponding well-mixed compartments for each nutrient or, to simplify the model, could be simply two compartments - one above and one below the feed.

Example 4.6 (Layered clarifier model 1) *The MATLAB program called* `Example_4_6.m` *implements the "robust layered model" of Jeppsson & Diehl (1995) and Diehl & Jeppsson (1998) with a simple exponential flux model (Equation 4.32 below). The left plot in Figure 4.20 shows the response to a period of higher feed flowrate, while the right plot in Figure 4.20 shows a period of decreased underflow. The responses show one important aspect of current layer models, they can predict at least qualitative behaviour in the sludge blanket but are seriously deficient in the clarification region and the compaction zone at the very bottom. This has to do with both the current flux models and the assumed mechanisms.*

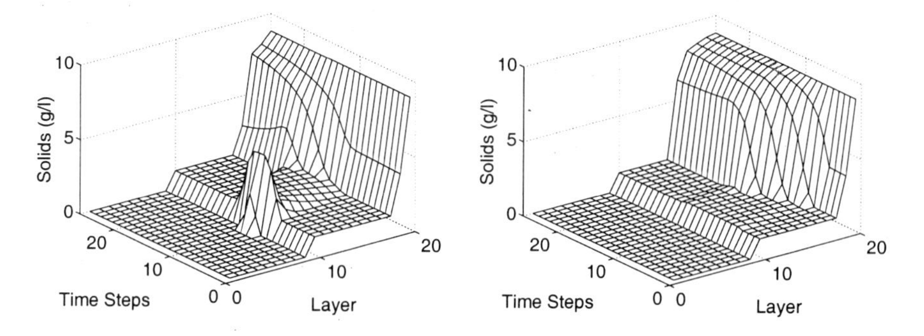

Figure 4.20: Feed flowrate increase and underflow decrease.

It has been demonstrated that more than 10 layers are needed in order to get a reasonable accuracy of the total mass in the settler and to better calculate the blanket height.

Example 4.7 (Layered clarifier model 2) *The MATLAB program called* `Example_4_6.m` *is used again to demonstrate the effect of using different numbers of layers (Figure 4.21). This is the same feed flowrate increase shown in Figure 4.20. Other layer models can show quite different profiles.*

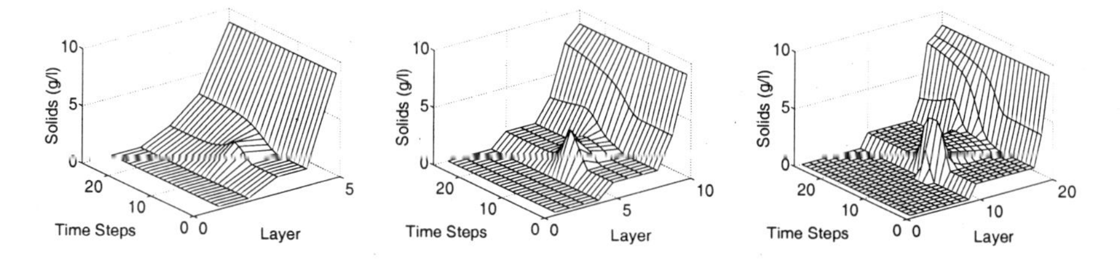

Figure 4.21: Profile with 5, 10 and 20 layers.

4.6.2 Settling Flux Models

Problems arise in defining a good settling flux model to define the settling flux in terms of the solids concentration. Popular models are of the form:

$$f_i = x_i v_i = x_i \, C_1 \, e^{-C_2 x_i} \tag{4.32}$$

where v_i is the settling velocity. Note that we consider the velocity a function of only the local concentration (the popular assumption of Kynch), demonstrated in Figure 4.22a.

The limitations of this form of model are:

- How should constants C_1 and C_2 be related to biomass characteristics? Generally they are empirically correlated to the sludge volume index (SVI).

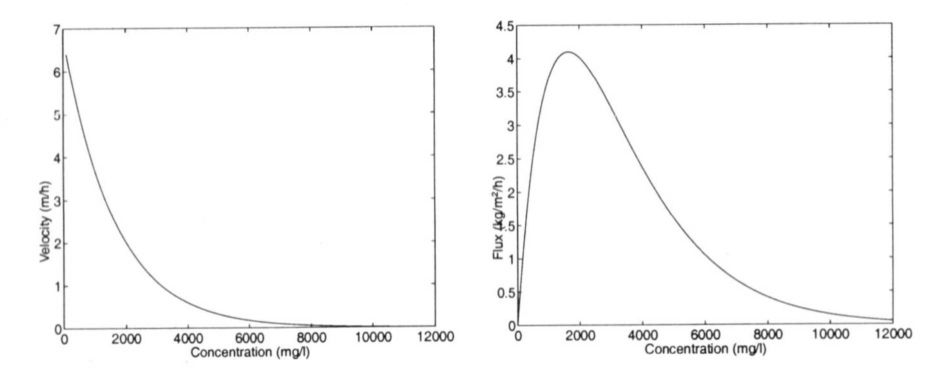

Figure 4.22: Settling flux model (a) velocity plot (b) flux plot.

- The settling flux does not go to zero at high solids concentrations so that an infinite solids concentration could result in the bottom layer. Various empirical *fixes* have been proposed such as the Omega function (defined in Härtel & Pöpel (1992)) and limiting the incoming flux to the slice.

- Most flux models do not describe the effluent concentration at all accurately. They are applicable to the region of zone settling within the sludge blanket. Clarification is discussed in the next section.

- Slice models may predict the solids concentration profile more accurately but sludge blanket level and sludge inventory are more complex to calculate.

4.6.3 Clarification

We mentioned above that most flux models apply within the sludge blanket. These flux models do not describe the effluent concentration at all accurately. The most used approaches to modelling clarification are:

- to use a double-exponential form of the flux model,

- to assume a constant settling velocity above the blanket level, or

- to assume a non-settlable fraction related to biomass characteristics (an empirical relationship exists for IAWQ AS Models).

A review of different models and their ability to fit several sets of experimental data recommends the multi-layer model of Takács *et al.* (1991) using a double-exponential form of the flux model (see Figure 4.23).

$$f_i = x_i v_i = x_i C_1 \left(e^{-C_2(x_i - x_I)} - e^{-C_3(x_i - x_I)} \right) \qquad (4.33)$$

where x_I represents the fraction that will not settle at all. Here the first exponential term represents settling in the sludge blanket and the second exponential term represents clarification.

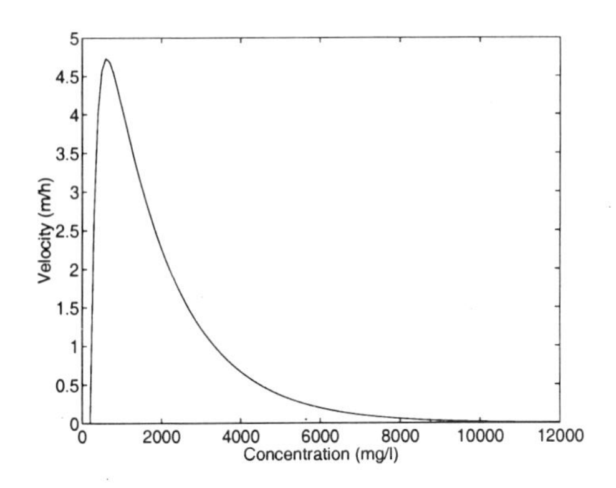

Figure 4.23: The double-exponential settling velocity model.

4.6.4 Compartment Models

A simpler approach considers two well-mixed compartments, one above and one below the sludge blanket level (Figure 4.24). From a control perspective the important parameters to be able to predict are the sludge blanket level, the sludge inventory, and the overflow and underflow concentrations. We will formulate a largely empirical model for predicting these variables. Consider first a mass bal-

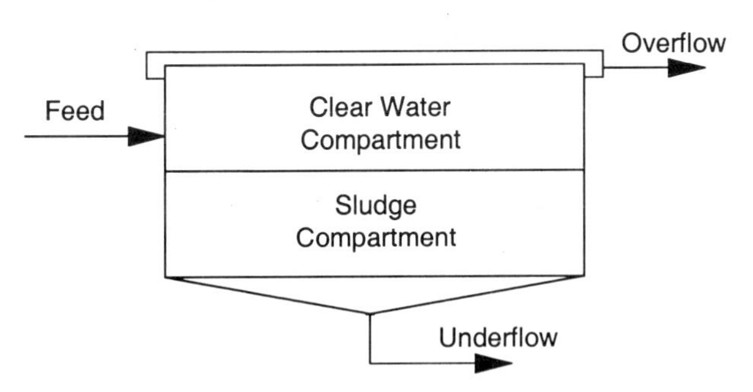

Figure 4.24: Compartment model.

ance on the clear water compartment.

$$\frac{dM_C}{dt} = \alpha q_F X_F - q_O X_O - m_S \tag{4.34}$$

where M_C is the sludge holdup, q the wastewater flowrate, X the suspended solids concentration, subscript F is the feed, subscript O is the overflow, α is the fraction of the feed entering the clear water compartment, and m_S is the rate of sludge settling into the sludge compartment. Assuming constant volume for the clarifier,

$$q_O = q_F - q_U \tag{4.35}$$

where subscript U is the underflow. Assuming the compartment is perfectly mixed (uniform composition),

$$X_O = \frac{M_C}{V - V_S} \tag{4.36}$$

where V is the total volume and V_S is the volume of the sludge compartment. The empirical settling model is,

$$m_S = K_S X_O \tag{4.37}$$

where K_S is a constant of proportionality. Other settling models could be readily substituted in place of Equation 4.37. Now consider a mass balance on the sludge compartment.

$$\frac{dM_S}{dt} = (1-\alpha)q_F X_F + m_S - q_U X_U \tag{4.38}$$

where M_S is the sludge holdup in the sludge compartment. Again assuming the compartment is perfectly mixed (uniform composition),

$$X_U = \frac{M_S}{V_S} \tag{4.39}$$

The empirical compaction model is,

$$X_U = K_C X_S \tag{4.40}$$

where K_C is a constant of proportionality. Other compaction models could be readily substituted in place of Equation 4.40. Finally we have an empirical thickening model which relates sludge blanket concentration X_S to the feed concentration X_F and the sludge mass M_S with a time lag.

$$\tau \frac{dX_S}{dt} = \left(1 + \frac{M_S}{K_T}\right) X_F - X_S \tag{4.41}$$

where K_T is a constant and the time constant was chosen as,

$$\tau = \frac{M_S}{(1-\alpha)q_F X_F} \tag{4.42}$$

Again, other thickening models could be postulated in place of Equations 4.41 and 4.42.

Example 4.8 (Compartmental clarifier model) *The MATLAB program called* `Example_4_8.m` *simulates the above compartmental model and Figure 4.25 show responses to periods of increased feedrate and decreased underflow respectively. These are similiar disturbances to those applied in Exercise 4.7 giving the responses in Figure 4.21.*

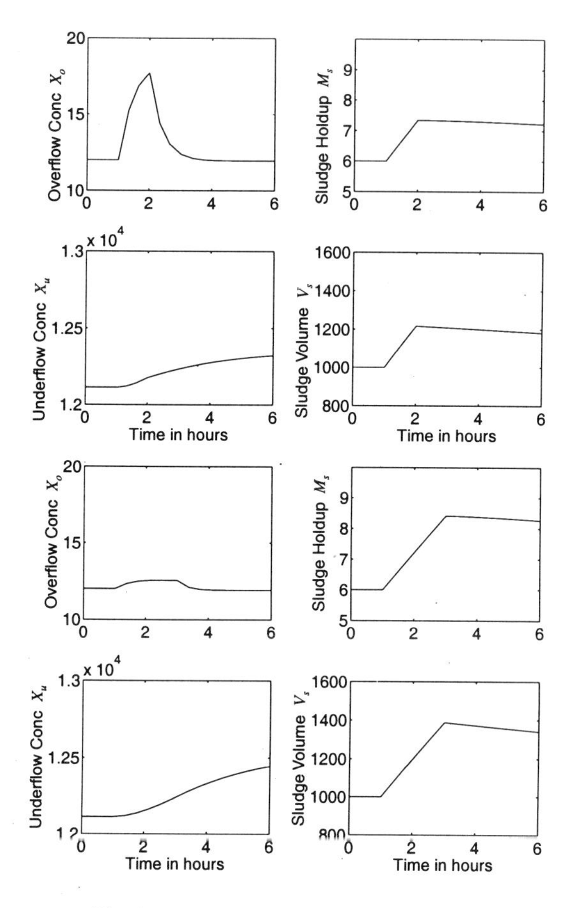

Figure 4.25: Feed flowrate increase and underflow decrease.

4.7 Modelling Batch Processes

Batch processes operate in a number of distinct time phases. For example, the fill, react, settle and drawoff phases of the typical activated sludge sequenced batch reactor (SBR) are shown in Figure 4.26.

In some cases, these phases can be further subdivided. The react phase of an SBR could be further divided into anaerobic, aerobic and anoxic phases. We must develop models for each time phase, as opposed to for each equipment item as we do in modelling continuous flow processes. Then we integrate these models in the time sequence, with the initial conditions of each phase related to the final conditions of the previous phase.

In principle, time phase process models are simply modifications of the continuous flow models developed above. The equipment is still a well mixed tank only we have different assumptions, sometimes variable volume and sometimes zero in-

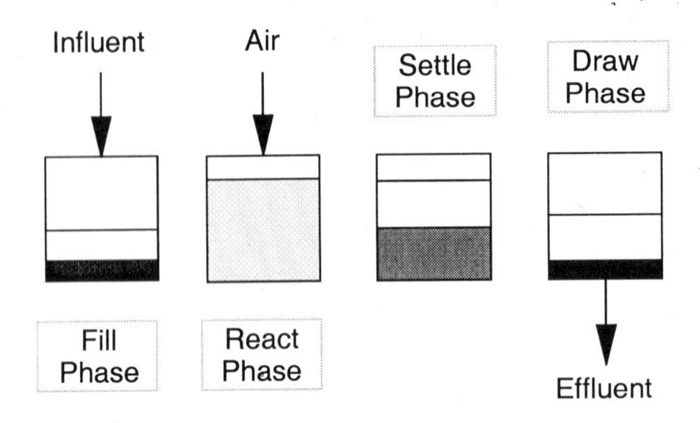

Figure 4.26: Main SBR time phases.

flows or outflows. For example, in a well mixed fill phase the overall mass balance and a component balance might look like:

$$\frac{dV}{dt} = q_{in} \tag{4.43}$$

$$\frac{d}{dt}(VS) = q_{in}S_{in} + rV \tag{4.44}$$

where V is the volume, q_{in} the feed flowrate, S the substrate composition, S_{in} the feed composition and r the reaction rate. There is no outflow in the fill phase. Notice that the volume must be included inside the derivative, as it is no longer constant. In order to accommodate most simulation software, the derivative must be expanded and the overall mass balance substituted to reduce the component balance to an explicit ordinary differential equation:

$$V\frac{dS}{dt} + S\frac{dV}{dt} = V\frac{dS}{dt} + S\left(q_{in}\right) = q_{in}S_{in} + rV \tag{4.45}$$

which simplifies to:

$$\frac{dS}{dt} = \frac{q_{in}}{V}(S_{in} - S) + r \tag{4.46}$$

The rate term can be obtained from the tables of stoichiometry and kinetics in the usual way. This looks the same as a continuous flow component balance, except that in this case the volume is a variable, determined from the simultaneous integration of Equation 4.43. In the constant volume react phase, component balances are simpler:

$$\frac{dS}{dt} = r \tag{4.47}$$

In fact, it is often very instructive to examine the behaviour of wastewater treatment reactions in a batch reactor. This is because you see the complete progress of the reaction from start to finish, whereas in a well mixed continuous flow reactor you see only the end conditions. The batch reactor solution, concentrations plotted against time, is also identical to the plug flow reactor solution, except that the time axis is replaced by the distance along the direction of flow (fluid velocity is the conversion factor from time to distance).

Example 4.9 (Carbon removal in a batch reactor) *Consider the simple carbon removal modelled in Chapter 3 by Equations 3.27 to 3.29 and with the kinetics in Table 3.4. The heterotrophic biomass mass balance will become:*

$$\frac{dX_H}{dt} = \hat{\mu}_H \left(\frac{S_S}{K_S + S_S}\right)\left(\frac{S_O}{K_{OH} + S_O}\right) X_H - b_H X_H \tag{4.48}$$

The carbon nutrient mass balance will become:

$$\frac{dS_S}{dt} = -\frac{1}{Y_H}\hat{\mu}_H \left(\frac{S_S}{K_S + S_S}\right)\left(\frac{S_O}{K_{OH} + S_O}\right) X_H + (1 - f_P) b_H X_H \tag{4.49}$$

The oxygen mass balance will become:

$$\frac{dS_O}{dt} = \frac{Y_H - 1}{Y_H}\hat{\mu}_H \left(\frac{S_S}{K_S + S_S}\right)\left(\frac{S_O}{K_{OH} + S_O}\right) X_H + K_L a (S_{O,sat} - S_O) \tag{4.50}$$

We assume that the air flowrate is adjusted to maintain a constant dissolved oxygen concentration, but with a maximum value. We obtain the concentration and air flowrate profiles shown in Figure 4.27, using the parameters and initial conditions from Table B.1 in Appendix B, and assuming the same exponential relationship for $K_L a$. The MATLAB program `Example_4_9.m` *was used for the calculations. The period with the air flowrate saturated illustrates the high initial OUR typical of batch (or plug flow) reactors.*

4.8 Summary

We have supplied you with basic information so that, together with the basic modelling techniques presented in Chapter 3, you will be able to configure quite complex process models for a variety of biological wastewater treatment systems. Even more importantly, we hope you will have learnt enough to ask intelligent and probing questions of the many model salespersons you will encounter. There is specific information on simulation and simulators in Chapter 9.

As we said at the end of the previous chapter, perhaps the more difficult task is obtaining reasonable estimates of the parameters, particularly those describing reaction rates and settling characteristics. The following Chapters 6 to 8 on model fitting discuss the problems of parameter estimation and model validation. Models which are not tailored to your processes properly are useless and potentially dangerous.

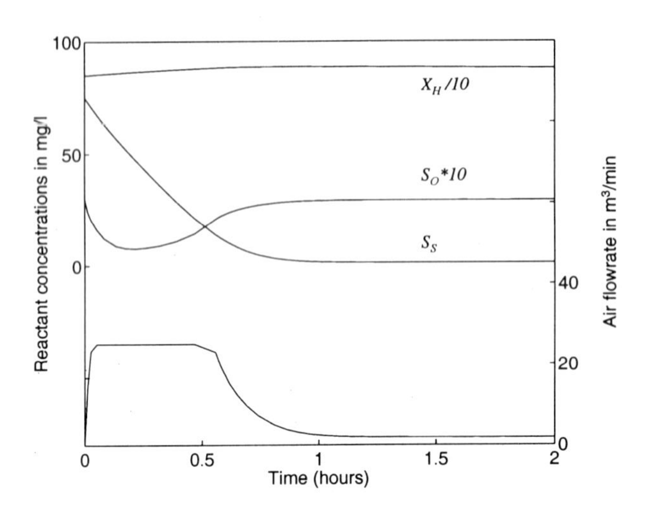

Figure 4.27: Batch reactor profiles.

4.9 Further Reading

- For more information on activated sludge modelling:
  1. Henze *et al.* (1995)
  2. Jeppsson (1996)
- An attempt to model biomass population dynamics was presented by Gujer & Kappeler (1992).
- For more information on anaerobic digestion modelling:
  1. McCarty & Mosey (1991), Costello (1989), Graef (1972), Graef & Andrews (1974), Heinzle *et al.* (1993), Hickey & Goodwin (1989), Hobson & Wheatly (1992), Jones *et al.* (1992), Keller (1992), Keller *et al.* (1994), Mosey (1983), Pavłostathis & Giraldo-Gomez (1991), Ryhiner *et al.* (1993), Sahm (1984) and Schink (1988),
  2. Costello *et al.* (1991a) and Costello *et al.* (1991b) for carbohydrates,
  3. Ramsay *et al.* (1994) for proteins,
  4. Batstone *et al.* (1997) for fats and oils,
  5. Gupta *et al.* (1994) for sulphate reduction.
- The flow propagation issue was first published in Olsson & Stephenson (1985) and further explored in Olsson (1985b).
- For more information on clarifier modelling:
  1. a review of different models and their ability to fit several sets of experimental data in Grijspeerdt *et al.* (1995),

2. Krebs (1991) and Krebs (1995), and
3. Jeppsson & Diehl (1995).

- Flux model parameters related to SVI in Härtel & Pöpel (1992).
- Various empirical *fixes* have been proposed such as the Omega function in Härtel & Pöpel (1992) and limiting the incoming flux to the slice in Vitasovic (1986), Takács *et al.* (1991) and Vitasovic *et al.* (1997).
- Clarifier non-settlable suspended solids related to bioreactor conditions in Dupont & Henze (1992).
- Modelling of abnormal clarifier problems such as scumming and bulking has also been attempted by Kappeler & Gujer (1994a), Kappeler & Gujer (1994c) and Kappeler & Gujer (1994b).

4.10 Exercises

1. Selecting just the appropriate stoichiometry and kinetics from Tables 4.1 and 4.2, write down the equations to model a continuous flow anoxic tank for post-denitrification. Using IAWQ parameter values and flowrates and volumes from your own process, write MATLAB code to simulate the process. You can find parameters in MATLAB program `Example_4_1.m`, and you can copy it as a template for your own program if you wish. Use your simulation to select an appropriate addition rate for COD and to examine the effect of hydraulic disturbances.

2. Using the stoichiometry and kinetics from Tables 4.6 and 4.7, write down the equations to model a continuous flow two stage anaerobic digester without recycle. Assume acidification only in the first stage and methanogenesis only in the second stage. Write MATLAB code to simulate the process. You can find parameters in MATLAB program `Example_4_3b.m`, and you can copy it as a template for your own program if you wish. Use your simulation to examine the effect of hydraulic disturbances. What is the very important mechanism that has not been modelled.

3. Write a hydraulic model for the major vessels and flows in your plant. You can use the MATLAB program `Example_4_5.m` as a template for your own program if you wish. Use your simulation to examine the effects of hydraulic disturbances.

4. Copy and modify MATLAB program `Example_4_6.m` to incorporate the following power law flux model in place of the exponential model currently used. Compare the solids profiles with the two flux models.

$$f_i = x_i \left(2000 x_i^{-0.9}\right) \tag{4.51}$$

5. Selecting just the appropriate stoichiometry and kinetics from Tables 4.3 and 4.4, write down the equations to model the two operational phases for a batch reactor to affect phosphorus removal. Using IAWQ parameter values and flowrates and reaction times equivalent to the volumes from your own process, write MATLAB code to simulate the two phases. You can find parameters in MATLAB program `Example_4_2.m`, and you can copy program `Example_4_9.m` as a template for your own program if you wish. Use your simulation to select an appropriate addition rate for VFA and to examine the effect of hydraulic disturbances.

Chapter 5

Empirical or Black-Box Models

The goal of this chapter is to introduce other types of models which are commonly used in the analysis of process data and in the design and implementation of diagnostic aids and schemes for optimisation and control.

5.1 Types of Models

Models may be classified in many different ways. One common classification is based on the source of the structural form of the model:

mechanistic which means based upon the actual or believed physics, chemistry and microbiology that governs the system,

empirical which means based upon a convenient mathematical function that reasonably represents available data from the system, also called black-box or input-output models, and

grey-box which means based upon both the mechanisms of the system and upon empirical functions where the mechanisms are complex or unknown.

In Chapters 3 and 4 we have developed what would normally be called mechanistic models for the nutrient removal processes. In truth all models of real processes are grey-box models. Rate expressions for heat and mass transfer and kinetics are empirical in nature and yet are built into fundamental mass and energy balances. The Monod form of the biological kinetic rate expressions is empirical in nature and could be replaced by a number of other forms with equally good results. Another classification is based upon the nature of the process inputs and outputs:

deterministic models have single valued inputs and outputs, which neglect any uncertainty,

uncertainty models have inputs and outputs with some empirical measure of uncertainty (interval or fuzzy models), and

stochastic models have at least some inputs or outputs described in terms of statistical probability distributions.

All models with parameters that have been fitted to real data will have parameter uncertainty, but this does not make them stochastic models. Only if this parameter uncertainty is built into the model as a statistical distribution does the model become stochastic. The influent and effluent characteristics in wastewater treatment plants contain both deterministic and random components as seen in the time series plot of influent total COD shown in Figure 5.1. However most modelling has been of the deterministic type.

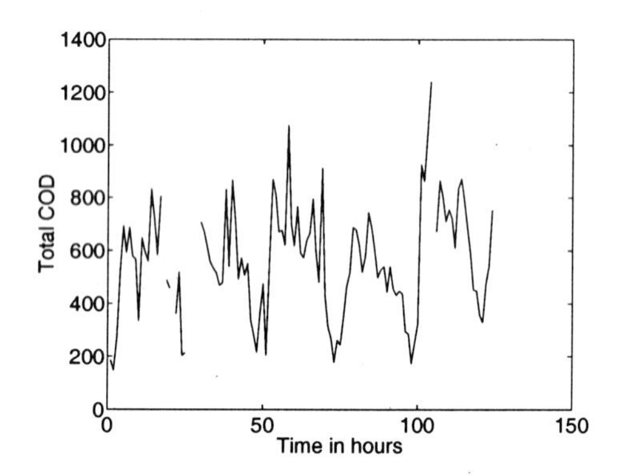

Figure 5.1: Typical influent total COD.

Stochastic models of mixed sewer systems have been used since the rainfall, which is the major input disturbance, is distinctly stochastic (despite what the weather forecast presenters will have us believe). Yet another classification depends on whether temporal or time dependency is incorporated:

static models represent the average or steady state behaviour of systems, and

dynamic models predict system behaviour as a function of time in response to disturbances.

Static models are frequently used for process design, though there are many instances where dynamics should be investigated to ensure designs are robust to disturbances or failures. Dynamic models are also used to investigate startup or for the design of control systems. Static models are simply a special case of dynamic models where the time derivatives are set to zero. Finally we could classify models according to the form of the mathematical functions used in formulating the model equations:

functional models use deterministic mathematical equations (differential, partial differential, and/or algebraic equations) usually with a simple structure but often with complex expressions,

neural network models use a complex structure of simple expressions based on one concept of how the human brain might work, and

qualitative models use a complex structure of simple qualitative expressions.

Functional and qualitative models may be derived mechanistically or empirically. All three classes can represent static or dynamic behaviour. In the following sections we will examine some common forms of empirical models in these three classes. But first we will define some basic terminology which the rest of the chapter will use. This follows on from the definitions in Section 3.1.

5.2 Basic Terminology

5.2.1 State and Input-Output Models

There are two basic forms of linear process models: state space models and input-output models. State space models contain much more information about the process interactions and dynamics, but also more parameters. For this reason state space models usually originate from physical or mechanistic modelling, as we have shown in Chapters 3 and 4, while input-output models are simpler and have fewer parameters that can be derived directly from plant data.

A state space model has the form:

$$\frac{dx}{dt} = f(x(t), u(t), d(t), p) \tag{5.1}$$

$$y(t) = g(x(t), u(t), d(t), p) \tag{5.2}$$

where $x(t)$ are the state variables (such as nutrient concentrations within the bioreactor zones), $u(t)$ are manipulated input variables (such as air and recycle flowrates), $d(t)$ are disturbance input variables (such as influent characteristics), $y(t)$ are output variables (such as effluent qualities) and p are parameters (such as volumes, oxygen transfer rates and biological growth rates). We assume throughout the chapter that these equations are in vector form representing any number of equations of the same form.

An input-output model has the simpler form:

$$\frac{dy}{dt} = h(y(t), u(t), d(t), p) \tag{5.3}$$

Input-output models can be derived from state space models by using the algebraic equations to eliminate $x(t)$. Although in principle there is seldom much sense in this, it is generally done because we know too little about our process to derive a mechanistic model or because we wish to avoid the modelling time and effort. The latter is possible because there are many techniques which enable us to derive the form and parameters of input-output models from plant data. There are a number of ways of writing an input-output model and the following are common forms you will see in many books and papers.

If we consider a time series for a process measurement or output (for example Figure 5.1), these data sets show time dependency, that is, measured values at time t are not completely independent of earlier measurements. There is a certain memory function in the system. The measurements are also known at discrete usually equally-spaced points in time. One way in which this dependency can be expressed is given by:

$$\begin{aligned} y(t) = a_1\, y(t-\Delta t) + a_2\, y(t-2\Delta t) + \ldots. \\ +a_n\, y(t-n\Delta t) + b\, u(t) + v(t) \end{aligned} \tag{5.4}$$

where the coefficients a_i and b are constants and Δt is the sampling interval, the time between adjacent data. In this traditional form, disturbances $d(t)$ are incorporated into either the inputs $u(t)$ or the disturbance term $v(t)$. For simplicity this is often written as:

$$\begin{aligned} y_t = a_1\, y_{t-1} + a_2\, y_{t-2} + \; \ldots \; + a_n\, y_{t-n} \\ +b\, u_t + v(t) \end{aligned} \tag{5.5}$$

Such a relationship is called an autoregressive (AR) process. The $v(t)$ term accounts for all the unmeasured or unknown disturbances affecting the process output. These equations are derived from equation 5.3 by linearising the right-hand side and approximating the derivatives by difference equations such as:

$$\frac{dy}{dt} = \frac{y_t - y_{t-1}}{\Delta t} \tag{5.6}$$

Another way of representing a time series, the moving average (MA) form, can be used to represent the dynamic effects of inputs giving us an ARMA model:

$$\begin{aligned} y_t = a_1\, y_{t-1} + a_2\, y_{t-2} + \; \ldots \; + a_n\, y_{t-n} \\ +b_1\, u_{t-1} + b_2\, u_{t-2} + \; \ldots \; + b_m\, u_{t-m} + v(t) \end{aligned} \tag{5.7}$$

The terminology y_{t-k} means the value of the variable y at the time k time intervals before time t. This form of equation is called a time series model which is the one we will often use to represent processes in subsequent chapters. Both of these forms are linear in both the inputs, the outputs and the parameters. We will discuss this further in the next section.

A time series model can also be used to represent a signal or single time series by omitting the input $u()$ terms. Signal models can be used for predicting or forecasting disturbances (see Chapter 15). Input-output models contain much less information than state space models, both because they tell us nothing about the behaviour of the system states, and because their parameters are derived from plant data with all its associated problems and inaccuracies. In addition, the parameters of input-output models are seldom simply related to our understanding of the physical phenomena.

5.2.2 Linear and Nonlinear Models

Process systems such as wastewater treatment processes are rarely inherently linear, so that a linear state space description usually comes about by the linearisation of a nonlinear model about some operating point. Why would we do that? Generally because most of the more advanced controller designs are based upon linear models. Of course we lose information when we linearise. However, we can argue that, because the controller keeps the process close to the operating point, the linear model contains sufficient information for the design. The problem arises when we move the operating point. Either we must accept some loss of performance, or we must relinearise and redesign our controller, or we may devise some form of automatic adaptation of the control law. Because of these problems, linear models are seldom satisfactory for purposes other than regulatory control.

Example 5.1 (Nonlinear DO - air relationship) *As an example, let us consider the aeration of a wastewater treatment tank. The balance of dissolved oxygen (DO) in the aerator of a wastewater treatment plant is a nonlinear dynamic system. Here the tank is assumed to be a batch reactor, there is no continuous flow of water in to and out of the tank (a simpler system than the carbon removal process of Section 3.7). Oxygen is supplied from a compressor with the air flowrate* q_a. *The transfer rate from gaseous oxygen to dissolved oxygen is determined by a transfer rate coefficient* K_La. *For simplicity, we will consider it proportional to the air flowrate, that is,*

$$K_La = K_a\, q_a \tag{5.8}$$

where K_a *is a proportionality constant. From the oxygen mass balance of Equation 3.35, with*

$$q_{in} = q_{out} = 0 \tag{5.9}$$

we obtain:

$$\frac{dS_O}{dt} = r_O + K_a q_a (S_{O,sat} - S_O) \tag{5.10}$$

Because of the product between q_a *and* S_O, *the system is nonlinear. The air supply* q_a *and the respiration rate* r_O *are assumed constant and the dissolved oxygen (DO) concentration is kept at a value of 3 mg/l. In Figure 5.2 it is shown that, when the air flowrate is changed by a step value (2 per cent, 4 per cent etc.), the DO concentration approaches a new steady-state value within an hour. A 4 per cent change will almost exactly double the concentration change with respect to a*

2 per cent variation. The behaviour looks quite linear. Note that these changes are symmetrical around the steady state value. But for an 8 per cent change, the asymmetry of the response becomes obvious. The upward and downward changes are not symmetrical, and furthermore are not two times larger than the 4 per cent change. These curves, produced by the MATLAB program `Example_5_1.m`, *show how nonlinearities appear in practice.*

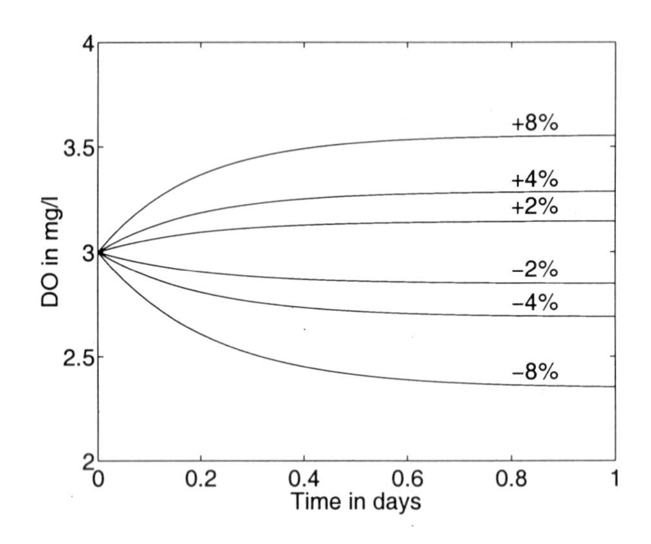

Figure 5.2: Changes of DO concentration with air flowrate.

A linear system has many attractive properties and obeys the **superposition principle**. This means that if the change in the output signal has a value of ΔY for a specific input amplitude ΔU, then the output will have the value $2\Delta Y$ if the input is doubled, or $4\Delta Y$ if the input is quadrupled. The previous example showed how this does not happen for a nonlinear system. Moreover, in linear systems the contributions from different input signals are additive. That is, if the input signal u_1 contributes with y_1 and u_2 with y_2 then the total output change will be $y_1 + y_2$. As a consequence the influence from the control input (the manipulating variable) and a disturbance signal can be analysed separately.

If you wish to talk of the degree of nonlinearity, then you have entered a research area. There is still no consensus on how we measure nonlinearity. Heuristically, we probably look at how much the gradient dy/du changes with u. However, this simplistic view is undoubtedly "necessary but not sufficient". A very important concept to remember is that a system may be nonlinear with respect to one variable, but still linear with respect to another.

Example 5.2 (Monod growth rate) *Consider the growth rate described by a* *simple Monod expression:*

$$\mu = \hat{\mu} \frac{S}{K+S} \tag{5.11}$$

where S is the substrate concentration. If the maximum growth rate is doubled, then the net growth rate will double, so the system is linear in the parameter $\hat{\mu}$. However, if K or S were doubled, then the net growth rate would not be doubled, so that the system is not linear in K or S.

The nonlinear state space models we developed in Chapters 3 and 4 can be linearised by a truncated Taylor expansion as follows:

$$\frac{d}{dt}(x - x_o) = \left.\frac{\partial f}{\partial x}\right|_{x_o,u_o} (x - x_o) + \left.\frac{\partial f}{\partial u}\right|_{x_o,u_o} (u - u_o) \tag{5.12}$$

$$y - y_o = \left.\frac{\partial g}{\partial x}\right|_{x_o,u_o} (x - x_o) + \left.\frac{\partial g}{\partial u}\right|_{x_o,u_o} (u - u_o) \tag{5.13}$$

where y_o, x_o, u_o is the operating point. The partial derivatives written in vector form are interpreted as:

$$\left.\frac{\partial f}{\partial x}\right|_{x_0,u_0} = \begin{pmatrix} \frac{\partial f_1}{\partial x_1} & \frac{\partial f_1}{\partial x_2} & .. & \frac{\partial f_1}{\partial x_n} \\ \frac{\partial f_2}{\partial x_1} & \frac{\partial f_2}{\partial x_2} & .. & \frac{\partial f_2}{\partial x_n} \\ .. & .. & .. & .. \\ \frac{\partial f_n}{\partial x_1} & \frac{\partial f_n}{\partial x_2} & .. & \frac{\partial f_n}{\partial x_n} \end{pmatrix} \tag{5.14}$$

where each partial derivative is evaluted at the operating point. We call this matrix A and its coefficients are renamed

$$[a_{ij}] = \left[\frac{\partial f_i}{\partial x_j}\right] \tag{5.15}$$

Similarly we define

$$\left.\frac{\partial f}{\partial u}\right|_{x_0,u_0} = \begin{pmatrix} \frac{\partial f_1}{\partial u_1} & \frac{\partial f_1}{\partial u_2} & .. & \frac{\partial f_1}{\partial u_r} \\ \frac{\partial f_2}{\partial u_1} & \frac{\partial f_2}{\partial u_2} & .. & \frac{\partial f_2}{\partial u_r} \\ .. & .. & .. & .. \\ \frac{\partial f_n}{\partial u_1} & \frac{\partial f_n}{\partial u_2} & .. & \frac{\partial f_n}{\partial u_r} \end{pmatrix} = \begin{pmatrix} b_{11} & b_{12} & .. & b_{1r} \\ b_{21} & b_{22} & .. & b_{2r} \\ .. & .. & .. & .. \\ b_{n1} & b_{n2} & .. & b_{nr} \end{pmatrix} = B \tag{5.16}$$

$$\left.\frac{\partial g}{\partial x}\right|_{x_0,u_0} = \begin{pmatrix} \frac{\partial g_1}{\partial x_1} & \frac{\partial g_1}{\partial x_2} & .. & \frac{\partial g_1}{\partial x_n} \\ \frac{\partial g_2}{\partial x_1} & \frac{\partial g_2}{\partial x_2} & .. & \frac{\partial g_2}{\partial x_n} \\ .. & .. & .. & .. \\ \frac{\partial g_p}{\partial x_1} & \frac{\partial g_p}{\partial x_2} & .. & \frac{\partial g_p}{\partial x_n} \end{pmatrix} = \begin{pmatrix} c_{11} & c_{12} & .. & c_{1n} \\ c_{21} & c_{22} & .. & c_{2n} \\ .. & .. & .. & .. \\ c_{p1} & c_{p2} & .. & c_{pn} \end{pmatrix} = C \tag{5.17}$$

$$\left.\frac{\partial g}{\partial u}\right|_{x_0,u_0} = \begin{pmatrix} \frac{\partial g_1}{\partial u_1} & \frac{\partial g_1}{\partial u_2} & .. & \frac{\partial g_1}{\partial u_r} \\ \frac{\partial g_2}{\partial u_1} & \frac{\partial g_2}{\partial u_2} & .. & \frac{\partial g_2}{\partial u_r} \\ .. & .. & .. & .. \\ \frac{\partial g_p}{\partial u_1} & \frac{\partial g_p}{\partial u_2} & .. & \frac{\partial g_p}{\partial u_r} \end{pmatrix} = \begin{pmatrix} d_{11} & d_{12} & .. & d_{1r} \\ d_{21} & d_{22} & .. & d_{2r} \\ .. & .. & .. & .. \\ d_{p1} & d_{p2} & .. & d_{pr} \end{pmatrix} = D \tag{5.18}$$

We can write this in the equivalent familiar linear state space form:

$$\frac{dx}{dt} = A\,x + B\,u \tag{5.19}$$

$$y = C\,x + D\,u \tag{5.20}$$

The matrix notation is interpreted as a set of linear differential equations,

$$\begin{aligned}
\tfrac{dx_1}{dt} &= a_{11}x_1 + a_{12}x_2 + .. + a_{1n}x_n \\
&\quad + b_{11}u_1 + .. + b_{1r}u_r \\
\tfrac{dx_2}{dt} &= a_{21}x_1 + a_{22}x_2 + .. + a_{2n}x_n \\
&\quad + b_{21}u_1 + .. + b_{2r}u_r \\
&.... \\
\tfrac{dx_n}{dt} &= a_{n1}x_1 + a_{n2}x_2 + .. + a_{nn}x_n \\
&\quad + b_{n1}u_1 + .. + b_{nr}u_r
\end{aligned} \tag{5.21}$$

and a set of linear algebraic output relations,

$$\begin{aligned}
y_1(t) &= c_{11}x_1(t) + c_{12}x_2(t) + .. + c_{1n}x_n(t) \\
&\quad + d_{11}u_1(t) + .. + d_{1r}u_r(t) \\
&.... \\
y_p(t) &= c_{p1}x_1(t) + c_{p2}x_2(t) + .. + c_{pn}x_n(t) \\
&\quad + d_{p1}u_1(t) + .. + d_{pr}u_r(t)
\end{aligned} \tag{5.22}$$

The coefficient matrices A, B, C, D are the partial derivatives and are constants. These coefficients will be combinations of our original model parameters and output, state and input variable values at the operating point. The vectors x, y, u are now perturbation variables, which are sometimes also normalised by dividing by the respective operating point values.

Example 5.3 (Linear state-space model) *Consider the simple carbon removal process detailed in Appendix B. Assume that the system of interest to us is substrate and oxygen dynamics. Further assume biomass concentration X_H is constant, air flowrate q_A is the input variable, and dissolved oxygen concentration S_O is the output variable. Our system is therefore described by the states S_S and S_O and the state equations:*

$$\begin{aligned}
\tfrac{dS_S}{dt} &= \tfrac{q_F}{V}(S_{SF} - S_S) \\
&\quad - \tfrac{\hat{\mu}_H}{Y_H}\left(\tfrac{S_S}{K_S + S_S}\right)\left(\tfrac{S_O}{K_{OH} + S_O}\right)X_H \\
&\quad + (1 - f_P)b_H X_H
\end{aligned} \tag{5.23}$$

$$\begin{aligned}
\tfrac{dS_O}{dt} &= \tfrac{q_F}{V}S_{OF} - \tfrac{q_F + q_R}{V}S_O \\
&\quad + \tfrac{Y_H - 1}{Y_H}\hat{\mu}_H\left(\tfrac{S_S}{K_S + S_S}\right)\left(\tfrac{S_O}{K_{OH} + S_O}\right)X_H \\
&\quad + a\left(1 - e^{-\frac{q_A}{b}}\right)(S_{O,sat} - S_O)
\end{aligned} \tag{5.24}$$

The first partial derivative is:

$$\begin{aligned}\frac{\partial f_1}{\partial x_1} &= \frac{\partial}{\partial S_S}\left(\frac{q_F}{V}(S_{SF}-S_S)\right)\\ &-\frac{\partial}{\partial S_S}\left(\frac{\hat{\mu}_H}{Y_H}\left(\frac{S_S}{K_S+S_S}\right)\left(\frac{S_O}{K_{OH}+S_O}\right)X_H\right)\\ &+\frac{\partial}{\partial S_S}\left((1-f_P)b_H X_H\right)\\ &= -\frac{q_F}{V}-\frac{\hat{\mu}_H}{Y_H}\left(\frac{K_S}{(K_S+\bar{S}_S)^2}\right)\left(\frac{\bar{S}_O}{K_{OH}+\bar{S}_O}\right)X_H\end{aligned} \tag{5.25}$$

where the bar over a variable means its steady state value. Similiarly:

$$\frac{\partial f_1}{\partial x_2} = -\frac{\hat{\mu}_H}{Y_H}\left(\frac{\bar{S}_S}{K_S+\bar{S}_S}\right)\left(\frac{K_{OH}}{(K_{OH}+\bar{S}_O)^2}\right)X_H \tag{5.26}$$

$$\frac{\partial f_2}{\partial x_1} = \frac{Y_H-1}{Y_H}\hat{\mu}_H\left(\frac{K_S}{(K_S+\bar{S}_S)^2}\right)\left(\frac{\bar{S}_O}{K_{OH}+\bar{S}_O}\right)X_H \tag{5.27}$$

$$\begin{aligned}\frac{\partial f_2}{\partial x_2} &= -\frac{q_F+q_R}{V}\\ &+\frac{Y_H-1}{Y_H}\hat{\mu}_H\left(\frac{\bar{S}_S}{K_S+\bar{S}_S}\right)\left(\frac{K_{OH}}{(K_{OH}+\bar{S}_O)^2}\right)X_H\\ &-a\left(1-e^{-\frac{\bar{q}_A}{b}}\right)\end{aligned} \tag{5.28}$$

The derivatives with respect to the control variable q_A are:

$$\frac{\partial f_1}{\partial u_1} = 0 \tag{5.29}$$

$$\frac{\partial f_2}{\partial u_1} = \frac{a}{b}e^{-\frac{\bar{q}_A}{b}}\left(S_{O,sat}-\bar{S}_O\right) \tag{5.30}$$

Substituting steady state values from Table B.1 in Appendix B, we can write the linear state equation as:

$$\frac{dx}{dt} = \begin{pmatrix} -170.6 & -18.6 \\ -66.6 & -69.5 \end{pmatrix} x + \begin{pmatrix} 0 \\ 49.0 \end{pmatrix} u \tag{5.31}$$

and the output equation is:

$$y = \begin{pmatrix} 0 & 1 \end{pmatrix} x \tag{5.32}$$

where:

$$x = \begin{pmatrix} S_S - 4.4 \\ S_O - 3.0 \end{pmatrix} \qquad u = (q_A - 6.3) \qquad y = (S_O - 3.0) \tag{5.33}$$

Equation 5.31 and 5.32 make up our linear model of the DO response to changes in air flowrate. Figure 5.3 shows DO responses to steps up and down of $1m^3/min$ in the air flowrate for both the linear and nonlinear models. The symmetrical linear responses approximate the asymmetrical nonlinear responses well close to the steady state, say within $0.5mg/l$, but are somewhat different further away.

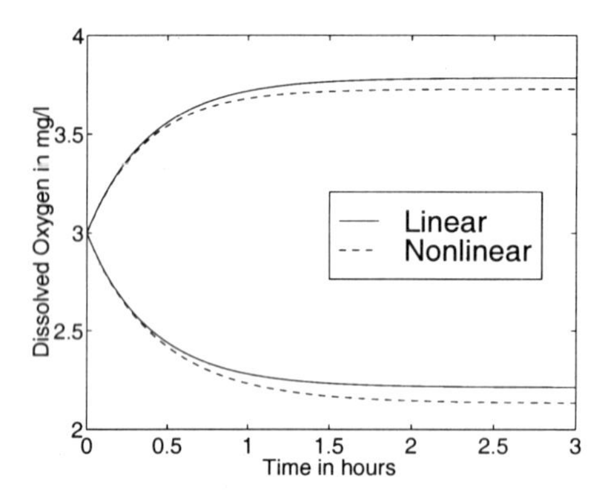

Figure 5.3: Linear and nonlinear DO responses.

Example 5.4 (Linear input-output model) *The input-output equation equivalent to Equations 5.31 and 5.32 in Example 5.3 for a sample time of 6 minutes is:*

$$y_t = 1.25\, y_{t-1} - 0.368\, y_{t-2} + 0.178\, u_t - 0.088\, u_{t-1} \tag{5.34}$$

This can be derived from Equation 5.31 by differentiating the state equation for x_2 *to obtain:*

$$\frac{d^2x_2}{dt^2} = -66.6\frac{dx_1}{dt} - 69.5\frac{dx_2}{dt} + 49.0\frac{du}{dt} \tag{5.35}$$

using the state equation for x_1 *to eliminate* dx_1/dt*, differentiating the output Equation 5.32 to eliminate* du/dt *and discretising the derivatives using for example backward differencing (Equation 5.6). Comparing Equation 5.34 with Equation 5.31, we can see that any information concerning the state* x_1 *has been lost. In fact, we only have information about* x_2 *because* $y = x_2$*, but this would not generally be the case. We have also lost information by discretising, in that Equation 5.34 only tells us what happens at each 6 minute sampling instant. We have no information about what happens in between the sampling instants.*

5.3 Functional Empirical Models

The most common empirical functional model used in science and engineering takes the form:

$$y_i = p_{i,0} + p_{i,1}u_1 + p_{i,2}u_2 + ... + p_{i,m}u_m \tag{5.36}$$

for each model output $y_i, i = 1,..,n$ where u_j are the model inputs and $p_{i,j}$ are the model parameters which take constant values. This classical linear model can be found throughout the statistical literature for experimental design models and regression models. It is used typically in time invariant (steady state) situations.

It is also clear that the dynamic input-output models presented in the previous section take the same form with the general u_i representing previous model outputs as well as current and previous model inputs.

Linear models are so popular because the superposition principle mentioned above makes the development of theory and algorithms a far simpler task. The ready availability of this theory and algorithms makes it very attractive for engineers to approximate their systems using linear models.

If the linear approximation has too small a region with acceptable errors, then one of three approaches must be taken:

- approximate the nonlinear system by a set of linear models each valid in small adjacent regions (Figure 5.4 left),
- transform the system variables such that the relationship between the transformed variables is linear or close to linear (Figure 5.4 right and also Example 7.1), and
- use a nonlinear model and find or develop the much more complex analysis tools.

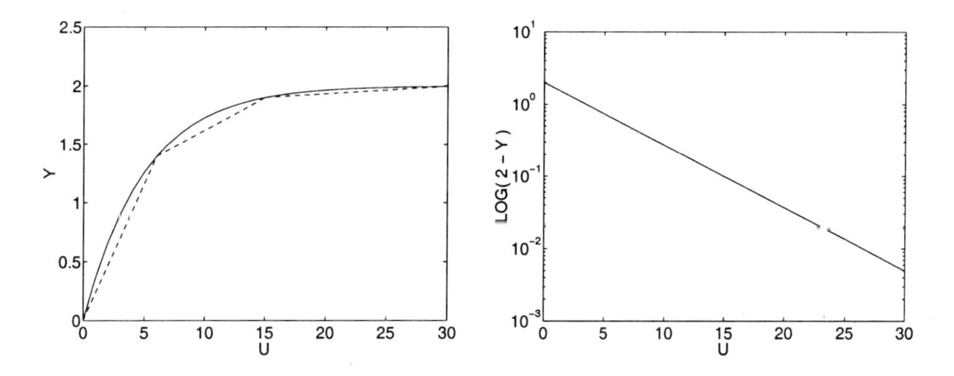

Figure 5.4: Linearisation approaches.

There are major problems with each approach. The use of many linear segments requires carrying out your design or analysis tasks many times, and the discontinuities between regions can cause major problems.

Transformation of variables is superficially attractive but also has its problems. The first is often to find the linearising transformation, there being no general theory to assist us. The second is that the transformations destroy many of the prior assumptions of the linear theory, for example normally distributed experimental errors are no longer normally distributed when that variable is transformed by a nonlinear function such as a square root or logarithm.

Finding nonlinear theory and analysis tools can be very difficult except for some very special and restricted types of nonlinearity. One special form which can be useful in process engineering is the bilinear equation:

$$\begin{aligned} y_i = p_{i,0} + p_{i,1}u_1 + p_{i,2}u_2 + p_{i,3}u_3 \ldots \\ +p_{i,12}u_1u_2 + p_{i,13}u_1u_3 + p_{i,23}u_2u_3 + \ldots \end{aligned} \tag{5.37}$$

This involves just the variables and second-order products of the variables. Process mass and energy balances frequently involve flowrate times concentration or flowrate times temperature terms which are bilinear in nature.

Example 5.5 (Models of settling velocity) *There are many functional empirical models for the settling velocity of biological flocs. It is one way to get your name remembered, though you may have trouble finding a functional form not already reserved. Examples include:*

$$V_s = k\ X^{-n} \tag{5.38}$$

$$V_s = V_o\ e^{-nX} \tag{5.39}$$

$$V_s = k\frac{(1-n_1X)^n}{X}e^{-n_2X} \tag{5.40}$$

Given that a basic functional form has been chosen for the model there remain two problems to solve:

identification is the problem of determining the model structure - the significant input variables, the significant output variables, and which inputs have a significant effect on which outputs, and

estimation is the problem of determining values for the model parameters from input and output data from the process. Aspects of the design of experiments and screening of data are presented in Chapter 6. The principles and practice of regression, determining the parameters, are presented in Chapter 7.

5.4 Neural Network Empirical Models

A quite different form of model has been developed and applied quite recently. These are the so-called **neural network** models. As with most new developments, neural network (NN) models seem to have very determined supporters and very determined detractors with not so many in between - at least not that are willing to step into the crossfire.

The NN model structure generally used is a simplified view of the structure of the brain. Rather than a fairly simple interaction between quite complex expressions, as we have seen in the mechanistic models developed in chapters 3 and 4, a NN model is a very complex interaction between fairly simple expressions called **neurons**.

It is able to represent linear and nonlinear relationships. This structure is illustrated in Figure 5.5.

There are a number of layers of neurons:

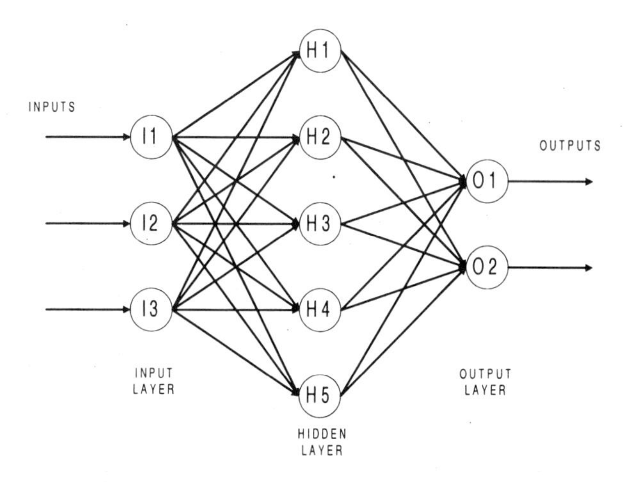

Figure 5.5: Neural network structure.

the input layer with one neuron for each model input which generally simply distributes the value of the input to every neuron in the next layer,

one or more hidden layers which perform a weighted sum of the inputs from each neuron in the previous layer, transform the sum according to some *activation function* and distribute the result to each neuron in the next layer, and

the output layer with one neuron for each model output which performs a similiar weighted sum but the transformation of the inputs from each neuron in the previous layer is usually much simpler.

There have been some variations on this basic structure but they are not as common.

Neural network models can represent static or dynamic relationships. Dynamic models are derived by including both previous process output values and the process inputs as network inputs. This configuration is similiar to the functional input-output AR or ARMA model form discussed previously. We will examine three types of neural network models, artificial neural networks, radial basis function networks and wavelet networks (wavenets).

5.4.1 Artificial Neural Networks

The original NN model was called an artificial neural network (ANN) with each neuron in the hidden layers computing according to the following expressions:

$$S_j = \sum_{i=1}^{n} w_{i,j} u_i + w_{n+1,j} \tag{5.41}$$

$$y_j = \frac{1}{1+e^{-S_j}} \tag{5.42}$$

where the u_i are the n inputs from the neurons in the previous layer. S_j is the weighted sum of the inputs to neuron j with weights $w_{i,j}$ and with a bias $w_{n+1,j}$. Inputs are typically normalised to the interval $[-1,+1]$.

The second equation is the so-called sigmoidal activation function where y_j is the output to each neuron in the next layer. Outputs are typically normalised to the interval [0.2,0.8]. The output layer generally has no activation function with the output y_k being simply equal to the weighted sum S_k of the inputs. All the weights $w_{i,j}$ for each layer constitute the parameters of the ANN model.

One disadvantage of the ANN is that there are a large number of parameters which have no particular physical significance and cannot be calculated directly from input-output data. There are a number of iterative "learning" algorithms to determine values for the parameters. The most common but perhaps the least efficient is the *back-propagation learning algorithm* outlined below:

1. The parameters are assigned random values typically in the interval $[-0.5,+0.5]$.

2. A *pattern*, a set of input and output values, is chosen randomly or sequentially from the *training* data.

3. The error in each output k is calculated:

$$\delta_k = y_k(1-y_k)(d_k-y_k) \tag{5.43}$$

   where d_k is the training data output value, y_k is the value calculated by the ANN from the training data inputs, and k is the output neuron.

4. The error for each hidden neuron is calculated layer by layer starting at the layer next to the output layer and working backwards towards the input layer:

$$\delta_j = y_j(1-y_j)\sum_k \delta_k w_{j,k} \tag{5.44}$$

   where $w_{j,k}$ is the weight from node j to node k, and k are all the nodes in the next layer.

5. The weights are adjusted according to:

$$\Delta_p w_{j,k} = \eta \delta_k y_j + \alpha \Delta_{p-1} w_{j,k} \tag{5.45}$$

   where Δ_p is the change for training pattern p, η is the learning rate and α is a momentum factor. These can be tuned to improve convergence.

6. The iteration returns to step 2 until some convergence criteria is satisfied. A typical convergence criteria is based on minimising:

$$J(w) = \sum_{p} \sum_{k=1}^{n} (d_k - y_k)_p^2 \tag{5.46}$$

where the squared errors in the n outputs are summed over the training data set $\{p\}$.

Example 5.6 (Artificial neural network model) *An ANN model with five hidden nodes was used to fit settling velocity data and the result is shown in Figure 5.6.*

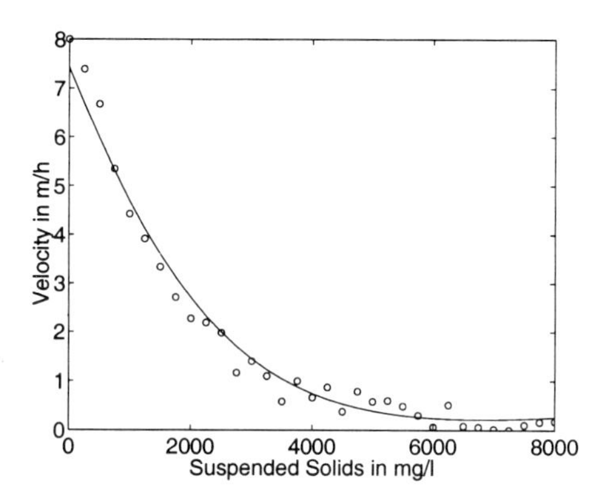

Figure 5.6: ANN model for settling velocity.

The fit was performed by the MATLAB program `Example_5_6.m` *and the WWT Toolbox functions* `annnet.m` *and* `nntrain.m` *which you can use or extend yourself. Several attempts with different random starting weights were usually required to get a stable convergence from the back-propagation algorithm.*

About 900 iterations through the training data were required to obtain the model fit shown.

ANN models have been used in modelling, identification, inferential or soft sensors, model-based control, and diagnosis or fault detection.

5.4.2 Radial Basis Function Networks

This type of neural network has been shown to be superior to the traditional ANN in fault detection applications. This is because the radial basis function is close to being a local approximator. Networks of local approximators also have much better convergence properties. The structure of the radial basis function (RBF) network is the same as a NN with only one hidden layer. The inputs are process measurements and the outputs each indicate a certain fault or operating condition. However, the neurons compute differently compared to the ANN. Hidden layer neurons compute according to:

$$y = exp\left(-\sum_i \frac{(\bar{x}_i - x_i)^2}{\sigma_i^2}\right) \tag{5.47}$$

where x_i are the i input values, $\bar{x}_i$ and σ_i are the mean and standard deviation of the radial basis functions for each neuron in the ith input direction. Output neurons compute according to:

$$y = \sum_h w_h a_h \tag{5.48}$$

where h are the hidden neurons and w_h are the weights for that output neuron.

The RBF network computes parameters $\bar{x}_i$ and σ_i for each hidden neuron and w_h for each output. They are also determined from sets of training data but the algorithm is quite different:

1. The RBF means $\bar{x}$ are first determined by the k-means clustering algorithm applied to the input data. The algorithm finds a local minimum in:

$$E_{k-means} = \sum_{h=1}^{H}\sum_{k=1}^{K} B_{hk} \sum_j (\bar{x}_h - x_k)^2 \tag{5.49}$$

   where B_{hk} is the cluster membership matrix with each column of K values represents a training set and each row represents one of H clusters. Each column has one 1 and the remainder 0s defining the cluster to which the point is assigned. There is one hidden neuron per cluster which must be determined iteratively. The algorithm is:

   (a) The centre $\bar{x}_h$ of each cluster is randomly assigned to a different test point.
   (b) Each training set is assigned to the nearest centre.
   (c) The average position of the sets in a cluster is made the new cluster centre.
   (d) The algorithm returns to step (b) until convergence is obtained (little or no change in the clusters).

   Some outer iteration may be required to choose the number of clusters.

2. The RBF standard deviations σ_h are determined next by the P-nearest neighbor heuristic:

$$\sigma_h = \left(\frac{1}{P}\sum_{j=1}^{P}\sum(\bar{x}_h - \bar{x}_j)^2\right)^{0.5} \tag{5.50}$$

   where P is typically 2.

3. The output weights w_h are lastly determined by linear least-squares regression.

Example 5.7 (Radial basis function model) *An RBF network was used to model the same settling velocity data as in Examples 5.6 and 5.8. The result that gave the best compromise between smoothness and minimisation of the residuals had 12 basis functions and a standard deviation of 0.15 m/h. The fit is shown in Figure 5.7, which can be compared to Figures 5.6 and 5.10.*

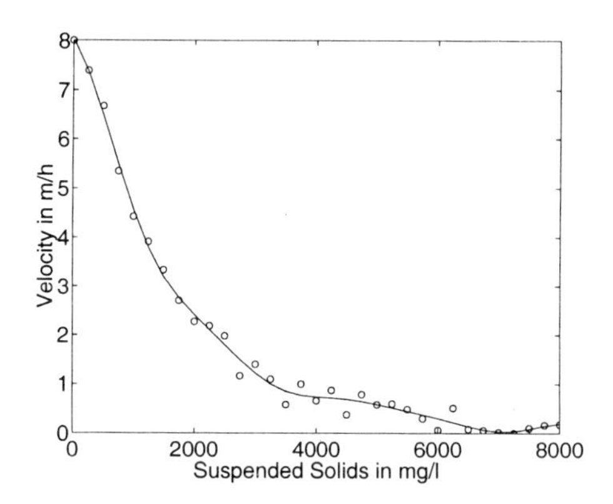

Figure 5.7: RBF model for settling velocity.

The fit was performed by the MATLAB program `Example_5_7.m` *and the WWT Toolbox functions* `rbfnet.m` *and* `fitrbfnt.m` *which you can use or extend yourself.*

5.4.3 Wavelets and Wavenets

Engineers know that any periodic signal can be portrayed as a weighted summation of sinusoids of different frequencies (Fourier analysis). In like fashion we can represent signals in terms of wavelets. Unlike sinusoids, a wavelet is a signal that is localised in time and hence gives a better representation of signals containing discontinuities which are quite common in process signals.

Wavenets are networks consisting of a single hidden layer whose activation functions are sets of wavelets. Adding additional wavelets to the sets increases the resolution of the representation, just as adding higher frequency sinusoids does in Fourier analysis.

Wavelets

The theory of functional analysis suggests that orthogonal basis functions are most effective as approximators. Wavelets are functions which can be orthonormal and can have good localisation properties. Families of wavelets are derived by translation (shifting along the axis) and dilation (amplification in the axis direction) according to the equation:

$$\frac{1}{\sqrt{s}}\psi\left(\frac{x-u}{s}\right) \tag{5.51}$$

where $\psi(x)$ is the base wavelet, s governs the dilation, u is the translation. Figure 5.8 shows a typical wavelet with dilated and translated forms.

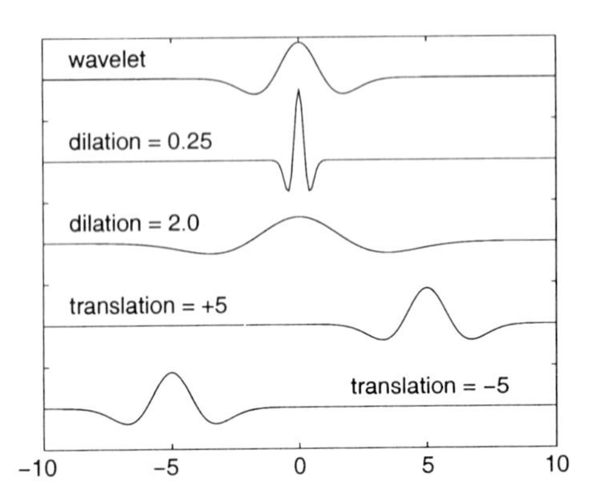

Figure 5.8: Wavelet with dilation and translation.

Discrete dyadic wavelet families are commonly used for data regression where the data is discrete and the dilation is always a factor of 2:

$$s = 2^m, \quad u = k2^m \quad for(m,k) \in Z^2 \tag{5.52}$$

so that the wavelet family is:

$$\sqrt{2^{-m}}\psi\left(2^{-m}x - k\right) \tag{5.53}$$

Wavenets

Figure 5.9 shows the centre points for a family of wavelets for $m = 0, 1, 2, .., L$ illustrating the progression of resolutions from the finest at $m = 0$ to the coarsest at $m = L$. The value of L is determined from the range and resolution of the data.

The wavenet hidden layer is made up of:

- scaling function nodes $\phi_{Lk}(x), k = 1, 2, .., n_L$ which approximate the unknown function at the coarsest resolution. For one-dimensional data $n_L = 2$ (Figure 5.9).

- wavelet nodes $\psi_{mk}(x), m = 1, 2, .., L$ *and* $k = 1, 2, ..n_m$ which form an orthonormal basis for approximating the detail of the unknown function at each level of resolution.

The wavenet weights are calculated by a series of linear least-squares regressions as follows:

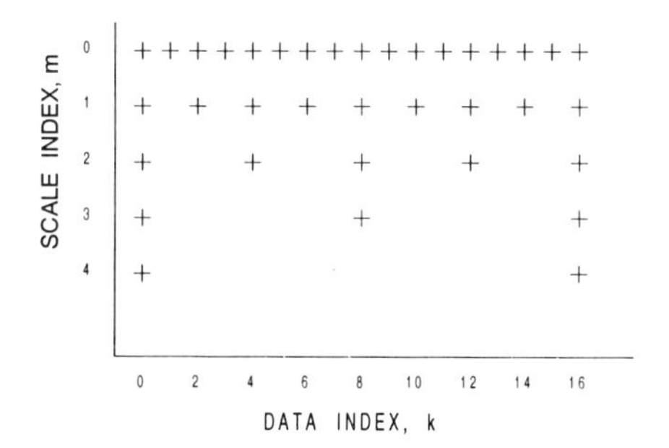

Figure 5.9: Dyadic wavelet family with $L = 4$.

1. calculate the n_L scaling function node weights by a fit to the input-output data,
2. set the resolution parameter $m = L$,
3. calculate the n_m resolution m wavelet node weights by a fit to the residuals between the previous fit and the data,
4. set the resolution parameter $m = m - 1$ and return to the previous step until $m < 0$.

Wavelet nodes can be included as required to approximate the level of detail required and the data available. In some cases more detailed data may be available in different parts of the range. The wavenet can also be optimised by omitting insignificant weights, either manually or by using a regression package which does so automatically.

Example 5.8 (Wavelet network model) *A wavenet model with a resolution of 32 (6 levels) was used to fit the same settling velocity data as in Examples 5.6 and 5.7. The result is shown in Figure 5.10 which can be compared to Figures 5.6 and 5.7. Although the standard deviation was 0.16 m/h, the wavenet fits more detail of the noise in the data giving a less smooth result.*

The fit was performed by the MATLAB program `Example_5_8.m` *and the WWT Toolbox functions* `wavenet.m` *and* `fitwavnt.m` *which you can use or extend yourself.*

5.5 Qualitative Models

There is a variety of reasons for using qualitative models. Sometimes we are only interested in whether things are related or not, for example in diagnostic reasoning. Often we glean model relationships from human experience, which is generally semi-quantitative and often expressed with a degree of uncertainty.

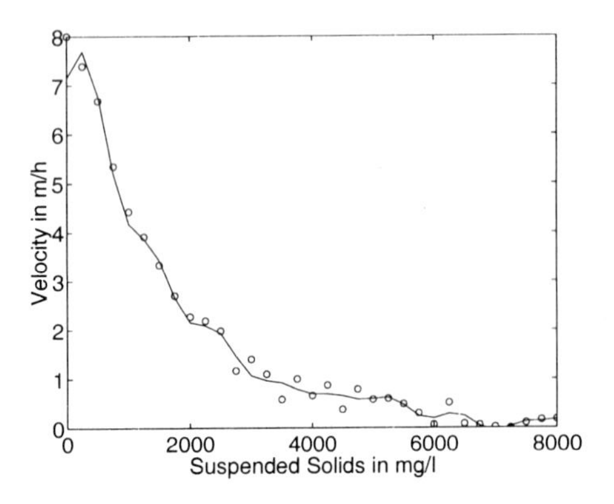

Figure 5.10: Wavenet model for settling velocity.

We will briefly introduce three of the more common forms of qualitative or semi-quantitative models.

5.5.1 Digraph Models

Digraphs and signed digraphs are particularly useful for expressing simple relationships between factors or variables. Diagnostic relationships can be displayed well by simple digraphs (Figure 5.11) which are related to the "fishbone" diagrams of statistical quality control (see Chapter 11).

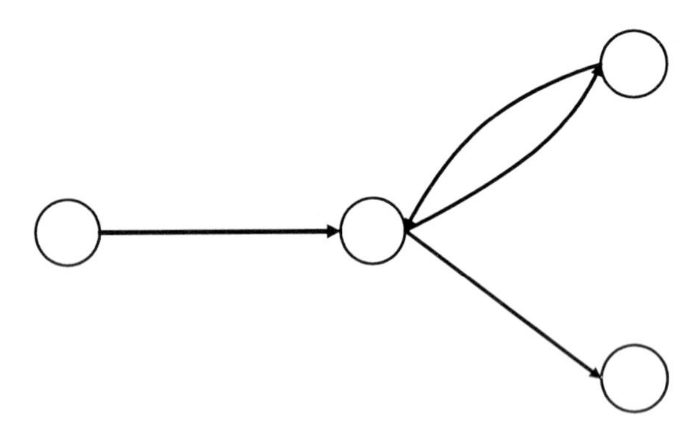

Figure 5.11: Digraph example.

Digraphs consist of nodes representing the factor or variable, represented as a dot or circle, and edges or branches which join the nodes and indicate the relationships, represented as lines or arrows. Digraphs are also used extensively to represent the structure of models or sets of equations and can be used to determine the order of solution for sets of algebraic equations. While the digraph simply indicates whether a relationship exists or does not exist, the signed digraph adds some quantitative directional information. A plus or minus sign beside the branch (or some similiar encoding) indicates whether the end variable increases or decreases if the starting variable increases.

Example 5.9 (Phosphorus release relationships) *The release of phosphorus in the anaerobic zone of the BNR process is governed by two processes relating nine variables (Tables 4.3 and 4.4 in Chapter 4). If we designate the nine variables* S_F, S_A, S_{O2}, S_{NO3}, S_{PO4}, X_H, X_{PAO}, X_{PP} *and* X_{PHA} *and the two rates* r_{FE} *and* r_{REL} *as nodes, the relationships governing the fermentation and phosphorus release processes can be represented by the signed digraph in Figure 5.12. The solid arrows represent positive (increase-increase) effects and the dotted lines represent negative (increase-decrease) effects.*

The loops are quite common in interacting systems and are stable if the product of the signs in the loop is negative. The input variables (S_{O2}, S_{NO3}, X_H, X_{PAO})*, the variables in balance* (S_F, S_A, X_{PP}) *and the output variables* (S_{PO4}, X_{PHA}) *are clearly seen.*

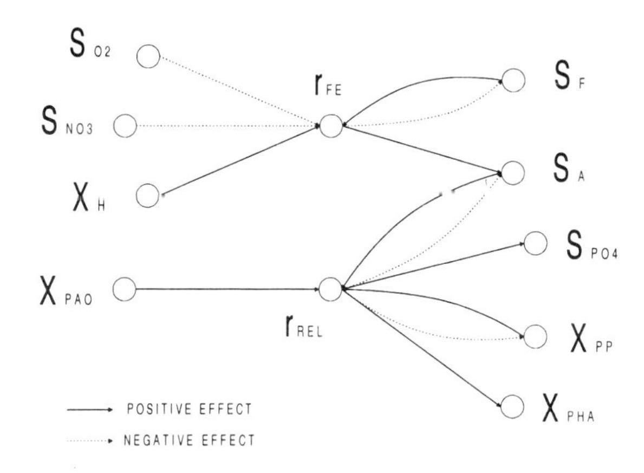

Figure 5.12: Phosphorus release digraph.

5.5.2 Rule-Based Models

Models based upon sets of rules received a boost in popularity in the early 1980's when simple rule-based expert system shells became available. The rules are based on simple predicate logic and take the form:

```
IF < conditional clause > THEN < action clause >
```

where conditional and action clauses frequently take an object-oriented form:

```
< attribute > OF < object > IS < value >
```

with a simple example being:

```
IF temperature OF water IS hot
 THEN motion OF hand IS withdraw
```

More complex conditional clauses can be constructed using the logical connectives AND and OR and the conditional prefix NOT. The OR connective can also be implemented by using two rules with the same action clause. Rule-based models are popular where the information is elicited from people with experience of the operation of the system. People generally find it much easier to verbalise relationships as rules.

Example 5.10 (Phosphorus release rule-based model) *The fermentation process in phosphorus release could be qualitatively modelled by the following rules:*

```
IF concentration OF oxygen IS increasing
 THEN rate OF fermentation IS decreasing
IF concentration OF nitrate IS increasing
 THEN rate OF fermentation IS decreasing
IF concentration OF heterotrophs IS increasing
 THEN rate OF fermentation IS increasing
IF concentration OF fermentables IS increasing
 THEN rate OF fermentation IS increasing
IF rate OF fermentation IS increasing
 THEN concentration OF fermentables IS decreasing
IF rate OF fermentation IS increasing
 THEN concentration OF acetate IS increasing
```

This can be compared to the top section of Figure 5.12.

There are a number of difficulties with rule-based models especially when they are derived from a number of sources. These difficulties include:

- detecting contradictions (similar conditions giving rise to opposite actions),
- determining completeness (are all possible conditions covered in the model), and
- defining verbalised attribute values such as "increasing", "decreasing", "low", "very low", and "OK".

The latter problem of defining linguistic values led to the popularity of so-called *fuzzy logic* (see also Section 13.3.3). More on rule-based models is found in Section 13.2.2

5.5.3 Fuzzy Models

We will introduce the use of fuzzy logic to define attribute values by an example:

Example 5.11 (Defining settleability) *The operators of an activated sludge plant may classify the performance of the clarifier as "poor", "ok" or "good". A quantitative measure of settleability frequently used is the Sludge Volume Index (SVI). We could define the terms using conventional or "crisp" logic as follows:*

```
good : SVI < 100 ml/g
ok : 200 ml/g > SVI > 100 ml/g
poor : SVI > 200 ml/g
```

but to say settleability is ok at 199 ml/g but poor at 201 ml/g is putting a bit too much faith in a relatively arbitrary boundary of 200 ml/g.

An alternative is to use the concept of a "fuzzy" boundary like that shown in Figure 5.13.

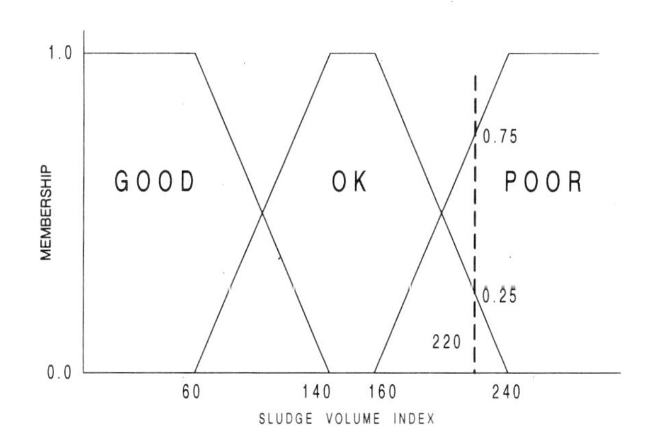

Figure 5.13: Fuzzy set definitions.

Each linguistic value is defined as a fuzzy set in terms of a "membership function" μ which takes values between zero and one (not just zero or one as with crisp sets). From the Figure 5.13 we can see that an SVI of 220 has memberships of $\mu(good) = 0.0$, $\mu(ok) = 0.25$ and $\mu(poor) = 0.75$. This more closely represents our uncertain definitions.

Fuzzy models can be represented in terms of rules, essentially rule-based models with fuzzy definitions of the attribute values. A popular alternative representation is in terms of possibility vectors and a relation matrix. Possibility vectors are simply the membership function values in the reference fuzzy sets arranged in vector form. Referring to Figure 5.13 the possibility vector p resulting from *fuzzifying* a SVI value of 220 would be:

$$p = [\, 0.0\ 0.25\ 0.75\,] \tag{5.54}$$

A fuzzy relation R between possibility vector p_y and p_u is evaluated by the max-min composition operation:

$$p_y = p_u \odot R \tag{5.55}$$

where the elements of the p_y vector are membership functions calculated from:

$$\mu(y) = max_y[min(\mu(u), \mu_R(u, y))] \tag{5.56}$$

We will discuss this representation further by looking at an example.

Example 5.12 (Fuzzy model for effluent suspended solids) *Suppose we wish to define a fuzzy model between clarifier settleability and effluent suspended solids. We would define reference fuzzy sets for both settleability and suspended solids like those shown in Figure 5.13. We would then define our model relation matrix which might look like:*

$$R = \begin{pmatrix} 0.7 & 0.2 & 0.1 \\ 0.1 & 0.8 & 0.0 \\ 0.0 & 0.2 & 0.6 \end{pmatrix} \tag{5.57}$$

For a settleability vector of $[\, 0.0\ 0.25\ 0.75\,]$ *the suspended solids vector would be* $[\, 0.2\ 0.25\ 0.6\,]$ *where for example the first membership value in the vector would be evaluated as follows:*

$$\mu_u(1) = max \begin{bmatrix} min(\ 0.0,\ 0.7\ ) \\ min(\ 0.25,\ 0.2\ ) \\ min(\ 0.75,\ 0.1\ ) \end{bmatrix} = 0.2 \tag{5.58}$$

Each element of the relation matrix is equivalent to a rule. For example the bottom right element would be a rule such as:

```
IF (SVI) is (poor) with membership = x
THEN (SS) is (high) with membership = min(x,0.6)
```

It is also possible to *defuzzify* a possibility vector by summing the elements of the vector weighted by some measure of the location of the respective reference fuzzy sets divided by the unweighted sum. One technique is called the *centre of area* technique where the weights are the centres of the reference sets on an area basis. Suppose the centres of area of the three suspended solids reference sets were 2, 10 and 25. Then defuzzifying $p = [\, 0.2\ 0.25\ 0.6\,]$ would give:

$$y = \frac{0.2 \cdot 2 + 0.25 \cdot 10 + 0.6 \cdot 25}{0.2 + 0.25 + 0.6} = \frac{17.9}{1.05} = 17.05 \tag{5.59}$$

In Section 13.3.3 we will discuss more on fuzzy sets as a way to represent knowledge. Section 18.8 will present applications of fuzzy control in wastewater treatment, where the controller is a fuzzy relation between the error signal and the adjustment to the manipulated variable.

5.6 Summary

We have briefly introduced the common forms of quantitative and qualitative black-box models. This has included both traditional functional models and network models which have proved popular for fitting nonlinear data. We have also introduced digraphs, rule-based and fuzzy set models. These models are semi-quantitative in that they are basically qualitative models of relationships between variables but do give quantitative information to varying degrees.

5.7 Further Reading

- identification and estimation in general in Ljung & Söderström (1983), Ljung (1987), Johansson (1993), Söderström & Stoica (1989), Stearns & David (1988) and in wastewater treatment systems Marsili-Libelli *et al.* (1978) and Beck (1986).
- artificial neural networks in Hoskins & Himmelblau (1988) and applications in Willis *et al.* (1991), Cote *et al.* (1995), Naidu *et al.* (1990), Sbarbaro-Hofer *et al.* (1992) and Kosko (1990).
- radial basis function networks in Leonard & Kramer (1991), Hofland *et al.* (1992), Kavuri & Ventkatasubramanian (1993), Haykin (1994) and Lee *et al.* (1998).
- wavelets are covered in the book Motard & Joseph (1994) and wavenets in Bakshi & Stephanopoulos (1993) and Bruce *et al.* (1996).
- digraphs are covered in Iri *et al.* (1979) and Shiozaki *et al.* (1979).
- rule-based systems in Lee *et al.* (1998).
- fuzzy systems in Zimmermann (1991), Drainkov *et al.* (1993), Jantzen (1994), Self (1990) and Klir & Folger (1988).

5.8 Exercises

1. How can you conclude that the dissolved oxygen dynamics is nonlinear? How would you see that in practice?
2. Sometimes pure oxygen is used instead of air for the aeration. The advantage is that the saturation concentration is about 4-5 times higher for the pure oxygen. What is the consequence from a control and operational point of view of using pure oxygen?
3. Linearise the following state space models and also convert them to input-output model form:

(a)

$$\frac{dx_1}{dt} = -2.5x_1^2 - 0.4x_1x_2 + 1.2u \quad (5.60)$$
$$\frac{dx_2}{dt} = 0.7x_1x_2 - 2.1x_2 \quad (5.61)$$
$$y = 0.1x_1 + 0.5x_2 \quad (5.62)$$

(b)

$$\frac{dx_1}{dt} = -0.4x_1x_2 - 2.3u^2 \quad (5.63)$$
$$\frac{dx_2}{dt} = 0.6x_1 - 0.2x_2 \quad (5.64)$$
$$\frac{dx_3}{dt} = 3.2x_2 - 0.1x_3 \quad (5.65)$$
$$y = x_3^{0.5} \quad (5.66)$$

4. The following Table 5.1 lists pump curve data at five settings of the variable speed drive. Graph the data in various ways and select the structure for an empirical functional model to represent discharge pressure as a function of flowrate and drive speed.

Flowrate	800 Hz	950 Hz	1200 Hz	1400 Hz	1750 Hz
0	80	95	120	140	175
25	76	92	117	138	173
50	64	82	110	131	168
75	45	65	97	120	159
90	29	52	86	111	152
105	11	37	74	101	144
120		19	60	89	134
130		6	50	80	127
140			38	70	119

Table 5.1: Pump curve data for different drive speeds.

5. The following data (Table 5.2) represents P release (mg/l) for different VFA concentrations (mg/l) in the influent. Compare fits of the data with empirical models based on artificial neural networks, radial basis functions and wavelets.

6. Construct a signed digraph model of the nitrification and pre-denitrification interactions. You can consult the Tables 4.1 and 4.2 in Chapter 4.

7. From your own experience or from a discussion with an operator, construct a rule-based model of the effect of hydraulic load on effluent suspended solids.

VFA	Release	VFA	Release
75	4.5	171	11.7
87	7.2	183	10.8
99	9.9	195	10.1
111	10.8	207	9.0
123	12.2	219	7.7
135	12.6	231	6.5
147	12.7	243	5.3
159	12.3	255	4.2

Table 5.2: P removal for different VFA concentrations.

8. Using the previous exercise as a basis, construct a fuzzy model of the relationship.

Chapter 6

Experiments and Data Screening

The goal of this chapter is to introduce you to techniques that will enable you to get good data, which will then give you good parameter estimates when you perform the model fitting discussed in the next two chapters. The techniques will enable you to choose the right sort of fitting approach, design the right experiments, and detect problems with and even fix up the data you do get.

6.1 Post-Modelling Tasks

Once a model has been developed and verified, there remain a number of tasks to be performed. These are summarised in the middle section of Figure 6.1.

We have previously addressed the first three tasks. Here we will address the issues of experiment design and preparing the data for model fitting and validation. The details of parameter estimation will be addressed in Chapter 7. Chapter 8 addresses strategies for fitting and validation and puts special emphasis on fitting for biological wastewater processes.

Experiment design is determining what experiments should be carried out to obtain data for parameter estimation and data validation. Parameter estimation is the fitting of the model to data from a particular process and validation is testing the fitted model. Related techniques are process identification and state estimation.

Process identification includes the problem to determine the right structure of a model. Such a structure can be an input-output model or a more elaborate model based on first principles as we discussed in Chapter 5. Once the structure is determined, then its parameters need to be determined, which can be performed by on-line estimation. State estimation is the on-line estimation of state variables which are not measured. For example, using state estimation techniques the

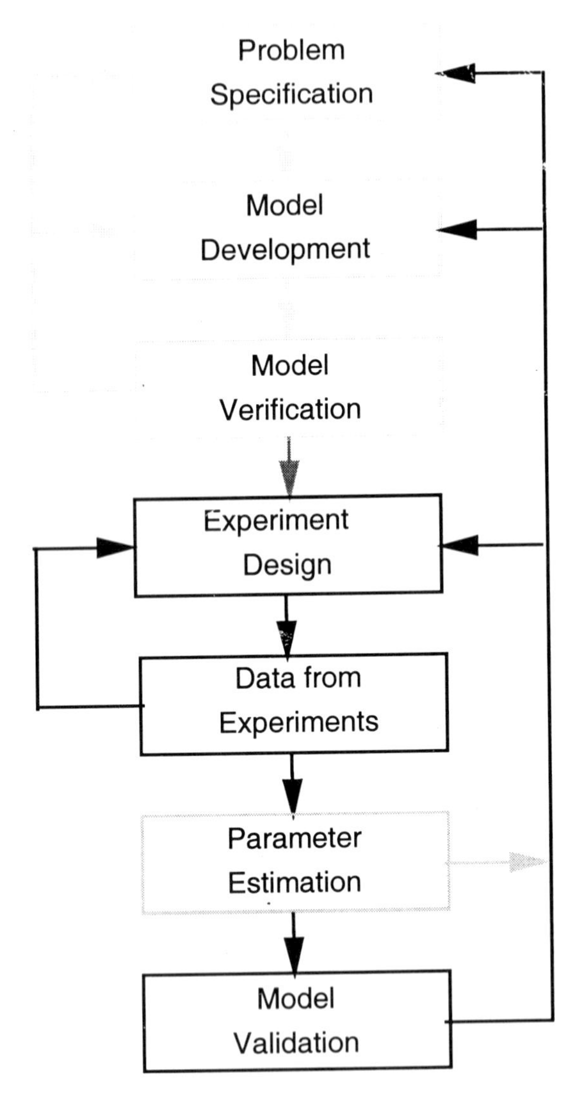

Figure 6.1: The model development process.

organism concentrations may be calculated, given measurements of various substrate concentrations. These two on-line techniques will be introduced in Part B, Diagnosis, and Part C, Control.

6.1.1 Experiment Design

The quality of real data is an important issue. What errors are involved in sampling and in measurement? Over what operating range is the data collected? Was the process really at *steady state*? If we are determining dynamic parameters, was there sufficient excitation or frequency content? We will address these issues in Section 6.4 below.

6.1.2 Parameter Estimation

Parameter estimation is the fitting of the model to particular process data by adjusting the model parameters to minimise some measure of plant-model mismatch. Let us assume that we have determined, for example by a sensitivity study and observability test, which parameters to adjust. Now we need to define a measure of plant-model mismatch, and to find a method to perform the minimisation. This task will be addressed in detail in Chapter 7 with a brief summary of the principles and dangers. Then Chapter 8 will discuss specifics on the fitting of biological wastewater treatment models.

6.2 The Model Fitting Problem

A key principle to follow in fitting models is "divide and conquer". Determining many parameters at the same time from experimental data will always be difficult. If you can divide the problem into two or more smaller fitting problems, you are much more likely to succeed. This can be done by concentrating on one piece of equipment in the flowsheet at a time. Sometimes, by keeping certain inputs constant, you can reduce the number of parameters. This depends on the structure of your model.

There are two basic approaches to parameter estimation. Firstly, you can fit a linear model to time series data - time series seldom have sufficient excitation magnitude to get a good fit for a nonlinear model. A time series or trend plot is simply values of the variable recorded at regular time intervals. This can be done off-line or on-line using recursive algorithms. Secondly, you can fit a linear or nonlinear model to steady state data from a number of different operating points using standard regression techniques.

Which fitting approach is used depends partly on the problem and partly on the inclination and expertise of the user. Figure 6.2 attempts to give you some guidance using the following factors:

- the size of the operating region for the process.

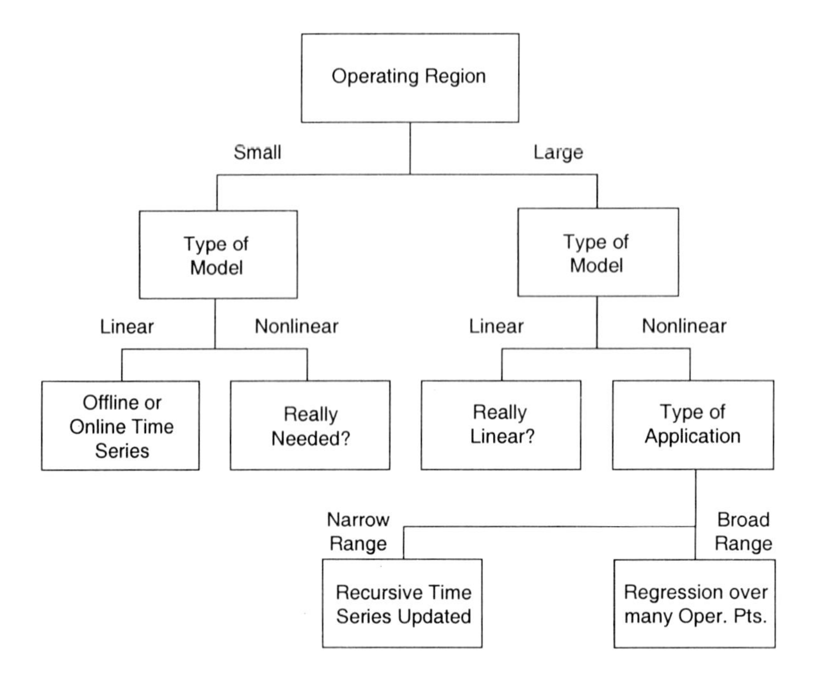

Figure 6.2: Model fitting approach.

- whether the model is linear or nonlinear in the parameters.
- the range of the application, a moving narrow range (such as regulatory control) or over the whole operating space (such as optimisation).

Time series parameter estimation and regression over operating points are introduced in Chapter 7. The first problem in the latter approach is to decide where the operating points should be located within the operating space.

6.3 Experiments for Steady State Data

Model parameters can be categorised into those identifiable from steady state operating data and those requiring dynamic tests. Most parameters fall into the first category since only the model time constants may require dynamic tests. This is also fortunate since steady state tests are less costly to perform, especially in terms of the number of measurements required. Of course we know that activated sludge plants are never at steady state, but averages over some multiple of 24 hours (daily or weekly) and composite samples will give a fair approximation. Mass balances can be used to check the steady state assumption, though in wastewater applications it can be expensive to obtain sufficient measurements.

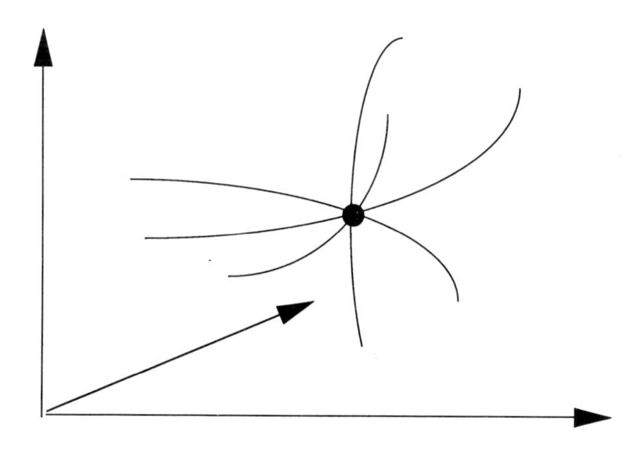

Figure 6.3: Which would you like?

Experiment design is deceptively simple in principle. We simply need to decide how many experiments to do, how many measurements to take, where the experiments should be, and in what order the experiments should be performed.

6.3.1 Number of Test Points and Measurements

This is an area of much confusion. We frequently see plant engineers perform a single test run on the plant, manipulate rather randomly the parameters in a complex nonlinear computer model to match, and happily proceed to extrapolate to new operating conditions or designs. What have they done? They have fitted a nonlinear relationship to a single point in a multidimensional space (Figure 6.3) - "not so difficult" they tell you - of course they have an infinite number of possibilities from which to choose just one.

We must distinguish carefully between the number of state variables and the number of parameters to be estimated. The state variables locate where the plant is operating within the state space. For one particular experiment we need to take sufficient measurements to locate ourselves in the state space, this alone can be a challenge. There are various observability tests to assist us in this regard. Sometimes we need to estimate some states together with the parameters.

Example 6.1 (Number of manipulated variables and parameters) *For example, lets look at the simple carbon removal process introduced in the verification section at the end of Chapter 3 and detailed in Appendix B. After the sensitivity study discussed in Chapter 3, we have the following situation:*

- *disturbances:* q_F, S_{SF}
- *manipulated variables:* q_R *or* q_W, q_A

- *states:* S_S, S_O, X_H
- *outputs:* S_S, S_O
- *known or assumed parameters:* V, $S_{O,sat}$, K_{OH}, f_P
- *unknown parameters:* $\hat{\mu}_H$, K_S, b_H, Y_H, a, b

The manipulated variables define a two dimensional operating space, there is one state X_H to estimate, and there are six parameters to fit.

We need to make sufficient experiments to be able to determine the unknown parameters (plus unknown states). The minimum is of course the number of parameters, but we generally need many more to have adequate confidence in the range of applicability and in the accuracy of the parameter estimates. We must also remember that, because of experimental error and model approximations, they are always only estimates.

6.3.2 Test Point Spacing

Having defined the operating region of interest and the desired accuracy, the next task is to select where within the region we should obtain validation data. This is called selection of test point spacing. A good selection of test point spacing will result in approximately equal plant-model mismatch throughout the chosen operating region.

The biggest pitfall at this stage is choosing too few experiments (after all they are expensive and bothersome to perform) over too small a region (production managers seldom wish to deviate very much from normal operation). This is despite the fact that we generally want to use our models to investigate abnormal operation. The modellers specification discussed in Chapter 3 needs careful thought and needs to be respected.

The selection depends on experimental errors and on the linearity of the model, both of which will vary throughout the region. Larger errors and higher nonlinearities require closer spacing of test points. With no measurement error and a linear model we would simply need $N+1$ points where N is the dimension of the operating space.. We would locate these at the extremes of the operating space.

Example 6.2 (Test point spacing) *Again let's look at the simple carbon removal process example described in Appendix B. Figure 6.4 is produced by the MATLAB program* `Example_6_2.m` *and shows plots of two of the three state variables across the operating region defined as:*

$(\quad 2 \leq q_A \leq 12, \quad 0.2 \leq q_W \leq 2.2 \quad)$

We can see that there are much greater changes at low air flowrates, so experimental points should be closer in this region. We will select the following grid of test points:

$q_A = (2.0, 3.0, 4.0, 5.0, 6.0, 8.0, 10.0, 12.0)$
$q_W = (0.4, 0.8, 1.2, 1.6, 2.0)$

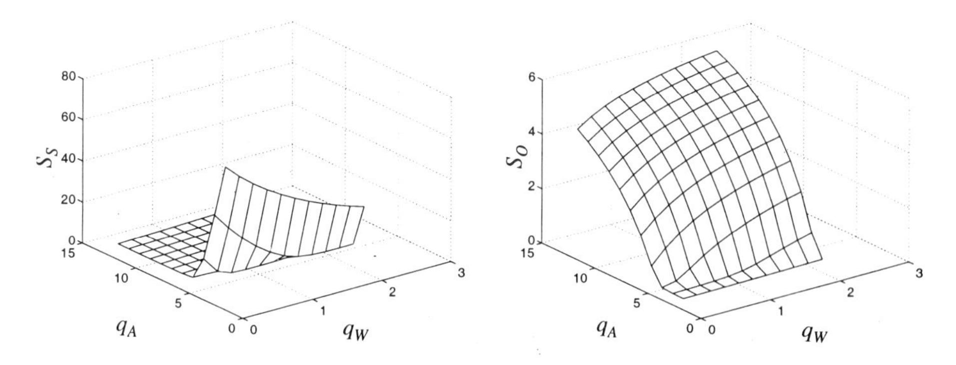

Figure 6.4: Surfaces for substrate and oxygen concentrations.

In order to confirm our estimates of experimental error, we will normally perform a number of replicate tests. As for the validation tests themselves, the selection of the number and placement of replicate tests aims to give us the best possible confidence in our estimate of experimental error, bearing in mind the desired accuracy of the fit. Experimental error is a composite of errors in setting input variables, disturbances over which we have no control, changing process characteristics and measurement errors. While we can estimate each of these and combine them into an estimate of experimental error which we can use for test point spacing, it is good to have a confirmation of our estimate with actual replicate experiments.

6.3.3 Ordering Experiments in Time

Having determined the experiments to be performed, we must then decide upon the order of the experiments in time. There are always a number of time dependent extraneous factors which affect the results of experiments. These factors include process characteristics which change slowly with time (very common with biological systems), drifts in the calibration of measuring instruments, and different operating styles and degrees of attention from the human operators. The perfect solution to the problem is to completely randomise the experiments to spread the effects of these factors equally across the test region. However this is usually also the most expensive both in time and operating expense. There are a number of techniques for block randomisation which can reach an acceptable compromise.

We have talked of experiment design in a technical sense, but as engineers we must also be concerned with the cost of data. Whether we are using a pilot plant or a real plant, the cost of obtaining good data is very high both in terms of time and real costs (analyses, materials, salaries, perhaps even effluent discharge penalties). Both in the specification of the desired accuracy and in the selection of test spacing and replicates, we must keep reminding ourselves of the costs and reach some acceptable compromise.

6.3.4 Optimal Sequential Design

An alternative to heuristic compromises, is an experiment design technique called optimal sequential design. This procedure begins at the point where you have selected your experiment tests. A representative subset of these experiments is performed and then the model is fit to the data giving a value for some lack of fit criterion (see Chapter 7). One further experiment is performed (or a small number of experiments), the model is refit, and the new lack-of-fit criterion is compared to the previous value. The whole procedure is terminated when the improvement is not significant. Some sequential design procedures also give guidance as to which experiment of a subset of experiments should be performed next. Like any optimisation technique, the success of the technique depends upon a well behaved response surface (how the lack-of-fit criterion changes over the experimental region).

6.4 Experiments for Dynamic Data

Regression techniques are discussed in Chapter 7 for identifying model parameters from *dynamic data.* Dynamic data is generally in the form of one or more time series. A time series is values of a certain variable collected at regular time intervals. A typical time series model is expressed in the form:

$$\begin{aligned} y_t = a_1 y_{t-1} + a_2 y_{t-2} + \quad .. \quad &+ a_n y_{t-n} \\ +b_1 u_{t-1} + b_2 u_{t-2} + \quad .. \quad &+ b_m u_{t-m} \end{aligned} \tag{6.1}$$

where y is the output variable at various times and u is the manipulated input. For example, in an aeration process y will denote the dissolved oxygen concentration and u the air flowrate. The data is normally collected at discrete times t, $t-1$, $t-2$, .. separated by a constant sampling interval.

The data may come from simple residence time tests (see Chapter 8), from special dynamic experiments, or sometimes from plant data logged during normal operation. Most of the experiment planning issues discussed above are just as relevent to determining dynamic parameters as they are to determining steady state parameters. There are also extra issues in planning dynamic tests - sufficiently rapid sampling, sufficient plant excitation of the plant outputs, and control loops.

6.4.1 Data Sampling Times

Sampling design involves sufficiently rapid sampling for a sufficient length of time. A very common problem with dynamic data, particularly data collected during normal operation, is that the sampling interval Δt is too large or that only averaged data was collected.

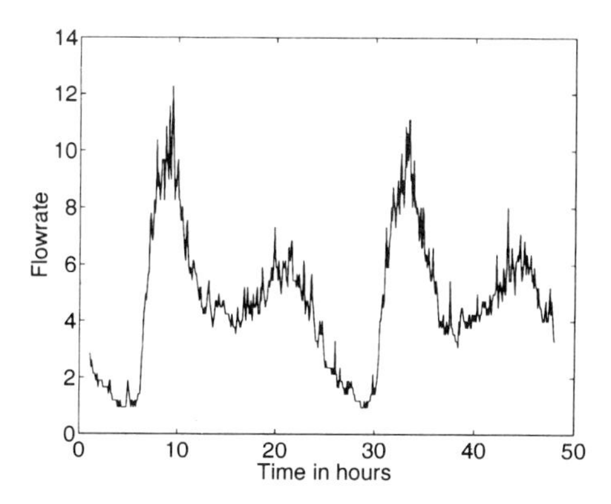

Figure 6.5: Influent flowrate data.

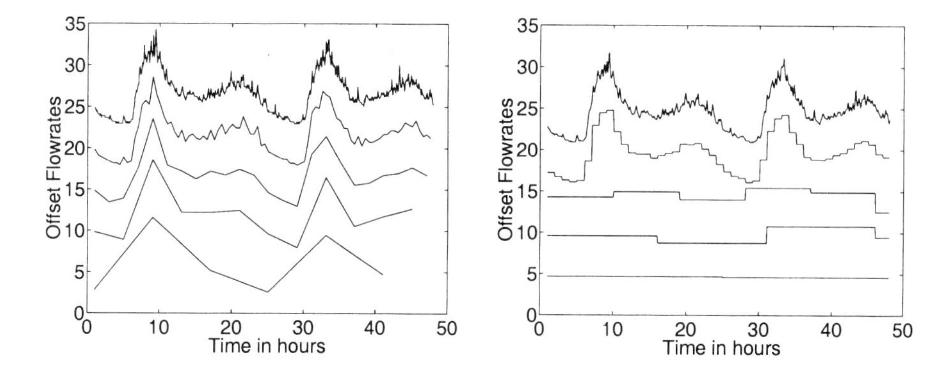

Figure 6.6: Effect of sampling interval and averaging period.

Example 6.3 (Intervals for sampling and averaging) *Some typical influent flowrate data to a municipal sewage plant is shown in Figure 6.5.*

One sample every eight hours is obviously not going to give you a reasonable characterisation of the disturbances. At least 2 hourly samples would be necessary to pick up the major (low frequency) changes. The left plot in Figure 6.6 shows sample times of 5 minutes, 30 minutes, 2 hours, 4 hours and 8 hours from top to bottom (the data has been offset vertically for clarity). Similiarly, since we have a periodic signal, composite samples or averages should be taken over an integer multiple of the disturbance period, for example 24 hours. The right plot in Figure 6.6 shows the data without averaging and with 1, 9, 15 and 24 hour averages. Averages over 9 and 15 hours bear little relationship to reality.

Figure 6.6 was produced by the MATLAB program `Example_6_3.m`.

If the data is to be used for parameter estimation, the sampling frequency must be high enough that the maximum frequency of interest (the inverse of the shortest time constant of interest) is one third to one quarter of the sampling frequency ($1/\Delta t$) in order to avoid the *aliasing* effects of sampled signals. Aliasing can also be reduced if the signals are pre-filtered to remove frequencies above the maximum frequency of interest. Section 6.5.5 discusses frequency content, sampling rates are discussed in 17.2.4 and in 24.2.4, and aliasing is discussed in Chapter 24.

Example 6.4 (Sampling interval selection) *For example, suppose we wish to fit a disturbance model to the influent data in Figure 6.5. The peaks of interest have a width of 3 hours, therefore the sampling interval should be around 30 to 60 minutes. It can be seen in Figure 6.6 that 30 minute sampling (second from the top) gives a reasonable representation of the flowrate.*

The length of the data record (hence the number of data points) must be long enough to reduce the error to acceptable values.

6.4.2 Plant Excitation

The dynamics of the plant must be sufficiently excited in order to estimate the dynamic parameters. While process plant inputs generally contain a certain amount of noise, this is very seldom of a high enough frequency or of sufficient magnitude for its effects to give a good signal-to-noise ratio on the plant measurements. It is generally necessary to apply some kind of test signal to excite the correct frequencies strongly enough. The degree of excitation can be determined by examining a power spectral density plot, a graph of signal intensity against frequency.

But what frequencies should be excited? Ideally we would like to excite all frequencies, but that would require an ideal impulse, infinite height and infinitesimal width, which is hardly practical. Our fall-back position is *reasonable* excitation up to the frequency corresponding to the smallest time constant of interest ($1/\tau$ in radians).

Example 6.5 (Normal excitation) *Figure 6.7 shows the frequency spectrum of the influent flowrate data shown in Figure 6.5. It was calculated and plotted by the MATLAB program* `Example_6_5.m` *calling the Toolbox function* `frspec.m`. *The dominant peaks correspond to 24 hours and 12 hours (sampling interval was 1 hour, so the frequency in radians for the 24 hour peak is* $2\pi/24 = 0.26$*).*

The time constants for nutrient removal are normally in the 30 minute to 2 hour range. To identify these dynamics we should have good excitation up to 2 radians/hour. It is clear from Figure 6.7 that normal influent disturbances would be inadequate.

Generally, as we have seen in Example 6.5, the desired excitation is generally not present in normal operation. We must apply some sort of stimulus. The most frequently applied stimulus is the pseudo-random binary sequence (PRBS) which can give reasonable excitation at very modest amplitudes though over a longer period. This allows the test to obtain good data without undue disturbance to the plant.

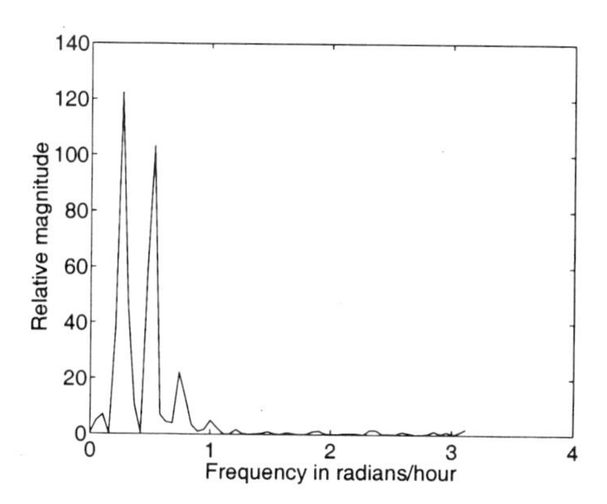

Figure 6.7: Frequency spectrum.

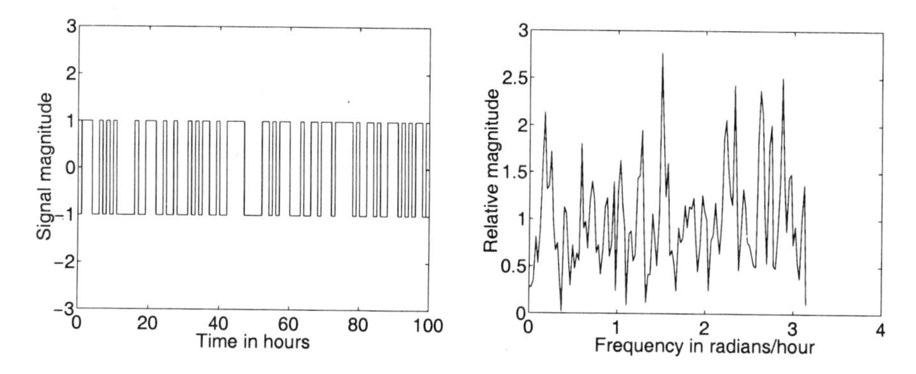

Figure 6.8: PRBS signal and spectrum.

Example 6.6 (PRBS excitation) *The left plot in Figure 6.8 shows part of a 10-bit PRBS generated by the MATLAB program* `Example_6_6.m` *calling the Toolbox function* `prbs.m`*. The right plot in Figure 6.8 shows the frequency spectrum of this signal. It can be seen that there is good but gentle excitation throughout the frequency range, unlike most signals which tail off rapidly to zero (Figure 6.7 for example).*

In designing a PRBS, the base time interval and sampling time should be about one fifth the smallest time constant you wish to identify. The sequence size in bits (n) should be chosen so that the sequence length $(2n - 1)\Delta t$ is about five times the major time constant of the system. Several sequences should be applied immediately following each other. The magnitude of the signal should be sufficient to get a reasonable signal-to-noise ratio in the measured output.

6.4.3 Testing Controlled Systems

If at all possible, place all control systems in manual operation when performing experiments on the process. This is not always possible or desirable, in which case great care must be taken or you will simply identify the controller rather than the plant. You must use a test signal to excite the controller setpoint to obtain a closed-loop model, and you must know the control algorithm and parameters so that you can back-calculate the open-loop model.

6.5 Data Screening

Regression techniques, whether they are fitting steady state or dynamic data, are particularly susceptible to poor quality data. The "garbage in - garbage out" principle certainly holds true for parameter estimation. Properly designed experiments, as we have discussed above, are a necessary but not sufficient condition for quality data. There are still plenty of opportunities for data corruption. Problems with excessive disturbances, equipment malfunction, and human error are very likely in any test programme.

Example 6.7 (Raw plant data) *Figure 6.9 demonstrates some of the problems of raw plant data. The plot on the left shows two obvious outliers (data values with very different values from the norm). Less obvious is that both are associated with a strange peak in the data and that that these strange peaks occur at the same time of the day but not every day. This would require some detective work to see if they are calibration abnormalities or some sort of regular yet irregular disturbance. The plot on the right shows some obvious outliers and possibly some less obvious ones, as well as a very obvious recalibration disturbance. Not obvious from the plots are many missing points, generally just one or two points randomly spread. All of these problems can play havoc with analysis techniques and considerable preprocessing of the data is required.*

So what can we do? We must carefully and very critically examine our data. It is best to do this as soon as possible after collection of the data. Early detection of problems can save much time, expense and frustration. There are several techniques and tests that we can apply to detect poor data and a few techniques for improving data quality, apart from repeating the experiment which is always best but not always convenient. Table 6.1 lists some of the techniques and a suggested hierarchical scheme for application.

The more commonly used techniques are discussed below. Statistical quality control (SQC) techniques are discussed in Chapter 11.

6.5.1 Data Visualisation

The first rule in getting a feel for your data is to plot it. Plot it against anything appropriate - dependent variables against independent variables, against time,

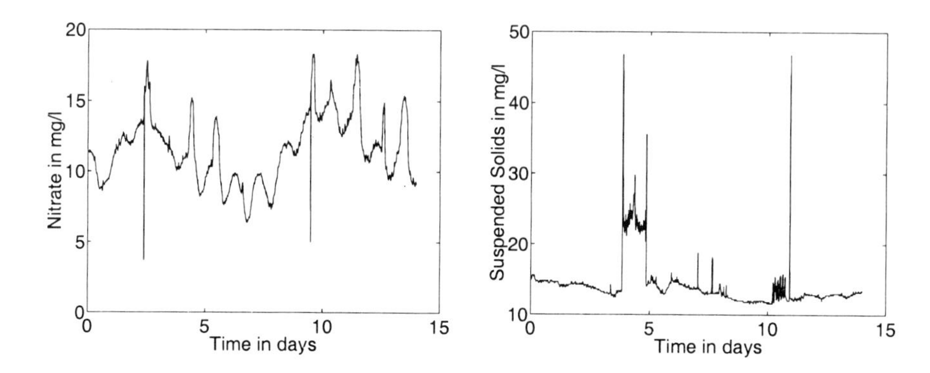

Figure 6.9: Raw data sets.

Category	Techniques
Phase 1 Basic Problems	Data visualisation Detecting and replacing missing points Detecting and fixing recalibrations Detecting and removing outliers
Phase 2 Feature Detection	Trend detection (statistical, SQC techniques) Detecting level changes (SQC techniques) Detecting variance changes (SQC techniques) Comparison to expected patterns Frequency content Cluster analysis
Phase 3a Fitting Steady State Data	Model based data reconciliation Detrending
Phase 3b Fitting Dynamic Data	Conversion to zero mean Detection of changes in variance Frequency content and coherence Pre-filtering
Phase 3c Feature Classification	Level classification with thresholds Gradient classification with thresholds Successive or bandpass filtering

Table 6.1: Data screening techniques.

against the operator, etc. There are several good books on data visualisation. There are also clustering techniques which can be used (see Chapter 11).

Simple extrapolation tests can be used if there are known constraints on the dependent variable at extremes of the independent variable. For example, pressure drop measurements across equipment should extrapolate to zero at zero flowrate.

Visualisation will help you to quickly pick up apparently abnormal data that does not conform to the general patterns. Of course, that does not mean it is bad data. Maybe you don't understand the process as well as you thought or maybe there is another factor or disturbance you did not anticipate (see fishbone diagrams in Chapter 11). If there are one or two data values that are still *strange* after considering all possible factors, they may be *outliers* resulting from inadvertant errors. Example 6.7 above illustrates the utility of data visualisation.

6.5.2 Outlier Tests

There are a few rules and a statistical test that should be used before apparent outliers are discarded. If they are outliers, they should definitely be discarded as they can lead to large errors in parameters fitted to the data.

One textbook example showed that one outlier in 100 points can give parameters of (1.16, 0.86) with apparently acceptable accuracy compared to (0.19, 0.98) without the outlier. This happens because the outlier increases the residual variance by a factor of four which makes parameter confidence tests insensitive. It points out not only the importance of rejecting outliers, but also the importance of having independent estimates of experimental error (the true residual variance). The following outlier rules are generally recommended:

- Do not reject if there are several neighbouring outliers. Chances are they are good points and something is happening you did not anticipate.
- Do not reject outliers at the edges of a data set. Take more data to move the edge outwards since the data may be the start of some change in direction.

Rosner's test can be used for univariate data such as a disturbance time series. Even quality control charts (see Chapter 11) can be used for univariate data. The MATLAB Toolbox program `outlier.m` implements Rosner's procedure.

Example 6.8 (Rosner's outlier test) *An example result is shown below where the software detected three outliers from the data set on the left in Figure 6.9 above. Detection is only possible if the data set is stationary. The best way to reliably detect outliers is to fit the data to a model or to lightly filter the data (left plot in Figure 6.10 below), and then to test the residuals (the differences between actual values and filtered values shown on the right in Figure 6.10 below), as was done in this case. The results and plots were produced by the MATLAB program* `Example_6_8.m`.

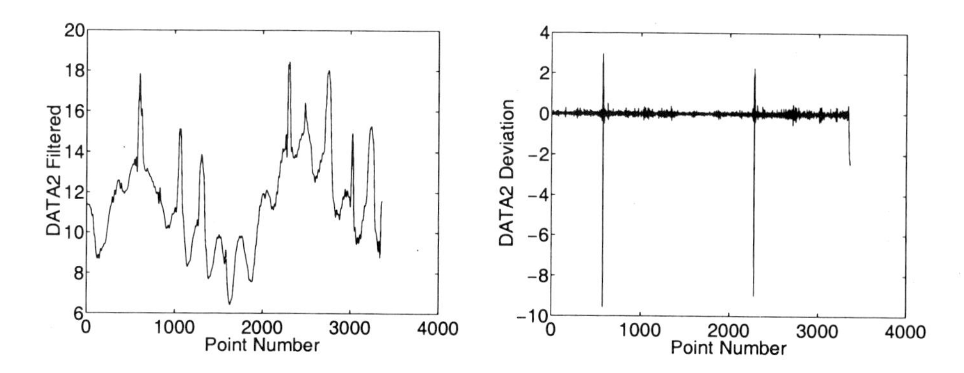

Figure 6.10: Filtered data and residuals after filtering.

```
Outliers detected = 3
Outlier indices are:
 575 2276 574
```

Outlier analysis is difficult and must be approached with a lot of care, as the following example will illustrate. It is easy to interpret points as outliers when they are a real effect.

Example 6.9 (Secondary clarifier dynamics) *Let us look at the dynamic behaviour of a clarifying process of a secondary settler. Consider the plant influent flow rate Q_{in} and the effluent suspended solids concentration C_{out}, and their interaction during normal conditions, Figure 6.11.*

The data were obtained from a full scale wastewater treatment plant. The suspended solids concentration is given as turbidity in FTU units (refer to section 22.4.1). The impact of the influent flowrate on the effluent suspended solids is obvious. It has been described in earlier work that there is an apparent dynamic relationship between Q_{in} and C_{out}. This is clearly demonstrated by Figure 6.11. The gain and the time constant of this relationship is strongly coupled to the floc properties and may change slowly with time. From such normal data representations it is straight-forward to derive dynamic models using time series analysis.

During periods of process failure the expected cause-effect relationship between the actual signals may get completely lost. There may be several reasons for this, both failing sensors and upsets in the settler operation, for example due to bulking sludge. Such a period is illustrated in Figure 6.12. Dramatic changes of the effluent suspended solids concentration appear. It may increase by a factor of 5 to 10 in a few minutes.

An obvious reason may be an overloaded settler, where the sludge blanket is close to the top of the settler unit. There are a number of data points here which appear to be outliers, but it would be foolish to assume so in a period of abnormal operation, even if they passed all the statistical tests.

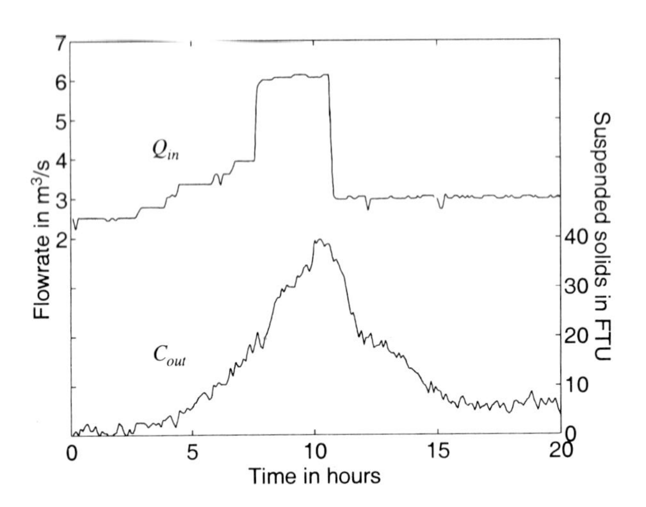

Figure 6.11: Typical variations of influent flowrate and effluent suspended solids.

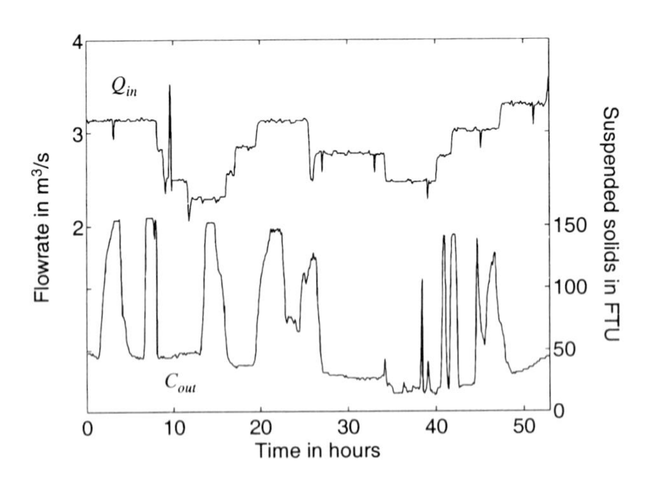

Figure 6.12: Flowrate and effluent suspended solids data during a period of settler failure.

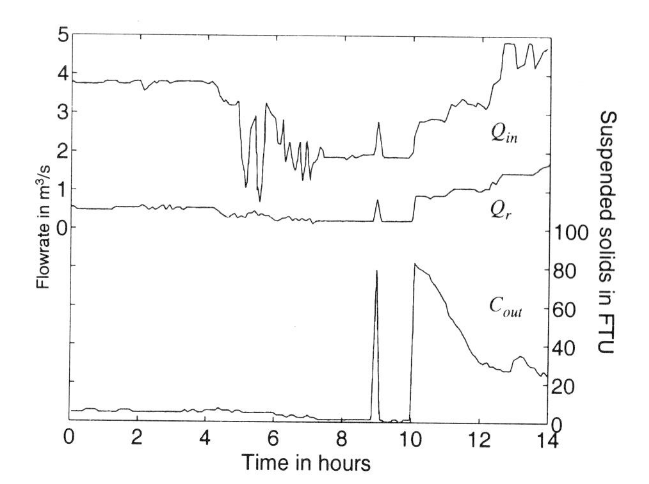

Figure 6.13: Flowrates and suspended solids with correlated peaks in the data.

Another situation is shown in Figure 6.13. At a time of around nine hours, there are obvious peaks in the signals. It would be natural to suspect that the peak in the suspended solids was an outlier, particularly if it is observed in isolation. When the flowrates are examined as well and the peaks are seen together, it can be seen that there is an obvious cause-effect relationship between the peaks.

6.5.3 Missing Data in Time Series

Many data analysis techniques, particularly those for time series of data, require equally spaced data without missing values. Sometimes that is more than we can expect. There are two situations we will discuss briefly, missing points and unequal spacing.

If there are relatively few missing points, we can postulate some model which we can use to calculate values to complete the series. Possible models are constant mean (replace missing values by the mean of the data we have), previous good value, constant trend (replace using linear regression of points on either side of the missing data) or more complex relations that can be used to interpolate (polynomials or splines). The MATLAB Toolbox program `missing.m` implements a selection of such models.

Example 6.10 (Missing data models) *Figure 6.14 shows treatment plant influent flowrate with four missing data regions showing three data reconstruction models, average, simple interpolation and a cubic spline through six points each side of the region (MATLAB program* `Example_6_10.m`*).*

It is clear that the average model is unsuitable for anything but steady state data. The simple interpolation and the multi-point spline interpolation produce similiar

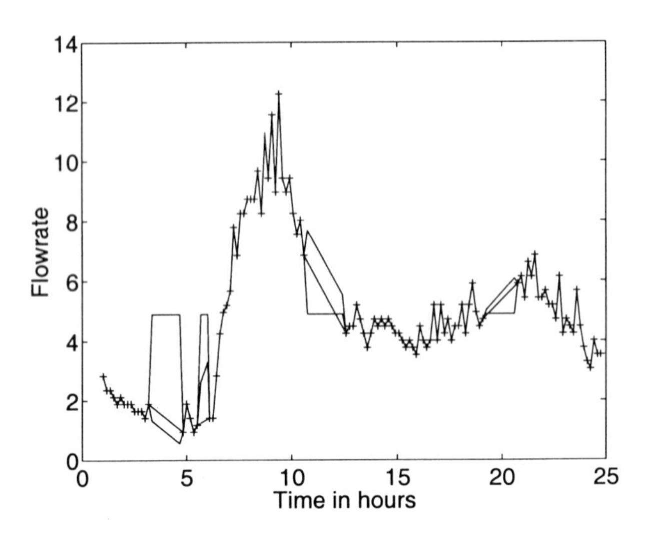

Figure 6.14: Replacing missing data.

results except for the second region which is very close to a sharp increase. In this case some of the points after the region have large values which influence the interpolation producing a spurious peak. In many respects the simple interpolation is the best solution.

Example 6.11 (Missing clarifier data) *Figure 6.15 illustrates missing points in data from a full scale activated sludge operation. It is apparent that missing data during a relatively smooth period can be successfully interpolated or even extrapolated on-line, while missing data during rapid transients are certainly not simple to replace.*

Correcting unequal spacing in a time series is done by fitting some relation to surrounding local data and then using that to calculate equally spaced points. The same MATLAB Toolbox function `missing.m` also handles this case.

6.5.4 Time Distribution

It is generally informative to examine the distribution of time series data about the mean. Unusual distributions, such as bimodal distributions, can indicate operating characteristics or faults. If the data is to be treated statistically, the distribution should approximate a normal distribution. The MATLAB Toolbox functions `resdist.m` and `normtest.m` can assist in examining distributions and checking normality respectively.

Example 6.12 (Data distributions) *Figure 6.16 illustrates the application of the MATLAB Toolbox functions. The influent flowrate data plotted in Figure 6.5*

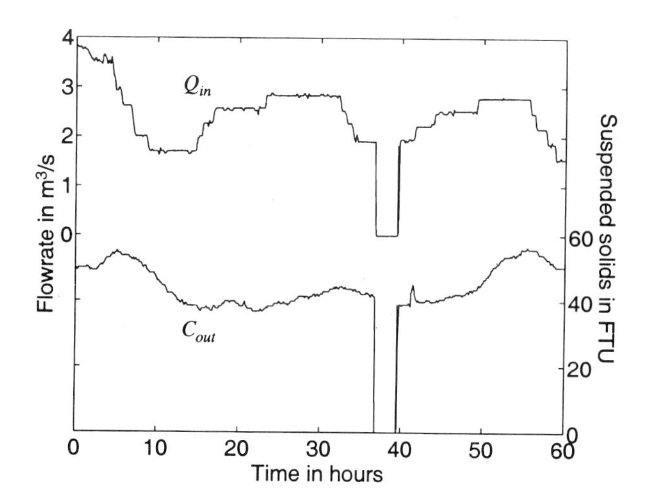

Figure 6.15: Illustration of missing clarifier data.

is tested. Both plots in Figure 6.16 show the abnormality in the distribution at low flowrates. The probability that the distribution is normal is 81% compared to the usual 95% required to accept a hypothesis.

6.5.5 Frequency Content of Time Series

We discussed frequency spectra in the previous section under the heading of excitation signals. The same technique can be used to detect some unexpected problems with data, particularly periodic effects which show up as peaks in the spectra. This may be expected as it was in the influent flowrate data (Figure 6.5 (data) and Figure 6.7 (spectrum)). The frequency content of interest also has an im-

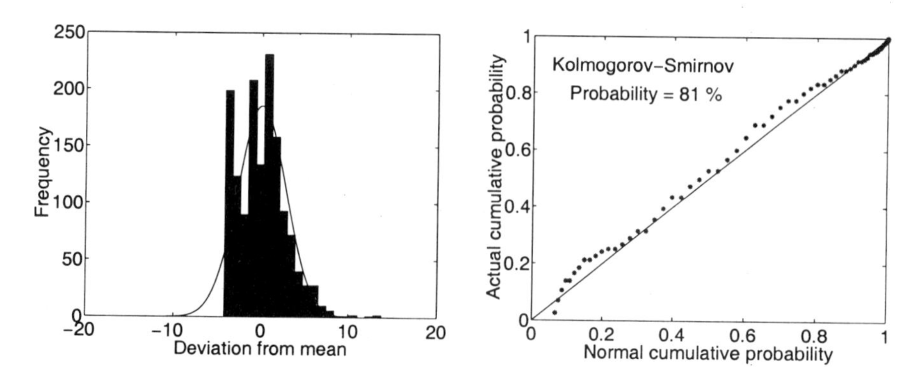

Figure 6.16: Data distribution tests.

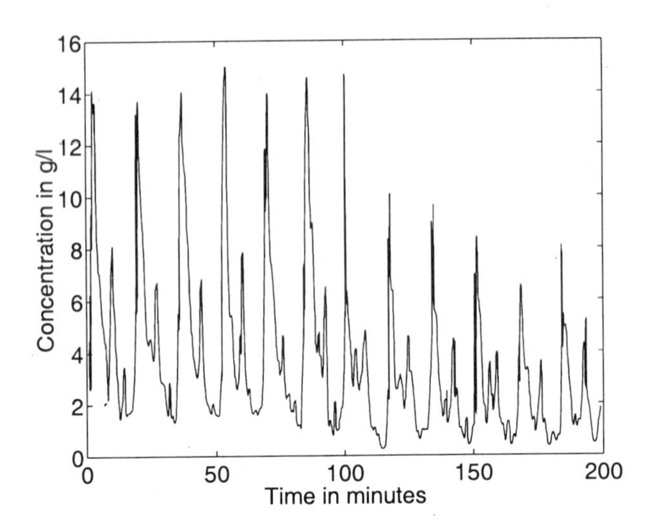

Figure 6.17: Settled sludge from a sedimentation unit.

plication on the sampling interval Δt as discussed in Section 6.4.1. Undesirable frequencies (high-frequency noise and low-frequency variations) can be removed by filtering, preferably using a high-order filter which contributes no phase shift such as `filt.m` in the WWT Toolbox. Filtering is discussed in more detail in Chapter 24.

Example 6.13 (Periodic effects) *For example, when sampling sludge concentration from a sedimentation unit there may be unexpected changes in the data. The example in Figure 6.17 illustrates the underflow concentration in a pilot scale settler. From a physical point of view it is expected to vary quite slowly, typically on an hourly timescale. Therefore a sampling interval of about 30 minutes seems to be adequate.*

The concentration has significant peaks every 12 minutes. They are caused by a sludge scraper that rotates along the bottom to remove sludge through the bottom valve. The rotation period is 12 minutes. Every time the scraper passes the bottom valve compacted sludge is brought past the suspended solids sensor. An adequate sampling time has to be in the order of minutes, and the proper process concentration is calculated as some 30 minute moving average value of the measurements.

Peaks in the spectra may also be unexpected, in which case it is important to identify the cause. This type of testing is used extensively for problem diagnosis (see Chapter 12), especially in vibration testing of running equipment.

6.5.6 Trends (Stationarity)

Stationarity, which includes the absence of trends, is a standard prerequisite for many time series data analysis techniques. Trend detection and removal is thus an important screening procedure. Again, it can also be an important diagnosis tool.

You could simply fit a straight line to the data and test whether the slope is significant. There are more powerful non-parametric tests less influenced by outliers, the Mann-Kendall and Sen's tests. The latter also gives an estimate of the slope, and is implemented in the MATLAB Toolbox function `trend.m`.

Example 6.14 (Trend detection) *Figure 6.18 shows influent TKN and NH_3 concentrations to an activated sludge plant. The output of the MATLAB program* `Example_6_14.m` *for this data is:*

```
NH3 Data:
Mann-Kendall Test:
 n = 48
 Mean Value = 40.85
 No significant trend
Sen's Nonparametric Estimator:
 Slope Estimate = -0.117
 Lower Confidence Limit = -0.3023
 Upper Confidence Limit = 0.008553

TKN Data:
Mann-Kondall Tcst:
 n = 48
 Mean Value = 72.61
 Downward trend detected
Sen's Nonparametric Estimator:
 Slope Estimate = -0.4453
 Lower Confidence Limit = -0.6162
 Upper Confidence Limit = -0.2607
```

Trends can be removed either by differencing or by postulating a straight line model and adjusting the data accordingly. The following equations give formulae for these two approaches.

$$\hat{y}_i = y_i - y_{i-1} \tag{6.2}$$

$$\hat{y}_i = y_i - s\,(t_i - \bar{t}) \tag{6.3}$$

where $\bar{t}$ is the mean of the independent variable and s is an estimate of the slope of the trend.

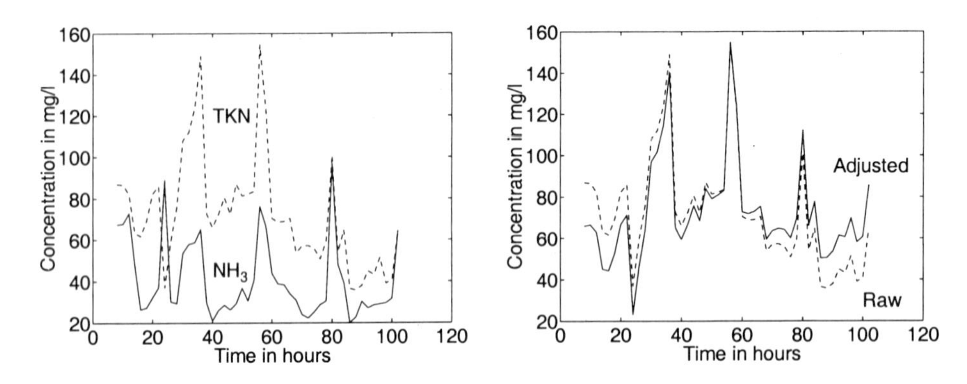

Figure 6.18: TKN and NH3 raw and adjusted data.

Example 6.15 (Trend removal) *The right plot in Figure 6.18 shows the TKN data before and after adjustment by the second technique. The output of the MATLAB Toolbox routine* `trend.m` *(called by program* `Example_6_15.m`*) on the adjusted data is:*

```
Adjusted TKN Data:
 Mann-Kendall Test:
 n = 48
 Mean Value = 72.61
 Z statistic = 0
 No significant trend
 Sen's Nonparametric Estimator:
 Slope Estimate = -2.307e-16
 Lower Confidence Limit = -0.1709
 Upper Confidence Limit = 0.1846
```

6.5.7 Seasonality

The absence of seasonality (cycling) is another requirement for many time series analysis techniques. You require data for a number of seasons greater than one to detect seasonality, the greater the number the more powerful the test. For equally spaced data, a simple frequency spectrum (see above) or an autocorrelation test can be used. MATLAB Toolbox program `season.m` will perform either or both of these tests.

Example 6.16 (Autocorrelation) *Figure 6.19 shows the autocorrelation against time lag (correlogram) for the influent flowrate data shown in Figure 6.5. The result can be compared to the frequency spectrum in Figure 6.7.*

The frequency spectrum is the more reliable as it does not show the multiples of the basic period that are evident in a correlogram, for example the 48 hour peak in

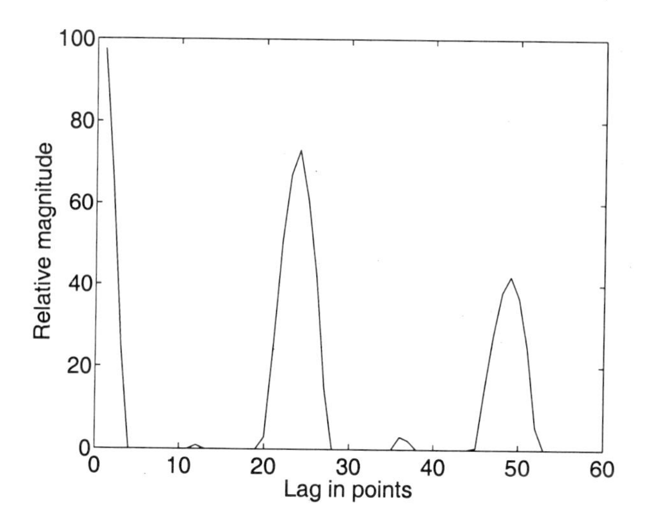

Figure 6.19: Autocorrelation test.

Figure 6.19. There would also be peaks at 72, 96, ... hours, if you had a sufficiently long time series.

Example 6.17 (Nitrification rates) *Figure 6.20 shows some interesting data on nitrification results, where the temperature and its seasonal influence on nitrification is apparent. This data is from an activated sludge plant in Sweden.*

6.5.8 Data Reconciliation

If you have complete or redundant data that is supposed to obey some conservation balance, then it is possible to adjust the data to exactly balance and to detect gross errors in the measurements. Mass balance data reconciliation is the most frequently applied example of this. It can also be applied to any data for which a known model exists. It is normally applied with steady state or averaged measurements, although it is possible to reconcile dynamic data. The data reconciliation problem is posed as follows:

$$\min_{y} (y - m)^T W (y - m) \tag{6.4}$$

subject to the constraint:

$$f(y) = 0 \tag{6.5}$$

where m are the measurements, y are the adjusted values, W is a usually diagonal weighting matrix with:

$$w_{ii} = \frac{1}{s_{e,i}^2} \tag{6.6}$$

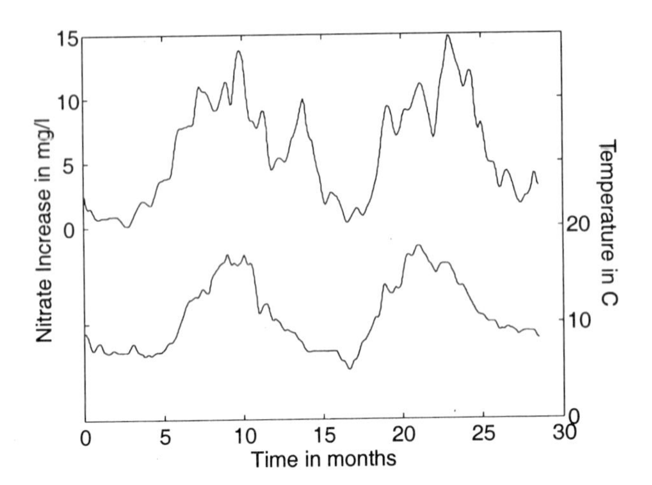

Figure 6.20: Seasonal temperature effects.

where s_e^2 are the error variances of the measurements, and $f(y)$ are some constraints, such as mass balances. The expression $(y-m)^T W(y-m)$ is interpreted as:

$$\begin{pmatrix} y_1 - m_1 & y_2 - m_2 & .. & y_n - m_n \end{pmatrix} \begin{pmatrix} w_{11} & .. & .. & w_{1n} \\ w_{21} & .. & .. & w_{2n} \\ .. & .. & .. & .. \\ w_{n1} & .. & .. & w_{nn} \end{pmatrix} \begin{pmatrix} y_1 - m_1 \\ y_2 - m_2 \\ .. \\ y_n - m_n \end{pmatrix} \tag{6.7}$$

which is a scalar number that has to be minimised. The function notation $f(y) = 0$ is a shorthand vector notation for n algebraic functions:

$$\begin{pmatrix} f_1(y_1, ..., y_n) = 0 \\ f_2(y_1, ..., y_n) = 0 \\ .. \\ f_n(y_1, ..., y_n) = 0 \end{pmatrix} \tag{6.8}$$

If the constraint $f(y)$ is linear in y, then it is a simple linear least squares problem. The MATLAB Toolbox program `datarec.m` will solve such problems. Even nonlinear problems such as energy balances are often solved by a two-step linear procedure, first reconciling the flowrates, and then the temperatures assuming the corrected flowrates are known constants.

Example 6.18 (Data reconciliation) *Consider the simple activated sludge plant in Figure 6.21 with five measured flowrates. The data file for* datarec.m *is:*

```
F1   F2   F3   F4   F5
1    1    -1   0    0    Aeration tank mass balance
```

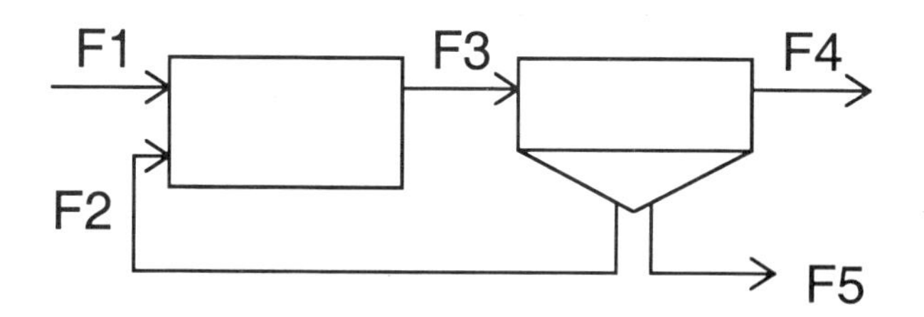

Figure 6.21: Data reconciliation example.

```
0    -1   1    -1   -1   Clarifier mass balance
0    0    0    0    0    No right hand side
0.1  0.1  0.1  0.1  0.2  Standard deviations
10.5 7.9  17.9 9.5  0.7  Measurements
```

The output of the program gives the following:

```
   New Value Adjustment Gross Error Test
F1 10.3176     0.1824     1.8235      0
F2 7.7412      0.1588     1.5882      0
F3 18.0588    -0.1588     1.5882      0
F4 9.5235     -0.0235     0.2353      0
F5 0.7941     -0.0941     0.4706      0
```

The new values of the flowrates now satisfy the mass balances. There were no gross errors detected, that is, no significantly excessive adjustments

6.6 Summary

There are three key requirements for good parameter estimates:

- good models, but if you did your homework on previous chapters, then you have good models.
- good data, which the techniques introduced in this chapter will help you get.
- good estimation techniques well used, which the next two chapters will address.

In this chapter we discussed the selection of the appropriate parameter estimation approach, good experiment planning practice, and finally data screening to diagnose and sometimes correct problems with the data. These techniques may not be the glamorous part of the job, but they can save time, cost and much frustration.

6.7 Further Reading

- For general information on experiment design see Walpole & Myers (1989) and Schenck (1968). For information on the parameter observability test see Ayesa *et al.* (1991).
- General information on data screening and analysis can be found in Gilbert (1987) which also has a good review and discussion of outlier tests.
- Information on the PRBS can be found in Eykhoff (1974) and Godfrey (1993).
- The textbook example on the effect of an outlier on regression can be found in Johansson (1993).
- The flowrate and effluent suspended solids data in Figures 6.11 to 6.13 are also published in Bergh & Olsson (1996).
- The data of settled sludge, Figure 6.17, from a sedimentation unit is also published in Olsson & Piani (1992).
- The seasonal temperature and nitrate data is also presented in the report Olsson (1985a).

6.8 Exercises

1. A simple model of the anaerobic phase of a biological SBR (batch reactor) is represented by the following equations:

$$\frac{dS_A}{dt} = -\mu_{fe} \quad (6.9)$$

$$\frac{dS_F}{dt} = \mu_{fe} - \mu_{rel} \quad (6.10)$$

$$\frac{dX_{PP}}{dt} = -Y_{PO4}\mu_{rel} \quad (6.11)$$

$$\frac{dS_{PO4}}{dt} = Y_{PO4}\mu_{rel} \quad (6.12)$$

where

$$\mu_{fe} = q_{fe}\frac{S_F}{K_{fe}+S_F}X_H \quad (6.13)$$

$$\mu_{rel} = q_{pha}\frac{S_A}{K_A+S_A}\frac{(X_{PP}/X_{PAO})}{K_{PP}+(X_{PP}/X_{PAO})}X_{PAO} \quad (6.14)$$

Design a series of experiments for collecting data to fit the parameters q_{fe}, K_{fe}, q_{pha}, K_A, K_{PP}, Y_{PO4}. The operating space is defined by initial values of S_A, S_F and the sludge concentration ($X_H + X_{PAO} + X_{PP}$). The proportions of X_H, X_{PAO} and X_{PP} in the sludge are known but not adjustable. The variables S_A, S_F, S_{PO4} can be measured with time.

2. An online controller tuning algorithm is being developed for dissolved oxygen. It requires a very simple model relating the oxygen transfer coefficient to the air flowrate which can be updated if the operating point (average air flowrate) changes.

   List the factors that must be considered, and give some quantitative estimates of experimental parameters where possible.

3. The data in the file `data0.dat` in the Exercises directory contains data of contaminants in river water (Figure 6.22). Use the MATLAB software we have presented in this Chapter to test for any trend and for seasonal effects.

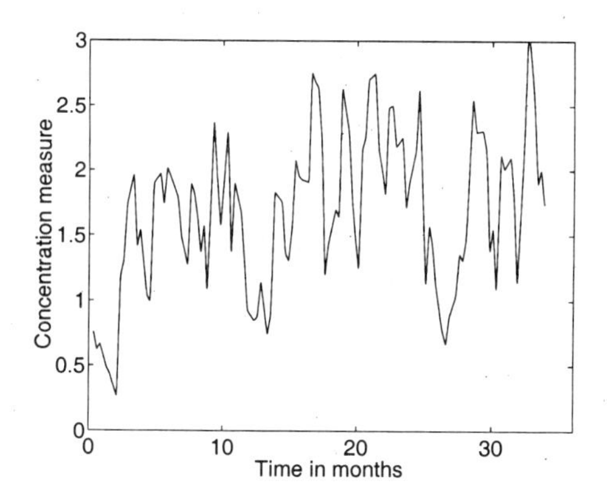

Figure 6.22: Contaminants in river water.

4. Obtain some time series of data from your process and examine the data thoroughly using the different techniques and MATLAB software we have presented in this Chapter (try the files `data1.dat` and `data2.dat` in the Exercises directory if you cannot get your own data).

Chapter 7

Principles of Parameter Estimation

Parameter estimation from experimental data is more commonly referred to as regression, and we will discuss some of the principles and pitfalls of both linear and nonlinear regression. Fitting typically involves searching for a set of parameter values which will minimise some function of the differences (or residuals) between real data values and corresponding values predicted by the model when it has been subjected to the same input conditions. Issues here are the criterion, the searching, the quality of the real data and the *sameness* of the input conditions.

7.1 Pitfalls

Great care must be taken when fitting data whether you are doing it qualitatively *by eye* or quantitatively by regression.

7.1.1 What is Best?

There are several criteria that can be used for parameter estimation to guage a *best fit* of the model to a set of data. The least-squares summation is perhaps the most popular, since the linear case has a single analytical solution.

$$J = (y - y_p)^T W (y - y_p) = \sum_i w_i (y_i - y_{p,i})^2 \tag{7.1}$$

where $y - y_p$ are the residuals and W is usually a diagonal weighting matrix with diagonal elements w_i. The least-squares criterion is a special case of the more general maximum-likelihood criterion. Maximum likelihood means the parameters are chosen such that the likelihood that the data can be predicted by the model is maximised.

Least-squares assumes the distribution of the data is Gaussian and the variance is known. With different assumptions the maximum-likelihood criterion will take other forms. Other forms of criteria are used when, for example, the parameters are probability distributions rather than deterministic.

7.1.2 Optimisation

More can be found about optimisation in Chapter 20. We will mention some factors here particularly relevant to parameter estimation.

Scaling

General optimisation techniques such as those typically used for nonlinear regression seldom include automatic scaling of the parameters or of individual step sizes. Whether the parameter is of order 1.0 or of order 1,000,000 the optimiser will apply the same step size when searching for the minimum value of the criterion. This means that small parameters will often change a lot but large parameters will hardly change at all. Scaling parameters within the model calculations is highly recommended.

Sensitivity

Parameter sensitivity is a different issue and was discussed under model verification in Chapter 3. Insensitive parameters, those which hardly affect the criterion no matter what value they take, are best omitted altogether.

Correlation

Parameter correlation is particularly troublesome with regression by optimisation and is especially common in wastewater model fitting. The product of a maximum growth rate and a biomass concentration are common and neither can be measured directly. It is therefore tempting to include both as parameters in the fitting. Unfortunately the two parameters often do not appear elsewhere or only where they are insensitive. The optimiser will get the same criterion with values of (1, 1), (1000, 0.001), (0.001, 1000) which have the same value when multiplied together. Which values the optimiser chooses will often depend on the initial estimates and it will frequently choose extreme values. Maximum growth rates and saturation constants are also highly correlated when concentrations are small relative to the saturation constant.

Response Surfaces

The shape of the response surface, criterion plotted against parameters, can have a dramatic effect on:

- the speed of convergence of the regression,
- whether an optimum is found, and
- which optimum is found.

If the surface is flat and irregularly shaped near the optimum the optimiser will often converge prematurely. A much worse situation can occur if the surface has more than one minimum. These problems frequently result in the wide ranges of parameter values often appearing in the literature. There is seldom a *magic* solution, so it is wise to run several regressions with widely differing starting values for the parameters and see if similiar final values are found. This is no guarantee but it does make you feel better.

7.1.3 Garbage In - Garbage Out

Parameter estimates can be severely affected by poor quality data. Tests of data quality such as data reconciliation, outlier checks, and simple extrapolation tests should be applied. These tests were discussed in Chapter 6. Regression is particularly sensitive to outliers, data points with some gross error involved, particularly when using the least-squares criterion. As well as quantitative tests there is nothing quite as effective as just plain common sense and experience, sometimes called *visualisation* to make it sound fancier. Plot it, look at it and think about it. Doubtful data can be discarded or used with a low weighting in the regression criterion.

7.1.4 Apples and Oranges

The *sameness* of the input conditions is particularly important if we are not to end up comparing *apples* with *oranges.* Except in very controlled conditions, there are many disturbances affecting the behaviour of the process, particularly with any real plant and even with laboratory pilot plants. Some of these disturbances were discussed in Chapter 2.

Care must be taken to monitor particularly those input conditions to which the model proved to be sensitive when we performed our model verification. Again this is an important outcome of a systematic model verification.

Of course we know that activated sludge plants (or any plant for that matter) are never at steady state. However, since the major disturbances are daily cycles, 24 hour averages and composite samples will give a fair approximation to steady state. Mass balances can be used to check the steady state assumption, though in wastewater applications it can be expensive to obtain sufficient measurements. While influent wastewater is subject to daily, weekly and seasonal variations, it would be foolish to compare *Mondays* or *Januarys* since random disturbances are frequent and biomass and settling characteristics change slowly and less regularly with time.

7.2 Linear Regression

The fitting task involves real data y, and predicted data y_p which has been generated by some process model of the general form:

$$y_p = m(\ u,\ p\ ) \tag{7.2}$$

where

$$u \quad = \quad \text{process input variables} = \begin{pmatrix} u_1 \\ .. \\ u_m \end{pmatrix} \tag{7.3}$$

$$p \quad = \quad \text{model parameters} = \begin{pmatrix} p_1 \\ .. \\ p_r \end{pmatrix} \tag{7.4}$$

$$m() \quad = \quad \text{the model relationship(s)} \tag{7.5}$$

The vector notation is a compact way of writing the relations:

$$\begin{pmatrix} y_{p,1} \\ y_{p,2} \\ .. \\ y_{p,n} \end{pmatrix} = \begin{pmatrix} m_1(u_1, u_2, ..., u_m, p_1, ..., p_r) \\ m_2(u_1, u_2, ..., u_m, p_1, ..., p_r) \\ .. \\ m_n(u_1, u_2, ..., u_m, p_1, ..., p_r) \end{pmatrix} \tag{7.6}$$

The fitting or regression problem is to find values for the parameters p such that some criterion is minimised, typically a least squares criterion (equation 7.1):

$$J = (y - y_p)^T W (y - y_p) = \sum_i w_i (y_i - y_{p,i})^2 \tag{7.7}$$

where $y - y_p$ are the residuals and W is a diagonal weighting matrix with diagonal elements w_i. The weights w_i given to individual data sets are theoretically the inverse of the variance estimate for that data. The more error in the data, the lower the weight that is assigned. Note, that in principle the observations have to be statistically independent of each other.

The solution to this optimisation problem is very simple if the model is linear in the parameters (not the input variables but the parameters). This means that the functions $m()$ could be rewritten in the form:

$$y_{p,i} = p_{0,i} + p_{1,i} u_1 + p_{2,i} u_2 + ... \tag{7.8}$$

for each predicted model output $y_{p,i}, i = 1, .., n$. It is often the case that the model is not linear, but can be linearised so that we have:

$$y\prime_{p,i}(y_p, u) = p\prime_{0,i}(p) + p\prime_{1,i}(p) u\prime_1(y_p, u) + p\prime_{2,i}(p) u\prime_2(y_p, u) + ... \tag{7.9}$$

where the new linear variables and parameters denoted by a prime are functions of the original variables and parameters respectively. The problem with this commonly practised approach is that experimental errors in the variables are no longer normally distributed or Gaussian, and variables and parameters are no longer uncorrelated. This means that the validity of the statistics for evaluating the fit is now in doubt. Of course, these conditions may not have been met originally in any case.

Example 7.1 (Monod linearization) *For an example of linearisation, suppose we have some laboratory data (simulated in this case) of growth rate μ for different substrate compositions S. Assuming we wish to fit this data to the Monod kinetic relationship:*

$$\mu = \mu max \frac{S}{K + S} \tag{7.10}$$

The parameters are μmax and K which makes the relationship nonlinear in the parameters (see Section 5.2.2). However, we can rearrange the equation algebraically to the equivalent linear form:

$$\mu = \mu max - K \left(\frac{\mu}{S} \right) \tag{7.11}$$

We can see this diagrammatically in Figure 7.1 which shows the data plotted as points. On the right, where the data has been transformed, the change in the variance at low rates is very obvious.

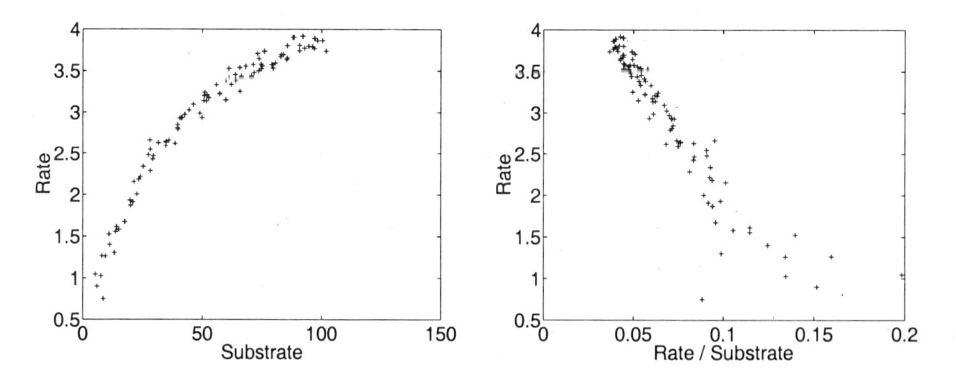

Figure 7.1: Nonlinear and linear forms.

The solution to this linear least squares optimisation, choosing p to minimize Equation 7.7, is simply:

$$\hat{p} = (X^T W X)^{-1} X^T W\, y \tag{7.12}$$

where X is the matrix $[1\ u_1\ u_2\ ...]$ with u_i being columns of data for the independent variables and $\hat{p}$ are the parameter estimates. The result $\hat{p}$ is sometimes called the BLUE (best linear unbiased estimate).

Example 7.2 (Monod fitting) *For our Monod example we have chosen to weight the data so that values at higher concentrations, where the variance is lower, have a higher weight:*

$$y = \begin{bmatrix} \mu_1 \\ \mu_2 \\ . \\ \mu_n \end{bmatrix} \quad X = \begin{bmatrix} 1 & -\mu_1/S_1 \\ 1 & -\mu_2/S_2 \\ . & . \\ 1 & -\mu_n/S_n \end{bmatrix} \tag{7.13}$$

$$W = \begin{bmatrix} S_1 & 0 & . & 0 \\ 0 & S_2 & . & 0 \\ . & . & . & . \\ 0 & 0 & . & S_n \end{bmatrix} \quad p = \begin{bmatrix} \mu max \\ K \end{bmatrix} \tag{7.14}$$

which gives a result such as $\hat{p} = [\,4.95 \quad 28.6\,]^T$. *The true values used to generate the simulated data were* $p = [\,5 \quad 30\,]^T$, *the difference being due to the 2-3 per cent normally distributed noise added to the true data and to the finite number of data points.*

7.3 Assessing a Fit

Obtaining a set of values for the model parameters is not enough. How do we know they are good values? How do we know the best fit is good enough? Perhaps our model is poor. Perhaps our data is poor. The assessment of the model fit that we have found is more important than the fitting task itself. We shall look at three common assessment procedures - residual tests, parameter confidence tests and the statistical lack-of-fit test.

7.3.1 Residuals Tests

If we have no plant-model mismatch, that is we have a perfect model, then the residuals, $y - y_p$, represent simply the experimental error. If we plot the residuals, they should be randomly distributed about zero with an average deviation of the same order as the experimental error that we estimated when we were planning the experiments. Residuals plots are simply graphs of the residuals $y - y_p$ against some other variable on the X-axis. It is good practise to plot the residuals against each of the input variables, against the predicted variables, and against extraneous variables such as the time of the experiment, the ambient temperature, and the process operator. If the model is a good one and the influence of extraneous variables is not significant, the residuals on every plot should be randomly spread about zero. If this is not the case then either the model is deficient or the real data is deficient. The plots can often give you indications of the likely problems.

One useful indication of the quality of the fit is to compare the difference between the sum-of-squares of the residuals about the regression line (unexplained

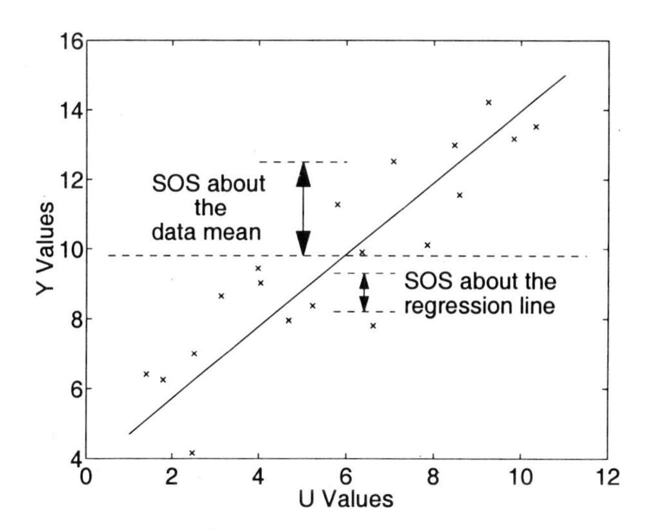

Figure 7.2: Measures of variation (SOS is the sum-of-squares of the residuals).

variation) and the sum-of-squares of the data about its own mean (total variation) as shown in Figure 7.2. This is usually expressed as a percentage of the total variation about the mean:

$$V_{\text{explained}} = 100 \frac{\sum_i (y_i - \bar{y})^2 - \sum_i (y_i - y_{p,i})^2}{\sum_i (y_i - \bar{y})^2} \tag{7.15}$$

This difference is a useful measure of the variation in the data that has been explained by the model.

Example 7.3 (Monod residual tests) *In the case of our Monod example the residuals and their distribution are shown in Figure 7.3. They are spread fairly evenly about zero, but show the changing magnitude caused by the data transformation. This can be seen by the distribution which is narrower than a normal distribution but with some outlying points.*

For the Monod example, 85 per cent of the variation was explained by the model, 15 per cent remaining in the residuals in Figure 7.3. This unexplained variation is quite high, considering we know that the noise is only 2-3 per cent, so the fit may not be so good or it may be an artifact of the linearisation.

7.3.2 Parameter Confidence Tests

Another test for the quality of the fit and the model is to examine measures of the quality of the parameters. These include the confidence limits and the confidence region. It is very important to look at both of these as we will show below.

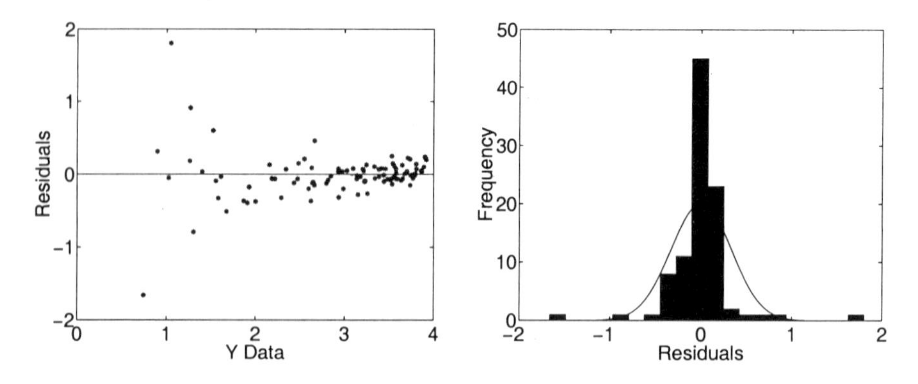

Figure 7.3: Monod fit residuals.

The parameters are 95 per cent certain to lie somewhere within the confidence region and, if the parameters are not correlated, within the confidence limits. Sophisticated fitting software will give you this information, possibly as an option. They can be calculated from the following formula:

$$p_i \quad \pm \quad t_{0.025,n-m} \sqrt{s_{ii}^2} \tag{7.16}$$

where t can be read from statistical tables, n is the number of data points, m is the number of parameters, s^2 is the variance of the parameters where

$$\left[s_{ij}^2\right] = \left(X^T X\right)^{-1} \frac{e^T e}{n-m} \tag{7.17}$$

where e is the vector of residuals $y - y_p$, $y_p = X\hat{p}$ and X and $\hat{p}$ were defined previously.

$$S_{95} = e^T e \left(1 + \frac{m}{n-m} F_{0.05,m,n-m}\right) \tag{7.18}$$

where S_{95} defines the sum-of-squares 95 % contour value and F can be read from statistical tables. The confidence limits can give an overly optimistic picture since they define a rectangle which can encompass quite a bit more area than the actual confidence region. This is particularly the case where there is a strong correlation between the parameters. Parameter correlation can be detected qualitatively by narrow sloping contours. The normalised correlation coefficient for parameters gives a quantitative measure, though the interpretation of the values can vary, particularly between disciplines,

$$r_{ij} = \frac{s_{ij}^2}{\sqrt{s_{ii}^2 \, s_{jj}^2}} \tag{7.19}$$

Example 7.4 (Monod confidence measures) *For our Monod example, $e^T e$ is a 2×2 matrix, and the confidence limits calculated from Equation 7.16 are:*

$$\mu_{max} = 4.95 \pm 0.16, \qquad K = 28.6 \pm 2.1 \tag{7.20}$$

The confidence region is shown in Figure 7.4 where the inner contour line defines the 95 per cent region, the square dot defines the parameter estimates and the diamond dot defines the true parameters. These points are off-centre due to the noise in the data. The square shows the area defined by the confidence limits.

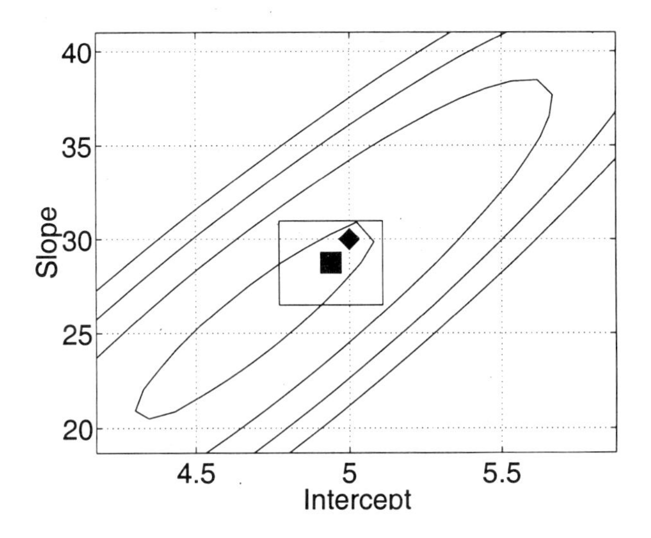

Figure 7.4: Linear confidence region.

The confidence limits give an overly optimistic picture since they define a rectangle 4.2 high and 0.32 wide which encompasses quite a different area than the actual confidence region shown. This is particularly the case where there is a strong correlation between the parameters, as the narrow 45 degree contours demonstrate. The normalised correlation coefficient between the parameters was 0.92, calculated from Equation 7.19 with $i = 1$ and $j = 2$.

7.3.3 Lack-of-Fit Test

If you have an independent estimate of the experimental error, you can perform a statistical lack-of-fit test. This is done by comparing the ratio of the variance of the residuals to the experimental error variance, and testing it against the statistic F with the appropriate degrees of freedom:

$$\frac{e^T e}{(n-m)s_E^2} \quad vs \quad F_{0.05}(n-m, n_E) \tag{7.21}$$

where n_E is the degrees of freedom of the experimental error variance s_E^2 (for example the number of replicates, duplicate experiments at exactly the same conditions, minus 1) and e the residuals defined following Equation 7.17.

Example 7.5 (Monod lack-of-fit test) *For the Monod example we would have:*

$$\frac{e^T e}{(n-m)s_E^2} = \frac{0.0976}{0.0133} = 7.34 \tag{7.22}$$

which is greater than the value of $F_{0.05,n-m,n_E} = 1.3$, *indicating that it was not a good fit.*

7.3.4 Precision of Predicted Values

Since there is some uncertainty about the estimated parameters, there is a corresponding uncertainty about responses predicted by the model. For the given x values the confidence interval for the response is:

$$\hat{y}_i \quad \pm \quad t_{0.025,n-m} \sqrt{s_{ii}^2} \tag{7.23}$$

where

$$\left[s_{ij}^2\right] = X^T \left(X^T X\right)^{-1} X \frac{e^T e}{n-m} \tag{7.24}$$

If it is desired to predict a response from the system, then both the uncertainty of $\hat{y}$ and the experimental error s_E^2 must be incorporated to give:

$$\hat{y}_i \quad \pm \quad t_{0.025,n-m} \sqrt{s_{ii}^2 + s_E^2} \tag{7.25}$$

7.3.5 Overfitting Data

There is a tendency, especially if a lot of data is available, to overfit, that is to use more parameters than is necessary. There is no quick way of simplifying the model. A new fit must be carried out and evaluated as discussed above. If two models A and B both satisfy the lack-of-fit criterion above, they can be compared using the statistic:

$$S = \frac{e_A^T e_A - e_B^T e_B}{(m_A - m_B)s_E^2} \tag{7.26}$$

where m_A and m_B are the numbers of parameters and e_A and e_B are the residuals for the A and B models respectively. If the statistic S is larger than the appropriate F value with $(m_A - m_B)$ and n_E degrees of freedom, model A is significantly better than model B. The general principle that should be used is that of *parsimony*, that is, use fewer parameters. Models with fewer parameters generalise better, that is, even if they are worse at predicting the given data, they will be better at interpolation and extrapolation.

7.4 Nonlinear Regression

If the form of the model relationships:

$$y_p = m(\ u,\ p\ ) \tag{7.27}$$

is not linear in the parameters p and cannot be linearised, then we have a nonlinear regression problem. The formulation of the problem is identical, but the solution becomes an optimisation without an analytical solution, which we had in the linear case. Search techniques must be used to find the minimum least squares criterion in the parameter space. This is generally numerically intensive.

Once a fit is obtained, the same general assessment procedures are used. Some extra care must be taken however. The parameter confidence regions and the lack-of-fit statistic given by good packages are mostly linearised around the solution point and are therefore only approximate indications, especially the confidence regions. While in the linear case confidence regions are guaranteed to be ellipses centred on the solution, in the nonlinear case the regions can be any shape in general. The more sophisticated packages will determine the confidence regions by a search procedure. Although this takes some time to compute, it is best to do this especially as a final check on the model.

Example 7.6 (Monod nonlinear regression) *We can fit our simple Monod example by nonlinear regression as well. If we do this we obtain parameter values of* $\hat{p} = [\ 5.06 \quad 30.7\ ]^T$. *This fit explained 98 per cent of the variation in the data (using Equation 7.15) and gave residuals which are plotted with their distribution in Figure 7.5.*

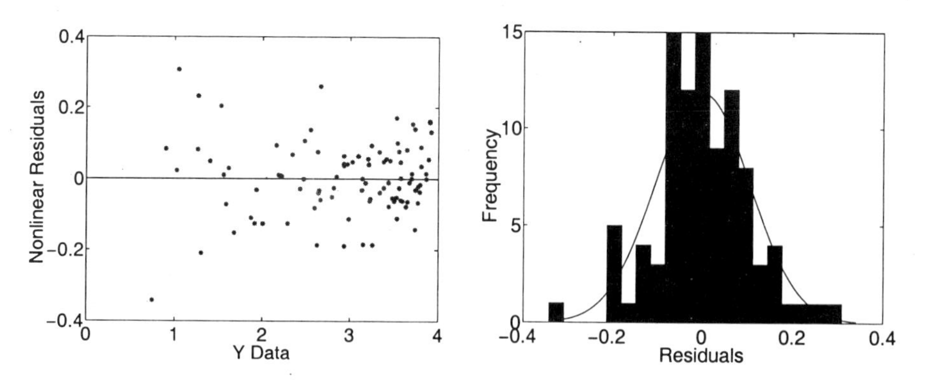

Figure 7.5: Nonlinear fit residuals.

Note that these residuals (Figure 7.5) have the even spread of experimental error you would expect from good data, unlike those in Figure 7.3 where a linearising transformation was applied. The distribution of the residuals is also much closer to a normal distribution, unlike the distribution of the transformed data in the previous Figure 7.3.

In the case of our Monod example, the nonlinear confidence region for the parameters shown in Figure 7.6 remains a similiar shape to that shown in Figure 7.4 above. For the Monod example the lack-of-fit test for the nonlinear fit has the

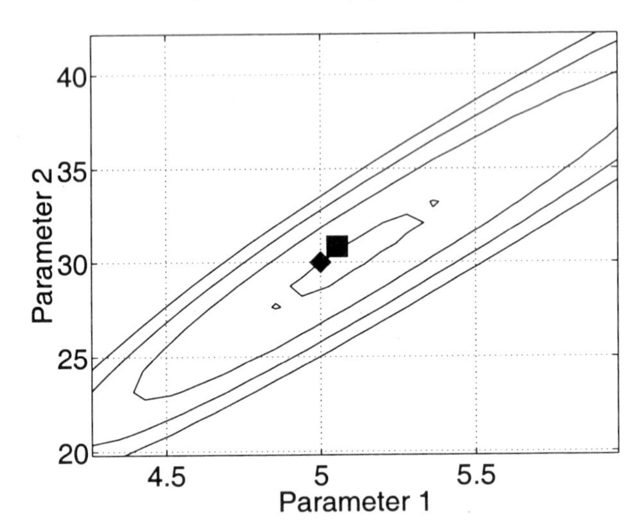

Figure 7.6: Nonlinear confidence region.

value:

$$\frac{e^T e}{(n-m)s_E^2} = \frac{0.0132}{0.0133} = 0.99 \tag{7.28}$$

which is less than the value of $F_{0.05,n-m,n_E} = 1.3$, *indicating that it was a good fit.*

Example 7.7 (Carbon removal process regression) *The simple carbon removal process example introduced in Chapter 5 and detailed in Appendix B shows the asymmetric confidence regions more typical of nonlinear regression models (see Figure 7.7).*

The left plot in Figure 7.7 shows the $\hat{\mu}_H$ *and* K_S *projection of the multidimensional surface (true values are 4.9 and 30). The right plot in Figure 7.7 shows the* b_H *and* Y_H *projection (true values are 0.4 and 0.6). Notice that the* valleys *are narrow and slanted showing a high correlation between these sets of parameters. The apparent irregularities in the contours are due to the coarseness of the calculated sum-of-squares data and the MATLAB contouring algorithm.*

In Figure 7.7, the *low point* is also at the most narrow end of the long narrow *valley*, which makes it very difficult to find. Most optimisation packages will find some point at the bottom of the valley, but may not find the true minimum. This contributes to the wide range of values often published for such highly correlated parameters. If correlation is suspected, several regressions should be performed with widely different starting estimates for the parameters. Only if several of these

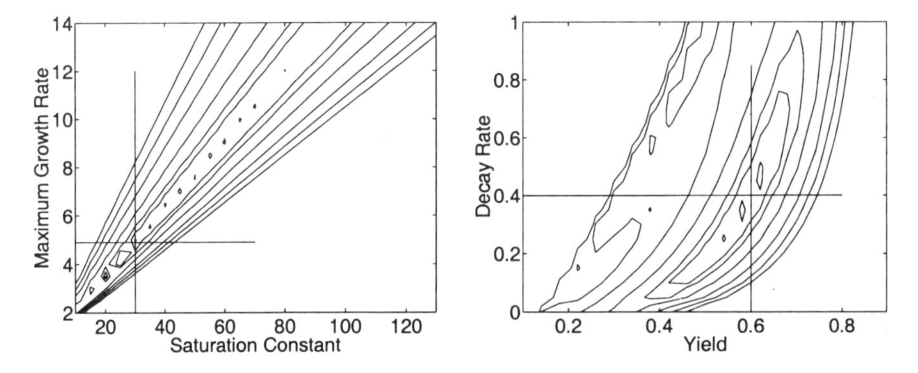

Figure 7.7: Confidence region projections.

converge to the same values can you be somewhat sure that the minimum has been located.

It is also important to have good quality data with such highly correlated parameters. Even modest amounts of noise can destroy any chance of finding the true values.

7.5 Fitting Simple Time Series Models

Time series models come under the category of dynamic input-output models which we discussed in detail in Chapter 5. We often want a quick and simple dynamic model of a process or some element of the process, for example a sensor or an actuator. In this case we can ignore the effects of unknown disturbances and choose a simple first or second order process model represented as follows:

$$y_t = a_1\, y_{t-1} + b_1\, u_{t-1} \tag{7.29}$$

$$y_t = a_1\, y_{t-1} + a_2\, y_{t-2} + b_1\, u_{t-1} \tag{7.30}$$

The order of the model is defined here as the number of previous y_{t-k} terms on the right-hand side of the equation. The dynamics of the process must be excited with a suitable stimulus as we discussed in Chapter 6. Typical test signals applied to fit the parameters for such simple process models are steps and impulses.

7.5.1 Step Signals

This is perhaps the most frequently used test signal in process systems. It does not have good excitation properties, but generally we are interested in determining only the gain and the major (largest) time constant or perhaps the two largest time constants. It is important to get as clean and sharp a step as you can. Typical step responses from processes are shown in Figure 7.8.

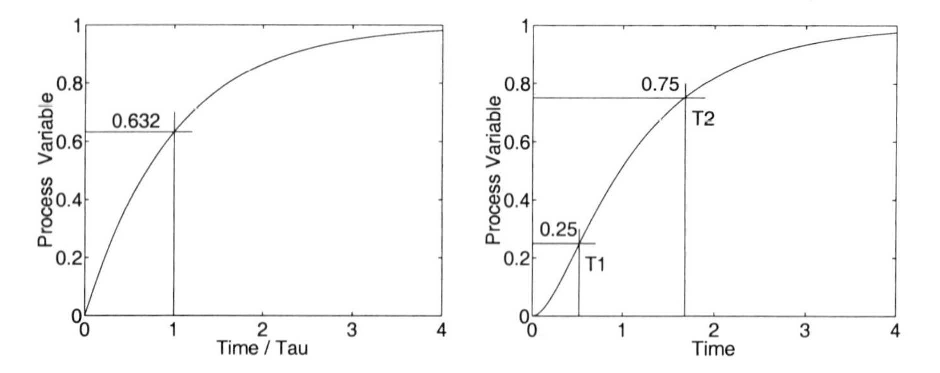

Figure 7.8: First and second order responses.

The gain K is the final change in output divided by the change in input (the magnitude of the step). Up and down steps and steps of different magnitudes can identify some nonlinear behaviour (give different gains). This was demonstrated in Chapter 5 in Figures 5.2 and 5.3.

As shown on the left in Figure 7.8, the time constant τ of a first order type response can be identified as the time at which 0.632 times the final change in the process variable has occurred. The model parameters are:

$$a_1 = e^{-\frac{\Delta t}{\tau}}, \quad b_1 = K(1 - e^{-\frac{\Delta t}{\tau}}) \tag{7.31}$$

where Δt is the time interval in the discrete time model.

Second order systems exhibit a range of responses depending upon the degree of damping inherent in the system. In Figure 7.9 the under-damped responses show an overshoot and some oscillation. These responses are uncommon in open-loop process systems though they do approximate the response of tightly-tuned feedback control loops (see Chapter 17). Open-loop process systems generally approximate an over-damped response rising gradually to the final value without an overshoot.

Fitting the under-damped second order system is a little more complex. As shown on the right in Figure 7.8, you must read two times, when for example 0.25 and 0.75 times the final change in process variable has occurred. Then you must perform a small optimization to search for the times at which two instances of the following equation are solved.

$$y_t = 1 + \frac{\tau_1}{\tau_2 - \tau_1} e^{-\frac{t}{\tau_1}} - \frac{\tau_2}{\tau_2 - \tau_1} e^{-\frac{t}{\tau_2}} \tag{7.32}$$

where y is the process variable, t is the time and τ_1 and τ_2 are the two unknown time constants. The accuracy of the parameters found when fitting a second order model is more susceptible to the effects of noise. The model parameters are:

$$a_1 = e^{-\frac{\Delta t}{\tau_1}} + e^{-\frac{\Delta t}{\tau_2}}, \quad a_2 = -e^{-\frac{\Delta t}{\tau_1}} e^{-\frac{\Delta t}{\tau_2}}, \quad b_1 = K(1 - e^{-\frac{\Delta t}{\tau_1}})(1 - e^{-\frac{\Delta t}{\tau_2}}) \tag{7.33}$$

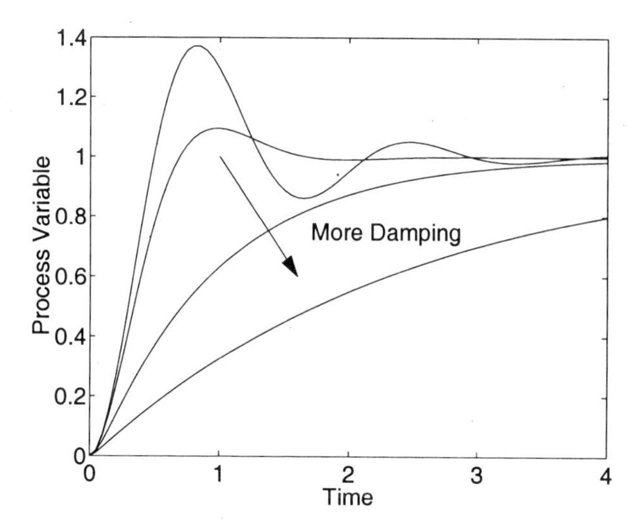

Figure 7.9: Over-damped and under-damped second order responses.

where Δt is the time interval in the discrete time model.

7.5.2 Impulse Signals

This is the usual test signal used for tracer studies, where a dye is injected into the inlet flow into a tank and the color intensity in the outlet is measured. There must be sufficient dye injected to achieve a reasonable signal to noise ratio in the outlet.

The response for first and n-th order responses can be calculated from the following equations:

$$y_t = e^{-\frac{t}{\tau}} \tag{7.34}$$

$$y_t = \frac{1}{(n-1)!}\left(\frac{t}{\tau}\right)^{n-1} e^{-\frac{t}{\tau}} \tag{7.35}$$

Typical impulse responses of the outlet concentration for an impulse in the inlet concentration for one to four tanks (first to fourth order processes) all with the same time constants are shown in Figure 7.10. Sampling times Δt and data record lengths T are dependent on the residence time or time constant τ and the order of the response. A knowledge of the system being tested and an examination of Figure 7.10 will give some guidance. An example of the use of impulse tests to identify the probable flow pattern in tanks is given in Chapter 8.

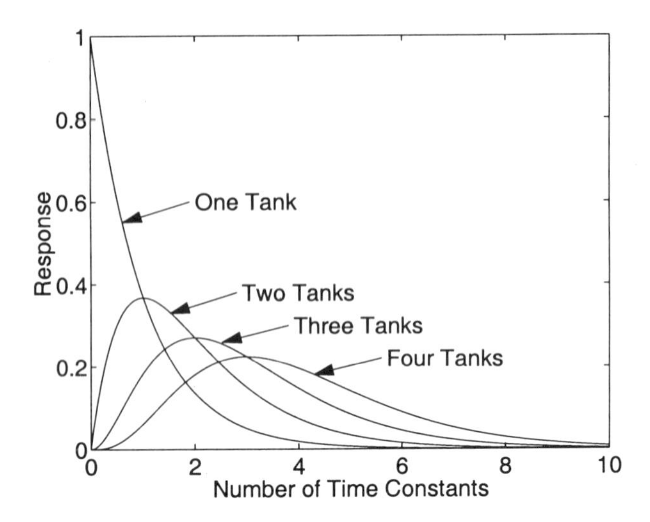

Figure 7.10: Impulse responses.

7.6 Time Series Analysis

The traditional technique called Time Series Analysis (TSA) is applied off-line to time series of input and output data. Chapter 6 discussed the problems of getting good time series data. The traditional TSA has two goals:

- to identify the form of the model (what is the minimum number of terms, or parameters, that will be required in the model to adequately describe the process), and
- to estimate values for the chosen parameters.

Time series analysis has been used for many years as a means for describing dynamic relations between inputs and outputs. It is important to preprocess data before TSA is applied in order to meet the assumptions that data is stationary and has a zero mean. This was discussed in detail in Chapter 6.

7.6.1 Parameter Estimation

We will only concern ourselves here with the second goal, parameter estimation. That is, we will assume you know the form of the model, the values of n, m, and q in the general time series model:

$$\begin{aligned} y_t = & \; a_1\, y_{t-1} + a_2\, y_{t-2} + \; ... \; + a_n\, y_{t-n} + \\ & \; b_1\, u_{t-1} + b_2\, u_{t-2} + \; ... \; + b_m\, u_{t-m} + \\ & \; c_1\, d_{t-1} + c_2\, d_{t-2} + \; ... \; + c_q\, d_{t-q} \end{aligned} \tag{7.36}$$

The a_i and b_i coefficients can be found from the experimental time series by linear regression in the same way we discussed in a previous section. We form a data vector and regression matrix:

$$y = \begin{bmatrix} y_t \\ y_{t-1} \\ y_{t-2} \\ .. \end{bmatrix} \tag{7.37}$$

$$X = \begin{bmatrix} y_{t-1} & \cdots & y_{t-n} & u_{t-1} & \cdots & u_{t-m} \\ y_{t-2} & \cdots & y_{t-n-1} & u_{t-2} & \cdots & u_{t-m-1} \\ y_{t-3} & \cdots & y_{t-n-2} & u_{t-3} & \cdots & u_{t-m-2} \\ .. & \cdots & .. & .. & \cdots & .. \end{bmatrix} \tag{7.38}$$

with the parameter vector:

$$p = \begin{bmatrix} a_1 & a_2 & \dots & a_n & b_1 & b_2 & \dots & b_m \end{bmatrix}^T \tag{7.39}$$

with the usual solution:

$$\hat{p} = (X^T W X)^{-1} X^T W y \tag{7.40}$$

where the generally diagonal weighting matrix W is seldom used in TSA but, as we will see in Chapter 12, can be used to progressively discount the value of older data when applied recursively for on-line parameter estimation. It is important once a fit is obtained to carefully examine the residuals as we did in Section 7.3.

We have thus far ignored the disturbance or noise terms with coefficients c_i. If the disturbances were random white noise, then the residuals will also pass the tests for randomness. However, if the disturbances are coloured noise (deterministic and white noise components), such as the cyclic influent disturbances to a wastewater treatment plant, then we must model this noise. Why bother to do this? Because the least squares regression only gives reliable parameter estimates if the residuals are random with zero mean and constant variance.

Hence we must remove the deterministic components from the residuals with a disturbance model. Typically, time series models have been developed for the input-output relationships between air flowrate and dissolved oxygen, flowrates and MLSS concentrations, flowrates and effluent suspended solids concentrations, and carbon source dosages and denitrification rate.

Example 7.8 (Dissolved oxygen time series model) *As an example we will consider determining a dynamic model for the response of dissolved oxygen concentration (output variable y_t) to air flowrate (input variable u_t) in an aerobic bioreactor where the air is controlled by the DO with an on-off controller. About ten minutes of data is shown on the left plot in Figure 7.11 labelled* y *and* u *respectively. The data shown is 60 data points of the 4153 points in the data file* `Example_7_8.dat`.

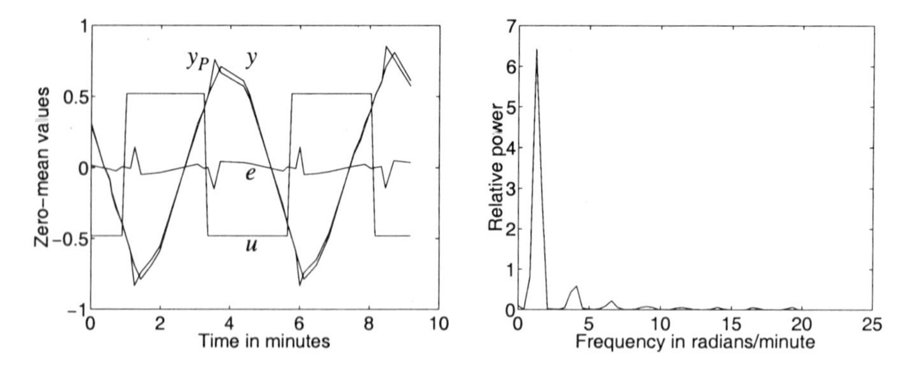

Figure 7.11: Plots from program `Example_7_8.m`.

The MATLAB program `Example_7_8.m` *was used to fit 100 data points to a time series model with the following structure:*

$$y_t = a_1\, y_{t-1} + a_2\, y_{t-2} + b_1\, u_{t-1} \tag{7.41}$$

resulting in the following parameters and confidence limits:

$$a_1 = 1.725 \pm 0.158 \tag{7.42}$$

$$a_2 = -0.791 \pm 0.143 \tag{7.43}$$

$$b_1 = -0.0367 \pm 0.0348 \tag{7.44}$$

Despite explaining 99.1 per cent of the total variance, it is not a great model, as can be seen from the large confidence limit on the parameter b_1 and the relatively large residuals following the switching of the air. The residuals are marked with e *on the left plot in Figure 7.11. This inability to fit the faster dynamics can be explained by examining the right plot in Figure 7.11 which shows poor frequency content (excitation) in the input signal u especially at higher frequencies.*

7.6.2 Disturbance Models

White noise is normally distributed uncorrelated random noise with a zero mean and a constant variance. If a signal is not white noise, then it is called coloured noise. A disturbance or noise model is a process or filter (Figure 7.12) that converts white noise into coloured noise. Such a model can be used to generate coloured noise. For example, Figure 7.13 shows a plot of white noise and coloured noises generated by the MATLAB Toolbox function `noise.m` from the following noise models:

$$v_t = 0.8\, v_{t-1} + w_t + 0.1\, w_{t-1} \tag{7.45}$$

$$v_t = 0.9\, v_{t-1} + 0.05\, v_{t-2} + w_t + 0.1\, w_{t-1} \tag{7.46}$$

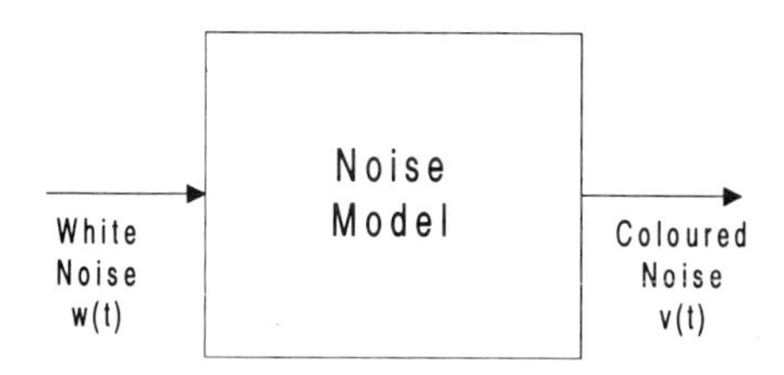

Figure 7.12: Noise or disturbance model.

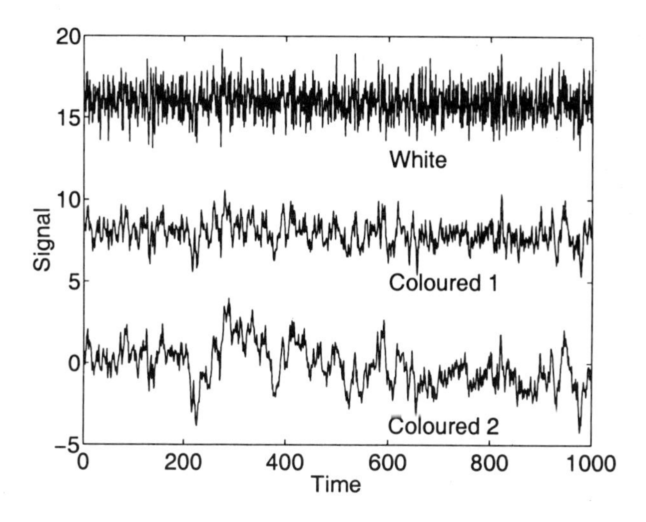

Figure 7.13: White and coloured noises.

It is also possible to determine a noise model that will generate a certain coloured noise signal or disturbance by regression. The MATLAB Toolbox function `fitnoise.m` will do this for low order models using a two-step regression. To fit more complex models, and to help identify the appropriate model, a more sophisticated procedure is needed. Several computer packages are available to do this, including one in the MATLAB Identification Toolbox.

In addition to improving the parameter estimates of our input-output model, a disturbance model can be used to predict disturbances (see Chapter 15) and take some feedforward control action (see Chapter 18). In Chapter 12 we will extend linear dynamic estimation further by discussing recursive online estimators.

7.7 Nonlinear Dynamic Estimation

If our dynamic model is nonlinear in the parameters, the task is much more complex. We now require a synergy between three components, as shown in Figure 7.14:

- the nonlinear dynamic model,
- a suitable integrator, and
- a nonlinear regression optimiser.

The first two components are often available in the form of a dynamic simulator. Indeed some simulators also have the optimiser available as an integrated or optional component.

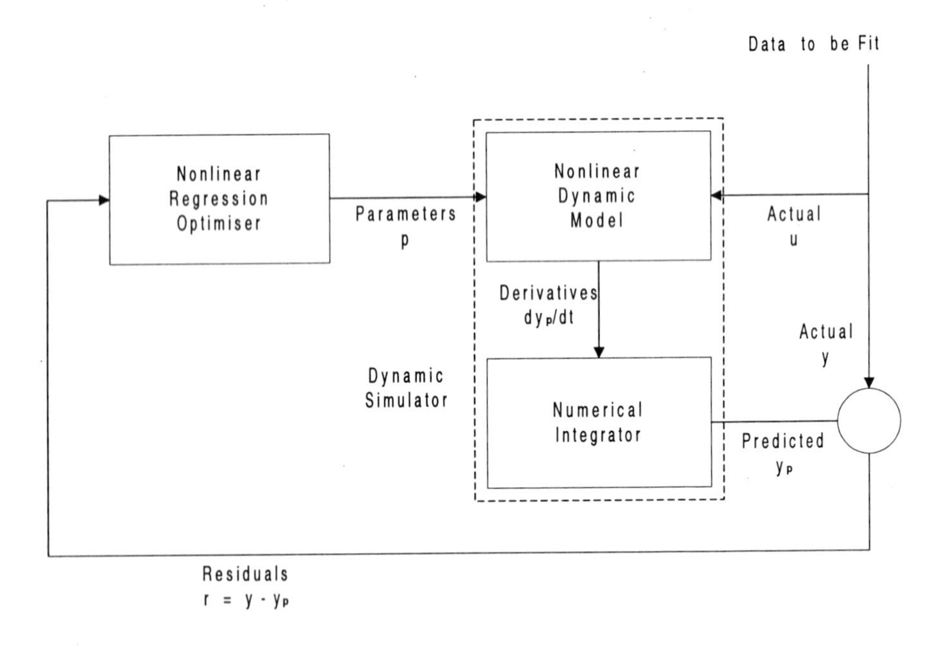

Figure 7.14: Nonlinear dynamic parameter estimation.

The regression criterion can be the usual least squares summation of the residuals which are again the difference between the actual and predicted output profiles over time. The predicted output profile is calculated by the simulator (model plus integrator) from the parameters supplied by the optimiser and the actual inputs. The optimiser will then choose new parameters attempting to minimise the criterion, the match between the actual and predicted profiles.

This type of regression suffers from all the same problems of standard nonlinear regression discussed previously in Section 7.4.

7.8 Summary

We have tried to give you a brief introduction to linear and nonlinear regression. Even more important than understanding regression principles, is understanding how to evaluate the results of the regression. If either the data or the model are poor, then the results will be nonsense. You must know how to decide whether you have good estimates. We have tried to emphasise this point. Now it is up to you. We have also shown how the parameters of dynamic input-output models might be determined from experimental data using simple test signals, using time series analysis for linear models and using nonlinear regression.

Experiment planning issues were covered in Chapter 6. Estimating parameters from dynamic data is always easier said than done. In real life it is difficult to get a system suitably excited, which we discussed in Chapter 6. It is also difficult to isolate your system from disturbances other than your input of interest.

There are many computer packages that will perform regression and time series analysis for you, though most only handle linear regression. Personal computer packages such as MINITAB are readily available at reasonable cost. You should look carefully for packages which give you lots of diagnostic information for performing that all important evaluation. The MATLAB Toolbox codes `linreg.m` and `nlreg.m` supplied with this book can be used as a basis for your own regression programs (see Appendix A). They supply more diagnostic information than most commercial packages.

7.9 Further Reading

- For more information on regression, there are a great many good books, again mostly restricted to linear regression. In our humble opinion, perhaps the best of all the books is still Draper & Smith (1981).

- For more information on time series analysis Box & Jenkins (1976) is the classic reference though others are perhaps more readable like Chatfield (1984), Cryer (1986) and Kendall & Ord (1990).

- Early examples of dissolved oxygen dynamics identification are shown in Olsson & Hansson (1976) and later in Ko *et al.* (1982), Sollfrank & Guyer (1990) and Carlsson & Wigren (1993). There is a comprehensive overview of many applications in Beck (1986) and an excellent survey of the potential of process identification in wastewater treatment in Beck (1989). Model identification experiences with wastewater applications are demonstrated in Olsson (1989), Marsili-Libelli *et al.* (1978), Marsili-Libelli (1989), Holmberg

(1982), Hiraoka & Tsumura (1989), Larrea *et al.* (1991), Carstensen (1994), van Dongen & Geuens (1998), Kabouris & Georgakakos (1996a), Kabouris & Georgakakos (1996b), Kabouris & Georgakakos (1996b) and Ossenbruggen *et al.* (1991). In particular, anaerobic processes are analysed in Jones *et al.* (1992) and alternating processes in Zhao *et al.* (1994a).

- A number of software packages are available such as MINITAB and the MATLAB Identification Toolbox (see MINITAB and MATLAB in the References).

- For more information on identification in general you can consult many good references such as Ljung (1987), Ljung & Söderström (1983), Johansson (1993), Söderström & Stoica (1989), and Stearns & David (1988).

7.10 Exercises

1. MATLAB Exercises file `kla_airflow_data.mat` contains K_La and air flowrate data. Plot the sum-of-squares response surfaces for the parameters and use linear regression to fit the following models. Evaluate your fits fully.

   a) $K_La = a\, q_A + b$

   b) $K_La = a\, \sqrt{q_A}$

   c) $K_La = a\, \left(1 - e^{-\frac{q_A}{b}}\right)$

   where q_A is the air flowrate and a and b are parameters.

2. MATLAB files `inactivation_data_a.mat` to `inactivation_data_c.mat` each contain three sets of concentration vs time data for the inactivation of bacteria by chlorine. The three files are data obtained using different contactors Pernitsky *et al.* (1995). The proposed model for the data is: $\log \frac{N_t}{N_0} = -a\, t^b$ where N_t is the concentration at time t and a and b are parameters. Plot the sum-of-squares response surfaces for the parameters and use linear and nonlinear regression to fit the model and evaluate your fits fully for each contactor.

3. The MATLAB file `acetate_data.mat` contains step test data for acetate concentration in the anaerobic zone of a BNR plant when the influent flowrate was increased from 70 to 140 (column one is time and column two is concentration). Use the techniques discussed in this chapter to determine parameters for first and second order models. Then simulate the step test with your model and compare it to the test data.

4. The MATLAB file `evaporator_data.mat` contains time series data from a forced circulation evaporator. The first column is time in minutes, the second column is the input data (calandria steam pressure), and the third column is

output data (product concentration). Either use the MATLAB Identification Toolbox, or the programs in the WWT Toolbox distributed with the book, or formulate your own least squares regression program, for example based on the program `Example_7_8.m`, to fit the following time series models to the data:

$$y_t = a_1 \, y_{t-1} + b_6 \, u_{t-6} \tag{7.47}$$

$$y_t = a_1 \, y_{t-1} + a_2 \, y_{t-2} + b_5 \, u_{t-5} \tag{7.48}$$

$$y_t = a_1 \, y_{t-1} + a_2 \, y_{t-2} + a_3 \, y_{t-3} + b_4 \, u_{t-4} \tag{7.49}$$

Then obtain your model prediction by feeding the input time series (second column) into your models. Compare the predicted and actual outputs and recommend a model.

5. If you are feeling brave, repeat the previous exercise using only half the data to fit your models. Then compare the model and actual outputs for the second half of the data not used for fitting.

Chapter 8

Fitting and Validating Models

In previous chapters we have introduced mechanistic and black-box models, designing experiments, preparing the data, and the principles of parameter estimation. It remains to perform the fitting task, or parameter estimation, and to validate the results. We will also address some specific issues related to strategies for fitting models of wastewater plants.

8.1 General Strategies

There are two competing validation strategies, *separate* fitting and validation and *integrated* fitting and validation. The fitting task obtains estimates of the model parameters. The validation task checks the predictive capability of the resulting model.

In effect it checks for *overfitting*, using too many parameters. Overfitting improves the fit of the fitting data, but reduces the ability of the model to generalise or predict other data sets. We will discuss both the separate and integrated strategies and their claimed advantages and disadvantages. The choice is up to you.

8.1.1 Separate Fitting and Validation

This involves dividing the available data into two data sets, one for fitting and one for validation. Regression is used to obtain parameter estimates from the fitting data set, usually one half to two thirds of the available data. Then the model is used to predict the validation data set. The validation residuals, the difference between the predicted and actual data, are then compared to the fitting residuals using an appropriate statistical test. An F-test of the variances can be used.

8.1.2 Integrated Fitting and Validation

There are a number of arguments against splitting the available data into two data sets. These arguments are especially relevant when the amount of data is limited.

- a smaller fitting data set results in less accurate parameter estimates,
- it is more difficult to ensure that smaller data sets are truly representative of the population,
- if the two data sets *are* representative, and if the fitting is done properly, then the validation test will pass by definition, and
- if the two data sets *are not* representative, then either the fitting or the validation test or both are meaningless.

Integration involves fitting the model to all the available data at least twice with models having different numbers of parameters. After the first fit one or more parameters are removed and the model is refit to all the data. The two models are compared by an F-test of the two sets of residuals. This is repeated until the models are significantly different.

Parameters to be removed are selected from the parameter confidence limits or from the correlation between parameters. If a parameter has large confidence limits, it suggests the parameter is insensitive and could be set to a constant value. If two parameters are highly correlated, either they should be combined into one parameter or one should be set to a constant value.

Example 8.1 (Fitting and validation strategies) *The MATLAB file called* `evaporator_data.mat` *contains 250 time series data points from a forced circulation evaporator. The MATLAB program* `Fit_Evaporator_Model.m` *performs a least squares regression to fit the following time series model:*

$$y_t = a_1\, y_{t-1} + a_2\, y_{t-2} + b_5\, u_{t-5} \tag{8.1}$$

to the data using a selected number of the data points. Table 8.1 shows the results for different numbers of points used for the fitting. It is clear that the parameter values vary and in particular the confidence limits increase as fewer points are used. All four model fits described 98.5% of the total variation.

8.1.3 Parameter Estimation Strategies

The "bull at the gate" strategy too often applied to parameter estimation is to derive a set of equations, gather some plant data, stuff both into a nonlinear least squares package and sit back and watch. Even worse of course is believing what eventually comes out.

So what is wrong with the bull's strategy? We have identified a number of issues in previous chapters that should be summarised here:

Points fitted	Parameter values	Confidence limits	Prediction error (mean %)
250	$a_1 = 1.673$ $a_2 = -0.825$ $b_5 = 0.0196$	0.053 (3.2%) 0.048 (5.8%) 0.0026 (13.1%)	0.43
150	$a_1 = 1.676$ $a_2 = -0.824$ $b_5 = 0.0188$	0.069 (4.1%) 0.062 (7.5%) 0.0032 (16.9%)	0.56
100	$a_1 = 1.690$ $a_2 = -0.834$ $b_5 = 0.0183$	0.091 (5.4%) 0.084 (10.0%) 0.0043 (23.5%)	0.70
50	$a_1 = 1.658$ $a_2 = -0.796$ $b_5 = 0.0179$	0.165 (9.9%) 0.147 (18.4%) 0.0081 (45.5%)	1.03

Table 8.1: Results of fitting with different amounts of data.

- Raw mathematical models frequently have too many state variables to be observable and too many parameters to be identifiable given the available plant measurements.

- Many model parameters are insensitive - over a wide range of values they have only a minor effect on state variable transients (see Chapter 3).

- Plant data frequently has excitation problems - it may not excite some state variables at all and may excite too narrow a range of frequencies (see Chapter 6).

- The quality of plant data can have profound effects on the results from least squares regression, the effects of outliers being particularly pronounced (again see Chapter 6).

- Regression algorithms suffer from convergence problems with large numbers of parameters (see Chapter 7).

The effects of these problems frequently results in unrealistic parameter estimates and very large confidence regions ("garbage in - garbage out"). The correct parameter estimation strategy involves a couple of simple principles - "divide and conquer" (examine and exploit the structure of the equations) and "a little forethought saves a lot of analysis" (design good experiments). Figure 8.1 shows some of the sources of parameter values available to the modeller.

The job of the modeller is to choose the appropriate source for the different parameters in his model. Some of the techniques which can help him or her make an informed choice are listed below.

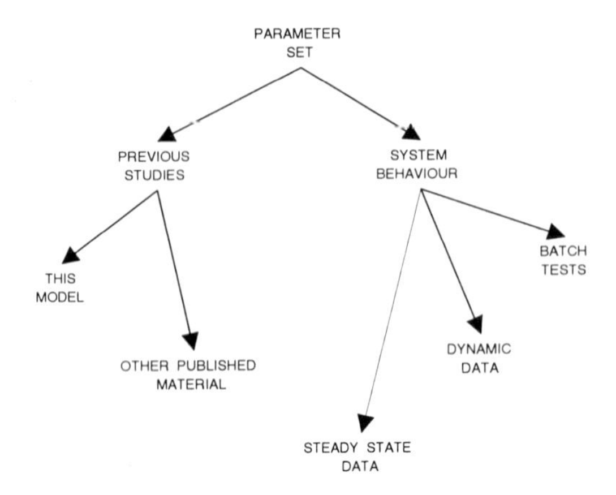

Figure 8.1: Some sources of parameter values.

- Divide parameters according to their sensitivity. Set insensitive parameters to reasonable literature values, and then simply ignore them.

- Divide state variables according to their response times and simplify the equation set. For very slow states simply set them constant and remove their equation. For very fast states convert their differential equation:

$$\frac{dx}{dt} = f(x, u, d, p) \tag{8.2}$$

  to an algebraic equation:

$$f(x, u, d, p) = 0 \tag{8.3}$$

- Examine the equations (using a sensitivity analysis where necessary) to determine the most sensitive connectivity between measured states and parameters.

- Where possible design experiments that excite subsets of the state equations to reduce the number of measurements required and the number of parameters to be fit.

- Be careful to measure or minimise the effect of variables external to the equation set that you are fitting.

Reducing the problem to a series of small fitting tasks and carefully designing experiments will almost always reward you with good parameter estimates with small confidence regions.

8.2 Specifics in Wastewater Fitting

For the activated sludge process, the model fitting can be divided into the following subtasks:

- wastewater characterisation
- fitting hydraulics
- fitting oxygen transfer
- fitting kinetic parameters
- fitting clarifier parameters
- fitting settling flux curves

These subtasks can be considered independent, either because they are independent, or because they react in significantly different timescales as discussed above and in Chapter 2. Some of them are also independent of the real plant as well.

8.3 Wastewater Characterisation

The selection of the fractions used to characterise the waste depends upon the stream descriptions used by the model and the practicalities of making measurements. The stream components used by the IAWQ AS Models are listed in Table 8.2 showing the magnitude of the problem.

Measuring all of these fractions in the laboratory is extremely difficult, let alone even contemplating online measurement. In addition, we have 5-10 streams of interest in a typical BNR treatment plant. This illustrates the identifiability problem with these models. Various simplified characterisations have been proposed with correspondingly simplified models. A typical feed characterisation for municipal wastewater is shown in Table 8.3.

Sensitivity studies have shown that a good characterisation of the COD, nitrogen and phosphorus components in the influent are very important to the performance of the IAWQ AS Models.

Characterising the feed to a plant is further complicated by the inherent disturbances. Daily, weekly and seasonal variations occur in many of the important component fractions. However, in many cases it can be shown that there is a high degree of correlation between the components (see Chapter 15) and as little as one measurement can be used to predict these regular variations. Shock loadings, typically from industrial sources, are by their nature impossible to predict and measurements must be used where early compensating control action is necessary or desirable.

IAWQ AS Model No1	IAWQ AS Model No 2
Soluble Components:	
Readily biodegradable RBCOD, S_S	Fermentable SBCOD, S_F
	Fermented RBCOD, S_A
Inert soluble COD, S_I	Inert soluble COD, S_I
Ammonium+ nitrogen, S_{NH}	Ammonium+ nitrogen, S_{NH4}
Nitrate+ nitrogen, S_{NO}	Nitrate+ nitrogen, S_{NO3}
Soluble organic nitrogen, S_{ND}	
	Nitrogen, S_{N2}
	Inorganic phosphorus, S_{PO4}
Dissolved oxygen, S_O	Dissolved oxygen, S_{O2}
Alkalinity, S_{ALK}	Alkalinity, S_{ALK}
Insoluble Components:	
Insoluble SBCOD, X_S	Insoluble SBCOD, X_S
	Internal PHA, X_{PHA}
Insoluble organic nitrogen, X_{ND}	
	Internal phosphorus, X_{PP}
Heterotrophic organisms, $X_{B,H}$	Heterotrophic organisms, X_H
Autotrophic organisms, $X_{B,A}$	Autotrophic organisms, X_A
	Phosphorous organisms, X_{PAO}
Unbiodegradable particulates, X_P	
Inert solids, X_I	Inert solids, X_I
	Total suspended solids, X_{TSS}

Table 8.2: IAWQ AS Model wastewater components.

Component	Value (mg/l)
Soluble Components:	
Fermentable SBCOD, S_F	20 - 250
Inert soluble COD, S_I	20 - 100
Fermented RBCOD, S_A	10 - 60
Ammonium+ nitrogen, S_{NH4}	10 - 100
Nitrate+ nitrogen, S_{NO3}	0 - 1
Nitrogen, S_{N2}	-
Inorganic phosphorus, S_{PO4}	2 - 20
Dissolved oxygen, S_{O2}	0 - 0.5
Alkalinity, S_{ALK}	-
Insoluble Components:	
Insoluble SBCOD, X_S	80 - 600
Internal PHA, X_{PHA}	0 - 1
Internal phosphorus, X_{PP}	0 - 0.5
Heterotrophic organisms, X_H	20 - 120
Autotrophic organisms, X_A	0 - 1
Phosphorous organisms, X_{PAO}	0 - 1
Inert solids, X_I	30 - 150
Total suspended solids, X_{TSS}	-

Table 8.3: Typical feed characterisation.

8.4 Fitting Hydraulics

A key question in the modelling of the hydraulics is whether well-mixed tanks are a reasonable representation of the real plant behaviour. We discussed this earlier in Chapter 3. If well-mixed tanks can be used, then do we need one, two or more in series or in some other configuration.

This question can only be answered by residence time tests on the real process. Typically a flourescent dye, such as Rhodamine, is injected into the vessel inlet stream and its concentration is measured in the vessel outlet stream. Figure 8.2 shows a typical result from a BNR pilot plant tracer test.

Fitting a residence time curve can be done independently of the biological processes and involves formulating a simple model and performing a parameter estimation. The simple model will involve some configuration of well-mixed tank models (Equation 8.4) and plug-flow tank models (Equation 8.5) together with recycle and bypass flows.

$$\tau \frac{dC_o}{dt} + C_o = C_i \tag{8.4}$$

$$C_o(t) = C_i(t - \tau) \tag{8.5}$$

where C_i is the inlet tracer concentration, C_o is the outlet tracer concentration and τ is the residence time.

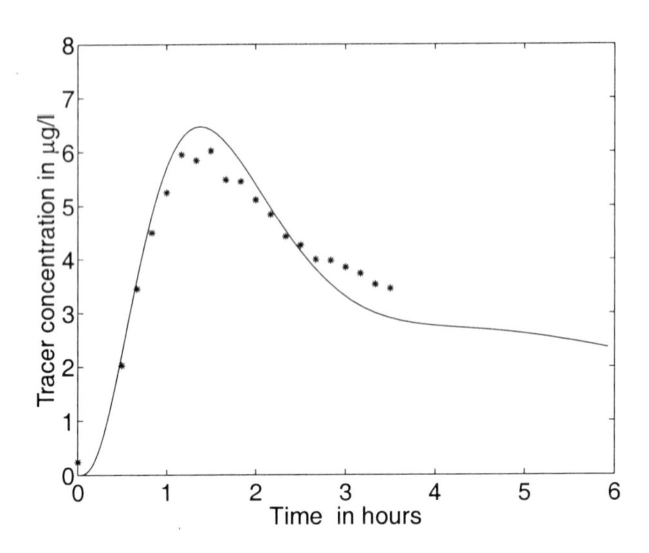

Figure 8.2: Aerobic zone fitted RTD test.

The configuration will be suggested by the process configuration of tanks and recycles, and the form of the residence time curve. The objective is to determine a configuration and set of residence times that will best match the measured residence time curve. Initial estimates of the residence time parameters can be obtained from known volumes and flowrates.

Without recycles, the number of stirred-tanks of equal size that best approximates an RTD curve can be roughly estimated from the peak time and the known residence time using the relation:

$$\text{peak time} = \frac{n-1}{n} \times \text{residence time} \tag{8.6}$$

Recycle flows destroy such simple relationships and have a major effect on the sensitivity of the residence time curve to the tank configurations (see Figure 8.3). If care is taken choosing the injection and sampling points, for example avoiding recycles, the analysis and interpretation of the data can be simplified. Small superimposed peaks generally mean some short circuiting or bypassing within the tanks. This can be represented by stirred tanks and a plug-flow vessel in a parallel configuration (see Figure 8.3).

These figures were prepared using a MATLAB special purpose simulator called `rtd`, developed specially for simulating different tank configurations to match experimental tracer curves (see Appendix A for a description of the program).

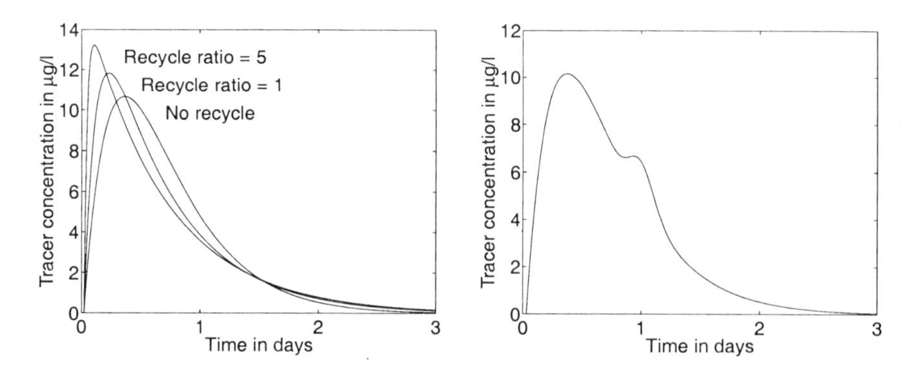

Figure 8.3: Recycle and bypass flow effects on RTD curves.

8.5 Fitting Oxygen Transfer

The mass balance for oxygen in a well-mixed region of a biological reactor was derived in Chapter 3 and can be written as:

$$V\frac{dS_O}{dt} = q(S_{O,in} - S_O) + K_La(S_{O,sat} - S_O)V + r_OV \tag{8.7}$$

where r_O is the negative of the oxygen uptake rate r_{OUR}. The first term on the right is small compared to the second and third terms and can be neglected. Then we can write expressions for the rate of change of DO concentration with the air turned "on" (Equation 8.8) and with the air turned "off" (Equation 8.9):

$$\frac{dS_O}{dt}|_{ON} = K_La(S_{O,sat} - S_O) - r_{OUR} \tag{8.8}$$

$$\frac{dS_O}{dt}|_{OFF} = r_{OUR} \tag{8.9}$$

From these, and assuming r_{OUR} is constant, we can determine K_La from the relation:

$$K_La = \frac{\frac{dS_O}{dt}|_{ON} - \frac{dS_O}{dt}|_{OFF}}{S_{O,sat} - S_O} \tag{8.10}$$

If we can measure or estimate the quantities on the right hand side, and if r_{OUR} is constant during the measurement period, then we have an independent way of determining the K_La parameter. The rates of change can be conveniently measured continuously online where the DO level is controlled by an on/off controller with a sufficient deadband. Perhaps the most uncertain value is $S_{O,sat}$, which is best determined by laboratory measurement and may require temperature compensation. The DO probe also needs to be well maintained.

Example 8.2 (OUR Estimation) *The left plot in Figure 8.4 shows DO responses from an aeration system with on/off control. The air came on at a DO reading of 2 mg/l and went off at a DO reading of 3 mg/l. The right plot in Figure 8.4 shows filtered and unfiltered estimates of OUR calculated according to Equation 8.9.*

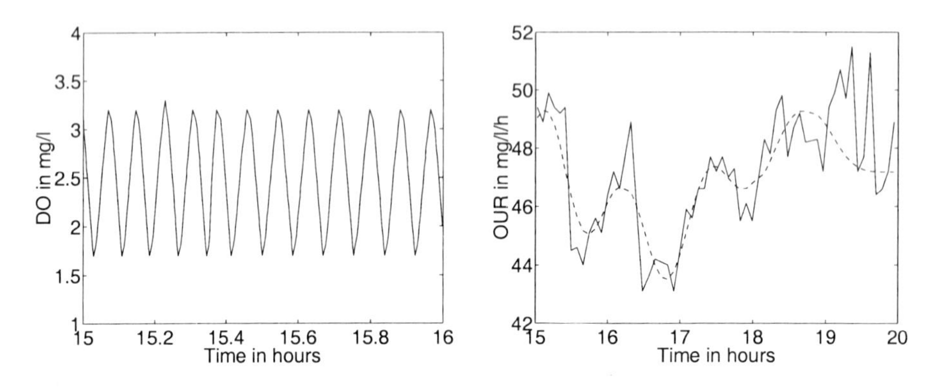

Figure 8.4: DO responses and OUR estimates.

If air flowrate measurements are also taken, then it is possible to develop an empirical relationship between the K_La parameter and air flowrate. Typical expressions used are:

$$K_La = a\, q_A + b \tag{8.11}$$

$$K_La = a\sqrt{q_A} \tag{8.12}$$

$$K_La = a\left(1 - e^{-\frac{q_A}{b}}\right) \tag{8.13}$$

Unless you need to estimate the air flowrate, for example for optimisation objective functions, it is possible to avoid the whole issue by assuming you have perfect DO controllers and using the DO setpoint as the model input variable. This can be a useful model simplification in many instances as the DO dynamics are relatively fast. The oxygen uptake rate r_{OUR} can be determined at the same time and can be used as a diagnostic indicator as discussed in Part B on Diagnosis. In carbon removal plants this can be related back to the heterotrophic growth rate, but in BNR plants the relationship is too complex to relate it to any one growth rate.

The estimation of oxygen uptake rate is further discussed in Example 12.9.

8.6 Fitting Kinetic Parameters

The observability problem is particularly severe in activated sludge plants, particularly when high-order models such as the IAWQ AS Models are used. These problems have generated interest in reduced order models such as those in Chapter 4. There have also been many proposals for independent experiments to determine some of the kinetic parameters, such as those in the IAWQ AS Model reports. However, there will always be the doubt and discussion as to whether the microorganisms behave the same in the environment of independent experiments compared to the *in situ* process environment.

In both independent and *in situ* environments, the parameter correlation problem inherent to the Monod kinetic model (discussed in Chapter 7) remains a serious problem. The problem can be alleviated, but not removed, if the right combination of measurements is made, in particular for heterotrophic parameters if OUR is included in the measurements. The problems can also be reduced if the sensitivity of parameters is used as a guide. Less sensitive parameters can be set to published values and more sensitive values are adjusted to achieve a fit to the data. This was discussed at the beginning of this chapter and in Chapter 3 Example 3.10 and some example rankings were given in Tables 3.6 and 3.7. In general terms:

- Saturation constants are frequently insensitive. Consider the form of the Monod kinetic expression:

$$r_S = \hat{\mu}\frac{S_S}{K_S + S_S}X_H \tag{8.14}$$

  If S_S is large relative to K_S then the Monod term will be "switched on", approximately equal to one, and changing the value of K_S will have a minimal effect. However $\hat{\mu}$ will have an effect.

  If S_S is small relative to K_S then the Monod term will be "switched off", approximately equal to zero, and changing the value of K_S will again have a minimal effect. In this case $\hat{\mu}$ will will also have a minimal effect.

  The value of S_S must be approximately equal to K_S before K_S has a significant effect.

- As mentioned in the previous point, maximum growth rates will be insensitive if any of the Monod terms in the rate expression are close to zero (very low substrate concentrations) or if any of the inhibition terms are close to zero (relatively high substrate concentrations).

- Saturation constants are highly correlated with maximum growth rates, particularly if they are large relative to the corresponding substrate concentration. The Monod kinetic expression then becomes:

$$r_S = \hat{\mu}\frac{S_S}{K_S + S_S}X_H \approx \frac{\hat{\mu}}{K_S}S_S X_H \tag{8.15}$$

- Maximum growth rates are highly correlated with biomass concentrations which are generally unknown. Both can not be fit together to either batch-test data or plant data.

- Maximum growth rates can also be highly correlated with yield constants, decay rates and oxygen transfer coefficients. For example, if we consider the following rate expression with growth and decay approximately in balance:

$$r_H = \hat{\mu}\frac{S_S}{K_S + S_S}X_H - bX_H \approx 0 \tag{8.16}$$

so that:

$$\frac{b}{\hat{\mu}} \approx \frac{S_S}{K_S + S_S} \tag{8.17}$$

- yield constants can also be correlated with oxygen transfer coefficients.

Sensitivity and correlation problems depend strongly on the experiments performed, the measurements taken, and the experimental conditions. Examining the component balances and careful experimental planning can reduce the size of parameter confidence regions markedly.

A factor that can compound the fitting of kinetic parameters using complete plant models is the frequently made assumption that there is no biological activity in the settler and return sludge line, whereas there is much evidence to the contrary.

8.7 Fitting Clarifier Data

Two aspects of clarifier modelling are difficult and often use empirical expressions fitted to experimental data. These are firstly predicting the effluent suspended solids, and secondly relating settling characteristics to biomass characteristics. We will discuss the first in this section and the second in the next section. Techniques for modelling the clarification section (above the sludge blanket) are less well advanced than those for modelling the settling and thickening sections (Chapter 4). Frequently, the effluent suspended solids concentration is empirically related to the clarifier influent flowrate.

Example 8.3 (Effluent Suspended Solids Prediction) *Experimental results shown in Figure 8.5 were collected from a pilot plant at the Wastewater Technology Centre, Burlington, Ontario, Canada. The influent flow was varied manually and the effluent suspended solids concentration measured.*

Initially, a model was fitted to the data assuming that the suspended solids concentration varied as a function of the flowrate in the same way for both increasing and decreasing flowrates. As shown on the left in Figure 8.6, there was poor agreement between the estimates and measurements. As suspended solids concentrations changed more quickly for an increase in flowrate than for a decrease, the need for two different models was indicated. On the right in Figure 8.6, one model has been fitted for an increasing flowrate and another for a decreasing flowrate. The results from the combined models gave satisfactory results.

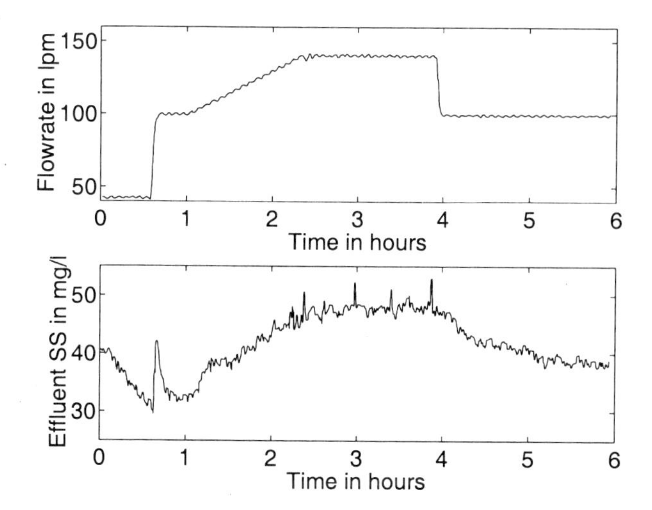

Figure 8.5: Clarifier flowrate and effluent suspended solids data.

The hydraulic model for clarification indicates that a dynamic relationship exists between the influent flowrate to the clarifier and the effluent SS concentration such that the effluent SS concentrations do not change instantaneously following a flowrate disturbance. This fact has important consequences for pumping practices; large and frequent changes in pumping rates will have detrimental effects on treatment efficiency.

8.8 Fitting Settling Flux Curves

In Section 4.6 we described the need for the flux curve in modelling the settling behaviour in the clarifier. The equation used to describe this curve was given as:

$$f_i = x_i C_1 e^{-C_2 x_i} \tag{8.18}$$

This equation has two constants C_1 and C_2, which must be determined. A series of batch settling tests, as outlined in Standard Methods (Greenberg *et al.* (1992)) are performed. These are performed at the concentration of the mixed liquor in the aeration basin, at one higher concentration and at several more dilute concentrations.

The height of the sludge liquor interface is plotted against time. The velocity of the fall of this interface at the beginning of the test, when it falls at constant velocity, can then be determined. In practise you need only enough points to get a good estimate of the initial settling velocity. The constants for the model can then be determined by plotting the natural logarithm of the velocity against the initial concentration. The constant C_2 is derived from the the slope of the line of best fit, and the constant C_1 is derived from the Y axis intercept.

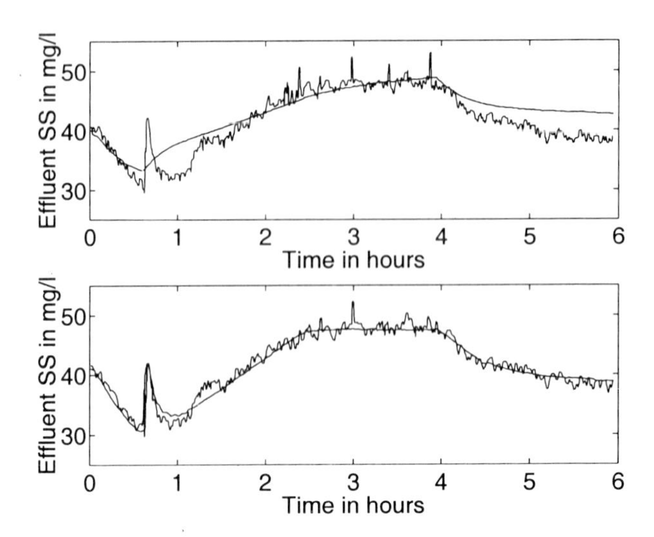

Figure 8.6: Single and dual models for effluent suspended solids.

Example 8.4 (Fitting settling test data) *Some data collected from a treatment plant is plotted in Figure 8.7. The concentration of the mixed liquor in the aeration basin was 4500 mg/l. This was concentrated to give the 6400 mg/l liquor. It was diluted with clarifier effluent to give the dilute concentration tests. The left plot in Figure 8.8 shows a line of best fit through the log-linear plot of initial settling velocity against initial concentration. The constants were derived from the slope:*

$$C_2 = -slope = 0.000604 \; l/mg \tag{8.19}$$

and the intercept:

$$C_1 = e^{intercept} = e^{1.9} = 6.71 \; m/h \tag{8.20}$$

The flux is the concentration of the liquor times the initial velocity of the fall of the sludge liquor interface. This is plotted against the initial sludge concentration. The points corresponding to the data in Figure 8.7 are plotted in Figure 8.8 together with the fitted flux curve.

Figures 8.7 and 8.8 were calculated by the MATLAB program `Fit_Flux_Curve.m` *(you can use it with your own data).*

Of course, conducting the required settling tests takes a great deal of time and it may not be convenient to do them as often as is necessary to keep them up to date with the current wastewater. To update these constants, an approximation involving an empirical correlation of the constants to the SVI has been used. The SVI is a much simpler settling test usually done on a regular basis.

It may then be possible to do occasional settling tests to check the correlation.

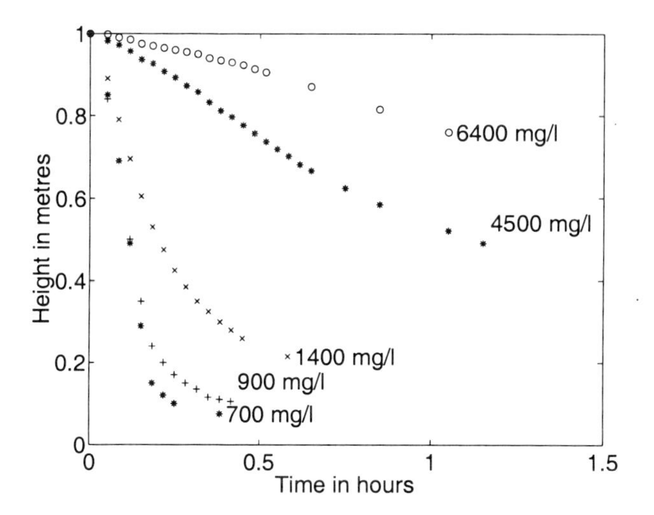

Figure 8.7: Settling test data.

8.9 Summary

In this chapter we have tried to give a pointer to some of the strategies used to obtain and validate parameter estimates. We have looked also at how we can obtain parameters which are at least partially independent of the operating process. If this can be done, it reduces the scale of the problems associated with fitting parameters to plant data.

Of course there is the uncertainty associated with whether things happen in the real plant in the same way that they happen in independent tests. But then fitting larger numbers of parameters to real plant data also increases the uncertainty in the parameter estimates. Our *gut feel* is that the latter uncertainty is greater, but others may feel otherwise.

After we have successfully validated the model, we can proceed to use it for the intended purpose. However, we must remember the limits set by the specification. We must particularly remember that we are modelling a biological system which has the disturbing habit of adapting to changing conditions without necessarily sending us a fax or e-mail. Periodic validation will be required and re-estimation of the parameters may be required. The frequency will depend upon the local circumstances and the purpose of the model. Continuous online adjustment of some parameters may be required and is discussed in Chapter 12.

Finally, you are "treading where angels fear to tread" if you use a model developed for one plant on another plant, no matter how similiar they seem to be.

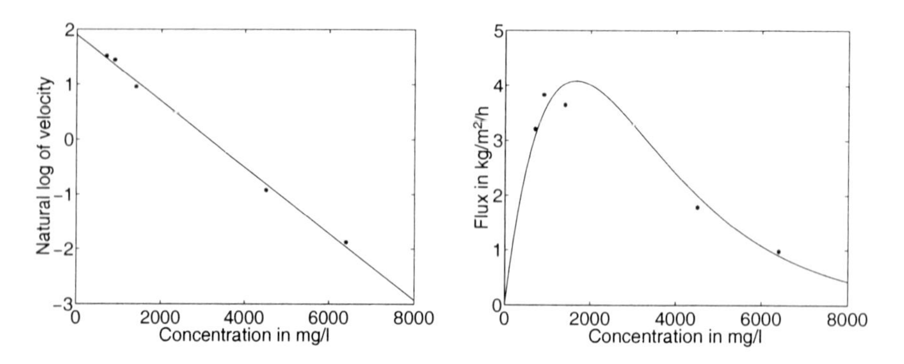

Figure 8.8: Regression plot and fitted flux curve.

8.10 Further Reading

- Basic books on statistics such as Schenck (1968), Cooper (1969), Berthouex & Hunter (1981), Draper & Smith (1981), Box *et al.* (1978) and Walpole & Myers (1989).

- For more information on wastewater characterisation see Henze (1992) and Naidoo *et al.* (1998).

- For fitting hydraulics see Levenspiel (1972) and Newell *et al.* (1997).

- For distributed parameter modelling of activated sludge tanks see Lee *et al.* (1999).

- For fitting of oxygen transfer see Lindberg & Carlsson (1996b), Carlsson *et al.* (1994), Holmberg (1986), Holmberg (1987) and Holmberg & Olsson (1985).

- For fitting of kinetic parameters see Henze *et al.* (1987a), Ayesa *et al.* (1993) and Henze *et al.* (1994). There are also many proposed techniques to determine the kinetic parameters such as Cech *et al.* (1985), Ekama & Marais (1979), Marais & Ekama (1976), Kappeler & Gujer (1992) and Kristensen *et al.* (1992).

- For fitting clarifier parameters see Krebs (1991), Krebs (1995) and Olsson & Chapman (1985). A function to relate clarifier non-settlable suspended solids to bioreactor conditions is found in Dupont & Henze (1992).

- For fitting settling flux curves see Greenberg *et al.* (1992) and Hasselblad & Xu (1996). For relating flux model parameters to SVI see Härtel & Pöpel (1992).

- The data in Figure 8.2, Figure 8.4 and Exercise 1 was collected from pilot plants at the Technology Development Centre of Sydney Water in Liverpool, Sydney.
- The RTD data can also be found in a paper by Newell *et al.* (1997).
- The settling curve data in Figure 8.7 was collected by Lisa Hopkins (Advanced Wastewater Management Centre, The University of Queensland).

8.11 Exercises

1. Use the MATLAB residence time distribution simulator `rtd` (for instructions see Appendix A) to fit a residence time model to the data in the text file `Exercise_8_1.dat`. The data was a tracer test on two anoxic/anaerobic tanks in series with suspected mixing problems. They each had a nominal volume of three cubic metres and the throughput was 100 litres per minute.

2. Use the MATLAB program `Fit_Kla_Model.m` to fit a variety of models for the relationship between K_La and valve position u (Table 8.4). The equation that is fit was shown above as Equation 8.7. The model is set up in the function in file `klaf.m` and the data files `do119.dat` and `vp119.dat` are required.

Model	Equation
Linear A	$K_La = p_2u$
Arctan	$K_La = tan^{-1}(p_2u)$
Square Root	$K_La = p_2\sqrt{u}$
Linear B	$K_La = p_2u + p_3$
Exponential	$K_La = p_2\left(1 - e^{-\frac{u}{p_3}}\right)$

Table 8.4: K_La vs. valve position models.

   The oxygen uptake rate is assumed constant and is estimated as the first parameter p_1.

3. The data file `anflow.dat` contains three columns of data (time in minutes, acetate in mg/l, phosphate in mg/l) collected from the anaerobic zone of a BNR plant. The influent wastewater flowrate increased from 70 to 140 kl/h at 10 minutes and back to 70 kl/h at 100 minutes.

   The volume of the anaerobic tank is 75 kl and 45 kl/h of return activated sludge is also fed to the tank. Influent wastewater contains 90 mg/l of fermentables (S_F), 110 mg/l of acetate (S_A) and 14 mg/l of soluble phosphate (S_{PO4}). Assume none of these nutrients are found in the RAS stream. At time zero the effluent from the tank contained 50 mg/l of fermentables, 24 mg/l of acetate and 50 mg/l of soluble phosphate.

(a) Write a simplified model with mass balances for the three nutrients (see Table 4.3 and 4.4) assuming a single well-mixed tank. Clearly state your assumptions.

(b) Considering the structure of the equations and the available data, select your parameters to be fit to the experimental data. Write down your reasoning and any calculations performed.

(c) Write a computer program to perform the parameter estimation (you can use program `Fit_Kla_Model.m` from the previous exercise as a template for the program structure).

4. Consider the aerobic zone of a BNR activated sludge process represented as a series of three well-mixed tanks.

   (a) Write the model equations for dissolved oxygen and the major nutrients, carefully noting the assumptions you make (refer to Chapter 4).

   (b) Assume typical values for the influent concentrations and the concentrations in each zone.

   (c) Examine the structure of the equations and the effective form of the kinetic expressions given the assumed concentrations (you could draw a digraph if you wished).

   (d) Propose a set of experiments specifying the parameters you aim to fit, the measurements to be taken, the location of the measurements (which tank), the excitation you propose to apply, the likely extraneous variables (disturbances) and how you propose to compensate for them.

5. The data file `settling.dat` contains the results of a series of settling tests. There are ten measurements for each test performed at initial biomass concentrations of 6, 4, 2, 1 and 0.5 g/l respectively. The first column in the file is the observation time in hours from the start of each experiment and the second column is the sludge blanket height in metres.

   Using the techniques illustrated above in Section 8.8 and in the MATLAB program `Fit_Flux_Curve.m`, fit the flux models in Table 8.5 to the data and select the best model. Discuss the reasons for your choice of model.

Model	Equation
Cho Model	$v_s = c_1 x^{-c_2}$
Vesilind Model	$v_s = c_1 e^{-c_2 x}$
Takacs Model	$v_s = c_1 \left(e^{-c_2(x-x_{ns})} - e^{-c_3(x-x_{ns})}\right)$

Table 8.5: Settling flux models.

Chapter 9

Simulators

We have seen that the knowledge of a plant's dynamics can be expressed in dynamic models. Such models can be packaged in special software platforms that we call simulators. Simulators have been used extensively in many disciplines over the years and are powerful tools both for design, planning, process analysis, operator guidance and education and training. Simulators can also be considered as packages of knowledge of the dynamics of the process.

Simulate means to mimic or imitate the behaviour of the system. The computer is used as the experimental platform. By using the simulator we can study the effects of different parameter values, model complexities and initial conditions. Simulation is also a good aid for training operators. For instance, pilots for aeroplanes are trained on a regular basis using simulators.

In this chapter we will explore some possibilities of simulators, but will also advise you to use them with care. They will never be better than the models and data they are using.

During the last ten years there has been a significant increase of the interest in simulators for wastewater treatment. Efforts in modelling as well as the general hardware and software development have together contributed to this. Moreover, the simulators are considered important tools to better understand the complexities of nutrient removal systems. Furthermore, the integration of several sub-processes within a plant makes it necessary to analyse all the interactions. In a larger scale, the interactions with the sewer network and receiving water are essential to understand, if the possibilities of the total wastewater system are to be utilised to their full extent.

Since a wastewater treatment system is truly dynamic, the time frame is crucial. This has been discussed in detail in Chapter 2.

In Section 9.1 we try to define the various categories of people who could use simulators. Various users have different expectations of the simulators. Fundamental features of simulators are outlined in Section 9.2. Since dynamic models are so essential for simulation, we once more return to some basic concepts in

Section 9.3. Any simulator user has to be aware of the model definitions and limitations. The general features of simulators which are important in assessing software are discussed in Section 9.4 and some selection criteria are listed in Section 9.5. In Section 9.6 we describe a few representative general purpose simulators. There are many more available in the market. Over the last few years, several application specific simulators for wastewater treatment systems have been developed. Some of them are discussed in Section 9.7. Finally in Section 9.8 we look at the dangers of packaged solutions and what we see as the future needs of simulators.

9.1 Who Needs Simulators?

Given appropriate models, simulators are powerful tools for process engineers or operators. They can give advice on possible control actions and can be used for making predictions. Probably the model accuracy in wastewater treatment systems will never be such that reliable quantitative predictions can be made over a long time (days to weeks), but still the couplings between different process units can be examined and the values or the trends will be of the right order of magnitude.

Simulation can be classified into *discrete event simulation* and *simulation of dynamic systems.* A typical example of discrete event simulation is simulation of queues, for example in manufacturing systems, or computer operating systems. Here we will exclusively talk about simulation of dynamic systems.

Simulators have been used for some time in many applications. Flight simulators for the training of pilots have been in use for many years. Full scale simulators of nuclear reactor operations and conventional power stations are used on a routine basis for training and for decision support. Over the last twenty years a number of software platforms have been developed to help researchers, designers, control engineers and many others to evaluate properties of dynamic systems in areas as wide apart as engineering sciences, macro and micro economy, natural sciences, biology, ecology, medicine and pharmacy.

In wastewater treatment the need for simulators is apparent for several groups of people:

- the *designer*, who wishes to explore not only the average properties of a planned plant, but also its robustness to dynamic disturbances. A simulator can be used to evaluate different design options without having to build prototypes. It may also include advanced control concepts;

- the *process engineer*, who wishes to explore different process configurations or operating strategies for an existing plant. Perhaps the most important benefit from an accurate simulator is the ability to project the plant's response to predicted future or catastrophic operating scenarios. Extreme flow and load projections can be simulated, and the simulator can identify process units that are stressed and those that have spare capacity. Failure mode analysis, such as predicting the consequences of the malfunction of a series of pumps or other key process components can be facilitated;

- the *control engineer*, who wishes to try different control structures and algorithms for the operation of a plant;
- the *operator*, who needs a decision support system, where he can explore different what-if situations. The simulator can predict the plant behaviour for various operations;
- the *educator*, who will use the simulator for teaching plant dynamics for different categories of people, ranging from operators to researchers;
- the *researcher*, who will use the simulator as a condensed version of his knowledge. This will give possibilities to further investigate different properties of the model.

It is quite apparent that the different users want to find different answers from simulators. Some of them have to calibrate the model to an existing plant, while others can use best available guesses of the plant parameters. A control engineer needs to have the flexibility to design various control structures and algorithms, while still the process model has to be available in the simulator. The different users are interested in different time scales and have different demands for simulator flexibility. While the researcher wants to be able to change almost everything in the model, the operator at the plant wants a user-friendly on-line style of interface.

9.2 Fundamental Features

Simulation programs are today in common use. With such programs the user has to formulate the model equations, along with the total integration period, numerical integration method, variables to be printed out or plotted, etc. The simulation program takes care of:

- checking if equations are consistent;
- sorting the equations into an appropriate sequence for iterative solution;
- integrating the equations;
- displaying the results in the desired formats (tables or graphical output).

Modern simulators have simple commands for parameter or initial value changes and have several integration routines to choose from. They also have advanced output features to present the results in easy readable graphic formats.

There are many aspects of a simulator. The modules are like links in a chain and each of them has to be reliable for the user. They include:

- *the dynamic model*: - does it truthfully represent the real system?
- *the model representation*: - how easy is it to build up a specific plant model?
- *real data*: - can it be included for model calibration?
- *the numerics*: - are the numerical integration routines sufficiently accurate?
- *the data representation*: - how flexibly can output data be presented and analysed?
- *the user interface*: - how easy is it to get started and to use the simulator?

We will discuss some of these aspects that are relevant for applications in wastewater treatment systems.

9.3 Models

All simulators are based on mathematical models describing the processes to be simulated. If these underlying models are not satisfactory, then neither are the results from the simulator. This obvious truth is a real problem for simulators in wastewater treatment systems.

9.3.1 Continuous Time Models

A simulator of a dynamic system assumes that the model is given in the form:

$$\frac{dx}{dt} = f(x(t), u(t), p, t) \tag{9.1}$$

as described in Section 3.1. The initial condition is assumed to be known:

$$x(t_0) = x_0 \tag{9.2}$$

The parameters p are indicated explicitly here to emphasise that they can be easily changed in a simulator. The functions $f()$ do not need to be simple equations. Sometimes they may include conditional expressions like *if-then-else*. For example: *if* the flow rate is increasing, *then* the dynamics for the suspended solids is described by one expression f_1, *else* it is described by f_2.

There are certainly dynamic systems that can not be described by the *explicit* form (Equation 9.1), that is where the derivatives are expressed explicitly on the left hand side. In the models we deal with here, however, this form of Equation 9.1 is adequate.

The outputs are defined by the expression:

$$y(t) = g(x(t), u(t), p, t) \tag{9.3}$$

A treatment plant consists of many unit processes, and each one of them may be described by the equations of the form of Equations 9.1 to 9.3. The simulator then treats each one of the unit processes as a module, or part of a model library, that will be connected to the other modules by connecting inputs and outputs.

The whole of Part A of this book has been devoted to models of different complexity. A wastewater treatment system has dynamics that cover several orders of magnitude, ranging from seconds to months. Naturally it is very difficult, and often not desirable, to have a model that incorporates the whole spectrum. Rather, different users may be interested in different time scales of the system. For example, an operator may test a number of what-if alternatives by simulation in the hourly time scale before any real action is taken. He may also examine different operating strategies for better long term operation. A designer usually wishes to examine the steady state behaviour in order to determine the plant capabilities, but may use a dynamic simulator as well to find out if the plant can tolerate disturbances in different time scales.

We have seen examples of widely different time scales:

- In Chapter 4 we studied hydraulics. A pump start will create a flow disturbance that propagates (within fractions of an hour) through the plant so that the clarifier effluent suspended solids concentration will be affected. The propagation time depends on the tank area. On the other hand, as was shown in Section 3.5, a concentration change caused by a varying flowrate is considerably slower (several hours) and depends on the tank volume.

- The transfer of gaseous to dissolved oxygen (DO) takes place within 15 to 30 minutes (Section 3.6). Consequently, a change in the air flowrate will not be noticed immediately in the DO concentration.

- Heterotrophic growth is noticeable in days, while nitrification and cell decay takes weeks.

- Annual seasonal variations produce a temperature influence on growth, particularly for nitrifiers.

Due to the widely different time scales, it is often possible to decouple the dynamics of a treatment plant. When considering hourly variations, the growth of organisms can be neglected and the biomass regarded as constant. Considering weekly and monthly changes, then much of the oxygen transfer, rapid degradation of carbon and hydraulics can be considered instantaneous.

The problem of time scale boils down to the question of the purpose of the simulation. The proper representation of the system is consequently not trivial. It is not realistic to look for the model that would represent the plant for all purposes. One has to remember that the different users need to choose from a whole spectrum of models.

A full simulation of a bio-P process can include around 20 state variables for each of the completely mixed compartments that make up the bioreactors. For a typical representation of two to four compartments per bioreactor the system includes about 160 state variables just for the biological part.

In a biofilm system, the concentration gradients into the film have to be represented by a number of layers. Each layer describes the concentration of different substrates and organisms, and the diffusion of oxygen and substrates through the layers have to be properly described. The number of states can easily be several hundred.

The weak link of many simulation models is the representation of thickening and clarification (see Section 4.6). The thickener is usually represented by a one-dimensional model of zone settling theory. Some simulators apply only a *point model* of the settler, that is, no spatial representation of the concentration at all, only an empirical thickening mechanism. Most simulation programs of treatment plants assume that the biological reactions in the final settler cease to occur (see Section 4.6). In order to get a reasonable approximation of the settler concentration dynamics, the spatial distribution has to be divided into typically 20-50 layers, each one with a homogeneous concentration. Adding biological activity models for each layer immediately brings the number of states to several thousand.

Models for rivers are mostly built up as a series of sections, each one of which is approximated as a completely mixed reactor. Apart from organism activities, such models usually include balances for algal growth. Similarly, lakes or ground aquifers can be represented by sections with a similar structure.

There is a vast literature on models for nutrient removal in aerated ponds or lagoons, and pond systems are common for both industrial and municipal use, see Vuillot & Boutin (1987) and Nie & Xu (1991). Some typical models are found in Preul & Wagner (1987) and in Sackellares *et al.* (1987). Still, an adequate description of the hydrodynamics is a major obstacle to obtaining good accuracy, and many activities are devoted to this problem, such as Wood *et al.* (1995) and Wood *et al.* (1998).

9.3.2 Discrete Time Models

Often controllers are implemented in a computer, and therefore they are time discrete. Therefore, in order to simulate a computer controlled system, the simulator should be able to describe both continuous processes, like Equation 9.1, and time-discrete descriptions of the regulators (or other parts of the process) that are time-discrete. The generic form of such a system is:

$$x(t_{k+1}) = f(x(t_k), u(t_k), p, t_k) \tag{9.4}$$

where x is the state vector and the kth sampling instant is denoted by t_k. The initial condition is given in the form of Equation 9.2. The output is calculated by:

$$y(t_k) = g(x(t_k), u(t_k), p, t_k) \tag{9.5}$$

and the next sampling instant is calculated by some function:

$$t_{k+1} = h(x(t_k), u(t_k), p, t_k) \tag{9.6}$$

In general, the sampling instants need not be equidistant and can be different in different subsystems. The outputs can only be defined at the sampling instants and not between them. To be able to interconnect continuous and time discrete systems the outputs of the time discrete systems are kept constant between the sampling instants.

9.3.3 Linearisation

Most wastewater treatment models are nonlinear, and we have seen many examples of this in the preceding chapters. Control engineers often want to linearise the plant to be controlled or at least a part of it. This enables them to perform a controller design in a simplified fashion, since there exist analytical controller design strategies for linear systems, whereas the nonlinear control system design would have to be done by trial and error. It also allows analysis of the system with respect to identifiability or parameter sensitivity.

We have seen techniques for linearisation of models in Chapter 5. It is highly desirable to do this in an automated fashion. Moreover, it needs to be done within the modelling environment, since the original nonlinear model needs to be interpreted in this process. It is crucial to be able to compare the linearised model with the original nonlinear model before designing the controller. Once the controller is designed, the control engineer wants to simulate the control behaviour of the controller when applied to the original nonlinear plant. One may want to use the linear control system design only as a first step on the way to determining an appropriate controller for the original nonlinear plant. Such a controller has the right structure, but the parameter values are only approximations.

Some of the currently available simulation environments, such as Simulink and ACSL, offer a limited model linearisation capability. A linear model of the type:

$$\frac{dx}{dt} = A\,x + B\,u \tag{9.7}$$

is obtained from the original nonlinear model (Equation 9.1) by approximating the two Jacobians:

$$A = \frac{\partial f}{\partial x} \quad B = \frac{\partial f}{\partial u} \tag{9.8}$$

through numerical differences. Since f, x and u are vectors, the Jacobians are interpreted as:

$$a_{i,j} = \frac{\partial f_i}{\partial x_j} \quad b_{i,j} = \frac{\partial f_i}{\partial u_j} \tag{9.9}$$

where $a_{i,j}$ and $b_{i,j}$ are the (i,j) elements of the A and B matrices respectively. The linearisation facility is limited at least in two ways:

- There is no control over the quality of the numerical difference approximation, and thereby over the linearisation. The problem can be arbitrarily poorly conditioned. A symbolic differentiation of the model to generate the

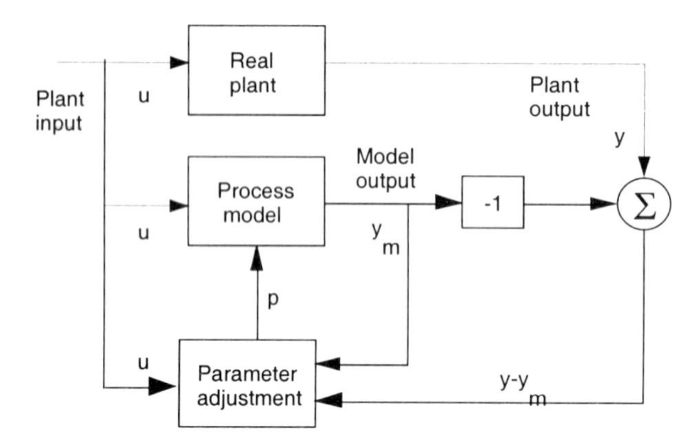

Figure 9.1: Principle for model updating, based on on-line measurements.

Jacobians may be more suitable and is entirely feasible. Particularly, in a wastewater treatment model, most of the nonlinearities can be readily differentiated.

- The approximation is necessarily local, that is, limited to a region close to an operation point $[x_o, u_o]$. If the solution leaves this region during the simulation the approximation may become meaningless.

9.3.4 Model Verification and Validation

Model verification and validation represents a large portion of Part A in this book. This is intentional, since they are such a crucial step in obtaining reliable simulations.

Accuracy can be a very difficult thing to define. Some of the models may be adequate to predict normal variations of nutrient removal. However, they may predict poorly if the process is gradually deteriorating towards a bulking condition. Similarly, they may describe normal hydraulic conditions well but be inaccurate during rain storms or other severe hydraulic disturbances.

A model for design has to capture the *average* behaviour. However, an operator may wish to use a model for advice on how to handle specific disturbances. This again may demand both a different time scale and a different kind of model verification and validation.

In some simulators there is a possibility to update the model parameters, based on available on-line measurements. The principle is illustrated in Figure 9.1 and will be further discussed in Section 12.4.

This feature may be used for different purposes. If the model is considered reliable, then any significant deviation between the model and the real process can be used as an indicator of an operating state that is undesirable. Such use is still not common in wastewater treatment systems, since the process knowledge is still insufficient. Moreover, many of the process parameters are truly time-varying, since the microbial culture will change according to the influent composition and other environmental factors.

A more natural way to use the updating feature is to adjust the model to changing conditions. Note, that some conditions, like bulking sludge, may very well be completely outside the definition area of the model (Chapter 3).

Since the model has to be able to be corrected according to the measurements, the identifiability is crucial. In other words, if the output of the system is insensitive to a specific parameter, then the model updating of that parameter is not successful. This may be a sign that this parameter is not identifiable in practice. A more serious problem is the use of too many parameters to describe the model. The effect of a change of one parameter may be cancelled by the change of another. The model may be a good fit to the measurements, but the interpretation may be completely wrong. This problem was discussed earlier in Chapters 7 and 8.

The parameter adjustment is not trivial. It consists of some kind of optimisation procedure, that has to converge in order to make the output error over a specific time horizon as small as possible.

If the simulator is connected to on-line sensors, then certain parameters may be recursively updated. This will be further discussed in Chapter 12.

9.3.5 Hierarchical Modelling

Control systems are frequently built like an onion. One control loop encompasses the other. For example, it is quite common that a dissolved oxygen controller is built around an air flow controller. The purpose of that is to make the air flow valve look simpler from the DO controller point of view. This system then becomes a block in another control system at a hierarchically higher level.

Also the plant may be modelled in the same hierarchical manner. One block may describe a well mixed zone. Several anoxic and aerated zones will together form a biological reactor. This reactor may be represented as a single block at a higher level, where the whole plant is represented.

Evidently, control and process engineers would like their simulators to behave in exactly the same fashion. One entire block diagram may become a single block at the next higher level. Many simulators currently on the market offer such a hierarchical decomposition facility.

9.4 Simulator Features

There is a wide range of simulators available today, both for general purpose simulation of dynamic systems and with specific application to wastewater treatment systems. The general simulator platforms are generally quite flexible. On the other hand they demand more knowledge of the user, since they do not have an interface specific to a wastewater treatment system. Usually both continuous-time and discrete-time systems can be simulated together. This is important if computer control algorithms should be implemented together with continuous process descriptions.

9.4.1 System Description

There are two principal ways to describe the dynamics of a system in a simulator:

- equation oriented descriptions, or
- block oriented descriptions.

By a simple example we will demonstrate the principal differences.

Equation Oriented Description

In equation oriented simulators (such as Simnon and ACSL), the dynamic equations are formulated as differential or difference equations like Equations 9.1 to 9.4, that is, the explicit form of the mass balances is formulated. Usually this is done in a specific language that allows nonlinear expressions. Some nonlinearities may be defined by tables or by Boolean expressions. The simulator compiles the equations and verifies that the variables and parameters are consistent.

Example 9.1 (Simulation of DO control) *Let us represent the dynamics of a Proportional-Integral (PI) controller of the dissolved oxygen (DO) concentration, as shown in Section 3.6, using a Simnon representation. The DO dynamics is described by Equation 3.19:*

$$V\frac{dS_O}{dt} = q_{in}(S_{O,in} - S_O) + K_La(S_{O,sat} - S_O)V + r_OV \tag{9.10}$$

The oxygen transfer rate is assumed to be linearly related to the air flowrate,

$$K_La = K_aq_a \tag{9.11}$$

where the air flowrate q_a is obtained from a PI controller. The PI controller is represented by (see Equation 3.23 to 3.24),

$$\begin{aligned} \tfrac{dz}{dt} &= e \\ u &= u_o + K_c\left(e + \tfrac{z}{T_I}\right) \end{aligned} \tag{9.12}$$

where the controller output u is the air flowrate q_a. The error e is obtained by comparing the DO sensor signal S_O to a desired value $S_{O,d}$:

$$e = y_d - y = S_{O,d} - S_O \tag{9.13}$$

where y is the DO concentration S_O. In Simnon the DO dynamic system description would look like:

```
Continuous system DOdyn
Input qa "air flow input signal, produced by the DO controller
Output so "output, DO measurement signal
State so "declaration of the state variable, DO
Der dso "declaration of the derivative names dDO/dt
dso = qin*(soin - so)/V + Ka*qa*(sosat - so) + ro
qin:1 "liquid flow rate
soin:0.5 "influent flow DO concentration
Ka:5 "oxygen transfer rate parameter
qa:1 "normalized air flowrate
sosat:10 "DO saturation concentration
ro:-30 "oxygen uptake rate
so:1 "initial condition of DO
V:1
end
```

The PI regulator is described by:

```
Continuous system PIreg
Input y "input, DO measurement signal
Output u "output, air flowrate signal
State z
Der dz
e = yd - y "calculation of the error
yd:2 "reference value
dz = e "integration of the error
u = uo + Ko*(e+z/Ti) "PI controller
uo:1 "steady state value of the controller
Ko:1 "controller gain
Ti:2 "controller integral time
z:0
end
```

Note that this code is just the basic formulation of a PI controller. A real controller has to be supplied with additional features, like anti-reset windup to make it useful in practice, see Section 17.2.2.

The system description is edited using a conventional text editor, and the equations can be changed simply by changing the text information. The initial values (`so` *and* `z`*) and the parameters (all other variables in front of a colon) can be changed by simple commands using the simulation command language. The coupling between the DO dynamics and the PI controller is defined by a* connecting system. *In this case it looks like:*

```
Connecting system DOcon
qa[DOdyn] = u[PIreg]
```

```
y[PIreg] = so[DOdyn]
end
```

Such a connecting facility allows unit processes to be interconnected in a straightforward way. Note that the connection is defined algebraically, and there is no graphical symbol to help the user to make this work easier. Furthermore, there is no support for hierarchical modelling, as described above. All modules are defined at the same level.

Block Oriented Descriptions

A process may also be described by general graphical symbols or block diagrams. A number of standard symbols are defined for elementary operations (such as summation and multiplication), different static nonlinearities, input generating functions, integrators. Each symbol is supplied with one or more input and output signals, so that the different symbols can be interconnected. In order to describe the dynamics of a process these symbols are connected into a flow diagram. Let us illustrate this with an example using Simulink.

Example 9.2 (Simulation of DO control revisited) *The dynamics of dissolved oxygen (Equations 9.10 to 9.11) is now represented in Figure 9.2 as a Simulink description.*

The dynamics is represented by connected symbols. The mass balance equation is hidden in the graphics. The equations can be changed with a graphical editor. Using the editor it is now possible to represent the whole DO dynamics with only one block, simply by marking the whole diagram and the resulting block is represented in Figure 9.3.

The input (air flowrate) and the output (DO sensor measurement) are marked. The PI regulator can be represented in a similar way. The graphical editor can be used to connect the blocks into the controlled DO process. The two blocks, DO dynamics and the PI regulator can be marked as a block on the next hierarchical level.

Hierarchical System Description

We have seen that Simnon supports only one level of modules in a model library, that can be coupled by connecting the inputs and outputs of the individual modules. Simulink makes it possible to go one step further, and several blocks can be redefined as only one block. In this way Simulink supports a hierarchical description of more complex plants.

An equation oriented system description is often desirable. The model is presented in a form that resembles the mass balance equations, and it can be changed in a flexible way. At the same time a block oriented description is an attractive way to present a plant, where the details may be hidden.

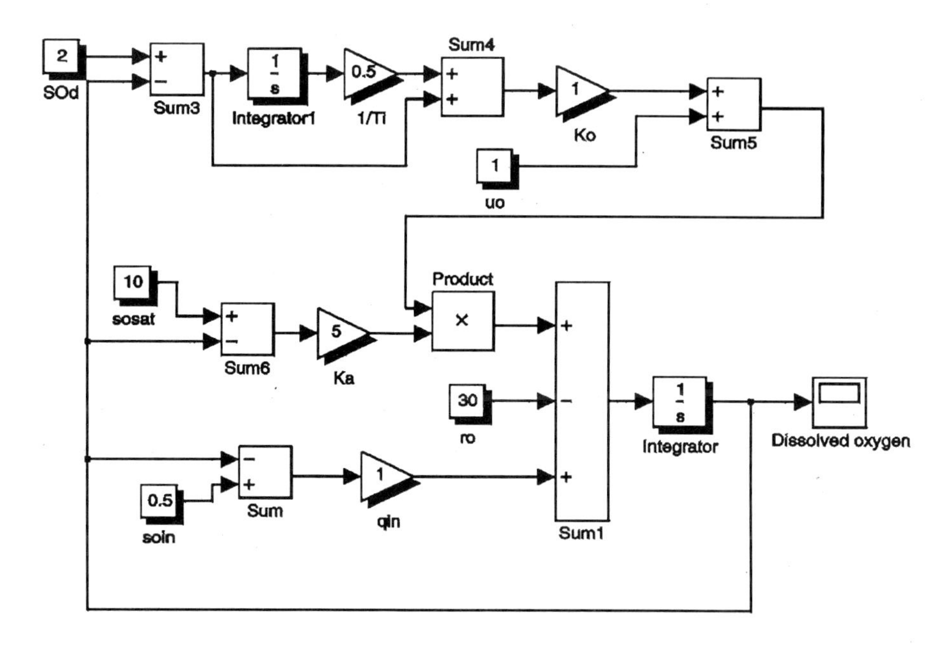

Figure 9.2: Simulink representation of a simple DO dynamics.

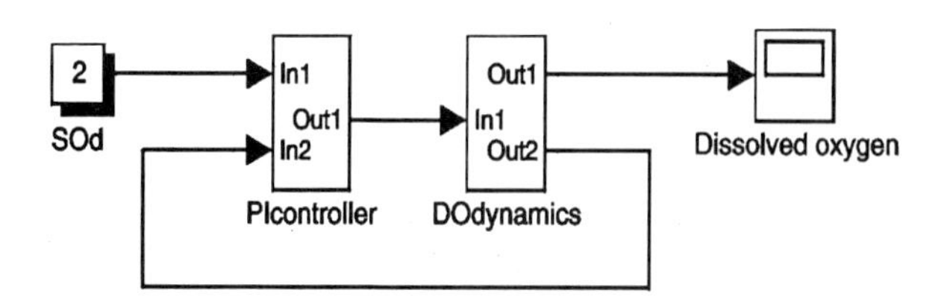

Figure 9.3: The DO dynamics of the previous figure represented as a single block.

There are simulation software packages that would allow a combination of equation oriented system description and block diagram presentation. Packages like SPEEDUP and NIMBUS can define the dynamics using equations like ACSL or Simnon. The equations of a unit operation may define a block. Such blocks can be interconnected as in Simulink.

Model Library

Many simulators have some component library available. System descriptions in Simnon or blocks in Simulink may be components in such a library. However, this is not a *true* component library concept in the sense used by a higher-level programming language. In the packages, the user can store model components in a (so-called) *library* and retrieve the component by *dragging* it from the library to the model area or copying the file. In this way the component is always being copied. Consequently, every change in the library requires manual repetition of the copying process, which is error prone and tedious.

Other simulators, like NIMBUS, work like higher-level programming languages, where a change in a library function just requires repeating the linking process between the various model components.

9.4.2 Equation Solving

Sorting the Equations

All the differential equations of a dynamic system are solved simultaneously by some sort of integration algorithm. This means that the derivatives, or the right hand sides of the equations, have to be evaluated and passed to the integrator. Usually the user does not need to bother about this. Equation oriented simulators will automatically sort the equations into such an order that the derivatives can be calculated. In block oriented packages, like Simulink, all the blocks are automatically sorted so that the derivatives can be evaluated properly.

When *recycles* of algebraic variables occur on the right hand side, in either equation oriented or block oriented simulators, one has to watch out for a proper solution. Such a property is called an algebraic loop and will create particular problems. Some simulators such as Simnon will simply give an error message and the user must reformulate the equations. Others such as Simulink will evaluate algebraic loops iteratively which can cause a marked degradation in speed.

There is another approach using more sophisticated solvers that can integrate differential equations and solve algebraic equations simultaneously. This is called a differential algebraic equation (DAE) system. Simulators which employ DAE solvers, such as SPEEDUP and NIMBUS, do not need to sort the equations and can inherently handle algebraic loops efficiently, in fact they don't even look for them.

Numerical Integration

Many users regard the numerical integration of the ordinary differential equations as a black box routine. Given a model of the form of Equation 9.1 with an initial condition given as Equation 9.2, the user expects the software to find an accurate solution $x(t)$ almost independently of the character of the function $f()$. Modern integration methods perform very well, but they do not yet arrive at this generality.

In all numerical methods the differential equation is approximated as a difference equation. In the simplest case the approximation of Equation 9.1 is obtained by replacing the time derivative with a simple difference equation (the *forward* Euler approximation):

$$x(t+h) = x(t) + hf(x(t), u(t), p, t) \tag{9.14}$$

where h is the step length. Knowing the initial conditions one wishes to compute the states that are close to the true solution at times h, $2h$, $3h$, etc. It is crucial to choose a step size h that is sufficiently small. Too short a step size will give unreasonably long solution times, while an h that is too large will cause numerical problems. These problems may be significant, particularly if the plant dynamics contain both fast and slow dynamics together. The choice of the step length will be illustrated by a simple example.

Example 9.3 (The problem of long step sizes) *To illustrate the problem of too long a step size consider the simple first order system:*

$$\frac{dx}{dt} = -a\, x \tag{9.15}$$

where $x(0) = 1$ and $a > 0$. The system has the analytical solution $x(t) = e^{-at}$. Let us solve the differential equation numerically by a forward Euler *approximation. Approximating the derivative with a finite difference:*

$$\frac{dx(t)}{dt} \approx \frac{x(t+h) - x(t)}{h} \tag{9.16}$$

we obtain:

$$x(t+h) \approx x(t) - h\, a\, x(t) = (1 - h\, a)x(t) \tag{9.17}$$

Figure 9.4 was produced by the MATLAB program `Example_9_3.m` *and shows what happens for different choices of the step size $h = [0.01, 0.25, 0.35, 0.4, 0.45]$ with $a = 5$. As h increases the integration error increases until for larger values of h such that $|1 - ha| > 1$, that is $h > 2/a$, the solution x oscillates with alternating sign and with an increasing amplitude.*

This instability has nothing to do with the system property but is only caused by a too crude an approximation in the solution method.

The problem of oscillations due to too long an integration step is called *numerical instability*. It illustrates, that the numerical integration of the differential

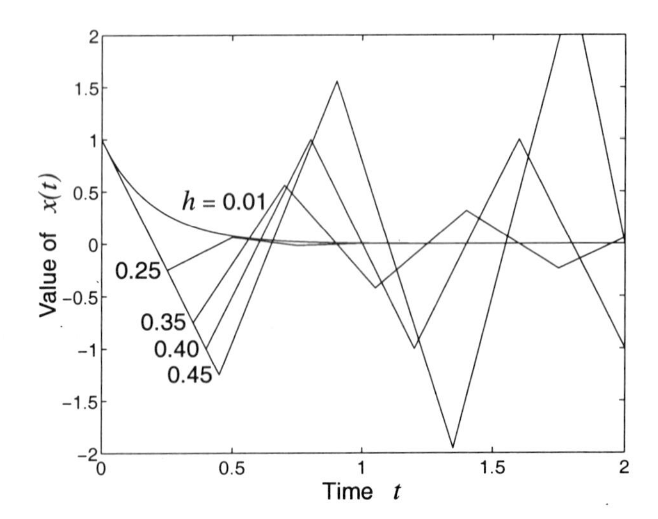

Figure 9.4: Numerical solutions of a simple first order equation.

equations has to be performed with some care. The numerical integration methods used by most commercial packages are usually more sophisticated than the simple Euler integration illustrated above, but they also suffer from the same stability problem.

Adequate numerical integration routines are crucial for most wastewater treatment systems. The function $f()$ in Equation 9.1 determines how sophisticated methods would be needed. For example, the wide time span of the dynamic equations make the system very stiff, which creates numerical difficulties. Likewise, the concentration profile of the settler may require many spatial divisions and special solution methods. Also the numerical solution of the partial differential equations describing biofilm processes may be quite cumbersome. Furthermore, many integration methods rely on the solution being smooth ($f()$ is assumed to be differentiable many times). A system that includes discontinuities needs special attention.

A numerical integration method approximates Equation 9.1 by a difference equation:

$$x_{n+1} = \bar{x}_n + h_n \, \bar{f}_n \quad n = 0, 1, 2, .. \tag{9.18}$$

where $\bar{x}_n$ is formed as a combination of past solution points and $\bar{f}_n$ is formed as a combination of function values evaluated at the solution points or in their neighbourhood. Note that the step length h_n is no longer assumed to be constant. Using Equation 9.18 we can compute $x_1, x_2, ..$ as approximations of $x(t_1), x(t_2), ...$ The accuracy of the integration method depends on how $\bar{x}_n$ and $\bar{f}_n$ are constructed.

We may categorise the methods as explicit or implicit, *one-step* or *multi-step*. In a one-step method x_n is the only input data to calculate $\bar{f}_n$. If any previous value x_k $(k < n)$ is involved in the calculation, the method is said to be multi-step.

A popular class of one-step methods are the *Runge-Kutta* methods. They used a number of iterated evaluations of f when forming $\bar{f}_n$.

An *explicit* method does not involve x_{n+1} in the calculation of $\bar{x}_n$ and/or $\bar{f}_n$. Both the Euler and the Runge-Kutta methods are explicit. An *implicit* method includes x_{n+1} in the construction of $\bar{x}_n$ and/or $\bar{f}_n$, and a nonlinear equation has to be solved in order to get the next solution point. As an example of an implicit solution, let us consider the simple model represented by Equation 9.15. Now the time derivative is approximated by a *backward* difference equation:

$$\frac{dx(t)}{dt} \approx \frac{x(t) - x(t-h)}{h} \tag{9.19}$$

so the corresponding difference equation becomes,

$$x(t) \approx x(t-h) - h\, a\, x(t) \tag{9.20}$$

In this case $x(t)$ can be solved easily:

$$x(t) = \frac{1}{1 + ah} x(t-h) \tag{9.21}$$

This is an *implicit* solution of $x(t)$. This method is more robust than the explicit Euler method, and the solution cannot get numerically unstable even for large values of h. For a nonlinear differential equation such as Equation 9.1, an implicit method can be formulated as the *trapezoidal* method, where Equation 9.1 is approximated as:

$$x(t_{n+1}) - x(t_n) = \frac{h}{2} \left[f(x(t_n), t_n) + f(x(t_{n+1}), t_{n+1}) \right] \tag{9.22}$$

Obviously $x(t_{n+1})$ cannot be solved directly. Instead iterative methods are often used, and predictor-corrector methods are common. There are several methods that are based on this scheme, such as Adams-Moulton's, Adams-Bashforth's and Adams-Gear's methods.

Generally, implicit methods are better when simulating problems with both fast and slow dynamics (stiff problems) like a complete wastewater treatment plant. This is especially important when trying to resolve what happens in steady state after fast transients have died out.

The accuracy of the numerical solution can be affected by varying the step size h_n. Normally the integration techniques have a variable step length that is automatically adjusted to fit an error criterion. The user defines an error tolerance which the integration method tries to fulfil by varying the step size in relation to the behaviour of the produced solution.

Having specified a value of the tolerance one may believe that the integration method will keep the global error of the numerical solution below this limit. What the software does is actually something very different. Each integration step leads to a local truncation error, perturbing the numerical solution from the true one. The local errors propagate through the differential equation and accumulate to form a global error $x_n - x(t_n)$. This is the fundamental measure of the quality of the numerical solution, but it is not computable, since the true solution is not known. In general, it is even difficult to estimate. Therefore the local error is kept below the defined tolerance in the hope that this will lead to a numerical solution with an acceptable global error. Do not count on tolerance proportionality! In a professional implementation, the global error is related to the local tolerance, but reducing the tolerance a factor of (say) 10 does not necessarily result in a similar reduction in global error.

Example 9.4 (Aerated biological reactor simulation) *Let us illustrate the importance of the numerical integration by simulating a simple system of carbon removal using only three states, living organisms (X_H), substrate (S_S) and dissolved oxygen (S_O). This is modelled in Example 3.7. It is a stiff system, since the DO dynamics is much faster than the growth and decay dynamics of the organisms. This is also reflected in the integration difficulties. The system has been simulated with the MATLAB program* `Example_9_4.m` *using Runge-Kutta integration with variable step length, the MATLAB* `ode45` *integrator. We have chosen different tolerances to demonstrate what can happen even in a very small wastewater treatment model. Figure 9.5 shows the results with relative tolerances of* 0.5 *and* 0.05. *This can be compared to Figure 3.17 which uses a tolerance of* 0.001 *which is the default.*

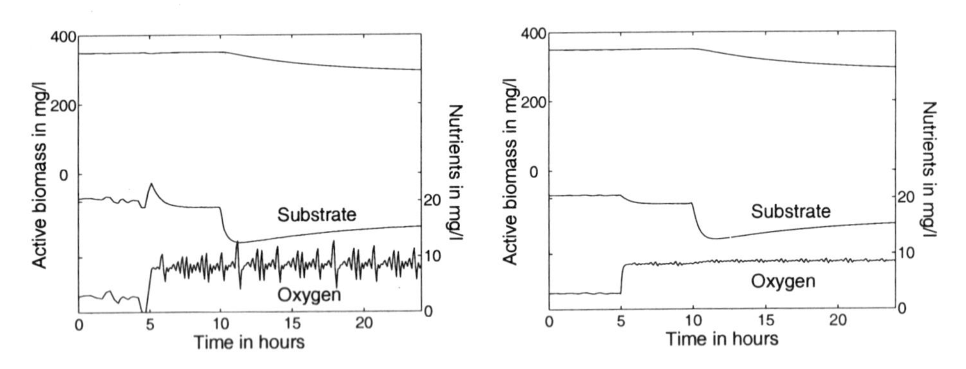

Figure 9.5: Simulations of aerated biological reactor.

While the biomass with very slow dynamics is close to the true solution, the substrate and oxygen show errors and even instability. While most problems occur with oxygen which has the faster dynamics, there is considerable interaction between the substrate and oxygen difficulties.

This experience should make us aware that, in the simulation of a large wastewater treatment plant with sophisticated models of the bioreactors and the settler, the numerical integration problem is of major interest and is far from trivial.

Which integration method to choose? There is no *best* method. Different integration methods perform well on different classes of problems. Repeat the simulation with different integration methods and different error tolerances to check that you get qualitatively similar results. Finally, never trust numerical simulation blindly, even if many of today's commercial tools will perform well on a large class of problems.

9.4.3 User Interface

System Definition

Some simulators have a graphical (or block diagram) representation of the models. However, nearly all block-diagram editors on the market, including Simulink, suffer from some drawbacks.

The equation oriented simulation packages usually have alphanumerical interfaces similar to those offered by general-purpose programming languages. Due to the success of the graphical simulation packages, more of the general-purpose simulation languages have been enhanced by graphical front ends as well. However, the text-oriented origin of these programs often remains clearly visible through their new interface.

The block oriented packages are usually supplied with some kind of graphical editor. This may offer a good intuitive user interface. Block diagram editors have the advantage that they are (usually) easy to master by even novice or occasional users, and this is the main reason for their great success. Symbols are picked from a *library* and *dragged* into the modelling area and then connected to each other. One was mentioned in the item *model library* in Section 9.4.1.

It is often the case that differential equations have to be incorporated directly in textual form, because the direct usage of block diagram components becomes tedious. For example, in Simulink and SystemBuild, the only reasonable choice is to program such parts directly in C or FORTRAN, that is by using a modelling technique from the 1960s. In this respect, the general-purpose simulation languages, like Simnon or ACSL, offer much better support, because differential equations can be specified directly, using user-defined variable names rather than indices into an array. Furthermore, the equations can be provided in an arbitrary order, since the modelling compiler will sort them prior to generating code. Even better support is given by a package like NIMBUS that can combine the advantages of an equation oriented and a block oriented package.

Having defined the model the user would need support in telling the software how to actually execute the simulation. The simulation time interval, the integration method and its tolerance have to be determined. The parameters and the initial conditions may be changed during simulation. In old simulation packages all of this is command driven, and the user has to consult the manual or help files to get the information. In a modern package the execution is often menu driven and more or less self-explanatory.

Naturally in a general simulation package, no support is given to find the proper parameters. However, some instruction can be built into the menus to help the user in a specific application simulation package.

It is highly desirable to get diagnostic information. During the model building it is crucial that the model is tested. Algebraic loops should be detected, missing parameters or other inadequate model behaviour should be detected. During the simulation, problems like numerical inaccuracies should be shown for the user.

Presenting Results

It is important to be able to effectively analyse output data. Therefore interfaces to programs for statistical analysis, parameter identification etc. are of great value (they may sometimes be included in the simulator). Moreover, by interfacing the simulator to spreadsheet programs and graphical packages the presentation can be made very versatile. Likewise, connecting the simulator to databases or directly to real-time input ports makes it possible to incorporate real data for off-line or on-line analysis or diagnosis.

9.4.4 Real Time Applications

Simulation of a complex plant should allow human operators to drive the simulation in just the same manner as they would drive the real system. This is useful for both system debugging and for operator training. However, since humans cannot be time-scaled easily, it is then important to perform the entire simulation in real-time.

Some simulators contain hardware in the loop. For example, flight simulators for pilot training are elaborate electromechanical devices by themselves. It is the purpose of these simulators to make the hardware components behave as closely as possible to those that would be encountered in the real system. Evidently, these simulations need to be performed in real-time as well.

A simulation of a wastewater treatment plant may describe the behaviour during long periods, weeks to months. This of course makes real-time simulation unrealistic and the simulation has to cover a long period in a relatively short computer time. However, it is important to be able to import real data from the process. Such data can drive the dynamics and may include such variables as the influent flowrate and concentrations.

9.4.5 Extensions of the Simulators

There are apparent needs for other extensions to simulators. Optimisation routines as well as identification algorithms are crucial for model verification and model reduction. A knowledge database can supply the user with guidelines about model validity, hints on numerical integration, model parameter values, or other training hints. Some general purpose simulators have the advantages (such as MATLAB/Simulink) that add-on toolboxes can easily be integrated with the simulator.

9.5 Simulator Selection Criteria

There is a large selection of different simulation systems on the market. They come in all shades and prices, specialised for different application areas, for different computing platforms, and embracing different modelling paradigms. Many of them are simply competitors of each other. It is not our purpose to survey all of them. A list of current products and vendors (available in 1994) is found in Rodrigues (1994).

Why are there many more simulation packages around than general-purpose languages? The answer is simple. The general-purpose programming market is much more competitive. Since millions of FORTRAN and C compilers are sold, these compilers are relatively cheap. It is almost impossible for a newcomer to penetrate this market, because they would have to work under cost for too long a period. Simulation software is sold in hundreds or thousands of copies, not millions. Thus, the software is comparatively more expensive, and those who sell ten copies may already have made a modest profit.

Rather than trying to be exhaustive we have decided to concentrate on a few of the more representative products to help explain the different philosophies embraced by these software tools.

9.5.1 Checklist for Selection

It is not an easy task to advise what simulator to choose, and it is not our purpose to favour any simulator over the others. It is too easy to be governed by want, instead of by need, so we give a few criteria to facilitate a rational selection.

- *Models available*: A useful simulator ought to have a library of useful unit process and equipment models. There has to be a good documentation of the assumptions and the range of validity of these models.

- *Model library*: In a previous section we noted that many simulators have some component library available. Such a library may be more or less advanced. Firstly, the number of processes described may be quite different. Secondly, the library structure may be more or less advanced from a software engineering point of view.

- *Modularity of the models*: The modularity of the models is not trivial. An anoxic zone can only be connected to an aerated zone that contains the same substrates and organisms. Similarly, one would like to have the same settler model for any reactor configuration. This is not easy, even if one agrees to use only one model for the thickening. If a nitrification aerator is connected to the settler, then all sludge components (organisms and stored substrate) have to be combined into one component before it enters the settler. If an aerator with only carbonaceous removal is added instead, then the sludge composition is different, but the settler needs to see an input of the same format.

- *Validated models*: This is of course crucial. What kind of support for the model validation is presented? Actually, part of the simulation might be a parameter fitting to real data. How easy is it then to incorporate real data together with the simulator? Can an optimisation like Figure 9.1 be implemented easily?

- *Flexibility to build larger plant models*: This is related both to available basic models in the model library and to the modularity of them. A full plant may contain a lot of blocks. Some kind of hierarchical modelling support can certainly help the user to deal with the complexity. For wastewater treatment simulators are there connections between plant models and sewer and receiving water models?

- *Graphical representations*: This has been commented on in Section 9.4. A package that can combine equation oriented and block oriented system description has a great advantage. Furthermore, intuitive user interfaces for the execution of the simulation are extremely important.

- *Numerics available*: One has to watch out for poor numerics. As described in Section 9.4.2 the integration of the equations has to be reliable and robust. The package ought to contain more than one standard integration algorithm, since the systems are often stiff. General purpose simulators often have superior numerical engines. Some simulators contain *steady-state solvers*. This makes it possible to obtain a steady-state solution much faster than by numerical integration.

- *Real time simulation*: Some simulators can be connected to a real-time environment (see Section 9.6.6).

- *Measurement data*: In a good simulator it is possible to include measurement data as input to the simulator. Since measurement data are not always represented with equidistant intervals, it is an advantage to be able to present time tags to each measurement point.

- *Output presentation*: Any decent simulator has some graphical output. Often one is interested in some form of post-processing facility. The output may be presented in a rich set of different formats. Furthermore, the output may be analysed, compared with experimental data, or simply exported to a word processor.

- *Optimisers*: The problem of model validation has been extensively discussed in Chapters 6 to 8. Referring to Figure 9.1, one has to be able to compare a model with a true process. One requirement is the import of real data. Another one is to find the right parameter adjustment. It is obvious that we try to make the simulator output as close as possible to the experimental output. The parameter adjustment can of course be made manually, but some optimising algorithm would be more realistic to use.

- *Operating systems*: There are simulators offered for all common operating systems. Naturally, only the most widespread simulators are offered for several operating systems.

What kind of simulator is the best? Naturally, there is no specific answer to this question. A researcher usually emphasises flexibility and can sacrifice some of the user-friendliness. The operator, on the other hand, usually wants well calibrated models with straight-forward user instructions.

9.6 General Purpose Simulators

Here we will discuss the most important packages of two kinds, block diagram simulators, and general-purpose simulation languages. All of them allow the simulation of continuous-time (differential equation) and discrete-time (difference equation) models and mixtures thereof.

Earlier we have categorised the simulators according to their system descriptions, block diagram or equation oriented. Some simulators, like SPEEDUP and NIMBUS, are a combination of both, and will be described separately.

Obviously there is a very fast development of the software, and here we can only describe the general features of the systems. For the various packages we also give web addresses for current information. Another source of information is a web based software directory

http://engineering.software-directory.com.

9.6.1 Block Diagram Simulators

The natural description for a control system structure is most often block diagrams. Most of the major simulation software producers offer a block-diagram editor as a graphical front end to their simulation engines.

Below we discuss three common packages on the market. All of them allow the simulation of continuous-time (differential equation) and discrete-time (difference equation) blocks and mixtures thereof. This is of particular importance to control engineers, since it allows them to model and simulate continuous processes controlled by discrete time regulators. The block diagrams can mostly be structured in a hierarchical fashion.

Simulink from the MathWorks Inc.

This is an easy-to-use point-and-click program. Simulink is an extension of MATLAB, the widely used program for interactive matrix manipulations and numerical computations in general. MATLAB is the basis for the simulations in all exercises in this book. MATLAB can be employed as a powerful pre- and post-processing facility for the simulation and can display the simulation result in a large number of different formats. Simulink and MATLAB are available for a broad range of computing platforms and operating systems (Windows, Macintosh, Unix workstations, VMS platforms). By default the equations of a Simulink model are pre-processed into an intermediate format, which is then interpreted. This has the advantage that the program is highly interactive, and simulations can run almost at once. Simulink enjoys a lot of popularity, especially in academia, where its highly intuitive and easily learnable user interface is particularly appreciated.

A large number of pre-defined building blocks is included in the program and it is easy to extend this library with user defined blocks. Hierarchical modelling is possible. Couplings to other MATLAB toolboxes (Signal Processing, Control System Design, Neural Network, Statistics, System Identification, Optimisation) makes Simulink powerful. By using the optimisation toolbox parameter estimation is readily available (see Section 7.7). The standard graphical output from Simulink can be extended by MATLAB graphics, such as 2D and 3D animation. Data can also be transferred between the program and a real process, connected to the I/O devices of the computer. This makes real-time analysis possible (see Section 9.6.6).

See MATLAB in the References for a source of more information. Current information about MATLAB and its toolboxes is found on the web address:

http://www.mathworks.com

SystemBuild from Integrated Systems Inc.

SystemBuild is a graphical workspace where a system is constructed using a pallet of primitive elements called blocks. Overall SystemBuild offers more powerful features than Simulink. It is designed to handle discrete events as well as complex structures like mechanical multi-body systems. They have to be described by so called differential-algebraic equations (DAE). Systems can be hierarchically structured using SuperBlocks which contain other blocks and SuperBlocks. Data is transferred between blocks using “wires”.

The price to be paid for this flexibility and generality is a more involved user interface that is a little more difficult to master. SystemBuild is an extension of Xmath (formerly $MATRIX_X$, the main MATLAB competitor). Xmath is similar to MATLAB, but supports more powerful data structures.

See $MATRIX_X$ in the References for a source of more information. Updated information about Systembuild and Xmath is found on the web address:

http://www.isi.com

EASY-5 from Boeing

EASY-5 is one of the oldest block-diagram simulators on the market, since 1981. It is designed for the simulation of very large systems. The tool is somewhat less easy to use than Simulink or SystemBuild. After a block diagram has been built, the models are translated into FORTRAN or C code and linked to the simulation run-time engine. This has the effect that the compilation of a block diagram into executable run-time code is rather slow. Yet the generated code executes generally faster than in the case of most other block diagram programs. As a comparison, Simulink and Systembuild are built on interpreted languages (MATLAB and $MATRIX_X$ respectively) which give a slower execution.

See EASY-5 in the References for a source of more information. Updated information about EASY-5 is found on the web address:

http://www.boeing.com/assocproducts/easy5

9.6.2 General-Purpose Simulation Languages

In the previous section we mentioned two limitations of the block diagram editors. Firstly, the aspect of a model library was discussed. Secondly, the incorporation of differential equations in textual form is often cumbersome. The general-purpose simulation software packages are usually based on a textual definition of the models, like the equation oriented systems. Later many of them have been enhanced by graphical front ends as well. Since they are command driven, they may be considered "high level" languages for simulation. Here we comment on a few common general-purpose simulation languages, Simnon, ACSL, SPEEDUP and NIMBUS.

Simnon from SSPA Systems

Simnon was an early direct-executing, fully digital simulation system on the market. It was designed initially at the Department of Automatic Control, Lund Institute of Technology by Hilding Elmquist in 1975. It is an easily manageable software system for the simulation of continuous-time and discrete-time systems. Simnon is command driven, which is often preferred by experienced users.

Initially Simnon offered a mixture of statement-oriented and block-oriented user interfaces. A graphical interface has been added. For years, Simnon has been a low cost alternative for simulation and has enjoyed a widespread acceptance. It is available in the Windows environment. Lately Simulink has been a serious competitor.

The system is usually built up by interconnecting subsystems, that may be both continuous-time and discrete-time. The user interacts with the simulator by typing commands (or menus in a graphical interface). Parameters and initial conditions can be changed by simple commands. A built-in macro facility allows the user to create his own set of commands. The program has real-time capabilities, that is data can be transferred directly from and to a real process, connected to the I/O devices of the computer on-line.

Simnon 3.0 can handle up to 10000 states and 100 subsystems. The integration routines include up-to-date algorithms.

See Simnon in the References for a source of more information. Updated information about Simnon is found on the web address:

http://www.sspa.se

ACSL from Mitchell & Gauthier Assoc. (MGA)

ACSL was made available in 1975 and has been a leading simulator on the market, where Simnon has been the low-cost alternative. Lately ACSL has lost a large percentage of their academic users to Simulink but is still fairly popular in industry. ACSL is now in release 11 and more than 3500 licenses have been sold. In order to achieve a platform independence, an ACSL program is pre-processed to FORTRAN. The FORTRAN code is then compiled further to machine code. As a consequence, ACSL simulations run efficiently, which is in contrast to the simulation code generated by most block-diagram simulators. User-defined FORTRAN, C, and Ada functions can be called from an ACSL model. ACSL can handle both ordinary differential equations and differential algebraic equations. Furthermore, ACSL can handle discrete events. Recently ACSL has been enhanced by a block-diagram front end.

A block in ACSLs block-diagram modeller can take any shape and the input/output points can be placed everywhere, contrary to the much more restricted graphical appearance of Simulink models. Consequently, it is easier to get a closer correspondence between reality and its graphical image. Still, ACSL is not truly modular. All variables stored in a block have global scope. This means that one has to be careful not to use the same variable name in different blocks. Furthermore, it is not possible to define a block once, and to use several copies of this block. As a result, it is not convenient to build up user-defined block libraries. ACSL is a fairly robust product, due to the more than 20 years of experience.

In 1994, MGA announced the introduction of the Graphic Modeller, a visual programming tool for creating simulations. Also in 1994, MGA released ACSL Vision, a suite of visualization and animation tools for Unix, Windows and VMS. Finally, MGA commercialized ACSL Realtime, a set of products for hardware-in-the-loop testing.

See ACSL in the References for a source of more information. Updated information about ACSL is found on the web address:

http://www.mga.com

9.6.3 Combined Block and Language Simulators

To the casual user these simulators appear like any other block diagram simulator. Blocks are selected and copied into a diagram, linked together and parameters set through pop-up menus. However, the more sophisticated user can program new blocks for the libraries in a high-level simulation language. In this respect they resemble general-purpose simulation languages, offering the flexibility that this offers yet with the ease of use of the graphical block diagram interface. The two products mentioned below are both based on sophisticated DAE solvers.

Both packages were developed specifically for process systems, but will simulate any system that can be described by differential and algebraic equations. NIMBUS has been used extensively for wastewater applications within the Advanced Wastewater Management Centre at The University of Queensland, from whom many models for wastewater treatment processes are available.

Aspen Dynamics from AspenTech

SPEEDUP was developed at Imperial College, London over many years, originally as an experimental process systems simulator of the simulation language type. More recently it has been commercialised by Aspen Technology. The software now offers a complete block diagram interface. SPEEDUP is a mature product offering many add-ons such as optimisers (with some pre-packaged applications), external data interfaces allowing off-line and on-line use, and physical properties libraries.

Today, combined systems represent the state of the art. With minimal effort, process designers can now switch from one type to the other, avoiding the rekeying of data, and accelerating the process by which an initial design (usually in steady-state form) acts as the base case for a dynamic optimisation.

The newest system is Aspen's DynaPlus, which is not so much a new program as a new interface. The program combines two existing products, AspenPlus, a steady-state package, and SPEEDUP, the dynamic package, with an "integrated environment" module that handles data flow from one system to the other.

Aspen Dynamics is AspenTech's new product for integrated steady-state and dynamic simulation. Unit operation models include typical chemical engineering units: mixers, two-phase and three-phase flows, heaters, pipes, valves, pumps, compressors, reactors of different kinds, distillation columns, etc.

The Aspen products are available on UNIX workstations and high-end personal computers running Windows NT. While it is a mature and fully supported package, the price is substantial.

See SPEEDUP in the References for a source of more information. Information about current products is found on the web address:

http : //www.aspentec.com

NIMBUS from The University of Queensland

This is a very recent package, first released in 1991 (Newell & Cameron (1991)). It was developed from the start as a combined block and high-level language simulator for process systems, with many of the features of SPEEDUP and some extras. It has some add-ons, such as an optimiser and a discrete event solver, currently both in prototype form. NIMBUS is a semi-commercial product with a number of industrial customers, but it is supported from a university department on a "best intentions but no promises" basis. It is available on UNIX workstations and high-end personal computers running OS/2 or Windows NT. Because of the low level of support, it is quite an economical package.

A source of more information is found in the References. Current information on NIMBUS is found on the web address:
http : //daisy.cheque.uq.edu.au/nimbus/.

9.6.4 Limitations of Current Simulation Software

Block diagram editors look deceptively modern and attractive, because they employ modern graphical input/output technology. However, the underlying concept is unnecessarily and unjustifiably limited. It is quite easy to offer block-diagram editing as a special case within a general-purpose object-diagram editor. However, to extend block-diagram editors to object-diagram editors is far from trivial. It is easy to predict that block-diagram editors will be replaced by object-diagram editors in the future in the same way as block-diagram editors have replaced the textual input of general-purpose simulation languages in the past.

There is still a problem with model component libraries. As previously explained the *library* technique supported by block-diagram systems, such as Simulink and SystemBuild, is only of limited use, because a modification in a component in a library cannot easily be incorporated into a model in which this component is being used. An object-oriented modelling system together with its object-diagram editors provide a much better and more satisfactory solution.

There is also a serious practical problem of organising and documenting simulation experiments. To organise the storage of the results of many simulation runs, possibly performed by different people, and to keep all the information about the simulation runs necessary in order to reproduce these runs in the future, that is, store the precise conditions under which the results have been produced, is a problem closely related to version control in general software development. At present, almost no support is provided for such tasks by available simulation systems.

9.6.5 Object-Oriented Modelling Languages

One reason for the limitation of the block-diagram simulators has to do with computational causality. It is always defined what is the input and what is the output of the equations. Unfortunately, causality is sometimes not defined. Simultaneous events are acausal. Modelling an electrical resistor, it is not evident ahead of time, whether an equation of the type:

$$u = R\,i \tag{9.23}$$

will be needed, or one of the form:

$$i = \frac{u}{R} \tag{9.24}$$

It depends on the environment in which the resistor is embedded. Consequently the modelling tool should relax the causality constraint that has been imposed on the modelling equations in the past. This concept has been coined the object-oriented modelling paradigm, since it provides a modelling language with a true correspondence between the physical objects and their counterparts inside the model. We will not present details of this, but two software products have been released recently. Dymola from Dynasim AB was created by Hilding Elmqvist (who also created Simnon) as part of his PhD dissertation (Elmqvist (1978)). The design of Dymola was influenced by the first object-oriented language Simula (Birtwistle *et al.* (1973)). Simula uses a class concept and Dymola introduced model classes. In 1992 the development of Dymola was resumed and Dymola was made a commercial product (Elmqvist (1995)).

Another product is Omola (Object-oriented Modeling Language) appeared in late 1988 from the Department of Automatic Control, Lund Institute of Technology. Omola is still a university product and is designed for generality and flexibility. Omola allows the models to be decomposed hierarchically with well defined interfaces that describe interaction. Omola also has primitives for discrete events. Omola supports its own simulator Omsim (Andersson (1994)). The features of Omola have now been included in Dymola.

In addition to Dymola and Omola there are many other languages with similar ideas implemented. In 1996 an international effort was initiated for the purpose of bringing together expertise in object-oriented physical modelling and defining a modern uniform modelling language. This language has got the name Modelica (which is now a trade mark of the Modelica Design Group). Modelica is intended for modeling within many applications, such as electric circuits, mechanical systems, hydraulics, chemical processes etc. It supports many different model descriptions, such as ordinary differential equations, differential-algebraic equations (DAE), bond graphs, Petri nets (for discrete events), etc. Modelica will serve as a standard format for the exchange of models between different users (Elmqvist *et al.* (1998)). Current information on Modelica is found on the web address:

http://www.modelica.org

Dynamic Modeling Library from Dynasim AB

Dynasim AB offers a Dynamic Modeling Laboratory, including four parts: the Dymola language that promotes reuse of library models, Dymodraw that allows graphical model composition, Dymosim for continuous/discrete simulation and Dymoview for 3D animation and plotting.

Dymola (Dynamic Modeling Language) is an object-oriented language and a program for modeling of large systems. Models are hierarchically decomposed into submodels. Reuse of modeling knowledge is supported by use of libraries containing model classes and by use of inheritance. Connections between submodels are described by defining cuts which model physical couplings. Special constructs are available for defining the connection topology of composed models.

Model details are given by ordinary differential equations and algebraic equations. The user need not convert the equations to assignment statements. Matrix equations facilitate convenient modeling of 3D mechanical systems, control systems, etc.

Discontinuous equations are properly handled by translation to discrete events as required by numerical integration routines. Dymola also supports instantaneous equations to model friction, impact and difference equations, etc. Dymola automatically generates the needed time and state events.

Dymola converts the differential-algebraic system of equations symbolically to state-space form if possible, that is, solves for the derivatives, or to reduced DAE form. Efficient graph-theoretical algorithms are used to determine which variable to solve for in each equation and to find minimal systems of equations that have to be solved simultaneously (algebraic loops). The equations are then, if possible, solved symbolically. Linear systems of equations can be solved symbolically or numerically. Code is generated to handle the nonlinear case iteratively. Higher index DAEs, typically obtained because of constraints between submodels, are handled by symbolically differentiating equations.

Dymola cooperates with several simulation packages. The converted equations can be output in different formats. The presently supported formats are: ACSL, Simnon, SIMULINK, C and FORTRAN according to Dymosim (DSblock). Dymola is available for Windows and Unix.

Dymosim (Dynamic Model Simulator) is a simulation program for solving ordinary differential equations (ODE) and differential-algebraic equations (DAE). It has a variety of numerical integration methods: one-step, multi-step and extrapolation methods. There are methods suitable for stiff dynamical systems and also methods that handle time- and state-events.

The model to be simulated is defined as a set of C functions according to the proposed standard DSblock (Dynamic System block). DSblock specifies when different routines are called, what the parameter lists are and what should be calculated. DSblock routines are generated by Dymola.

See Dymosim in the References for a source of more information. Current information about the software is found on the web address:

http://www.dynasim.se

9.6.6 From Simulation to Real-Time Implementation

Once an advanced controller is built up in a simulation environment it has to be transferred to a real-time environment. It is obvious that if the code were to be written again, there are many chances for human mistakes. Therefore, automatic code generation from the simulator to a real-time environment is highly desirable. Three platforms are described below:

Simnon from SSPA Systems

There is a real-time feature in Simnon, so that the simulation can be tested in a real-time environment. Since Simnon can import and export variables it can be connected to physical hardware as well. Via DDE links Simnon can receive commands and return results during an ongoing simulation. For example, Simnon can be connected to a man-machine interface created in InTouch (Wonderware, Irvine, California). This gives an environment for operator training. See Simnon in the References for a source of more information.

Real-Time Workshop from the MathWorks Inc.

Simulink has an extension feature, that makes it possible to automatically generate C code directly from Simulink block diagrams. This allows the execution of continuous-time, discrete-time and hybrid system models on a wide range of platforms, including real-time hardware. The product is called Real-Time Workshop.

The Real-Time Workshop can be used for embedded real-time control. The simulation code for real-time controllers is generated, cross-compiled, linked and downloaded onto a target processor, such as a DSP board or a variety of commerical hardware, such as 68xxx, MIPS, SPARC and Intel. Other applications include "hardware-in-the-loop" simulations. Typical applications are training simulators. See MATLAB in the References for a source of more information.

RealSim from Integrated Systems Inc.

This product also has the same purpose, to translate from simulations in SystemBuild to real-time execution with some connected physical process. RealSim includes both a software environment and a class of computers for the implementation. Models built in SystemBuild model are implemented in the RealSim computer by using a software development environment. The most advanced RealSim computer can accomodate anywhere from 1 to 11 Intel i486 CISC or i860 RISC processors, or Texas Instruments DSP. RealSim also provides graphical user interfaces for the real-time application. See $MATRIX_X$ in the References for a source of more information.

9.7 Application Specific Simulators

There is a large number of applications reported in the literature where simulators have been used for wastewater treatment design or operation studies. The references are not intended to be complete, but a few examples are presented. Results from a working group within the European Union research program COST 682 concerning the use of simulation software for wastewater treatment processes have recently been published by Jeppsson (1994) and Vanrolleghem & Jeppsson (1994).

There are quite a lot of simulators available for wastewater treatment systems, although they have a lot of different features.

An early example of the application of a general purpose simulator is reported by Andrews & Graef (1971). A model library for simulation based on a general simulation package was reported in Olsson *et al.* (1985a). Simnon was used as the simulation platform. The work to improve and verify further unit processes has continued. Other simulation applications, mostly for research purposes, are reported by Barton & McKeown (1986), Bidstrup & Grady Jr. (1988), Gujer & Henze (1990), Font & Ruiz (1993), Pons *et al.* (1993), Angelidaki *et al.* (1993) and Ryhiner *et al.* (1993).

The research at McMaster University, Hamilton, Ontario, Canada (Patry & Takács (1990)) has led to the GPS-X software, while model developments at the Swiss Federal Institute of Technology (Reichert (1994)) have been packaged in ASIM and AquaSim. BioWin is a product of bio-P research at the University of Cape Town, South Africa and at McMaster University, see Dold (1990), Dold (1992) and Barker & Dold (1997). EFOR is a result of Danish development work at Krüger Systems. Simba has been developed by Institut für Automation und Kommunikation (IFAK) in Germany and is based on Simulink as the general simulation platform.

9.7.1 Models

Simulators dedicated to wastewater treatment processes have already defined a number of elementary unit processes relevant for wastewater treatment or water quality modelling. Such unit processes may define static components like flow splitters, mixer boxes and short pipe connections as well as complex biological descriptions of aerated, anoxic or anaerobic zones. Each one of these zones is defined as a complete mix reactor, and several reactors can be connected in series. Also, more or less complex settler models have been defined as modules.

A complete plant can be defined simply by connecting a number of units representing reactors, settlers and their connections. Likewise, some of the simulators (such as GPS-X) offer model units for primary settlers, sludge handling, deep bed filtering as well as anaerobic digesters.

Of course, any general purpose simulator can be supplied with modules describing elements of wastewater treatment systems. From a simulation point of view this will give exactly the same outputs. The main difference to the dedicated simulators is the human interface and the guidance for the user during the execution.

Almost all the simulators below have models for the biological part of the activated sludge system based on the IAWQ AS Models No 1 and No 2. As already mentioned in Section 4.1 the first model describes C and N removal and has proven to be very reliable. At the same time it suffers from too many parameters describing the system, so there is no unique way of updating the parameters of the model. This has been thoroughly examined by Jeppsson (1996), and reduced order models have been derived for operational purposes. The IAWQ AS Model No 2 also includes the P removal mechanisms, as described in Section 4.1. This model is extremely complex and has not yet been tested as thoroughly as the IAWQ AS Model No 1. There are also other "competing" models for P removal, one described by Dold (1992) and Barker & Dold (1997) and another by Johansson (1994).

The settler models are far from satisfactory. In some of the simulators the settler is represented as a simple ideal "point settler", that is the solids/liquid separation is simply represented by a simple algebraic expression. Other settler models are based on the flux models (see Section 4.6) and allow the calculation of the sludge concentration profile in one dimension. The GPS-X simulator has a model which includes a complete AS model within every concentration layer of the settler. This makes it possible to predict DO and nitrate concentration profiles in the settler as well as the amount of nitrogen gas generated (which in turn may lead to problems with rising sludge).

9.7.2 Some Simulators on the Market

Here we describe some wide-spread simulators on the market. All of them are being used both in academia and in industry. Some general remarks may first be made with respect to the models available. Most of the simulators have included the IAWQ AS Model no 1 (for nitrification and denitrification) which is still the best available model for the nitrogen removal. The IAWQ AS Model No 2 for bio-P removal is still not sufficiently tested, and there are other similar models, like Dold (1992) which has been tested to a greater extent. Furthermore, these models are only as good as the wastewater characterisation and the determination of the nitrifier growth rate.

Whatever the final simulator will be it may be advisable to borrow or rent for a few months. It should be tested on some process that is well known. All the relevant mass balances should be tested properly. Furthermore, do not buy a simulator that doesn't let you build your own configurations. Every plant is slightly different and one has to think about storm flows, sludge liquor returns, industrial wastewater equalisation, hybrid treatment, etc.

GPS-X from Hydromantis

GPS-X is a very powerful simulator for dynamic simulation of wastewater treatment systems. GPS-X is supplied with a large number of models covering virtually all of the unit processes found in wastewater treatment plants, including advanced nutrient removal models, fixed-film operation, anaerobic reactors, secondary settler models, primary settler models and several units for sludge operation. The user builds a layout of the plant by manipulating graphical icons on a drawing board and assigns the appropriate process models and parameter values.

The simulator is built on the ACSL simulator, that provides powerful integration and general simulator features. A particularly interesting feature in GPS-X is the fast steady state calculation, which is done iteratively. This makes it very easy to find new initial conditions for dynamic simulation.

On top of the off-line capabilities GPS-X can download on-line data automatically from a SCADA system. This makes model calibration as well as parameter estimation (see Section 7.7) much less cumbersome. In particular there is a module, called Respeval, that automatically derives kinetic and stoichiometric data from properly developed respirograms (see Section 23.9).

GPS-X is also linked to MATLAB and can use the extra data analysis capabilities of that package. The MATLAB based controller design module GMI links GPS-X to MATLAB for the design and off-line testing of controllers.

GPS-X has an optimiser that can be used for model calibration. Historial data can readily be included in the simulations. There is a sensitivity analysis module, that allows automatic execution of many simulations, where certain parameter values are varied within certain limits.

GPS-X is available for Unix and Windows NT platforms.

See GPS-X in the References for a source of more information. Updated information is found on the web page:

http://www.hydromantis.com

SIMBA (Otterpohl Wasserkonzepte)

SIMBA (SIMulationsprogramms für die Biologische Abwasserreinigung) has been developed at IFAK, Institut für Automation und Kommunikation e.V., Magdeburg, Germany. It can be considered a custom made version of Simulink for wastewater treatment applications. Using Simulink as the general platform, it allows the simulator vendor to concentrate all the development work on modelling and an attractive and intuitive user interface. Therefore SIMBA makes use of all the features of Simulink and also can use MATLAB for all the post-processing. SIMBA also includes facilities for software sensors and supervision of sensors.

For biological reactors, the IAWQ AS Models No 1 and No 2 are implemented. Furthermore several settler models as well as influent flow data series are available. The model library can be readily extended, since many users can document their model developments in Simulink format. Thus, academic users make valuable contributions to the modelling efforts, while SIMBA enjoys an increasing popularity among industrial users. SIMBA can readily import historical data and can also be connected on-line to a SCADA system for data acquisition into the simulator.

SIMBA also offers a model library called SIMBA Sewer, which is a set of model blocks to simulate the upstream sewer network of a treatment plant. This makes it possible to simulate integrated operations. The SIMBA Sewer does not claim to replace conventional sewer simulation software. For example, it does not describe the catchment of rain water.

See SIMBA in the References for a source of more information. Current information about SIMBA is found on the web address:

http://www.ifak-system.com

AQUASIM

AQUASIM is an interactive system for simulating and analysing the dynamics of aquatic systems. It is a graphical, mouse driven program that includes a number of pre-defined compartment models (mixed reactors, biofilm reactors, river sections). Using compartments limits the generality of the program, but allows the selection of efficient numerical algorithms, depending on the type of partial differential equations used to describe the transport mechanisms. The internal dynamic processes within a compartment are formulated by the user without any restrictions. Compartments can be combined by different types of links. The program contains built-in tools for identifiability analysis (by sensitivity analysis), parameter estimation and uncertainty analysis. Results are presented as traditional two-dimensional plots (not on-line) or saved as ASCII files. The program does not have any real-time capabilities.

AQUASIM was developed mainly for internal use, but it is also available to interested scientists. Technical support for program usage is not provided, but an electronic user group is established to facilitate communication among program users and AQUASIM courses are organized regularly. AQUASIM is available for numerous platforms, including Solaris, AIX, HP-UX, Apple PowerMac, DEC-VMS, DEC-Unix, Windows 95, 98 and NT.

Further information is found in Reichert (1994). Current information is found on the web address:

http://www.eawag.ch/soft/aquasim/shortdesrc.htm

BioWin from Envirosim Associates Ltd. of Oakville, Ontario, in partnership with Reid Crowther & Partners Ltd., Vancouver, B.C. Canada

Biowin is made up of a number of linked processes to simulate a wastewater treatment plant. The cornerstone of Biowin is the model of the biological treatment component. The IAWQ AS Model No 1 has been extended by Professor Peter Dold (Dold (1990), Dold (1992)) modifying the basic package to include biological phosphorus removal. The models contain 20 differential equations for each bioreactor in the configuration.

The secondary settler is modelled as either an ideal settler or a one-dimensional settler. It also can include biological reactions. Biowin also incorporates sludge handling, such as anaerobic digestion and dewatering. The plant hydraulics are modelled as flows in pipes, channels or tank inlets and outlets.

Biowin incorporates two modules, a steady state analyser, assuming constant influent flowrates and compositions, and a dynamic simulator, using time-varying inputs. Biowin has been implemented in a Windows environment.

See BioWin in the References for a source of more information. Current information on BioWin can be found on the web address:
$http://www.envirosim.com$

EFOR from Krüger Systems, Denmark

EFOR is a stand-alone software package for the simulation of complete wastewater treatment works. The models include organic as well as biological N and P removal. The IAWQ AS Model No 1 is included among the modules. The IAWQ AS Model No 2 for biological P removal has been modified to include bio-P activity for denitrifying organisms as well as some hydrolysis processes. Professor Mogens Henze at the Danish Technical University has been a partner in this work. There are three types of settlers included, having 0, 2 or many layers.

EFOR allows not only continuous flow processes but also sequential batch and alternating processes. EFOR is available in the Windows environment. See EFOR in the References for a source of more information. Current information on EFOR is found on the web address:
$http://www.efor.dk$

STOAT from WRc, Software Services, Swindon, Wiltshire, UK

STOAT has been developed in the UK at the Water Research Centre (a stoat is the symbol of the WRC) and is now a commercial product. It can simulate entire sewage works and has been used for design and operation of wastewater treatment plants. The software can be used together with commerical sewerage and river quality models such as MOUSETRAP and MIKE 11 (from the Danish Hydraulics Institute) and HydroWorks-QM (Wallingford Software, UK). STOAT also contains a simple sewer model to assist integrated modelling of sewers and sewerage works.

STOAT includes modules for many process treatment configurations, such as primary settlement, storm tanks and detention tanks, wet-wells, activated sludge systems, including oxidation ditch, sequencing batch reactor, and Deep Shaft systems (Deep Shaft is a trademark of Kvaerner Engineering), activated sludge settlement, biofilm systems, such as trickling filters, BAFs, RBCs and fluidised beds, humus tanks (biofilm settlement), disinfection, anaerobic and thermophilic aerobic (ATAD) sludge digestion, sludge consolidation, dewatering, drying and incineration. There are also models for simulating popular process control options such as PID and PLC systems. User-defined or written models can also be added, should a novel treatment process be required.

The biological models include both BOD and COD based models. The COD models include the IAWQ AS Models Nos 1, 2, 2d and 3 implementing N and P removal in activated sludge systems. The biofilm models are based on the model of Wanner, Gujer and Reichert, with the kinetics taken from the IAWQ AS Model No 1 COD and the Jones BOD models.

Recently an Integrated Catchment Simulator (ICS) has been developed that runs MOUSE, STOAT and MIKE 11 in parallel, allowing the control element of the programs to operate across all three programs, simulating the effect of control in the sewer based on the state of the sewerage works. The ICS will also support integration with SCADA systems on a site-specific basis.

The Water Research Centre have validated the models in the UK on many full-scale sewage works. STOAT has been used for the design of many plants in Europe, America and the Middle East.

See STOAT in the References for a source of more information. Updated information is found on the web address:

http://www.wrcplc.co.uk

WEST from Hemmis n.v., Belgium

WEST (Wastewater treatment plant Engine for Simulation and Training) is an interactive dynamic simulator for wastewater treatment plants. It has been developed in collaboration with Epas n.v., Belgium, Aquafin n.v., Belgium and the University of Ghent, Department of Applied Mathematics, Biometrics and Process Control. The simulator includes carbon removal as well as nitrogen and bio-P removal. Sludge settling, recirculation and effluent turbidity are modelled. It also includes controllers and has a modern graphical interface for the user. The second generation software package is called West^{++}. In contrast to WEST, which only runs on IBM RS/6000 machines, West^{++} is a multi-platform program now running on UNIX (IRIX, AIX and Linux) and Windows NT platforms.

For the model building there is a Hierarchical Graphical Editor (HGE). User support is provided so as to minimise modelling errors and maximise reuse of existing knowledge. To this end a user extendable Modelbase has been created for wastewater treatment systems. Other Modelbases exist also for electrical systems, system dynamics, etc.

A powerful feature of the modelling environment is that different model libraries can be accessed and models can be specified in a Model Specification Language (MSL). In this language previous modelling work can be reused. In view of the widespread use of the matrix representation of conversion (see Chapters 3 and 4) models within the wastewater treatment modelling community, MSL also supports the definition of models in this format.

See WEST in the References for a source of more information. Current information about West^{++} is found on the web address:

http://hobbes.rug.ac.be/ peter/WESTpp.html.

SSSP from Clemson University, S.C., USA

SSSP (Simulation of Single Sludge Processes) is a university product for the interactive simulation of activated sludge systems, developed at the Clemson University, South Carolina, USA. The software, often called the Clemson model, includes models for carbon oxidation, nitrification and denitrification, and is basically the IAWQ AS Model No 1. The settler is highly simplified and is modelled strictly as a separation point with all solids entering it being returned in the recycle stream and none being accumulated within the clarifier nor lost in the effluent. Consequently SSSP does not calculate a true dynamic situation. The simulator allows time-varying inputs. The treatment plant is composed of up to nine completely mixed reactors. Currently SSSP is implemented in the DOS environment.

Further information on SSSP is found in Bidstrup & Grady Jr. (1987). Current information and a user manual is found on the web address:

http://www.eng.rpi.edu/dept/...
env-energy-eng/WWW/CLEMSON/sssp.html

DSP from The University of Queensland, Australia

DSP (Dynamic Simulator for Prefermenters) has been developed in a joint project between the Cooperative Research Centre for Waste Management and Pollution Control Ltd, the Advanced Wastewater Management Centre at The University of Queensland and the software developer Science Traveller International (STI). It is the first commercially available prefermenter simulator. The prefermenter model has been developed and verified by Elisabeth v. Münch at The University of Queensland. The prefermenters produce the VFA that bacteria would need, in particular for the P release in the anaerobic reactor (see Section 4.2).

DSP takes the influent characteristics to a prefermenter (VFA, SCOD, TCOD, TKN, and ammonia-N concentrations) and the operating conditions (prefermenter hydraulic and solids residence times) and uses these to predict the prefermenter effluent characteristics and mass of VFA. This output can be used as input for subsequent biological nutrient removal processes.

DSP is developed for Windows and is meant to assess the costs and merits of prefermenter designs and the addition of prefermenters to wastewater treatment plants. It can also be used to analyse operation of existing prefermenters. It contains a calibration facility so that a model can be calibrated based on appropriate data.

Current information on DSP can be found on the web address:

http://www.scitrav.com/wwater

9.8 Buyer Beware!

In Section 9.5 a checklist for the selection of simulators was presented. Let us now critically consider what are the important issues.

9.8.1 Buying Solutions

Available models is a key issue. All models have to be verified and validated. Even if a simulator seems to offer a large number of models, it is not the same as an accurate simulator.

Different users need different levels of flexibility. There are seldom ready solutions. A researcher or designer may appreciate flexible modularity and advanced model building facilities. A process engineer may instead emphasize validated models and will probably not make major changes to the models. Still model validation tools are useful. However, as we have seen in Chapters 5 to 8 model verification and validation is not trivial.

All modern simulators are supplied with more or less graphical features. This is of course helpful, both for model building, system set-up and for the presentation of results. However, graphics is not everything. Many other issues have to be considered, such as:

- What does the interface look like? How easy is it to change parameters of the model? How easily can modules be changed in the model?
- What kind of numerical integration is available?
- Are there optimisers available?
- How easily can the output results be analysed and transferred to other software platforms?
- Which operating systems are available?
- What kind of software support is available?
- Does the software manufacturer have any model knowledge?

Finally, the prize is the bottom line. It is not always best to buy the lowest cost software. If a number of man-months work can be saved by buying more expensive software, it may still be advantageous.

9.8.2 Future Needs

Several difficulties remain to be solved. First of all, there is still a considerable lack of knowledge of many reactions and mechanisms within a treatment plant. Secondly, the high complexity of the mathematical models used today creates problems in calibration and identification, which hinder the practical implementation of simulators at WWT plants. Due to the complexity of the wastewater treatment systems there is a need for several improvements and developments in future simulators. The most important ones discussed in the European COST 682 working group as reported by Vanrolleghem & Jeppsson (1994) have been identified to be:

- help for model calibration
  - identifiability analysis;
  - parameter reliability analysis;
  - required measurements and experimental design;
  - variable conversions;
  - input data manipulation (outliers, filtering, etc.);
  - parameter sensitivity analysis;

  - parameter correlation and importance;

- tools for model structure selection and model databases;
- models including pH variations and chemical precipitation as well as settling and hydrolysis;
- better handling of hydrodynamics phenomena;
- combinations of WWT simulators with

  - sewer system simulators;
  - receiving water models;

- possibility for symbolic equation manipulation and analysis;
- user guidance for solutions to numerical problems;
- better internal checks for "obvious" modelling errors (reliability vs. flexibility);
- on-line help systems with current knowledge and history of the processes;
- knowledge based system for process diagnosis;
- support for both graphical and symbolic model descriptions;
- multiple low-level programming language support;
- drivers to external programs

  - better numerical routines;
  - data acquisition and real-time control systems (SCADA);

- optimisation tools (cost functions for investments vs. operational costs).

It is obvious from the list above that much work and research still remain until fully operational simulators are a standard tool at wastewater treatment plants.

Many of the items above are specific to wastewater system simulators. However, there are some developments in general purpose simulators that can also help specific simulators. In particular, we can see great progress being made in model descriptions, better numerical routines and optimisation tools.

9.9 Summary

The chapter has given a survey of simulation and simulation tools. It is a necessary ingredient for both process design, operation and control. The range of software available to the process engineer or control engineer is enormous. One of the key issues is not so much the facilities (although important) but often the level of support provided. Those interested in pursuing more information should refer to the web sites of the major vendors.

9.10 Further Reading

- A more elaborate discussion on modelling and simulation software in general (not only on control systems) can be found in Cellier (1993).
- The survey by Åström *et al.* (1998) gives a very good insight into continuous-time modelling and simulation.
- There are many good textbooks on numerical methods, like Butcher (1987), Gear (1971) and Hairer *et al.* (1987).
- Gustafsson (1994) gives an excellent survey of common pitfalls in integration of differential equations.
- There is a special treatment of stiff differential equations in Hairer & Wanner (1991).
- The concentration profile of the settler may require many spatial divisions and special solution methods Jeppsson (1996) and Diehl *et al.* (1990).

9.11 Exercises

1. What is the difference between discrete event simulation and dynamic system simulation?
2. Give some system where it is of interest to mix difference equation models with differential equation models.
3. Compare advantages and disadvantages of a block oriented (graphical) system description and an equation oriented description.
4. What is the problem of a stiff system? How can this be overcome?
5. Why is the Euler approximation insufficient in many cases of numerical integration, in particular in wastewater treatment systems?
6. What is the advantage of hierarchical modelling?
7. What is the advantage of object oriented modelling?
8. When does causality have to be particularly considered?
9. What is the difference between explicit and implicit numerical integration?
10. What is the advantage of having a steady-state solver?
11. Explain numerical instability. How can it be overcome?

12. What is the difference between a general purpose simulator and an application specific simulator?

13. What is the difference between a real-time simulator and an off line simulator?

14. Make a specification of the simulator you would need for your purpose (operator, plant engineer, researcher, educator). Give reasons for including or excluding the items in the simulator checklist in Section 9.5.

Part B

Diagnostics

Chapter 10

Diagnosis - an Introduction

The goal of this chapter is to introduce this part of the book on diagnosis. A wastewater treatment plant is a very complex system, including a lot of equipment and complex processes. As we have seen the system can be monitored by on-line sensors, analytical instruments, laboratory analyses and can be observed by the operator. All these measurements and observations make up the basis for control or operational actions.

In a large wastewater treatment plant there are literally hundreds of sensors available. Only a few of them will measure the quality of the wastewater, while many of them monitor various physical parameters. Some of the basic plant instrumentation will be described in Chapter 24. Sensors for primary physical properties of wastewater are discussed in Chapter 22, while water quality analysers are described in Chapter 23 . In the discussion on diagnosis we will simply assume that certain measurements have been made available in some way.

Data acquisition systems may collect a huge amount of data, but there are relatively few significant events taking place. Therefore the data from all the measurements must be mapped into meaningful descriptions of what is going on in the plant. As previously remarked a plant is often "data rich but information poor", so it is easy to get fooled by the huge amount of data available from a plant data acquisition system.

In a biological wastewater treatment plant the major challenge is to find out the current status of the process. The measurement task is far from trivial. For a long time lack of instrumentation has been a stumbling block for improved wastewater treatment operation. However, we can see encouraging instrumentation developments. Furthermore, the knowledge of the microbial behaviour is limited. Despite the complex models described in the Modelling part of the book, we have only a very limited understanding of the true microbial behaviour of the system. Therefore any detection of changes in the plant behaviour will be of significant value in the operation.

The problems of diagnosis have many faces and basic definitions are made in Section 10.1. Detection, to extract adequate information from sensors and observations, is discussed in Section 10.2. Part of this is signal filtering, Section 10.2.5. Diagnosis, to find likely causes of disturbances or faults, is described in Section 10.3. Once the problem is diagnosed, then consequence analysis, is the next natural step, Section 10.4.

The whole chain of diagnosis work requires adequate information about the plant. This is discussed in Section 10.5.

10.1 Defining the Problem

More and more wastewater treatment systems are getting supplied with advanced instrumentation, not only for physical parameters, but also for water quality analysis. This will give a mine of information, if only the important and relevant information could be extracted and interpreted, especially in real-time.

Any action taken on a plant relies on an explicit or implicit description of the state of the operation or events that are occurring. The action may be a manual correction or an automatic controller responding to some measurement. The information is mostly coming from on-line sensors that are connected to a computer, but often some operator observations are important complements to the on-line sensor information.

The diagnosis problem can be described by three phases:

- *data rectification*, that is, screening the available data to remove irrelevant information (this problem was discussed in Section 6.5);
- *detection*, to examine the available data from on-line sensors or from observations to detect abnormal features or "effects"; and
- *diagnosis*, to analyse the "effects" and find possible "causes".

Below we will discuss this in further detail.

10.1.1 Hierarchical Structure

Earlier we have illustrated the various time scales of a wastewater treatment plant, ranging from weeks to minutes and seconds. It is natural to separate the various plant operating problems into at least three layers, as illustrated in Figure 10.1. The diagnosis problems are defined in the supervisory control level, above the on-line control system and below the planning layer. The general goal at the supervisory control level is to assure the proper operation of the plant by monitoring the performance of the plant and its automatic control systems. At a minimum this implies keeping operators, managers and maintenance people better informed about what is going on in the process. The design and implementation of computerised fault diagnosis must be driven by the needs of these user groups.

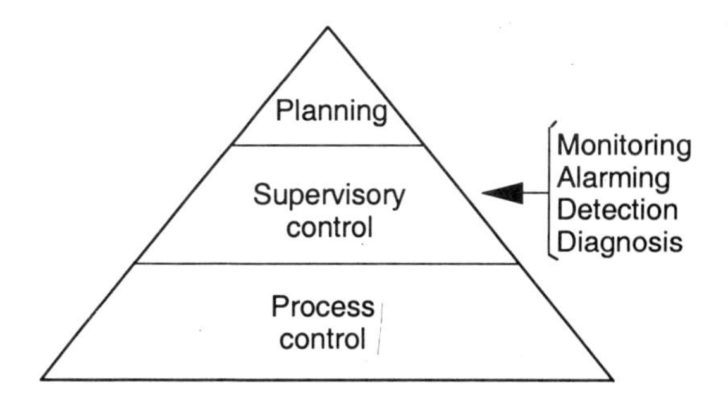

Figure 10.1: A hierarchical structure of the plant operation.

10.1.2 Detection and Diagnosis

Diagnosis is part of a feedback scheme for the plant, basically for manually implemented control, as illustrated by Figure 10.2. The plant is influenced by various kinds of disturbances, and we will discuss these in more detail in Chapter 15. We assume that there is some computer data acquisition system that is supervised by the operator.

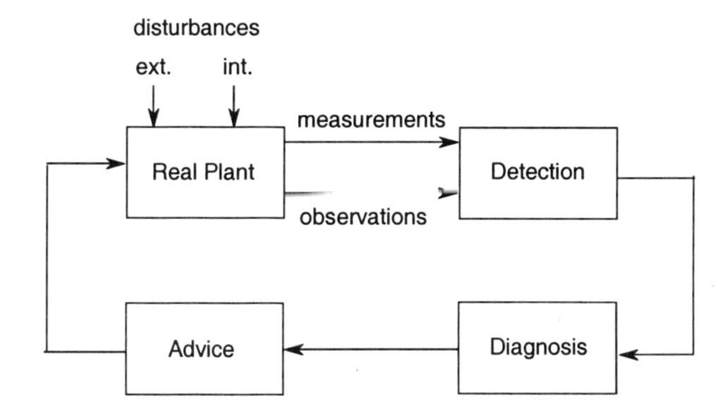

Figure 10.2: Diagnosis as part of the plant operation.

In future chapters we will discuss more "normal" behaviour. The disturbances will cause various state variables to deviate from their desired values. For "normal" behaviour the detection and diagnosis problems are unambiguous, and the cause-effect relationship is well defined. The detection problem is obvious (even if the measurement may be advanced). Other variables may be indirectly calculated, based on the primary measurements.

During "normal" operation the structure of the control action is well defined. Even if the variable deviations are large, the manipulated variable is defined uniquely. For the supervisory control, the system will give some advice for some correction of a valve, a set-point of a controller, etc.

During an "abnormal" situation there are many more choices. It is not apparent which manipulated variables to use, or which control loops to use. For example, if the problem is caused by equipment failure, completely different actions are needed. Furthermore, if a major process upset is observed, like bulking sludge, there may be many different actions needed in order to return the process to normal operation.

Data analysis and interpretation is an important part of all plant operation. Data analysis is a term used to describe how data are manipulated and processed to produce the features of interest. Data interpretation describes mechanisms to put labels on the data observed. Here we will define *detection* as the combination of process observations and measurements, data analysis and interpretation to detect abnormal features or "effects" and the isolation of faults. In this context there are many labels of interest, including state descriptions (low, high, normal), trends (increasing, decreasing, steady), process changes, shape descriptions and fault descriptions. Detection does not involve any explanations or causes for the behaviour. *Diagnosis* involves the analysis of "effects" to identify and rank likely "causes". This generally involves iteratively proposing and testing hypotheses based on known cause-and-effect relationships.

The *advice* involves the problem of synthesising strategies that can be implemented by plant operators to eliminate the "causes" and return the process to normal operating conditions.

The ultimate reason for detection and diagnosis is to find early warning systems for changes in the process. This is particularly important in a biological system, where some of the changes are not very obvious and may grow gradually until they become a serious operational problem. One example is sludge bulking. If an early warning is possible, then the chances for successful corrections become much greater.

In practice the diagnosis and response tasks are most often characterised by manual, ill-documented or *ad hoc* operator procedures. There is a tremendous scope for improvement by adopting computer-based fault detection, diagnosis and advisory systems. Some of the benefits of this are:

- reduced costs by vigilant monitoring of key parameters;
- consistent monitoring of the water quality;
- increased consistency by rapid detection and correction of disturbances; and
- reductions in human error due to misassessment of the process condition or failure to follow standard procedures.

These incentives have stimulated a large number of academic and industrial activities over the last decade. Still there is no widely accepted diagnostic technique and there are few standard vendor-supported tools available to support off-the-shelf solutions. Below we will illustrate some of the proposed diagnostic approaches, including artificial intelligence (sometimes called reasoning systems or knowledge based systems (KBS), probability and statistics (such as statistical process control), control theory (for example, estimators) and safety and reliability (such as fault trees and alarm trees).

10.1.3 Faults and Failures

In detection and diagnosis one speaks about faults and failures. Here a *fault* is defined as an unexpected component or process change that might be serious or tolerable. A *failure* suggests a complete breakdown of a process or a process component. Sources of faults or failures may be actuators and sensors. Typical faults are bias or drifting values for a sensor or sticking for an actuator. In general speedy and accurate correction of faults is required. Some of the hazards of slow or incorrect treatment are:

- faults not treated early enough may propagate catastrophically;

Example 10.1 (Fault propagation) *Figure 10.3 shows a gradual change of the effluent suspended solids concentration in a secondary settler. If the gradual increase of the concentration can be detected early, then the chance for some countermeasure would increase.*

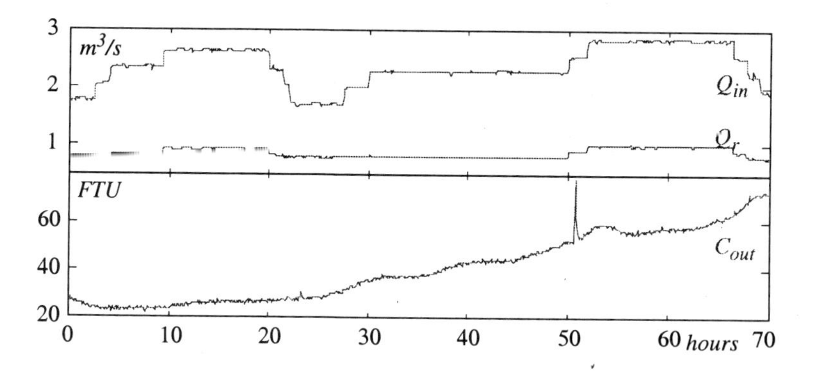

Figure 10.3: The floc settling properties have deteriorated and are causing the secondary settler effluent suspended solids concentration to increase.

- dormant faults undetected or left untreated may greatly affect the overall reliability and maintainability of the system;

Example 10.2 (DO membrane fault) *The membrane of a dissolved oxygen (DO) sensor was broken. The probe showed a too high DO level. Consequently the DO control system supplied the aerator with too little air. Actually, this was difficult to detect during almost steady state operation.*

A dynamic experiment, however, showed the fault. The air flowrate was changed stepwize. For a normal probe the response time for the DO concentration should have been in the order of 15-20 minutes. However, the sensor had a significant response within the order of 10-15 seconds (Figure 10.4), which is much too fast for a normal probe. Thus the broken membrane was readily detected.

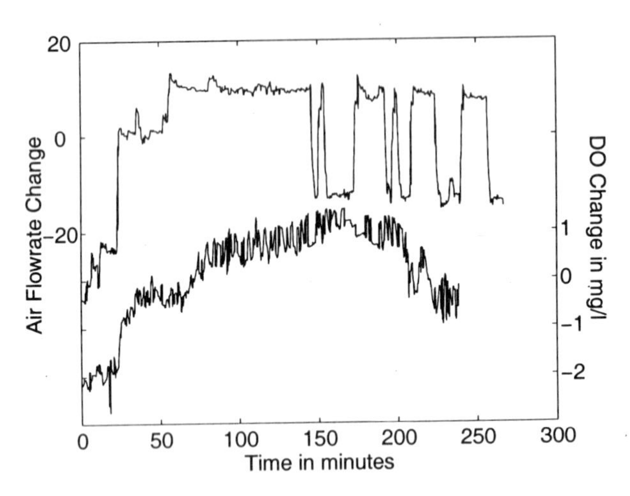

Figure 10.4: Dissolved oxygen sensor output for a stepwize change in the air flowrate. The response time is unreasonably short due to a broken membrane of the DO probe.

- the existing control system may itself be prone to failure and thus may mask the true cause of misoperation;
- wrongly identified faults and consequential repair actions may make the matter worse.

Disturbances are the main reason for control. In Chapter 15 we will see that some of them are caused by external sources. Most of them are related to the influent flow. Internal disturbances are created by various kinds of control actions, like recirculating flows, inadequate dosages or human mistakes. Other disturbances are caused by equipment failures, like breakdowns of motors, valves, communication, electronics, etc. Various instrumentation and sensor failures are of course also considered disturbances in this context.

10.2 Detection

Fault detection and diagnosis has been practised for many years in certain disciplines by using redundant hardware. A critical component has been duplicated or triplicated and then using a majority decision rule. Hardware redundancy is fast and easy to implement but has several drawbacks. It introduces more complexity in the system and can be very expensive. The attitude taken in this and the following chapters is that analytical redundancy can be utilised to reduce or avoid the need for hardware redundancy.

10.2.1 Observations and Laboratory Tests

Important information for detection comes from observations and laboratory tests. The quantity of scum or the colour and odour of activated sludge are useful information. Various analytical laboratory tests are also important. Microbial data can be classified according to the species, floc forming or filamentous organisms, protozoa or metazoa. The number of microorganisms per millilitre is another measure. To know the overall activity of different organisms is important.

Three classes of observation will need to be defined and evaluated:

- Initial intensive observation is necessary to build a database of normal ranges and daily (sometimes weekly, sometimes seasonal) baselines. An occasional intensive observation would also be required to update the database to account for changes in influent characteristics, in the plant flow sheet and in operating practises.
- Regular (daily and historical) observations of key variables would be made to confirm normal operation or locate abnormal events or trends.
- Special observations to assist in problem diagnosis and in evaluation of advice. This may involve laboratory analyses or on-line monitoring not normally carried out.

10.2.2 On-line Measurements and Data Screening

On-line data is obtained for all kinds of physical variables, like flowrates, levels, pressures. Turbidity, colour, pH, dissolved oxygen, alkalinity or oxygen reduction potential (ORP) are other interesting on-line variables. Sensors for these variables will be dicussed in Chapters 22 and 23.

In Section 6.5 we discussed some data screening problems, such as loss of data and extreme data points. Before any analysis can be made, the data screening is crucial, so that false conclusions are avoided.

10.2.3 Elements of Detection

Detection involves the examination of process observations or variables calculated from them, for example estimated parameters to detect abnormal features or "effects".

It involves at least:

- normal range comparisons for set variables: High and low limits are defined. Often *very* high or *very* low limits are used as a complement.
- rate of change: The trend is usually quite informative. Again it has to be averaged over a defined time horizon. Noisy data may corrupt the information.
- change limits: Most commonly some sudden changes (steps) appear in variables. It is obvious to detect a positive or negative step. Likewise a "pulse" or rectangular disturbance is readily detected. Again noisy data may corrupt the information.
- variance: The variation around some average value may give a good indication of the properties of the signal. Sometimes a large variance is a sign of poor sensor performance, sometimes of abnormal process behaviour.

Example 10.3 (Detection of settling problems) *Figure 10.5 illustrates the plant influent flowrate and the secondary settler effluent turbidity in a large municipal treatment plant. The influent flowrate contains both significant peaks (that are real) and a trend. It is obvious that the rate of change has to be calculated with some care, especially around the time 30 hours. The effluent suspended solids concentration is certainly very unsatisfactory at around 30 hours. This may be detected both by its absolute value and its variability. The problem of poor settleability is easily detected.*

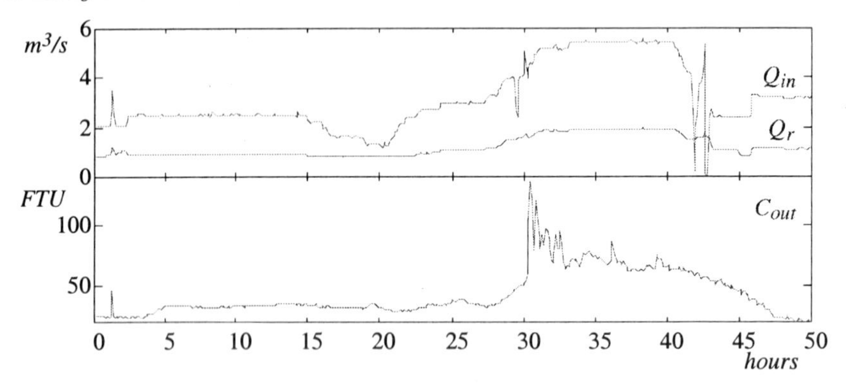

Figure 10.5: A rain event causing serious effluent suspended solids. The settling properties are quite poor, and the rain will cause a washout. Data records of influent flowrate (Q_{in}), return sludge flowrate (Q_r) and the effluent suspended solids concentration of the secondary settler (FTU). The data is recorded at a full scale plant during 50 hours.

Example 10.4 (Detection of rapidly changing air flow demand) *In an aerator with DO control the air flowrate approximately corresponds to the biodegradable load to the system. This was demonstrated in Example 3.9. Assume that* *the air flowrate demand to keep the DO concentration around the setpoint value falls rapidly. The DO control system can only detect that there is a fast decrease of the plant load. The reason can be either some toxic material killing or inhibiting the organisms or it can be a true rapid decrease in substrate concentration. In other words, there is a need to find the true cause of the effect, since the actions have to be quite different.*

Example 10.5 (Detection of an abnormal flowrate) *Since flowrates are measured continuously (for water or air) it is straight-forward to detect a flowrate change that is abnormal. What is the cause for it?*

We have noticed before, that the time scales in a wastewater treatment plant are widely varying, from seconds to months. For a detection this has to be taken into consideration. When comparing trends and variations one has to carefully consider the time scale, and the detection may be classified into:

- baseline comparisons for rapidly changing variables, mostly related to various equipment;
- baseline comparisons for variables changing in the minute-to-hour time scale (flows and nutrients); and
- normal range comparisons and trend detection for slowly changing variables (biomass and settling characteristics).

Variables can also be characterised by their frequency content. The appearance of certain frequencies in a signal can be an indicator of something special. For example, frequency analysis of sensor signals often reveals sensor problems. The signal from a sensor with a clean probe will look different than that of a dirty probe, Figure 10.6. Both the noise level (variance of the signal) and the frequency characteristics may give an early warning to clean the sensor.

In a plant there may be several binary signals indicating process alarms, equipment failures, and limit indicators. Such signals may give complete information by themselves. Other times they have to be combined with other information to get the full picture.

10.2.4 Numerical Methods for Analysis

Before we discuss diagnosis in more detail, we will mention a number of levels of analysis that are part of a detection problem. The basis for the diagnosis is the quantitative information obtained from on-line sensors or laboratory analysis. We have already mentioned that data reconciliation, as described in Section 6.5.8 is a prerequisite for all the data analysis. Some further considerations are:

- *binary information.* This may give the status of individual variables or equipment. If a pump has stopped, a flowrate has exceeded a limit or a pressure

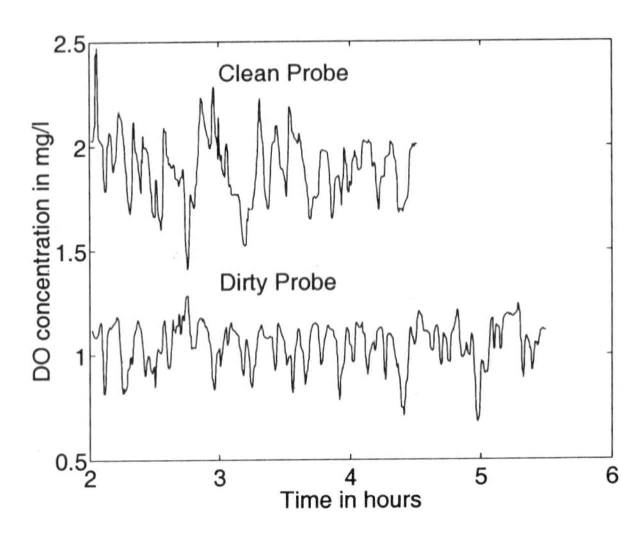

Figure 10.6: Clean and dirty sensor.

has become too small or too large, we will have some alarm or binary information. Such a signal may be used to combine with other alarms and to qualitatively predict what may happen as a result of the alarm.

- *analog signals.* A measurement signal has to be screened with respect to missing values, extreme values, trends, pulses and sudden steps (see Chapter 6). The screening in itself can give useful information. A continued signal analysis can add further information. Important features are of course the mean value and the standard deviation (or variance).

- *static relationships.* In Chapter 7 we discussed the problem of fitting static data. This is often an interesting first step for further analysis. For example, a continuous updating of the ratio between the settler underflow suspended solids concentration and the mixed liquor suspended solids may readily monitor the settler performance.

- *dynamical relationships.* One important aspect of feature extraction is to compare the behaviour of an explicit model with the obtained data set. The parameters in the model may be changed in order to drive the differences between the predicted and observed data towards zero. In Chapter 7 we described how parameters can be fitted to a series of data. The dynamical models may have the structure of the physical mass balances, or they may be in a time series form. When the whole data set is used simultaneously for the analysis we call it an *off-line* technique. In Chapter 12 we will reconsider the parameter estimation problem for an *on-line* situation. This means, that certain parameter values of a model are adjusted *recursively*, whenever a new measurement value is obtained. This is a useful technique to detect if a parameter is drifting.

A comprehensive interpretation system will map numeric data from the process into useful labels. Such a *feature extraction* is intended to produce the features that are resolved into the labels. As indicated above data pre-processing is necessary to remove irrelevant information and condition the input data to allow easier subsequent feature extraction.

Explicit (physically based) models are not always available for the feature extraction. Rather than focusing on comparisons between observed data and predicted plant behaviour, those approaches focus on distinguishing patterns such as shapes or characteristic signatures as well as movement or changes in patterns relative to a reference set. Earlier in Section 7.2 we described the least squares method as a way to extract features of a model. Empirical network or rule-based models (Chapter 5) are frequently used. Another method, multivariate analysis, will be described in Chapter 11.

10.2.5 Filtering

The ultimate purpose of filtering is to extract as much information as possible from a noisy signal. Here we consider digital filters, which means that we use a numerical algorithm to reduce the noise in the signal.

Digital Low Pass Filters

In order to examine slow variations it is necessary to remove individual spikes in the measurement data and other quick disturbances which do not contain relevant information. This is done with a digital low pass filter. Constructing a filter which effectively removes the quick variations and at the same time does not affect the slow variations is always a compromise. As for the analog filters, a higher order filter is more efficient in removing undesirable high frequencies. The two most important types of low pass filters are *moving averages* and *exponential smoothing.* Low pass filters in the process industry are almost always implemented with these types of simple filters.

Moving Average - the Simplest Low Pass Filter

A simple moving average filter is obtained summing the last n measurements and dividing the sum by n. Such a filter is written:

$$\hat{y}_i = \frac{1}{n}\left(y_i + y_{i-1} + \ldots + y_{i-n+1}\right) \tag{10.1}$$

where $\hat{y}_i$ is the calculated average at time i and y_i is the true measurement at time i. Note that in on-line filtering the calculation has to be based only on old values. Such a filter is called a *causal filter*, since it cannot be based on future measurements.

Example 10.6 (Moving average filters) *Figure 10.7 shows a measurement series together with moving averages of order 3, 5 and 10. Notice, that the moving averages are lagged behind the raw measurement values.*

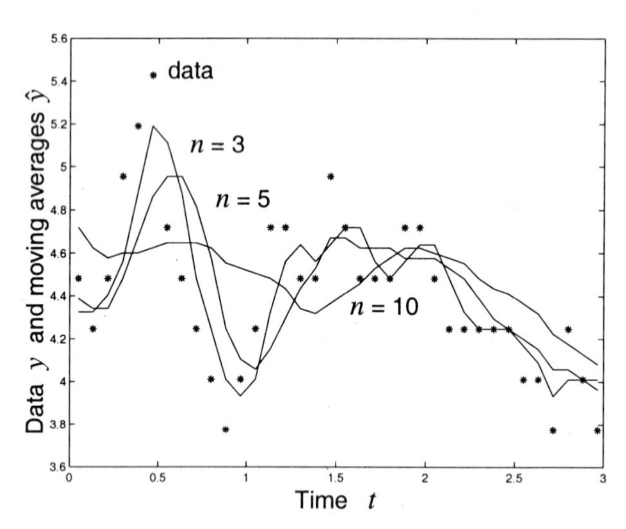

Figure 10.7: Measurement series and moving averages of order 3, 5, and 10.

If the output is the average of the input over the last n samples, it is also shifted $1 + n/2$ cycles. For increasing n, the filter output becomes smoother but more delayed. In off-line analysis, however, the filter can be non-causal and a moving average can be calculated using measurements both before and after the current time. Then the filtered value is not lagged in relation to the input values. For example, a non-causal moving average for five values is:

$$\hat{y}_i = \frac{1}{5}\left(y_{i+2} + y_{i+1} + y_i + y_{i-1} + y_{i-2}\right) \tag{10.2}$$

The moving average is a simple method but has certain limitations. If equal coefficients are used, the filter can be unnecessarily sluggish and not react adequately to real changes. On the other hand, if the coefficients taper off, it is difficult to analyse how the magnitude of the filter coefficients are related to the filter properties.

Exponential Filter

A very common low pass filter is the so called exponential filter. The filter output $\hat{y}$ is a combination of the last filter output and the last real measurement y. The filter is a first order system and is defined by:

$$\hat{y}_i = \alpha \, \hat{y}_{i-1} + (1 - \alpha) \, y_i \tag{10.3}$$

The filtered value $\hat{y}_i$ is computed by adding a weighted version of the earlier value of the filtered signal $\hat{y}_{i-1}$ to the latest measurement value y_i. The coefficient α has a value between 0 and 1. Equation 10.3 can be rewritten in the form:

$$\hat{y}_i = \hat{y}_{i-1} + (1 - \alpha) \, (y_i - \hat{y}_{i-1}) \tag{10.4}$$

which reveals another interpretation. The exponential filter corrects the filtered value as soon as a new measurement arrives. The correction has little gain and becomes small if α is near 1, which means that the filter is sluggish. This will greatly reduce the noise but at a cost of poor agreement with real changes in the measurement signal. If α is small, near 0, the correction gain is large. Consequently there is a poorer reduction of the noise level but the filter will track real signal changes more easily. For α =0 the filter output is identical with the measurement value.

Example 10.7 (Exponential filters) *A signal that is changed stepwise and is corrupted by noise, Figure 10.8, is used to illustrate the consequences of different choices of α.*

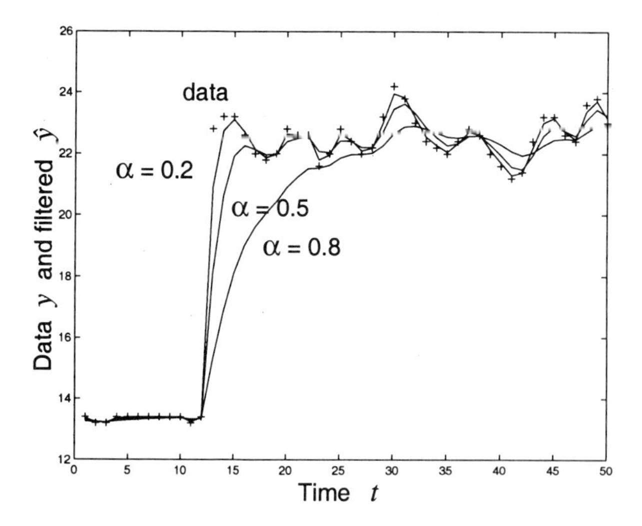

Figure 10.8: Effect of a first order filter with exponential smoothing.

The Equation 10.3 can be rewritten in another form, which will give still another interpretation of what the filter does. The right hand side can be further evaluated:

$$\begin{aligned}
\hat{y}_i &= \alpha\, \hat{y}_{i-1} + (1-\alpha)\, y_i && (10.5)\\
&= (1-\alpha)\, y_i + \alpha\, ((1-\alpha)\, y_{i-1} + \alpha\, \hat{y}_{i-2}) && (10.6)\\
&= \ldots && (10.7)\\
&= (1-\alpha)\, y_i + \alpha\, (1-\alpha)\, y_{i-1} + \alpha^2\, (1-\alpha)\, y_{i-2} + \ldots && (10.8)
\end{aligned}$$

Since α is less than 1 the coefficients will grow smaller for older measurement values. This is the reason that the filter is called exponential. Note, that for α small the "memory" is short, and for α close to 1 the memory can be made very long.

The tuning of the parameter α becomes a compromise. There are methods to overcome this dilemma, and α can in principle be variable. For more advanced filters we refer to the specialised literature.

Digital High Pass Filters

In some instances it is desirable to highlight the higher frequencies instead of the slow variations. A *difference builder* is a simple example of a digital high pass filter:

$$\hat{y}_i = \Delta y = y_i - y_{i-1} \tag{10.9}$$

The output differs from zero only when a change of the signal occurs. This can be an indicator for a fast change that should be detected readily. Thus high pass filters can be useful diagnosis tools.

10.3 Basic Elements of Diagnosis

Diagnosis involves the analysis of "effects" to identify and rank likely "causes". This generally involves iteratively proposing and testing hypotheses based on known cause-and-effect relationships.

A basic idea of diagnosis is to make use of all possible information in order to find out how the process is operating. On-line measurements can be combined with observations and more sophisticated laboratory analyses. Furthermore, measurement and observations at the actual instant can be combined with the previous history of measurements and of parameters. During the diagnosis there may be new questions coming up and new analysis or measurements will be required. This is similar to what happens when your medical doctor makes a check-up. If some of the information is contradictory or insufficient he wants to make still another test.

In the previous section we assumed that the information is given in a numerical form. This is not always the case. For example, an operator observation may be non-numeric (linguistic or symbolic). For example: "there is a lot of (little; as much as last Friday) foam on top of the aerator", "the sludge is very dark", "there are some dead spots in the aerator", "the smell is earthy today".

Knowledge-based systems (KBS), sometimes called *reasoning systems*, provide a technique for interpretation, when the operating histories do not provide sufficient information to develop an adequate numerical calculation from on-line input data. KBS have been shown to be particularly useful for fault and malfunction description where the data input relating the malfunctions directly to the measured data are very limited and in many cases non-existent. In such a system we will combine measurements with operator observations. It is intuitively obvious, that a competent combination of various observations will give a more reliable diagnosis. A KBS approach constitutes a model description of how the presence or absence of various features (with degrees of certainty) will map to a set of desired labels. There are different computer support systems for such an analysis, and we will discuss them further in Chapter 12.

In order to illustrate different diagnosis strategies we will present some scenarios.

Example 10.8 (Diagnosis of settling problems I) *Settling problems can be further analysed by relating flowrates and turbidity measurements. Firstly, let us assume that the turbidity measurement has indicated a settler problem. The high effluent turbidity may be caused by a too high sludge blanket or an extremely high flowrate. Then the relationships between various variables have to be tested. Since the diagnosis is completely driven by available data we have to test the various cause-effect relationships. This may be done in the form of IF .. THEN.. ELSE rules, fault trees, etc.*

Example 10.9 (Diagnosis of a too small flowrate) *An abnormally low flowrate can now be analysed together with other measurements. If the flowrate measurement is compared with an electrical current measurement at the relevant pump motor a relationship has been established that can diagnose the problem. The pump motor has failed.*

These examples are all diagnosis based on given data. Such a diagnosis is called *data driven* or *forward chaining*. It contains two elements, a data base including relationships, models, etc. and a number of rules. The procedures are a set of rules of the type:

```
IF (condition) THEN (conclusion)
```

and a control structure determines what rule is tried next. It is often called a rule interpreter or inference engine. A rule "fires" then the *if* condition is satisfied.

Another situation is when some simple observation has been made, and the cause for it has to be found. Compare the case: you have got a serious stomachache and the doctor has to find the cause for it. The problem is easy to detect, but far from trivial to diagnose.

Example 10.10 (Diagnosis of settling problems II) *An operator discovers a special condition in the settler, and he hypothesises "bulking sludge". The detection of the basic phenomenon is quite straightforward. This hypothesis now has to be confirmed. He may look at some of the recorded data. This leads to a few different conclusions. An analytical test or a microscope test may reduce the number of possible causes, and by adding further evidence he may confirm his hypothesis.*

Example 10.11 (Diagnosis of rapidly changing air demand) *A diagnosis system has to find the true cause for an abnormally decreasing air flow demand. Further evidence has to be established. There may be a hypothesis that there is a toxic load. Then an independent toxicity test has to be made, since the DO control system could not unambiguously determine the cause. The toxicity test may or may not confirm the hypothesis.*

The examples 10.10 and 10.11 are *hypothesis driven* diagnoses or *backward chaining*. From the observation or measurement one has to backtrack in the chain of events. The hypothesis driven diagnosis (backward chaining problem) requires information on how possible faults relate to observable symptoms or features. In the parameter identification problems that were discussed in Chapter 7 we looked at the natural causal direction of dynamic systems. In the diagnosis we consider the inverse problem. This creates a new problem such as uniqueness. A certain feature may be caused by several events. If such a problem is solved numerically there may be several solutions. In optimisation terms we may say that there are several local optima. Parameter estimation may converge arbitrarily to one of the local optima and miss other solutions that may be more likely. It may be almost prohibitive to derive all possible numerical models of various fault models. Therefore qualitative models are often tried. Many qualitative methods used in diagnosis are designed specifically to yield a ranked list of possible faults.

Some diagnosis problems use a combination of data driven and hypothesis driven reasoning. A data driven diagnosis may have analysed a number of relationships between variables and found two different possible abnormal conditions. This creates different hypotheses that have to be tested. The hypothesis driven diagnosis now has to ask for further evidence to reach a final conclusion.

10.4 Consequence Analysis and Advice

Having made the detection and diagnosis it is often important to make some consequence analysis of the abnormal situation before any advice can be given to the operator (Figure 10.2). Let us illustrate the point by some examples.

Example 10.12 (Consequence of pump failure) *Consider the pump motor failure from Example 10.9. Once the problem has been diagnosed it is important to find out the consequences of the failure, so that the proper operation can be performed before the equipment is replaced. This may involve some prediction* *or simulation of the plant having too low a flowrate in one channel. Once the consequences are derived it is time for an operating decision. Once again the computer can act as a decision support for the operator.*

Example 10.13 (Consequence of a toxic load) *In Example 10.11 we diagnosed a toxic load to the system. A consequence analysis can be calculated by simulating the plant behaviour and predict the various adequate concentrations. Sometimes, if such a disturbance would have been anticipated, there may be some pre-calculated consequences to show the operator. The system may then advise the operator what to do. For example, step feed control might be a reasonable choice in some cases (see Chapter 18).*

The consequence analysis is well defined in the cause-effect structure. Often it includes some prediction, so dynamical models may be required. Sometimes the consequences can be tabled in *if..then* rules. When the diagnosis of the abnormal behaviour and its consequences have been obtained, then there is time for the operator action. Of course the system may support the operator decision by various knowledge bases. We will illustrate some possible strategies in Chapter 13.

10.5 Information Requirements

Diagnosis is sometimes very complex and suffers from some inherent problems which current techniques do not entirely overcome. A major reason is simply that we know too little about the detailed plant dynamics. Some of the problems are:

- *incompleteness* in the lists of known effects and causes and in the cause-and-effect relationships;
- *multiple "causes"* in close proximity (in a time or spatial sense);

  Example 10.14 (Effluent turbidity increase) *Assume that the effluent turbidity increases significantly. The reason may be an increasing flowrate, too large mixed liquor suspended solids concentration, or changing floc properties. The time scales for these causes are quite different, and the necessary corrective actions consequently have to be different.*

- *"chain reactions"* where one effect triggers other causes and sometimes one alarm may trigger another alarm.

  Example 10.15 (Three Mile Island) *The nuclear reactor disaster that occurred at Three Mile Island in the USA in 1979 has been widely publicised and analysed. The initial alarm caused a whole series of alarms to be triggered. Within minutes the operators were overwhelmed with alarms and did not know where to start. This event triggered large research efforts in* *alarm analysis, and the purpose was to automatically filter out the secondary alarms, so that the proper alarm could be noted. Fortunately, in a biological wastewater treatment plant the time scale is kinder to the operator, but still the basic problem of secondary alarms is the same.*

In Chapter 13 we will further discuss alarm analysis.

- *temporal or dynamic relationships*, particularly important in the case of biomass changes.

Current diagnosis techniques can be of great assistance to a human "expert", but are not sufficiently advanced to allow automatic diagnosis. The operators are responsible for the day-to-day operation of a plant. In large plants there are of course several operators responsible for different sections of the plant. Both the operator - working closest to the plant - and the process engineer require valid information about the plant. This information will be a pre-requisite for good monitoring, diagnosis and control.

Each diagnostic function is characterised by an information flow to the operator. There are several considerations to be made in a diagnostic system, such as:

- which diagnostic function should be performed autonomously by the system, which on an on-demand basis, and which should be left to the users?
- what information should the system display on a continuous basis, and what information should be accessible on demand?
- how should status information be presented so the user always knows the status of the system at a glance? How should the system focus user attention?
- how should the system alert the operators when a deviation is detected, for example via text, colour coding, or sound?

The information presented below is usually not provided by a conventional system, but should be considered possibilities for a modern plant control system. All of these features have the purpose to guide the operator through the diagnostic analysis. Graphical information is of great value, and innovative computer displays would enhance the possibilities for better interaction. In the following chapters we will then describe various tools that will assist the operator in the difficult tasks of detection and diagnosis.

Geographical Plant Overview

Physically and isometrically correct pictures of the complete plant, such as photos and 3D drawings. The main function of these pictures is to find a link to the operator's reality. Every pump, tank, valve, motor and sensor that is of interest for the operator should be included and positioned in the actual plant. The pictures should probably be organised hierarchically. Data sheets that relate to the pictures will improve the information. Such displays are frequently used for design purposes and could be utilised in operation as well.

Topological Process Overview

Process variables, such as flowrates, levels, flow directions, concentrations, active/nonactive units should be presented and indicated in a picture similar to the geographical overview or in the form of a process schematic. It should be possible to move between different descriptions of the system. Also here, the descriptions should be hierarchical. An example is given in Figure 10.9. The figure indicates the structure of the information in the plant. In a practical implementation the graphical view can be more attractive.

There should be a possibility to focus on subsystems of the plant. In many cases a unit process has to be further divided into further subsystems, like a single pump and how it functions. Of course, data sheets, maintenance manuals and hints for operation can be added to the process unit and equipment information.

Presentation of Process Data

A continuous presentation of sensor values or calculated process values has to be available. There should be possibilities to access other values, such as alarm limits and control parameters. Dynamic values should be updated in real time. Historical data should be available as graphs.

Especially to help inexperienced operators to interpret the information from the data acquisition and control system, there should be a presentation of how things are going with regard to quality. This could consist of, for example, a Chernoff's face, that smiles when the process is going well and looks sad when things are going bad, as shown in Figure 10.10. This information should be available on every picture of the system.

Focusing on an Event

It should be possible to focus on events, and not only on topological components. Depending on the operator's choice, a new picture might be built automatically.

Example 10.16 (DO control) *The dissolved oxygen (DO) control system may not be able to maintain a concentration close to the set-point during a certain time. Then a presentation of the air flowrates, respiration rate, wastewater flowrate, and mixed liquor suspended solids could present a relevant data background for such an event.*

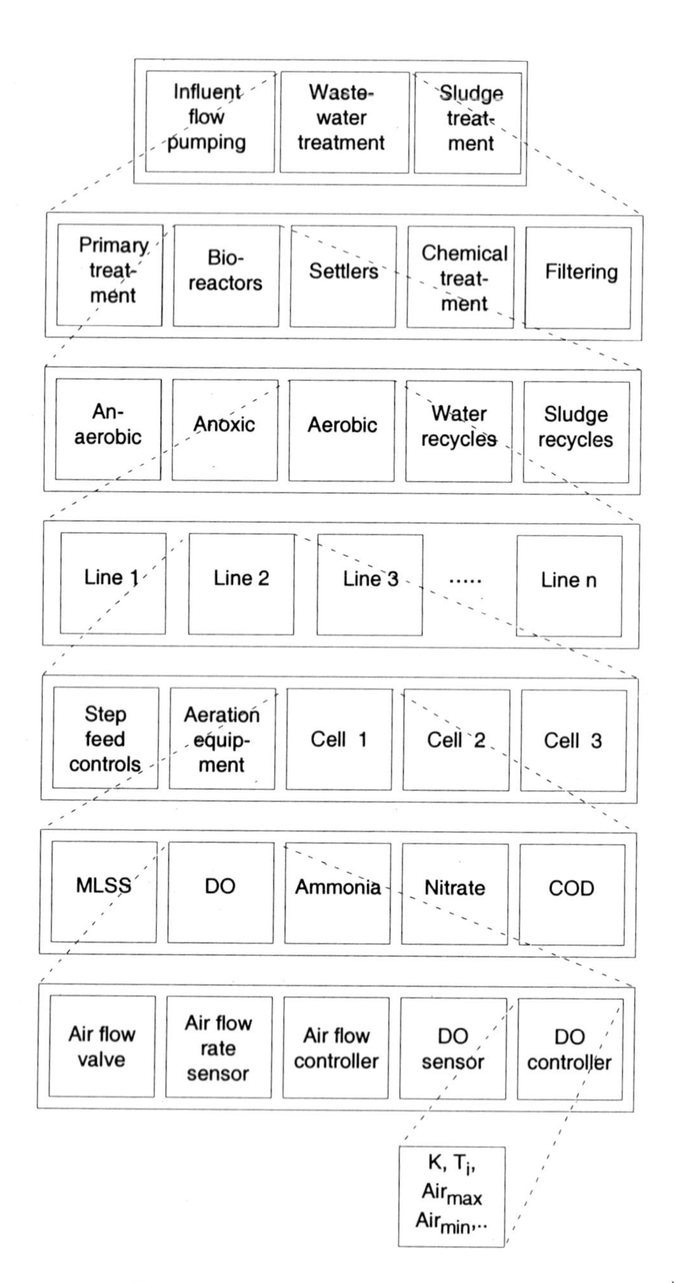

Figure 10.9: Topological view of a municipal treatment plant.

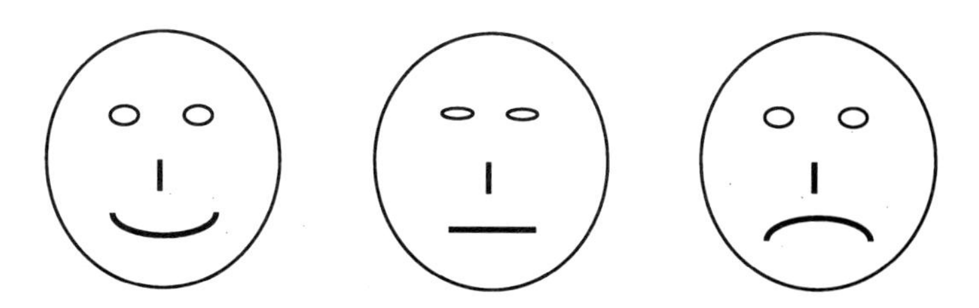

Figure 10.10: A Chernoff's face showing the process state.

Process Alarms

The alarms should be analysed before being presented, so that misleading or irrelevant parts of the information should be discarded, and conclusions based on what is left should be presented as alarms of a higher level.

Alarm signals could be complemented with audible and visual signals. However, one has to be very careful not to overdo it. The messages should require an acknowledgement, after which only the display remains. Alarms could be presented in the following forms:

- every picture in the system has a specific box for alarm messages;
- in the overview pictures the components relevant to the alarm should be marked;
- a special alarm list, where all alarms are shown, including activation time, value, limits, unit, time when acknowledged, and priority.

Trends indicating probable future alarms should also be shown, for example as graphs.

Process Warnings

Warnings differ from alarms as they are activated before something has gone wrong, and there is still time to avoid malfunctions. The activation of a warning is normally based on the value or the rate of change of process data. When one of these passes a pre-set limit, a warning is given. Warnings should be handled in a similiar manner to alarms.

Quality Monitoring

The system should provide a possibility to monitor the process with regards to the quality of the product. A list of important quality parameters for each unit process could be mapped every so many minutes to a list of effluent quality parameters. When some effluent quality is observed the system (or the operator) could inspect the corresponding process parameters and in this way get closer to the cause of the problem.

Condition Monitoring

Performance of individual sensors or components of the plant can be observed in order to find signs of wear among the equipment. This is a complement to pre-scheduled maintenance.

Example 10.17 (DO sensor noise) *The variance of the noise of a DO sensor is different for a new and for an old membrane. This can be used as extra information for preventive maintenance (Figure 10.6).*

Example 10.18 (Motor bearings) *Inductive or acoustical sensors are sometimes used in large motors to detect any imbalance of bearings.*

Example 10.19 (Wrong measurement) *One of the lessons from Three Mile Island: The operators failed for over two hours to recognise that a pilot-operated relief valve, which had supposedly opened and then closed, had in fact remained stuck open. A light on a panel told them that it was shut. However, the light was not operated by the valve position but by the signal to the valve. The operators either did not know this or had forgotten. The lesson is that whenever possible instruments should measure directly what we want to know, not some other property from which it can be inferred. If a direct measurement is impossible, then the label on the panel should tell us what is measured (in this case, "signal to valve" and not "valve position").*

Prediction

In the case of alarms and warnings some prediction (forward chaining) ought to be presented. One may want to ask more explicitly about the future, like "What will the dissolved oxygen concentration become in 20 minutes, if compressor 2 is shut down?" or "How much will the sludge blanket rise, if the return sludge pumping is decreased by 10%?"

Standard Operating Procedures

For many standard operations there are manuals of practice. Of course, it is of great value to have standard procedures documented on the process computer. They should of course include normal operating procedures for unit processes as well as for the complete plant. Furthermore, standard procedures for equipment ought to be added. Records should be kept on maintenance history, running times and of special problems or corrective actions.

Operating Costs

The cost information is not only for the managing people, but equally important for the operator. Various costs should be presented, like electric energy, chemical dosages, total air flow per day or week. This will supply the operator with better chances to follow up the plant operation and achieve cost reductions.

Software for Data Presentation

We have indicated that there is a wide variety of data that ought to be presented to the operator in a plant. Today there are several commercial software packages that offer an operator interface in process control systems. There are a lot of features that ought to be considered for such software:

- The software has to be able to communicate with the database of the control system. Such a database is the heart of the whole plant operation. It will contain both static data (such as volumes, yield coefficients, control gains, etc.) and dynamic data (such as time series of flowrates, concentrations, etc.)
- New sensor information should be easily added to the system.
- It should be easy to change the graphics, both the static and the dynamic information. The software has to give support so that the operator or the process engineer can implement minor changes.
- Many of the packages support sophisticated static information, such as photos and complete maintenance manuals for the equipment.
- Signal processing, such as data screening and filtering should be readily implemented.
- Alarm handling is also based on the primary information in a database. More or less "intelligent" alarm handlings can be implemented.
- Of course the operator interface has to be able to communicate with all the different controllers of the plant system.

The commercial market is huge, and it is difficult to pick only a few manufacturers. Still we indicate a few large players in the operator interface market, such as:

- Rockwell Automation (http://www.software.rockwell.com) with RSview,
- Intellution (http://www.intellution.com) with Fix, and
- Wonderware (http://www.wonderware.com) with InTouch and InSQL.

Large automation manufacturers like ABB, Siemens, Hitachi, Fuji Electric, GE Fanuc and Mitsubishi also have their own operation information systems.

The best way to find current information about plant control systems is to regularly follow the trade journals of automation.

10.6 Summary

We have described some of the basic elements of detection and diagnosis. In the following chapters we will discuss some specific methods used for detection and diagnosis. Methods based on numerical information from sensors and observations will be discussed in the next two chapters. In the subsequent chapter methods based on qualitative information and reasoning systems will be further discussed.

We will illustrate, that there is not an all-encompassing approach to diagnosis. Contributions from artificial intelligence, statistics and control theory all play important roles in solving different aspects of the problem. While individual elements of the problem may be fairly well understood, integrating these elements into an overall operating system presents significant engineering challenges.

10.7 Further Reading

- Many interesting ideas on knowledge based control systems are presented by Årzèn (1989) and Årzèn (1990).
- Good tutorial articles on fault analysis and diagnosis have been presented in an International Conference on Intelligent Systems in Process Engineering, see Davis *et al.* (1996) and Willsky (1976).
- The Three Mile Island disaster has been documented extensively. The Congressional hearings US Congress Subcommittee on Energy and the Environment (1979) present interesting evidence given by the operators. They showed that there was a serious problem of fault diagnosis and that aggravating factors were the large number of alarms, the lack of direct measurement on some critical indicators of state (reactor water level, pressure relief valve opening) and the tendency of the operators to give undue weight to particular items of information.
- More on low pass and high pass filters can be found in standard textbooks on control, see Olsson & Piani (1992).
- An early application of diagnosis in wastewater treatment systems is found in Beck *et al.* (1978). Applications in activated sludge systems are found in Gall & Patry (1989), Chan & Koe (1991), Ischikawa *et al.* (1993), Barnett & Gall (1996) , Bergh & Olsson (1996) and Bergh (1996). An example of detection of settling problems due to hydraulics is described in Olsson *et al.* (1986). Disturbance detection is treated in Rosén & Olsson (1997) and in Rosén (1998)
- Application of diagnosis methods in the chemical industry is found in Iri *et al.* (1979) and Shiozaki *et al.* (1979).
- Fault detection using neural networks is discussed by Naidu *et al.* (1990).

10.8 Exercises

1. Detections can be made by different kinds of measurements and observations. Give some examples of binary information, of analog measurements and of human observations.

2. We have talked about 'data rich' and 'information poor'. Suppose that you see the recording of a variable as a straight line. Is that data information rich?

3. What is the advantage of an exponential filter compared to a moving average filter?

4. Consider an exponential filter. The filter can be rewritten as a moving average of an infinite number of past measurements. Write the first 4 terms in that series. Compare how fast the size of the coefficients decrease for two cases, α=0.9 and α=0.1. Discuss the result.

5. The file `influent_data.mat` contains influent flowrate, COD and ammonia data to a full-scale wastewater treatment plant with a sampling interval of six minutes. The file `effluent_data.mat` contains suspended solids, ammonia and nitrate data to the sand filtration unit in the plant. The MATLAB program `exponential_filter.m` as its name implies is an exponential filter. Apply this filter to some of the experimental data and try different values of α. For example, try α=0.5,0.9 and 0.95.

6. The MATLAB program `moving_average_filter.m` as its name implies is a moving average filter. Apply this filter to the same experimental signals as the previous example. The order n of the moving average filter has to be given. Try the values n=3, 5, and 20.

7. List some possible applications for a high pass filter.

8. Give some examples of possible backward chaining procedures in a plant.

9. Noise is defined as the component that does not contain any information. Still there are cases when noise will contain information that helps to detect faults. Explain!

10. Suppose that bulking sludge is detected. How would you go ahead and test its causes? What would you test?

11. How could you detect that the membrane of a DO probe is failing?

12. Suppose that a too low value of the dissolved oxygen level is detected. What kind of causes can you look for?

13. A toxic load enters an aerator with automatic DO control. How is that detected by the DO control system? Can the DO control system distinguish the toxic load from other load changes?

14. A too high effluent turbidity is detected. What are the possible causes? Think of possible actions to take.

15. It is discovered that the DO level rises to about 9-10 mg/l from the normal operating range of 1-3 mg/l within a minute. What could be the reason?

Chapter 11

Quality Management

Quality is a term that means different things to different people. To the operations manager quality is the degree to which the product meets specifications. Quality control has become one of the most important precepts of international business. This is also true for environmental processes. In Section 11.1 we will discuss quality definition and its meaning in wastewater treatment operation to some extent. Statistical process control is then described in 11.2 and the standard tools of SPC are discussed in 11.3. It is an integrated part of the overall control of a plant. In 11.4 we will describe some more advanced tools for the analysis of large amounts of data.

11.1 Defining Quality

The quality of a wastewater treatment plant operation is usually only defined in terms of the effluent quality, measured as BOD, suspended solids, total nitrogen and total phosphorus content. Other components may also be included. This is of course the "product" of the operation, so the definition is good, but not sufficient in all situations. For example, if the effluent quality limit is BOD $= 10\ mg/l$, is a value of 5 a better quality of the effluent water? Here we define that the quality of the effluent is the degree to which the effluent meets the specifications. This suggests that the wastewater treatment operation has to set a goal relating effluent water quality and costs providing the specific level of effluent quality. Differently stated: quality cannot be inspected only in the product, the effluent water.

Quality has many dimensions. One is of course the standards defined by the regulatory agency. The annual average is one value, the monthly, daily averages are something else. This reflects what the receiving water is assumed to be able to accept. Then there is another more subtle quality, that has to be considered. The residents living close to a wastewater treatment plant may have strong opinions about odours from the plant, or the colour of the plant effluent entering the stream close to the buildings. Such a perceived quality measure is also part of the total quality definition and has to influence the way the plant is operated. Still another quality dimension is the level of maintenance. Is the maintenance of equipment and competence of people such that the quality will be kept even a year from now?

Quality has to be related to the specifications. We can define quality as how the plant operation will meet the specification. Stated differently: *quality is satisfying the customer requirements.* Here we note, that the "customer" for the treatment plant is not only the regulatory agency. It is also its neighbours, its personnel and its management. It is common to define total quality as quality *at the lowest cost.* Total quality control has been defined as *managing the entire organisation so that it excels in all dimensions of products and services that are important to the customer.*

An interesting definition of the concept quality has been presented by Myron Tribus [1]:

Quality is what makes it possible for a customer to have a love affair with your product or service. Telling lies, decreasing the price or adding features can create a temporary infatuation. It takes quality to sustain a love affair.

Love is always fickle. Therefore, it is necessary to remain close to the person whose loyalty you wish to retain. You must be ever on the alert to understand what pleases the customer, for only customers define what constitutes quality. The wooing of the customer is never done.

11.1.1 How Quality Control has Evolved

It is apparent that any wastewater treatment plant is like any process industry. In the early days of wastewater treatment the professional skill was not emphasised. The plant operation emphasised the mechanical parts, where pumping, valving, etc. were the dominating problems, while the process part of the operation were considered more on an *ad hoc* basis. It was considered normal that the effluent concentrations exceeded the limits on a regular basis. Bypassing during large flow was considered necessary. Today the operator skill in general is significantly higher than in the old days. Also the plant complexity has increased and usually there are very few persons in a large plant that have an overview of all the operations. This has an impact on quality control in the plant. In the industry in general the importance of quality grew during and after World War II and companies recognised that more than just inspection was needed to make a quality product. The same is true for the wastewater industry. Inspection of the effluent quality is not enough. Instead quality needs to be built into the whole treatment process. This requires the involvement of design engineers, process engineers, operators, maintenance people *and* the support of the management.

It may be interesting to note that some dominating teachers of modern quality engineering were Americans. W. Edwards Deming went to Japan soon after the

[1] ASQC Statistics Division Newsletter, 1990, No 3, page 2

World War II to teach quality. The Japanese learned. Today *the* control award of Japan is called the Deming prize, and is awarded once a year. The Western industry were a little slower to adopt these quality ideas. Deming's principles have been adopted for example by the Central Cosa Nostra Sanitary District in Martinez, California (Zayac *et al.* (1996)). Teams of employees are used to analyse work processes and determine how to do them better, faster and cheaper. From a utility's point of view the Deming principles are:

- know your mission;
- let the customer define quality;
- do it right the first time;
- eliminate variability;
- emphasise processes;
- focus on teams, not individuals;
- increase employee involvement;
- change organisational culture;
- avoid blaming;
- improve continuously.

Some other philosophical elements may be added:

- leadership;
- employee training (in particular: training to obtain greater operator skill);
- quick response;
- management commitment;
- management leading by example;

Employee involvement has to be encouraged by a good communication between everybody in the organisation. Furthermore employee education gives a base for comprehension and implementation through a common language and the application of special skills. To quote Albert Schweizer [2]: "Example is not the main thing in influencing others, it is the only thing."

In the quality management literature it is emphasised that a quality organisation:

- is dedicated to improvement (does not delegate quality);
- values people (does not value only numbers); and
- recognises that quality improvement reduces cost (does not trade quality against cost);

[2](1875-1965) German theologian, philosopher, organist, and mission doctor in equatorial Africa, who received the 1952 Nobel Peace Prize for his efforts in behalf of "the Brotherhood of Nations"

11.1.2 Quality Assurance

More and more companies have developed a quality measure of the implementation of quality control by being certified to the international standard ISO 9000. ISO has different names in different countries, such as ANSI/ASQC Q90 in the USA and BS 5750 in England, SS-ISO 9000 in Sweden, but the contents are almost identical. Another standard being completed is ISO 10012 concerning quality assurance in measurement and calibration systems.

The standard ISO 9000 "Quality management and quality assurance standards" is actually a collection of five standards (ISO 9000 through 9004), in which it is defined how controls and verifications should be carried out in the different phases of work organization for project development. These standards specify also in what measure manufacturer and customer carry the main responsibility for different project phases and define also the type of contractual regulations. ISO 9000 does not only refer to the technical production or the construction of a plant, but includes also the related services and organizational measures on the part of the contractor. In other words, the quality-conscious contractor can follow the indications of ISO 9000 in its production and in this way show his customers a concrete quality certification.

The five standards within ISO 9000 are:

- *ISO 9000* - Basic definitions, summary of the use of the four other standards.
- *ISO 9001* - Quality systems - Model for quality assurance in product design, development, production, installation and servicing.
- *ISO 9002* - Quality systems - Model for quality assurance in production and installation.
- *ISO 9003* - Quality systems - Model for quality assurance in final inspection and test.
- *ISO 9004* - Quality management and quality system elements - Guidelines.

The European Committee for Standardisation (CEN) has developed a special version of ISO 9000 for the European Union (EU), called EN 29000. Every member nation within the EU has to accept this standard. From 1993 it is required that a company is certified with EN 29000 to be considered for purchasing orders by public organisations. The certifying process is an official audit process where the company is evaluated according to the ISO 9000 series. One result of this certification is that the company expects to be more readily accepted by customers, without any extra evaluation by them.

Some remarks about ISO 9000 ought to be made:

- ISO 9000 is not an absolute measure of product quality. It represents a minimum of requirements to ensure good product quality. This standard does not define how quality can be ensured; to this purpose there are thousands of other standards, and in the end this problem can be solved only with human know-how. Note that the *documentation* of each product is essential. Any product raw material should be possible to *trace.*

- The system is quite defensive and is more directed towards products than towards processes.
- The system does not encourage more progressive quality improvement work. In that sense the ISO standard is conservative.
- The system includes a lot of documentation. Therefore, at least in small companies, the paper work seems to dominate and the real quality improvement work is easily pushed into the background.

There is sometimes a tendency to satisfy the ISO system instead of working with constant improvements of the processes and the products.

11.2 Statistical Process Control

Statistical process control (SPC) is recognised as a generic tool for quality management. It is used to monitor processes and to identify causes of lost quality. As described in Chapter 10 measurements and observations are the basis for all quality control. Again we look at the basic problems of detection and diagnosis. The term SPC is often confused with process control, which will be the subject of Part C of the book. SPC, however, is more related to monitoring the process, and therefore the term statistical process monitoring (SPM) is used synomously with SPC.

SPC is widely used to ensure that processes are meeting standards. All processes are subject to a certain degree of variability. It is useful to make a distinction between *common* and *special* causes of variation. Many people refer to these variations as *natural* and *assignable* causes. A process is said to be operating in statistical control when the only source of variation is common (natural) causes. The process must first be brought into statistical control by detecting and eliminating special (assignable) causes of variation. Then its performance is predictable, and its ability to meet the requirements can be assessed.

Example 11.1 (An assignable cause of variation) *An activated sludge plant had a very unsatisfactory performance, and the suspended solids concentration out of the secondary settler was too high. It was recognised that the influent flowrate had large peaks every hour. Therefore it was decided to try out some step feed control in order to attenuate the effect of the large flow disturbances. Analysing the reason for the large flow peaks revealed that the primary pumping was the whole reason for the problem. The pump was located almost one kilometre away and was not operated by the plant people. Furthermore, the pump was largely overdesigned. Since it could only operate on/off it was normally turned on only 5 minutes every hour. Consequently the plant received a high peak flow every hour.*

In this case it is apparent that a more suitable size of the primary pump, probably supplied with variable speed, could solve most of the variability problems in the plant. Thus the pump was an assignable cause of variation.

In SPC there are some pre-conditions that should be emphasised. First we will discuss benchmarking, as we need some measure or objective so that quality can be measured and compared. Secondly, we will assume that data represents steady state conditions. Transient conditions have to be taken care of differently.

11.2.1 Benchmarking

Benchmarking can be defined as "...the ongoing structured and objective process of measuring and improving products, services, practices and processes against the best that can be identified world-wide in order to achieve and sustain competitive advantage..." (Grinyer & Goldsmith (1995)). The idea of benchmarking is to:

- continually compare the current performance of the organisation against its historical performance, and
- continually compare the performance of the organisation against the performance of the "best in class" organisation(s).

Once again we notice, that we have to measure in order to know. Measuring is here defined as inspection and tracking the performance of the overall operation. With benchmarking some consequences can be noted, as described in Table 11.1.

There is a disturbing tendency in industry to measure "productivity" in terms of inputs rather than outputs, particularly long term outputs. If manpower is reduced to "best practise", but long term outputs such as equipment reliability deteriorates and accident rates increase, then we seldom have a more productive system.

11.2.2 Inspection and Detection

To make sure the processes are producing at the right quality level expected, inspection is needed. In other words, performance can be measured. It can involve measurements and visual inspection (odour, noise, colour, etc.). Its goal is to detect unacceptable behaviour, but inspection does not correct the deficiencies. There are three basic issues relating to inspection:

- how much and how often to inspect;
- when to inspect;
- where to inspect.

Let us consider the possibility of toxic material entering the plant. The issue of how much and how often to inspect and analyse is a matter of economics. This

Aim	Without Benchmarking	With Benchmarking
Becoming competitive	- internal focus - slow, evolutionary change - low commitment	- external focus - acceptance of new ideas and methods - higher enthusiasm and commitment
Adopting best-of-class techniques	- not-invented-here syndrome - myopic vision - self-imposed limits	- active search for new ideas - expanded horizons into other industries
Understanding "customer" requirements	- based on history and intuition - no interest in exploring new trends	- constant search for significant trends
Establishing clear goals and incentives	- reactive to historical standards - based upon catching up with industry	- aimed at leadership
Improving "productivity"	- acceptance of status quo - low receptivity of new initiatives	- passion for continuous improvement - no tolerance for coasting on past success

Table 11.1: Benchmarking vs. non-benchmarking approach.

trade-off is illustrated in Figure 11.1. Often the measurement or inspection costs are directly proportional to the number of measurements, due to labour or chemical costs. On the other hand, there is a cost for not detecting serious disturbances. These costs may be related either to a direct problem associated with the killing of organisms in the sludge, or to costs related to exceeding the effluent limits.

Deciding when and where to measure or inspect depends on the type of unit process and its importance for the total operation. Again, inspection can take place if the cost of inspection is less than the likely loss from not inspecting. "When" is a risk assessment involving:

- frequency of the disturbance;
- consequences of not detecting the disturbance;
- the definition of what is an acceptable risk.

"Where" can be performed in two ways. If a product is inspected against specifications we talk about *feedback*. If the disturbances are inspected, it is a *feedforward* structure. In order to determine the cost of quality, a cost/benefit analysis is usually performed. Firstly costs are identified and then benefits, both

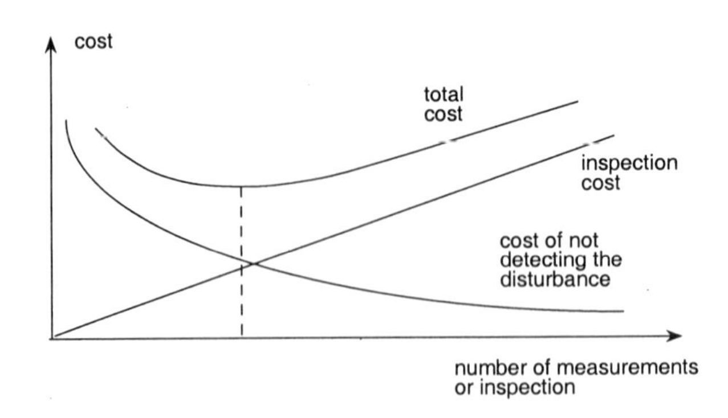

Figure 11.1: The trade-off between the cost of inspection and the cost of not detecting the disturbance

intangible and tangible, and each are assigned a value. The following can be attributed to the cost of quality:

- failure costs - cost of not getting it right the first time;
- appraisal costs - cost of making sure it was right the first time;
- prevention costs - cost of making sure it is right the first time.

If a process variable is controlled at a certain setpoint, then costs will be incurred when the variable deviates from the desired value. The tighter the control of the variable, the more cost savings are achieved. If a process variable operates against a constraint, then costs are incurred for operating further from the constraint and especially if the constraint is violated. Again, the tighter the control, the closer you can operate to the constraint without violating it, and the more are the cost savings. This is detailed in Chapter 21.

When inspections take place, quality characteristics may be measured as either attributes or variables. *Attribute inspection* classifies items as being either good or defective. It does not worry about the degree of failure or performance. For example, a broken membrane in a dissolved oxygen sensor is useless and has to be replaced. This is a kind of binary information, mentioned in Chapter 10. *Variable inspection* measures quantitatively like concentration, flowrate, speed to see if the value falls within an acceptable range. In other words, this is an analog signal. Naturally, the statistical quality control approaches are different for attribute and variable inspections.

11.2.3 Diagnosis

One of many available tools helpful in listing possible locations of quality problems is the fish-bone chart, also known as the Ishikawa diagram and as the *cause and effect* diagram, illustrated by Figure 11.2. Note the shape resembling the bones of a fish. Each bone represents a possible source of error. When such a chart is systematically developed, possible quality problems and inspection points can be determined.

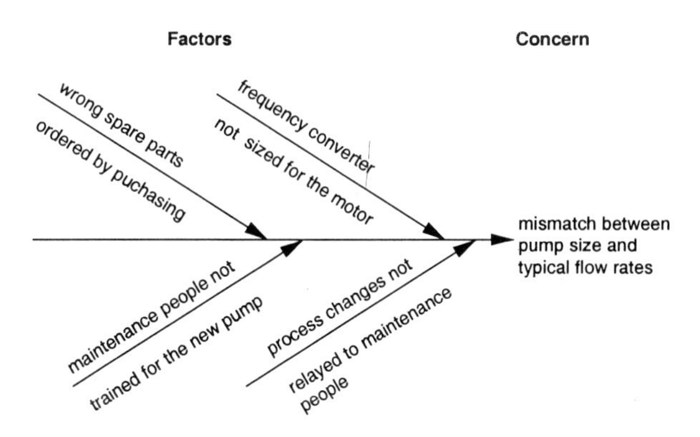

Figure 11.2: Simple fish-bone chart (Ishikawa diagram) for a mismatch of a pump.

Example 11.2 (Fishbone diagram on nitrification) *Some of the factors influencing nitrification are shown as an Ishikawa diagram or fish-bone chart in Figure 11.3.*

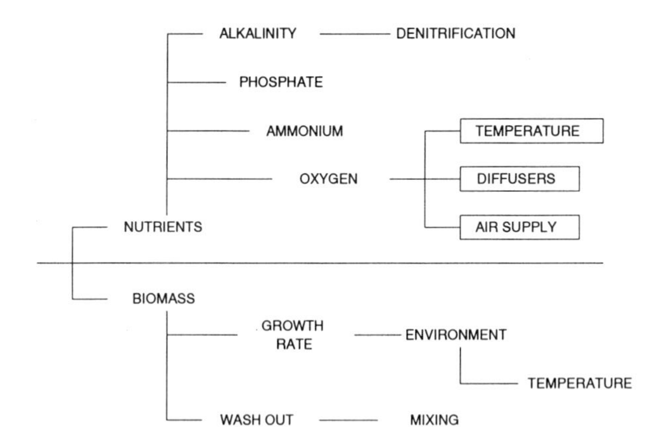

Figure 11.3: Ishikawa diagram or fish-bone chart for nitrification.

11.2.4 The Meaning of Stationary Processes

A random signal (or stochastic process) can be characterised by its probability distribution. At each time instant there is a probability distribution of the signal amplitude, as illustrated in Figure 11.4. If the probability distribution of the signal does not change with time it is said to be *stationary.* In most cases it is sufficient to characterise the probability distribution by the mean value and the variance (or standard deviation). In a stationary process the mean value and the standard deviation are constant in time. The probability function of such a signal is shown in Figure 11.4 and the corresponding analog signal is illustrated in Figure 11.6(a).

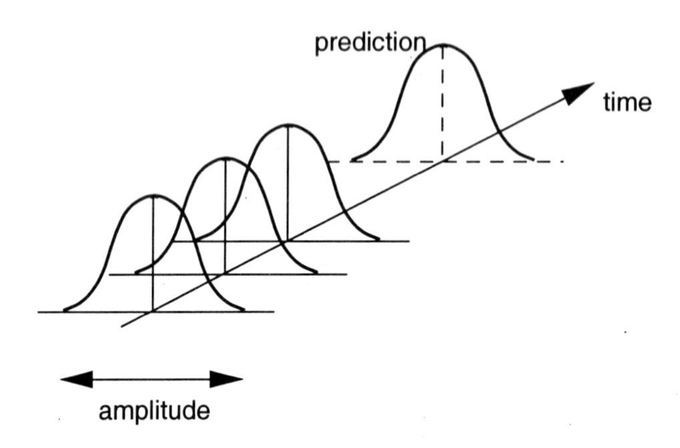

Figure 11.4: In most cases the measurements can be described by a stationary time series, that has a constant average and standard deviation.

If the operating conditions of a plant have not changed, then it is probable that the signals can be represented by stationary processes. In such a case it is reasonable to predict a future value of the signal. Here we assume that the signals are stationary. If the probability distribution is not constant in time it may look like Figure 11.5. This may happen if the operating conditions change or if unexpected disturbances occur. The process is said to be non-stationary. Its mean value and/or its variance are not constant. A simulation of non-stationary processes is also shown in Figure 11.6(b)-(d).

If there is no knowledge about the time-variability of the mean value and the variance, then there is no way to predict the future behaviour of the variable.

11.3 Basic Statistical Process Control Tools

Statistical process control (SPC) is concerned with monitoring standards, making measurements, and taking corrective action. The basic structure is like that of any process control system. As remarked already there is sometimes a confusion between the aims and objectives of SPC and (automatic) process control which we will talk about in Part C. This is not helped by the word "control" in both terms. We will highlight some differences:

- SPC and automatic control usually work in different timescales. Usually the time scale is much faster for automatic process control, compared with SPC.

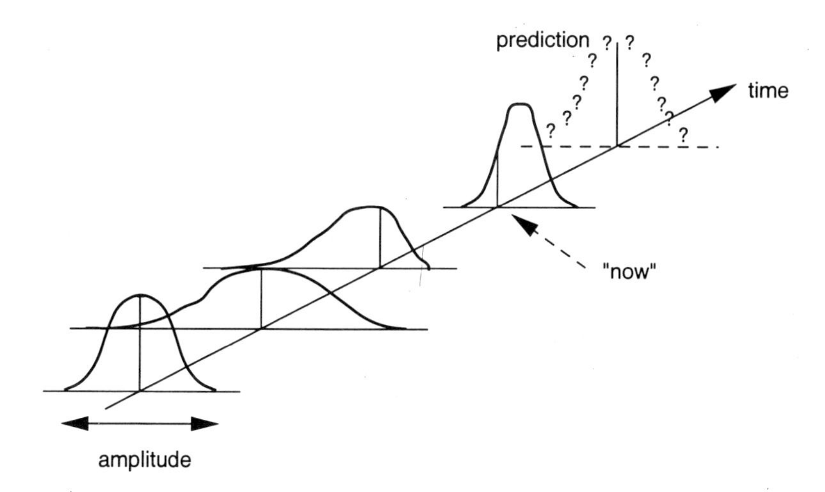

Figure 11.5: If assignable causes of variation are present, the process output is not stationary (the mean and/or the variance are not constant) and it is difficult to make predictions.

- In SPC both the measurements and the corrections may be made by a human operator, while an automatic controller depends on on-line instrumentation and controllers. Therefore SPC may be called supervisory control.

- Automatic process control takes control action to alleviate the effects of disturbances acting on the process. Process control does not try to eliminate the disturbance from recurring in the future. SPC on the other hand tries to identify and eliminate the disturbance altogether. In the short term, SPC does not alleviate the effects of the disturbance.

- The costs may be different. Process control actions are usually relatively inexpensive. The manipulation of flowrates is cheap, while of course chemical precipitation is more costly. SPC actions may involve changes to the process or control system design. This is also why SPC control actions are less frequent than process control actions.

- Process control, at least the more advanced forms (see Part C) require the use of empirical as well as phenomenological information. SPC mostly uses empirical information about the process.

- SPC will not respond to random variation at all. SPC separates random variation (common causes) from unusual occurrences (special causes) and responds only to the latter. Automatic process control responds to all variation. Thus in some circumstances, particularly if the controller gain is too large, the automatic controller can actually "amplify" the noise, creating more variation in the process (see Chapter 17).

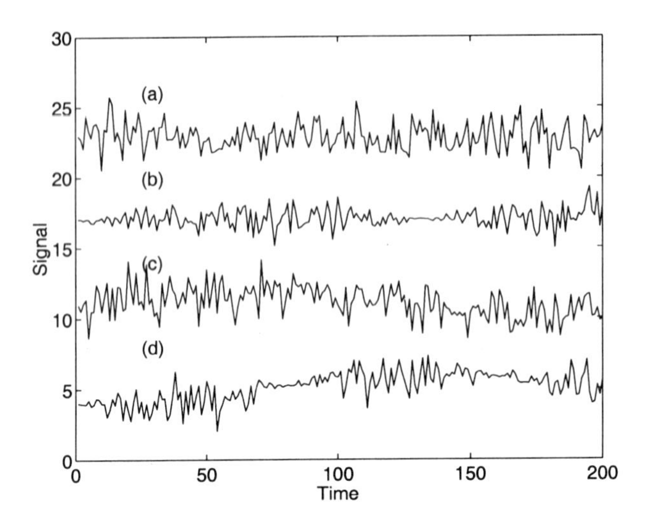

Figure 11.6: Simulation of a stochastic process. (a) stationary, constant mean and variance, (b) non-stationary, constant mean but time-varying variance, (c) non-stationary, time-varying mean and constant variance, (d) non-stationary, time-varying mean and variance.

We will now consider some simple tools used to display the performance of the plant.

The problems discussed in this Section are not widely studied by the academic control community, but are not any less important than other issues of control that will be described in the Control part of the book. The reason for the neglect may be the dominant interest in problems with interesting dynamics. In the situations considered here, dynamics are generally not important or can be addressed in a straight forward and *ad-hoc* manner.

The diagram in Figure 11.7 gives an overview of various steps involved in the SPC detection and diagnosis strategy.

11.3.1 Pareto Analysis

The Pareto analysis was named after the Italian economist Vilfredo Pareto (1848-1923). He made the famous observation that 80% of the worlds wealth was owned by only 20% of the people. This principle is valid in many situations, and can be equated to process operations as 80% of the process problems are due to 20% of the causes. A Pareto analysis can therefore determine the important few causes from the trivial many causes. The analysis can be developed easily by the following method:

1. identify the causes of a particular problem;

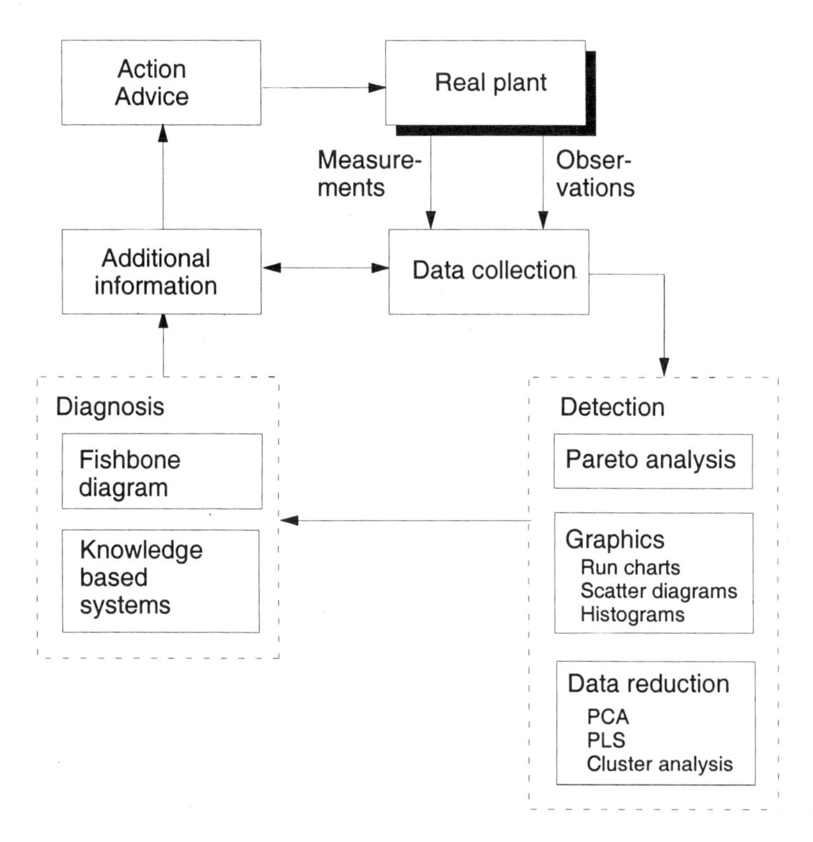

Figure 11.7: Flow chart of various tools for detection, diagnosis and action in process analysis.

2. determine the frequency of each cause;
3. list the causes in descending order of frequency;
4. calculate the percentage of each cause from the total frequency;
5. plot the frequency and percentage together.

The choice of frequency variable determines the resultant Pareto chart. This form of Pareto chart draws attention to the "squeaky wheel". It is often *better to multiply frequency by the cost consequence* before ordering and graphing. This form prioritises problems that cost the most.

Example 11.3 (Pareto chart) *The chart on the left in Figure 11.8 illustrates the "squeaky wheel" problem where frequent but not costly problems distort the priorities. The chart on the right ranks the costly problems. The data is shown in Table 11.2.*

On the basis of annoyance factor (frequency) we would attend to problems B, A and C, while on the basis of net cost over one year we should attend to problems

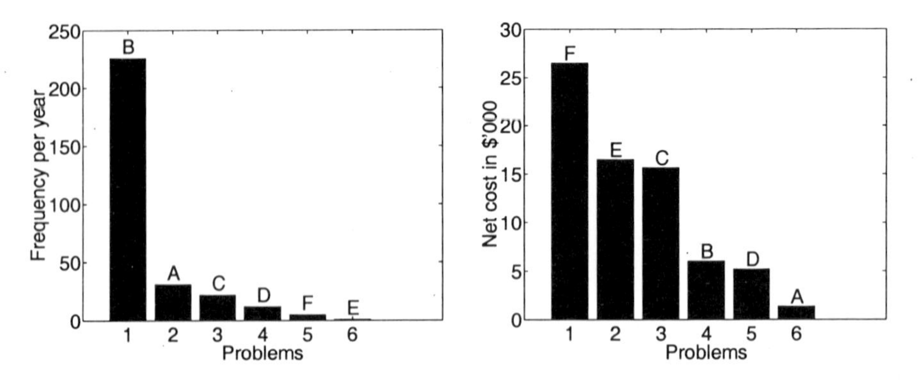

Figure 11.8: Pareto charts of frequency (left) and net cost (right).

Problem	Frequency (per year)	Lost Revenue ($ per time)	Elimination Cost ($)	Net Cost over 1 year
A	31	125	2,500	1,375
B	226	50	5,300	6,000
C	22	780	1,500	15,660
D	12	1,250	9,800	5,200
E	1	28,500	12,000	16,500
F	5	9,000	18,500	26,500

Table 11.2: Problem data for Pareto analysis.

F, E and C. Of course we would get yet another answer on the basis of percentage return on investment. Of course it is not unusual that infrequent problems are often costly. This tends to happen because of the way we design systems.

11.3.2 Graphical Tools

Graphics is usually quite powerful to illustrate what is going on. There are several graphical representations that are useful in SPC and will help the operator to understand what is going on in the plant. We will describe the following:

- flow charts;
- histograms;
- scatter diagrams;
- run charts;
- control charts.

Flow charts have been described in Chapter 10. They are not limited to material flow. They can be used to graphically display the decisions operators make in controlling the process. Flow charts are used to give a graphical representation of the flow of all aspects of the process, and reveals the relationship between each process step.

Histograms are used to identify the distribution of the data being monitored. If the distribution is not close to normally distributed, then there is a major process problem. Furthermore, one of the assumptions of control chart analysis is violated (see below). Histograms are also used in capability analyses to determine the amount of data within the specification limits of a process.

Scatter diagrams are sometimes useful to illustrate the relationship between two variables. The pattern in which the points group together may reveal whether some correlation exists between the variables. However, caution is required when interpreting scatter diagrams. They do not tell about the mechanisms of the causes and effects, and the effects of different causes can be mixed, which can confuse or hide relationships.

Example 11.4 (Scatter diagram of effluent suspended solids) *Scatter diagrams will not always easily reveal relationships. Figure 11.9 illustrates the plotting of flowrate against effluent suspended solids produced by the MATLAB program* `Clarifier_SPC_Charts.m`.

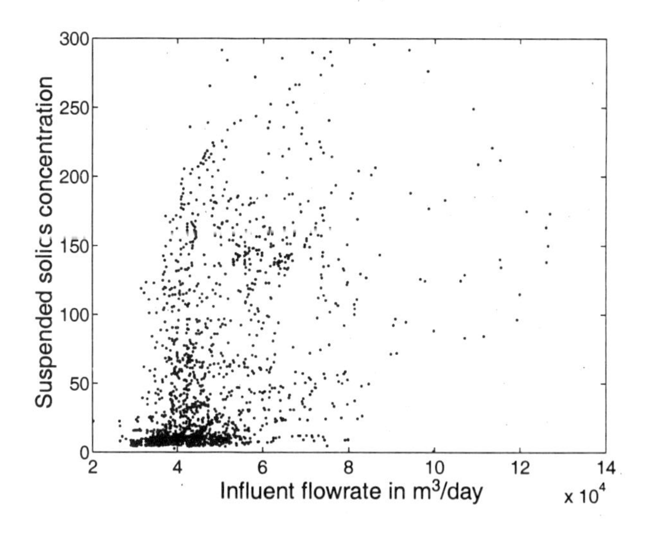

Figure 11.9: Scatter diagram of daily data of flowrate vs. effluent suspended solids.

While a relationship does appear to exist, there is a broad scatter. The output below from the MATLAB Toolbox function `linreg` *confirms the impression - perhaps some relationship but not strong.*

```
LINEAR REGRESSION
Fitted Parameters:
 Intercept = -18.67
```

```
 Slope = 18.39
Per cent Variance Explained by Fit = 72
Correlation between Parameters = -0.9319
Confidence Limits:
 Intercept = 7.833
 Slope = 2.255
```

A **run chart** is a plot of variables as a function of time, and is always strongly recommended.

Example 11.5 (Run chart of effluent suspended solids) *Let us plot the flowrate and the effluent suspended solids concentration in Figure 11.9 as run charts in Figure 11.10. This will reveal obvious peaks that appear in the variables. Sometimes they coincide in time, sometimes not. In fact, this is an example of inadequate sampling rate (compare Section 6.4.1).*

The flowrates and the suspended solids measurements are daily averages. It is intuitive that a large flowrate will cause the effluent suspended solids concentration to be large. As shown in the diagrams, however, this is not always the case. Sometimes the peaks appear at the same time, other times they do not. The reason is the dynamics. Actually, there is quite a short time lag from a flowrate increase to an increase in the effluent suspended solids (less than an hour). That means, that daily averages will completely hide the true dynamic behaviour. When there is a high flowrate peak it does not say anything about the intensity of the rain. It may be quite a light rain during 24 hours or a heavy storm during an hour. The resulting effluent suspended solids may look quite different. For some flowrate peaks there is no apparent increase in the suspended solids concentration. At other times, there is a high suspended solids peak and no apparent flowrate increase. The reason may be a short rain storm. This has caused the effluent solids concentration to increase significantly. However, the daily flowrate average has been quite moderate.

A run chart display of a variable is always recommended. New data have to quickly be compared with past performance.

The idea behind a **control chart** is to plot the data as a function of time and to add upper and lower (control) limits. Stress is placed on the fact that the limits are not the same as the specification limits. Instead, action limits are determined from the data output from the process. In Figure 11.11 we have indicated the existence of such limits. Upper and lower limits are displayed, so that the operator easily can judge the performance. Whenever the measurements approach or exceed the limits there is a reason for action.

The data can be tested for various properties as indicated in Figure 11.12. It is important to detect random variations around the mean value or to identify the existence of trends or oscillations of the variable. There are various tests to do that. Often an operator can judge from his experience if the data is "normal" or not. We may define standard tests for "out-of-control" signals such as:

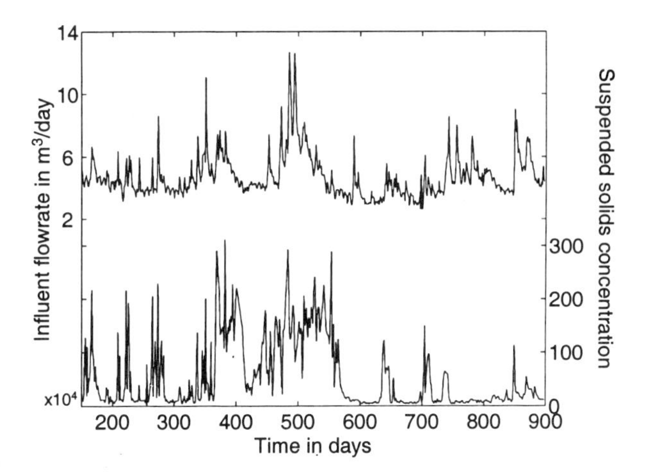

Figure 11.10: Run charts of daily data of the flowrate and the effluent suspended solids.

- one point beyond the upper or lower limit (out of control);
- nine points in a row above the mean value (sustained poor performance);
- six points in a row, all increasing or decreasing (trend);
- fourteen points in a row, alternating up and down (erratic or noisy behaviour).

Many measurements tend to follow a Gaussian (normal) distribution, that is described by its mean value ($\bar{x}$) and its variance (σ^2) or standard deviation (σ). The upper and lower limits shown in Figure 11.12 are usually $\bar{x} \pm 2\sigma$ or $\bar{x} \pm 3\sigma$. For a Gaussian distribution 68% of all the samples are expected to lie within $\bar{x} \pm \sigma$. During 95.4% of the time the samples will fall within $\bar{x} \pm 2\sigma$. If a point on the chart falls outside $\bar{x} \pm 3\sigma$ then we are 99.7% sure that the process has changed [3]. Such a value is usually called an *outlier*.

Often group averages (see Chapter 6) are plotted. Typically an on-line sensor in a wastewater treatment plant may be sampled once every 6 seconds. A one minute average (that is the average of 10 values) is then shown on the computer screen. Such a value will have much less variance, since some of the sensor noise will be filtered out. The control limits for an averaged value (often called an X-bar chart) have to be smaller than those of individual measurements. The more values that are averaged, the closer the limits have to be set.

Another useful parameter to display is the difference between two consecutive measurements $|X_i - X_{i-1}|$. Too large values of the difference may indicate erratic behaviour.

[3] In manufacturing, some companies like Motorola and the Ford Motor Company have published the 6σ goal, meaning an almost perfect production. Quantitatively this means 3.4 defects per million parts are accepted.

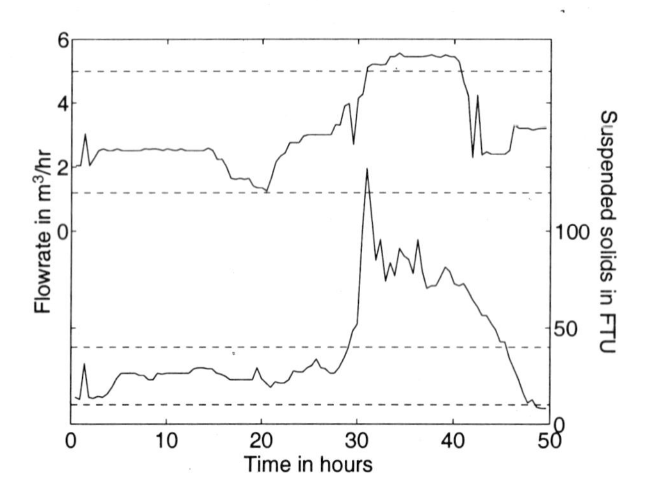

Figure 11.11: Control charts of on-line measurements of the flowrate and the effluent suspended solids. The sampling rate is 6 minutes.

The control charts may sometimes display filtered data. One filtering method is exponential filtering, described in Section 10.2.5. Other charts may show cumulative sums. This is informative if some accumulated mass needs to be displayed, such as total dosage or total effluent mass of some component. Cumulative sums are also very sensitive for detecting small changes in the mean value.

11.4 Advanced Tools

Unfortunately the techniques described in 11.3 are woefully inadequate to deal with many situations in complex plants, because they are inherently univariate in nature. Techniques which have gained acceptance to deal with multivariate problems are principal component analysis (PCA) and partial least squares or projection to latent structures (PLS). The underlying idea of these techniques is that although a large number of variables may be measured in most process situations, these variables are often highly correlated. The true dimension of the space in which the process moves is almost always much lower than the number of measurements. The methods shown below are multivariate statistical methods which consider all the noisy correlated measurements and project this information down onto lower dimensional subspaces which (hopefully) contain all the relevant information. They are examples of "factor analysis" commonly used in statistics. PCA was introduced already in 1901 by K. Pearson and is now a standard tool in regression analysis. The development of PLS is more recent and largely due to the work of Harald Wold between the mid 1960s and the early 1980s. It has become a standard in chemometrics. Since both PCA and PLS are based on projecting the

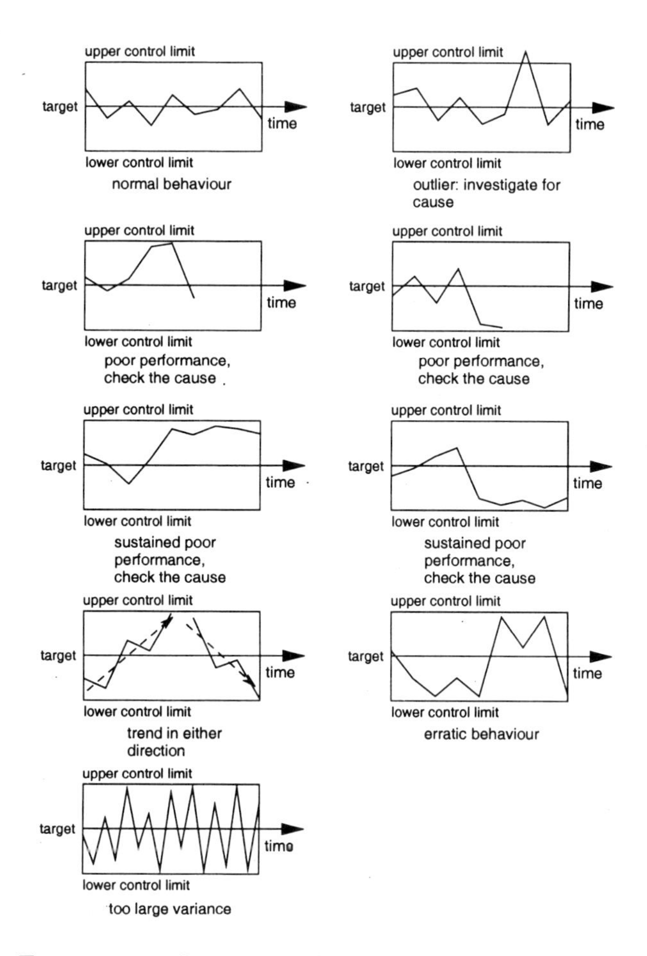

Figure 11.12: Patterns to look for in measurements.

information contained in high dimensional data spaces down into low dimensional spaces they are also called *projection methods*.

The geometric approach in PCA has been rediscovered in many diverse scientific fields, resulting in, amongst other things, an abundance of redundant terminology. In the 1930s the development of factor analysis (FA) was started by psychologists. FA is closely related to PCA and often the two methods are confused and the two names are used interchangeably. Singular value decomposition (SVD) used in numerical analysis and the Karhunen-Louve expansion used in electrical engineering are also similiar techniques. In image analysis the term Hotelling transformation is often used for a principal component projection. In chemistry PCA was introduced around 1960 under the name principal factor analysis. In geology PCA has lived a more secluded life, partly overshadowed by its twin brother

FA.

11.4.1 Principal Component Analysis

Principal component analysis (PCA) is used as an aid when investigating data sets with many variables. It is essentially investigating the linear relationships within data. PCA reduces the dimensions of the data set, whilst retaining as much of the variations described by the data as possible. A process computer may be collecting hundreds of data points every minute on flows, pressures, concentrations etc. For PCA these measurements are arranged into a $(n \times k)$ matrix X consisting of n observations on k variables.

A major difficulty with analysing such data is the shear size of the data set. There is an obvious "data overload" and, despite the fact that we have computers to store such data, the usual reaction to the problem has been to look at only a few key variables and to ignore the rest. The result will then be that many aspects of the process may be lost. Another problem is the colinearity of some variables, that is, they are highly correlated with one another. When we measure a large number of variables they are almost never independent of each other. Consequently, the true dimension of the space in which the process moves is usually much smaller than the number of variables. A human often has large difficulties interpreting such highly correlated data. We are often trained to think of the independent effects of each variable.

The fact that most of the measurements are noisy and that not all measurements are available at each sampling interval creates further challenges, and makes it even more difficult to interpret the data. Many statistical methods can not easily handle missing data.

We now assume that data is available and is represented by one large process data matrix X. The matrix X is first mean centred by subtracting the mean value of each variable from the respective column. It is then common to scale all variables (columns) to unit variance.

Principal component analysis is a method for explaining the variance of this matrix in terms of a number of new latent variables called principal components (PCs). Now consider Figure 11.13 with three variables plotted against each other in a three-dimensional space. The first *loading* vector p_1 is defined as the direction of greatest variability, as indicated in Figure 11.13(b) (p_1 has the dimension k). The *score* vector t_1 is defined as the projection of each observation vector onto p_1, as shown in Figure 11.13(b) (t_1 has the dimension n). The first principal component is expressed as the linear combination of the k measured variables that explain the greatest variability and is expressed algebraically as:

$$t_1 = Xp_1 \tag{11.1}$$

Using linear algebra we can express p_1 as the first (largest) eigenvector of the matrix X^TX and t_1 the first (largest) eigenvector of XX^T. The second PC has to be orthogonal to the first PC. It is the linear combination of the variable explaining the next greatest variability. This can be expressed algebraically as:

$$t_2 = E_1 p_2 \quad (11.2)$$

where the matrix E_1 is a so called *residual* matrix left after the predictions of the first PC have been removed. It is calculated from:

$$E_1 = X - t_1 p_1^T \quad (11.3)$$

which is another $(n \times k)$ matrix. It is possible to continue in this fashion until k PCs have been found. In general we will have:

$$E_m = X - t_1 p_1^T - t_2 p_2^T - ... - t_m p_m^T \quad (11.4)$$

$$t_m = E_{m-1} p_m \quad (11.5)$$

where $1 \leq m \leq k$.

The idea of the PCA is to reduce the dimensionality of the problem. Fortunately one often finds that it is possible to explain most of the variation in the data matrix X with a lot fewer than k PCs, say after r PCs. By stopping at this point it is to say that the k-dimensional observation space has been approximated by the projection of the observations into an r-dimensional space. In Figure 11.13(c) we have found that most of the variation can be explained by the plane defined by the first two PCs. For most process data, depending on the correlation between the measurements, the number of PCs necessary for capturing the data variance is significally smaller than the number of measured variables, in other words $r << k$.

Each observation is projected onto the plane of PCs via its scores. The score is the distance from the origin of the plane along each PC and is calculated from Equations 11.1 or 11.2. The perpendicular distance from each observation to the plane is the residual for that observation. Algebraically the X matrix has been approximated by:

$$X = t_1 p_1^T + t_2 p_2^T + ... + t_r p_r^T + E \quad (11.6)$$

where E is the residual matrix. In the ideal situation the dimension r is chosen so that there is very little significant process information left in E. Then E will represent only some random noise, and adding another PC would not add any new information. There are various methods proposed for testing that E is sufficiently random. All the t vectors can be columns of a matrix T, and the p vectors are columns of a matrix P. The expressions in Equations 11.1 and 11.2 can then be described by the matrix operation:

$$T = X\ P \quad (11.7)$$

and the calculation of the PCs can also be expressed by:

$$X^T X = P^T \Sigma P \quad (11.8)$$

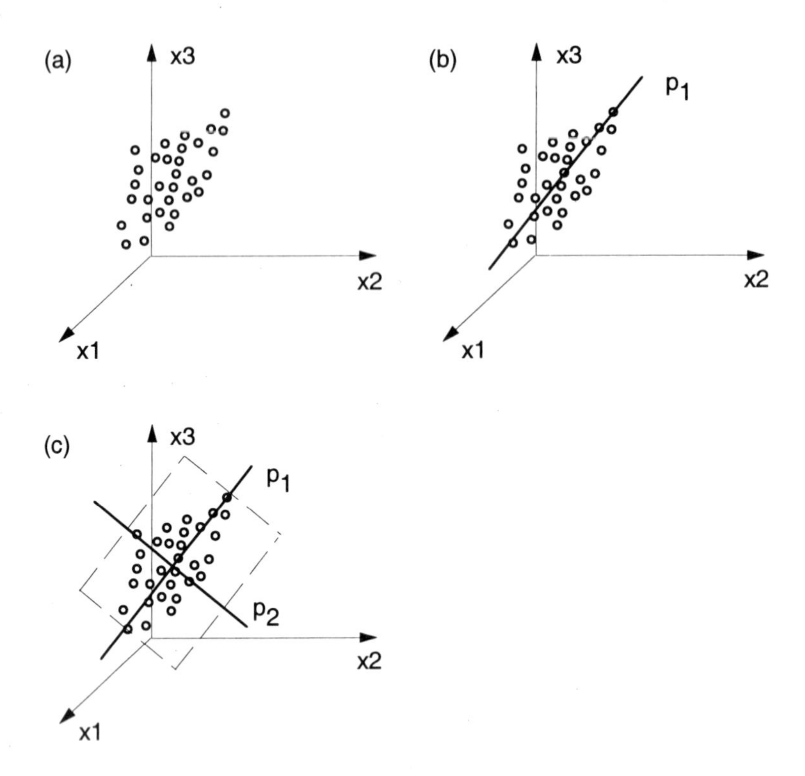

Figure 11.13: Geometric representation of PCA. When the original variables are correlated, their scatter is enclosed in an ellipse which has axes which are at an angle to the variable axis. The axes of the ellipse become the principal components. (a) observation of three variables plotted in the space of the measurements; (b) the first principal component; (c) the plane defined by the first two principal components.

This transformation is called singular value decomposition (SVD). The decomposition acts on the covariance matrix of X and produces the PCs for a change of coordinates. We remember that the PCs are orthogonal. The covariance matrix Σ of the independent variables in the new coordinate system is diagonal. The elements of Σ indicate how much of the overall variation in X occurs in the corresponding direction in P.

Using principal components a model that describes normal operating conditions (NOC) can be described. The model can then be used to detect outliers as excessive variation from normal target and as unusual patterns of variation. The envelope of the NOC can be described as in Figure 11.14. For data using two PCs the observations can be displayed with a control ellipse to monitor process performance. Points located inside the ellipse are considered in-control with respect to the PCs. The third axis denotes the residual, orthogonal to the PCs. Points in the residual direction can also be considered outliers. They denote the possible loss of information by using fewer PCs than original variables.

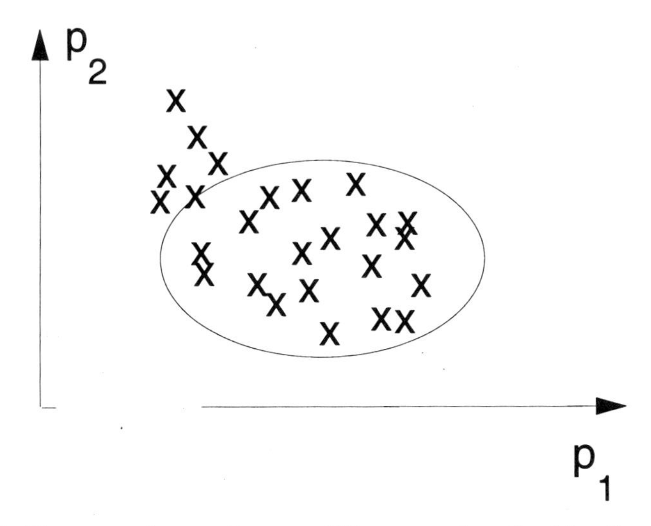

Figure 11.14: The normal operating conditions are described by the ellipse in the plane of the two PCs. When an upset occurs the observations leave the envelope. Upset data from different source causes movement to different regions.

For data requiring more than two dimensions the same statistical measures can be used for quantitative testing of points in a hyper-dimensional elliptical envelope. A system having such a detection of measurements outside a normal operating condition is sometimes called a "smart alarm".

Example 11.6 (Monitoring influent wastewater characteristics) *PCA is applied in the monitoring of the influent wastewater characteristics of the Ronneby plant, Sweden. Available on-line measurements on the influent wastewater are; temperature (1), conductivity (2), ammonia (3), pH (4) and flowrate (5). The detection model is built and trained from more than 6000 samples (about 21 days) of normal conditions. Already two principal components describe 75 % of the variance. In other words, by projecting the variables from dimension 5 to dimension 2 we manage to capture 75 % of the variance. The Figure 11.15 describes the projected values. Normal data remain within the ellipses, and the elliptic boundaries correspond to the 95 and 99 percentage significance level of the original training data.*

Two periods of data deviate significantly, the first between samples 141 to 165 and the other from 336 to 352. From the plotted data one can see certain deviations in some of the variables at these intervals. However, the deviations are not always caused by what is believed from the simple time series plots. Instead, the contributions from every measurement variable can be calculated, and a much more obvious cause for the deviation can be detected.

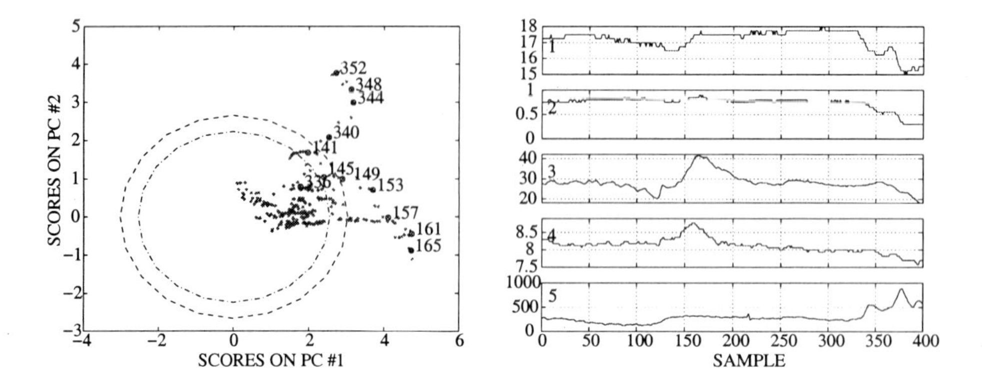

Figure 11.15: Monitoring the influent characteristics with PCA (left) and traditional time series charts (right).

11.4.2 Multiple Linear Regression

In the principal component analysis we tried to find the reduced space that would sufficiently explain the variability of a single data matrix. Often we can identify two groups of variables. Above we considered the process variables X, on which observations are always available. Measurements are also made, often much less frequently, on a number of process output or product quality variables that are the key indicators of process performance. Some of these variables may be measured on-line with special sensors (such as phosphorus or nitrogen content), but many are often the result of sample analyses made off-line in a quality control laboratory (such as SVI or floc shape test). Arrange the n observations of these m variables into an $(n \times m)$ matrix Y.

The problem is now not only to explain the variability in the process variables x, but to use them to predict the output variables y.

Multiple linear regression (MLR) is the most common method for developing multivariable statistical models.

Traditional multiple linear regression has severe problems dealing with large sets of highly correlated variables. This leads to imprecise parameter estimates and poor predictions.

A least squares solution usually gives very imprecise parameter estimates and may often be singular (see also Section 7.1). To overcome this problem many of the variables in x can be eliminated either by subjective means or by step-wise linear regression routines. Unfortunately, these methods are often unsatisfactory. Furthermore, the aim of the analysis is to extract information from all measurements available, so there is a need to focus on methods that retain the entire data set.

11.4.3 Partial Least Squares Regression

The Projection to Latent Structures (PLS) or Partial Least Squares method addresses the problem when the variables in both x and y are highly correlated. Conceptually PLS is similar to PCA except that it simultaneously reduces the dimensions of the X and the Y spaces to find latent vectors for the X space which are more predictive of the Y space. In effect the plane in the X space is tilted so that it is more predictive of Y. The PLS algorithm simultaneously computes the latent vectors (called t and u) of X and Y respectively in an iterative manner. An empirical model is developed to relate process variables (X) to product (or quality) variables (Y) under normal operating conditions. Then, by monitoring the process variables only (X) and projecting them to the reduced dimensional space defined by PLS, we monitor the variation in the process variables that are more influential on the product quality variables.

Example 11.7 (Using PLS to detect cause-effect relationships) *PLS analysis has been applied to an activated sludge plant. The latent variables (LV) of the X space (16 variables) are calculated to maximise the correlation between the input matrix X and an output matrix Y, containing the product quality variables (the effluent turbidity is chosen here). The model is built from 2500 samples (about 9 days) of normal conditions. The new measurements are transformed into a two-dimensional space defined by the first two latent variables.*

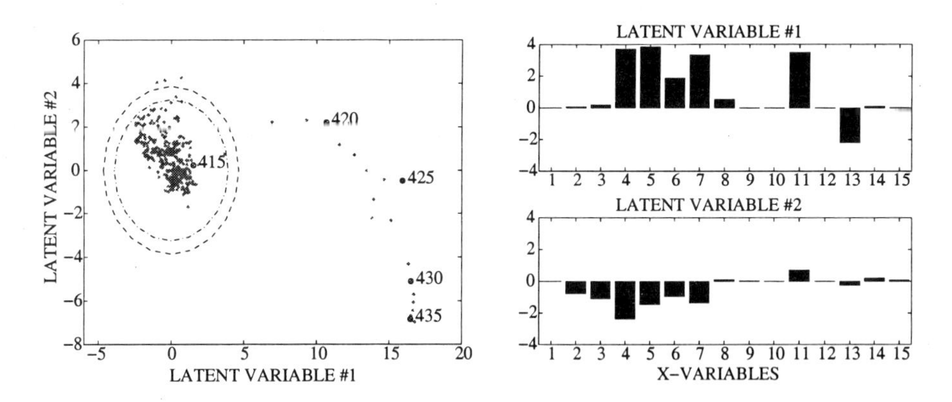

Figure 11.16: Monitoring process with PLS (right) and diagnosis of the event at sample 415-435 (left).

The left graph in Figure 11.16 shows that the process is well inside the boundaries until a disturbance occurs at about sample 415. In a few samples the process has drifted far outside the boundaries until a maximum deviation is reached at sample 439. In Figure 11.16 (right) the major contributors to the deviations are shown. Along the number 1 direction the major inputs are the air valves (3-7) and the flowrate (11).

11.4.4 Cluster Analysis

Processes will often operate in a number of more or less distinct operating regimes. These may be related to different feedstocks or products, for example in wastewater treatment influent may sometimes contain toxic elements for some populations of microorganisms or could be deficient in some nutrients.

They might also be due to changes within the process itself due to process materials or equipment. Wastewater examples could be again a change in the bacterial population due to washout, a change in settleability, or it could be a DO probe diaphragm or other sensor fault.

Sometimes these changes in the process can be detected by the traditional trending of individual process variables. Clustering is another way of analysing and displaying process data that is often found to aid the operator in detecting changes in the process. Figure 11.17 illustrates a simple two dimensional cluster display. Three dimensional displays are also possible. The dimensions can be measured or calculated process variables or could be components from a principal components analysis.

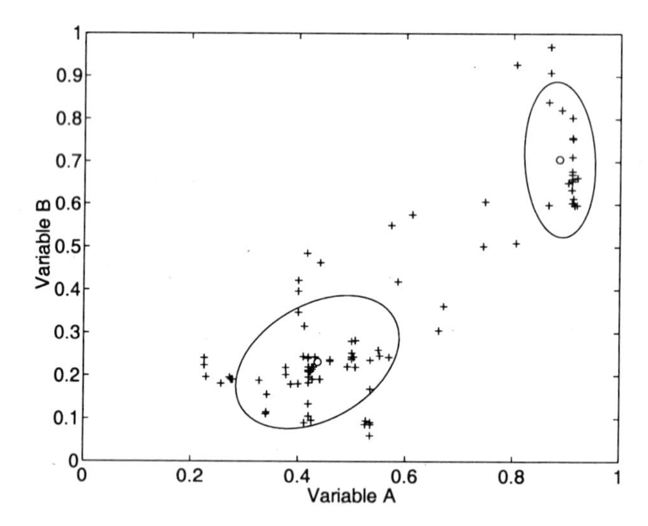

Figure 11.17: Two-dimensional cluster display.

If the most recent operating points are distinguished, it may be possible to identify the problem from the cluster in which the data is located. It may also be possible to detect problems early from the general direction in which the process is moving.

Partition type clustering algorithms are most appropriate for the often large amounts of process data where there may be no *a priori* knowledge of the operating regimes. Three algorithms will be discussed here, the popular K-means method, an ellipsoid method developed for mineral processing and the FCM fuzzy C-means technique.

K-means clustering

K centres (or means) must be chosen to start the algorithm. This may be done randomly or from prior process operating knowledge.

Then each data point is assigned to the closest centre, using a Euclidean distance or 2-norm. The centres are then recalculated as the means of the points assigned to that cluster. The assignment and recalculation of the means is repeated until there is no significant movement of the means.

The major disadvantage of K-means clustering is the need to specify the number of clusters *a priori* and the lack of a boundary definition. Function `kmeans.m` in the MATLAB Toolbox implements this algorithm.

Ellipsoid Clustering

This algorithm defines clusters as multidimensional ellipsoids whose shapes and sizes are defined by the eigenvectors and eigenvalues of their sample covariance matrices. This is a mouthful that really only the algorithm needs to worry about too much.

The allocation of points to clusters is performed by selecting an agglomeration distance and an unassigned data point, and then agglomerating all the other points within this distance. Then the covariance matrix is calculated defining the cluster shape and size. Unassigned points are again tested to see if they are within the agglomeration distance. This procedure continues until no further points can be added.

Then another unassigned point is chosen as a centre and another cluster is grown. This continues until all but a few "outlier" points remain.

The agglomeration distance can also be increased by a factor up to some limit to ensure all but outlier points are clustered.

Function `ellclust.m` in the MATLAB Toolbox implements this algorithm.

FCM Fuzzy Clustering

The fuzzy C-means (FCM) algorithm is basically the K-means algorithm where the points have a weighted membership value in each cluster rather than a definite assignment.

The weighted membership value takes the form $u_{k,i}^m$ where k is the cluster, i is the point and the power m is a measure of the "fuzziness" of the cluster boundaries. The membership is calculated from distance measures:

$$u_{k,i} = 1/\sum_{j=1}^{C}\left(\frac{d_{k,i}}{d_{k,j}}\right)^{2/m-1}. \tag{11.9}$$

The disadvantage of needing to nominate the number of clusters remains. However a measure of size can be obtained by mapping the boundaries for some membership function value, say 0.9. Function `fuzclust.m` in the MATLAB Toolbox implements this algorithm.

Example 11.8 (Comparison of clustering algorithms) *These three algorithms have been applied to a set of data containing two operational clusters and an outlier. The results are shown in Figure 11.18.*

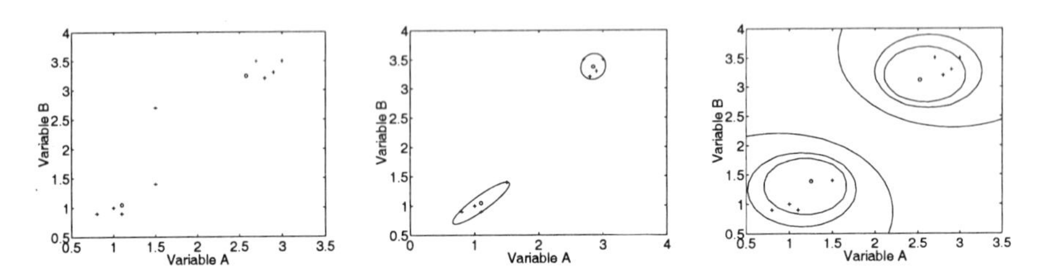

Figure 11.18: K-means, ellipsoidal and fuzzy C-means clustering.

The K-means algorithm will not reject outliers and allocates it to one of the clusters depending on the randomised starting points. On the left in Figure 11.18 it allocated the outlier to the lower cluster pulling the centre of that cluster up to the right. The ellipsoidal technique rejected the outlier and formed two compact cluster boundaries shown in the centre plot in Figure 11.18. The right plot in Figure 11.18 shows the results of the fuzzy C-means clustering technique with $m = 0.5$ *and showing contours with membership values of 0.9, 0.98 and 0.99. In this case the outlier affects both clusters pulling the centres towards it.*

Example 11.9 (Detection using multiple operational classes) *We have seen how the operational data can be plotted in the plane of the first two principal components, as in Figures 11.15 and 11.16. Points that form a cluster represent similar process behaviour. Any deviations from normal operating behaviour will be seen as "snakes" of points moving outside the cluster. Here a PCA model was identified from about 10,000 measurements (about 36 days)of 11 variables. The model was able to catch about 85 per cent of the variability using the two PCs and is shown in Figure 11.19. We note that most of the measurements are located around the origin. There are, however, some measurements that deviate more significantly from the central cluster a of measurements. From the plottings of the individual measurements it was possible to distinguish some distinct features. The area d represents periods of high influent flowrate and large oxygen demand (shown by large air flowrate demand). In cluster b there is just a high influent flowrate. Peaks in the effluent suspended solids concentration are represented by c. The central cluster a is essentially a dry weather period. The boundaries between the clusters have been defined simply from these empirical observations.*

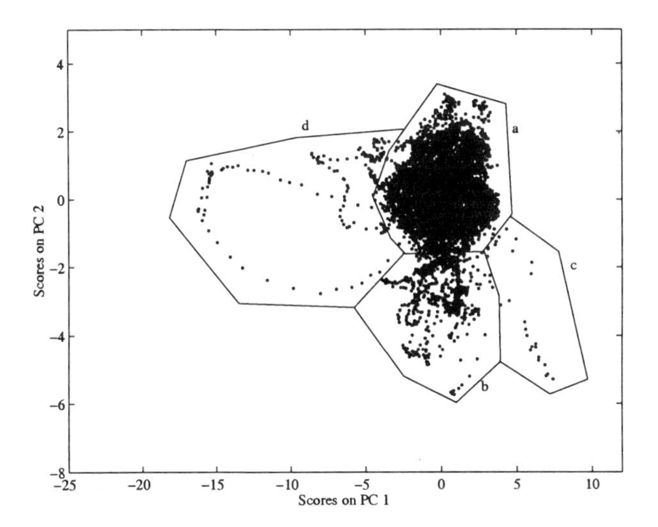

Figure 11.19: Two dimensional cluster display, defined from some 10,000 measurements. Clusters have been defined from the operating data.

The model with the principal components displayed in Figure 11.19 is now used for new data. Another period of 10,000 data is displayed in Figure 11.20. Many of the data can be directly classified as the previously defined operating conditions. However, some data deviate from the previous definitions. Actually this sequence of data has a combination of high flowrate and high oxygen demand and apparently falls in between the two previously defined classes b and d.

11.4.5 Summary

It ought to be emphasised that in all inferential approaches one has to assume that, in order for the methods to work, the operational conditions have to be comparable and the events have to be observable. The first assumption states that the method is valid as long as the reference database is representative of the process operation. If something changes in the process (such as a completely new composition of the influent water) then one has to build a new database which embodies the change and then re-apply the method. The second assumption says that the events which one wishes to detect must be observable from the measurements that are being collected. No monitoring procedure can detect events that do not affect the measurements.

11.5 Further Reading

- A nice introduction into the ideas of quality is found in Garvin (1984).

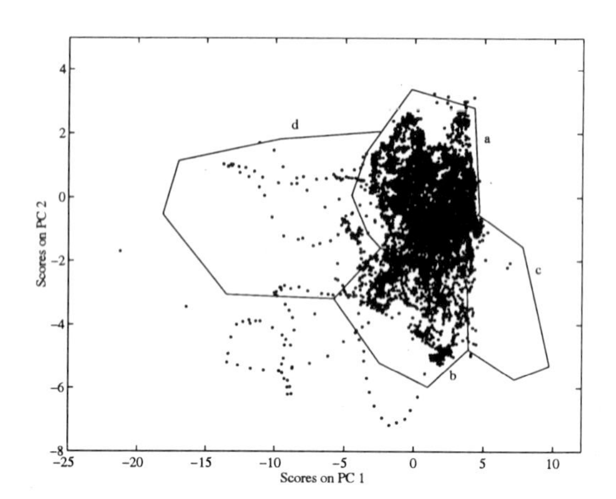

Figure 11.20: Two-dimensional cluster display. New measurements will reveal deviations from previous clusters.

- A classic in the quality control area is Crosby (1989). His basic idea is that "with management and employee commitment great strides can be made in improving quality". Other pioneering books in quality are Deming (1982), Ishikawa (1989), Juran & Gryna (1980) and Sullivan (1986).

- Another quality control expert, Feigenbaum (1961) on page 17 delivered a fundamental message, "Make it right the first time....".

- The most important international standardization organization is ISO (International Organization for Standardization). The acronym is not misspelled, but derives from the Greek work "isos" ("equal"). ISO standards cover all possible technical fields and are therefore the main reference for the industry, together with the applicable national standards. Current information on ISO standards can be found on the web address:

  http://www.iso.ch/

- The June 1996 issue of the journal *Quality Progress* discusses and critically reviews the *ISO 9000* series, its history and its future (Anon. (1996)).

- Statistical tools are essential for detection and diagnosis. Box *et al.* (1978) is an excellent reference for the practical user. Other standard textbooks are Cryer (1986) and Kendall & Ord (1990).

- Experiences of time series analysis in wastewater treatment systems are found in Adeyemi *et al.* (1979), Novotny *et al.* (1991), Beck (1986), Beck (1989) and Carstensen (1994).

- PCA was first described in Pearson (1901). It is described in many textbooks, such as Anderson (1984) and Johnson & Wichern (1992). The PLS development is described in Wold (1982) and further refinements are introduced in Wold *et al.* (1984). The methods are further described in the textbook by Martens & Naes (1989). Good overviews are found in Wold *et al.* (1987), Kresta *et al.* (1989) and MacGregor *et al.* (1991).
- Three cluster algorithms were discussed: the K-means method by Hartigan (1975), an ellipsoid method by Ginsberg & Whiten (1991) and the FCM fuzzy C-means technique by Bezdek (1981). A recursive and adaptive form of the latter was developed and applied to an anaerobic digester by Marsili-Libelli & Müller (1996). Marsili-Libelli (1991) also applied clustering to ecological data.
- The figures 11.15 and 11.16 are published in Rosén & Olsson (1997) and the figures 11.19 and 11.20 are published in Rosén (1998).
- There are many useful software packages. For the MATLAB user there is a Statistics Toolbox available. Another commercial package for multivariate analysis is SIMCA from Erisoft and Umetri (see SIMCA in the References).
- There is a MATLAB PLS toolbox available from Eigenvector Technologies. Barry Wise, Neal Gallaghan, Manson WA 98831. 73633.2451@compuserve.com, fax +1 509 967 3973.

11.6 Exercises

1. Provide your own definition of quality in a wastewater treatment plant.
2. What are the major components of a quality control system in a treatment plant?
3. Describe the difference between high quality and consistent quality.
4. Normally the effluent requirements are given as fixed limits of BOD, total nitrogen, suspended solids and phosphorus. Why should the plant operation just meet the effluent requirements and not try to minimise the effluent concentrations?
5. In some countries the effluent requirements are changing. The treatment plant has to pay a fee proportional to the accumulated effluent mass of BOD, nitrogen and phosphorus. How will this change the operating strategy of a plant?
6. Which variable in a closed loop should you monitor, the manipulated variable or the controlled variable? Why?

7. List and describe the key differences between on-line process control and statistical process control.

8. How would you define the "product liability" for a wastewater treatment system?

9. Discuss different ways to define lower and upper limits. Discuss the reason for introducing a low and very low limit, and a high and very high limit. What would be reasonable values?

10. Often signals are sampled every 6 seconds in a large treatment plant. Still the measurements are stored only every 6 minutes, and hardly any control action takes place more often than every 6 minutes. What is the advantage of sampling the sensors so fast?

11. Why is the Gaussian distribution used so often to describe measurement deviations around an average?

12. How would you determine if a signal in a real plant is stationary?

13. Describe the underlying purpose of principal component analysis (PCA).

14. The data file `cluster_data.dat` in the exercises directory contains clarifier data, flowrate and suspended solids. Use the Toolbox functions `kmeans.m`, `ellclust.m` and `fcmeans.m` to cluster this data and compare the results.

Chapter 12

Model Based Diagnosis

Many characteristics of a treatment process slowly change with time. For instance, at a wastewater treatment plant (WWTP) using biological treatment the rate at which the mixed liquor settles changes from day to day. How then can mathematical models be used in the operation of WWTPs? The use of on-line estimation enables process models to be updated. Regular or continuous updating can account for time varying characteristics of the process and it can also update linear models to take account of a nonlinear process. The knowledge reflected by the updated models can be used for improved control or on-line diagnosis of operating problems.

12.1 Introduction

Detection and diagnosis has been discussed in Chapter 10 and important methods for detection as well as quality control have been presented in Chapter 11. In this chapter we will add another dimension to detection and diagnosis. By using steady state or dynamic models there are more refined methods available in order to find faults or operational problems.

The main idea is to combine the information obtained from sensors and observations with the knowledge expressed in models. This makes it possible to detect deviations in certain process parameters as well as in non-measureable state variables. This is illustrated in Figure 12.1.

The idea of using models for diagnosis can be used to check instruments. Two examples are given below.

Example 12.1 (MLSS concentration check) *By making use of redundant measurements of suspended solids it is possible to check two sensors against each other. We will consider the mixed liquor suspended solids (MLSS) concentration measurements x in an aerator and suspended solids concentration x_r in the return sludge. The two measurements can be checked against each other by calculating a*

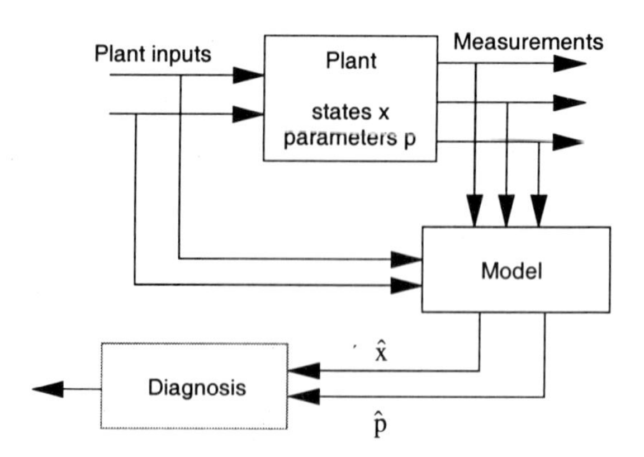

Figure 12.1: Elements of detection, using dynamic models.

concentration from the mass balance of the sludge and comparing it with the measurement. In a short time scale (hourly) the growth of the sludge can be neglected. Referring to Figure 12.2 for the notation, the mass balance for the sludge is:

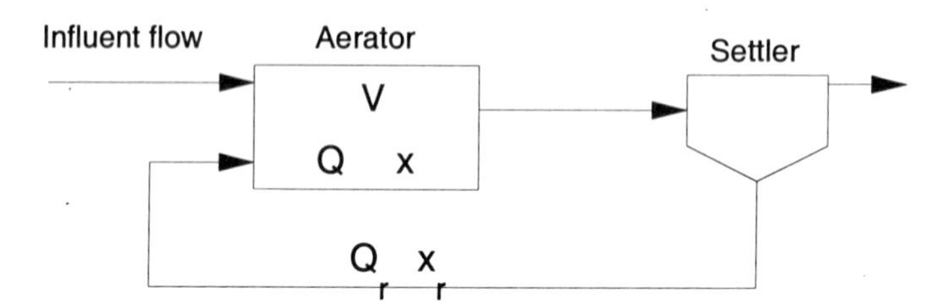

Figure 12.2: Notation for the variables of Example 12.1.

$$\frac{d\hat{x}}{dt} = \frac{Q_r}{V}\, x_r(t) + \frac{Q + Q_r}{V}\, x(t) \tag{12.1}$$

By comparing $\hat{x}(t)$ with the measurement value x it is possible to check the sensors. If there is a significant difference, then one of the sensors has to be recalibrated. Figure 12.3 shows a practical measurement from a full-scale plant. The measured MLSS and the estimated MLSS are close to each other until about 12 hrs, when they start to deviate. The problem was found to be the calibration of the return sludge concentration sensor. Taking the calibrated signal into consideration the estimated and measured values become much closer to each other.

Example 12.2 (Ammonium measurement check) *Data quality is routinely checked against high, low and slope limits. The next level of data quality check is by using experience about the way in which the measurements normally behave and by comparing trends in different parameters that are known to be interlinked.*

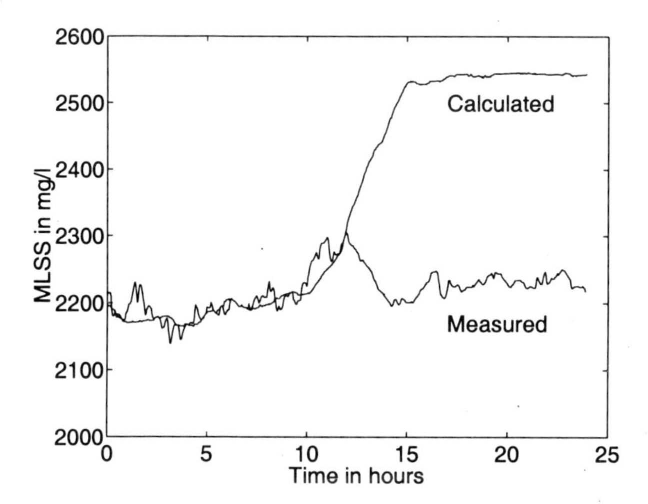

Figure 12.3: Measurement of MLSS from a full-scale plant.

One example is the link between ammonium load and the oxygen rate (OUR). The OUR can be calculated automatically, based on monitoring of the aeration system (see Section 18.5). If a sudden unexpected increase in the ammonium occurs this should normally be seen in the OUR estimates as well. If no sudden change is detected in the OUR, the measurement of ammonium is likely to be erroneous.

12.2 Steady State Models

By making relatively simple calculations various parameters can be checked on a routine basis. Here we give some examples of steady state models where some parameters can be indirectly calculated.

In Example 3.9 we showed that there is a simple relationship between the air flowrate and the oxygen uptake rate if the dissolved oxygen is kept constant. Therefore the air flowrate is a good indicator for changing biodegradable load to the plant. The OUR can of course be more accurately estimated using more sophisticated models, as in Section 18.5, but the air flowrate is a good first approximation.

The sludge settling properties can be defined in various ways, like initial settling velocity and sludge volume index (SVI) (see Chapter 22). One first check on the thickening properties is to calculate the thickening ratio x_r/x, the ratio between the return sludge concentration and the MLSS concentration. If this ratio (averaged over a few hours) becomes too small, then the floc properties need to be checked further.

Another reason for a too small return sludge concentration may be the return sludge flowrate. If this is too high, then there will be a diluted flow. Therefore, a good way to check that the return sludge flowrate can be increased is that the dry mass $Q_r \cdot x_r$ is increasing for an increasing flowrate Q_r. In other words, the ratio

$$\frac{Q_r(2) \cdot x_r(2) - Q_r(1) \cdot x_r(1)}{Q_r(2) - Q_r(1)} \tag{12.2}$$

where "2" and "1" are two operating points, has to be positive and sufficiently large.

There is a close relationship between the steady state influent flowrate Q and the effluent suspended solids concentration x_e. It is advisable to regularly check the average ratio between the two variables. During transients the relationship is different, and will be discussed in Section 12.4.3.

12.3 State Estimation

In Chapter 7 we discussed estimation methods, how to find unknown parameters in steady state or in dynamic models. We will now proceed to describe estimation methods for systems described by state equations.

In Section 5.2 we defined state space models. We repeat the linear representations (Equations 5.20),

$$\frac{dx}{dt} = A\, x(t) + B\, u(t) \tag{12.3}$$

$$y(t) = C\, x(t) + D\, u(t) \tag{12.4}$$

Generally the state variables cannot be measured directly. Instead the process is observed via the output vector y, which is usually of a lower dimension than x. In other words, there are fewer sensors available than the number of states. For example, in the activated sludge model the organism concentrations are not directly measureable, while we may measure variables like dissolved oxygen, suspended solids, COD, or TOC. The question is now whether the non-measureable state variables can be indirectly calculated from the model and the measurements.

12.3.1 Observers

We can reconstruct the state variables from the measurements by an *observer* (when no disturbances are modelled) or a *filter* (when disturbances are modelled as stochastic processes). In many situations it is interesting to know all the states x even if adequate sensors are not available or simply their cost is prohibitive. To denote that the estimated value of x differs from the true value we use the notation $\hat{x}$ for the estimate.

Linear Systems

The structure of the estimator is similar to the real system description (Equation 12.3). A correction term is added that takes the real measurements y into consideration:

$$\frac{d\hat{x}}{dt} = A\ \hat{x}(t) + B\ u(t) + K\ [y - C\ \hat{x}] \tag{12.5}$$

If there is only one sensor then K is a vector, otherwise it is a matrix, that has to be chosen appropriately. If x were equal to $\hat{x}$ then the last term in Equation 12.5 would be zero, since $y = Cx$. The estimated value then would obey the same dynamic equation as the true state Equation 12.3. As long as $\hat{x}$ is different from x, the last term in Equation 12.5 is an error correction term, the difference between the true measurement y and the estimated measurement $C\hat{x}$. The error corrects the equation and compensates the estimated value.

It is possible to show that the estimate $\hat{x}$ approaches the true state x given certain conditions. Equation 12.5 can be subtracted from Equation 12.3 and we obtain:

$$\frac{d\tilde{x}}{dt} = A\ \tilde{x} - K[y - C\hat{x}] \tag{12.6}$$

where $\tilde{x} = x - \hat{x}$. Furthermore we know that $y = Cx$, so the result is:

$$\frac{d\tilde{x}}{dt} = [A\ - KC]\tilde{x} \tag{12.7}$$

Note that $\tilde{x}$ goes to zero once the matrix $A - KC$ has certain properties. If the system is *observable*, then we can always find a K such that $\tilde{x}$ goes to zero.

The estimator structure is illustrated in Figure 12.1. Note that the model output $C\hat{x}$ is compared all the time with the true measurement y. If K is properly chosen, then $\hat{x}$ can approach x.

The *plant* (upper box in Figure 12.1) represents the physical process, while the estimator is an algorithm that implements Equation 12.5 in a computer. In other words, the estimator is nothing more than a computer simulation of the physical process, continuously corrected with the measured real data y. If K can be chosen so that $\hat{x}(t)$ converges to $x(t)$ sufficiently fast, then the observer is satisfactory. This means that $\hat{x}(t)$ will converge to $x(t)$ regardless of the initial guess.

The fact that K can be chosen so that the estimator works in the desired way depends on the system property called **observability**. This property depends only on the A and C matrices. Roughly speaking observability refers to the ability to deduce information on all the states x of the system by monitoring the measured outputs y. It also guarantees that K can be found so that $\hat{x}(t)$ converges to $x(t)$ *arbitrarily* fast. Unobservability results from some state or subsystem being disconnected physically from the output and therefore not appearing in the measurements.

Example 12.3 (State observer) *Consider a mechanical system of second order, describing the position and velocity of a mass connected to a spring and a damper. We use high school mechanics with Newton's law to model the position as:*

$$m\frac{d^2z}{dt^2} = F - b\frac{dz}{dt} - kz \tag{12.8}$$

where z is the position, F an external force, m the mass, b a viscous damping coefficient, and k a spring constant.

We assume that the position can be measured exactly. *The system is rewritten in a state form, where we define:*

$$z = x_1 \qquad \text{and} \qquad \frac{dz}{dt} = x_2 \tag{12.9}$$

so that:

$$\frac{dx_1}{dt} = x_2 \tag{12.10}$$

$$\frac{dx_2}{dt} = -\frac{k}{m}x_1 - \frac{b}{m}x_2 + \frac{1}{m}F \tag{12.11}$$

where x_1 is the position and x_2 the velocity. The measurement is simply:

$$y = x_1 \tag{12.12}$$

Figure 12.4 shows the measured position (which is the true one), the true velocity and various estimates of the velocity when $m = b = k = 1$ and $F = \pm 5$.

First the estimator Equation 12.5 is simulated with $K^T = [0, 0]$. The true velocity has an initial value of 5, while the best guess for the initial value of the estimate is 0. The estimate converges towards the true velocity, once the influence of the initial condition has disappeared. The model is perfect and it is driven by the same force as the real system (compare Figure 12.1). If the value of K is increased then the estimated velocity approaches the true one much faster. In theory the estimation can be made infinitely fast, as long as the model is correct and the measurement perfect.

Nonlinear and Time Discrete Systems

Let us now look at a state estimator for a nonlinear and time discrete system. The state model of the system (compare the linear continuous system in Equation 12.3) is described by

$$x_{k+1} = f(x_k, u_k) \tag{12.13}$$

while the measurement relationships to the states (compare the linear form in Equation 12.4) are modelled by the algebraic relation,

$$y_{k+1} = g(x_{k+1}) \tag{12.14}$$

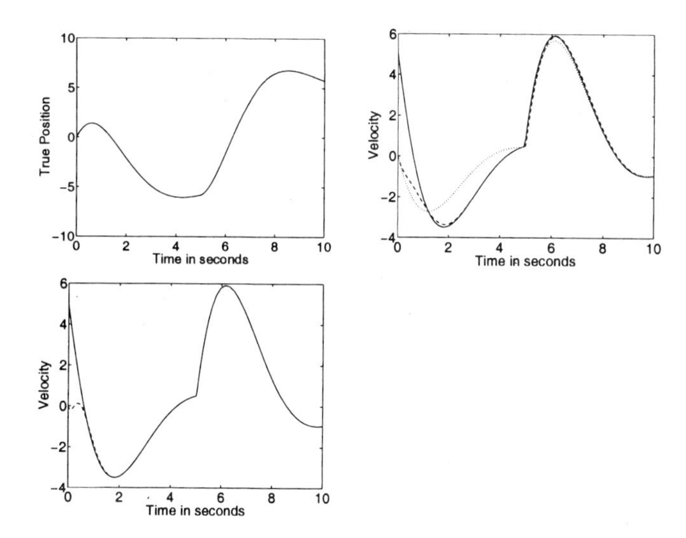

Figure 12.4: An observer for the velocity of the mechanical system: (a) the measured position, (b) estimation of the velocity, true velocity (solid line), estimated velocity with $K^T = [0, 0]$ (dotted line), with $K^T = [3, 3]$ (dashed line), (c) estimation of the velocity, true velocity (solid line), estimated velocity with $K^T = [6, 15]$ (dashed line).

The mathematical form of the state estimator consists of three parts. The first one *predicts* the next state, based on the current estimate and Equation 12.13:

$$\hat{x}_{k+1|k} = f(\hat{x}_{k|k}, u_k) \tag{12.15}$$

where $\hat{x}_{k|k}$ denotes the estimate at time k, given measurements up to time k, $\hat{x}_{k+1|k}$ is the estimate at time $k+1$, given the same measurement data and u_k are the process inputs. The estimation filter makes a *correction* based on the latest measurements at time $k+1$:

$$\hat{x}_{k+1|k+1} = \hat{x}_{k+1|k} + K_{k+1}\left(y_{k+1} - g(\hat{x}_{k+1|k})\right) \tag{12.16}$$

where y_{k+1} are the actual process measurements. The function:

$$\hat{y}_{k+1|k} = g(\hat{x}_{k+1|k}) \tag{12.17}$$

corresponds to the estimated output at time $k+1$ given the measurements up to time k. The filter parameter K_k is a function of time for an optimal filter, but is usually set to a constant K making the filter sub-optimal but possible to implement.

The structure of the estimator of states and/or outputs is shown in Figure 12.5 and involves the predictor model Equation 12.15, and a corrector filter Equation 12.16. Note that the plant may have many inputs and outputs. The predictor and corrector outputs are vectors containing the estimated state at time $k+1$, given measurements up to time k and $k+1$ respectively.

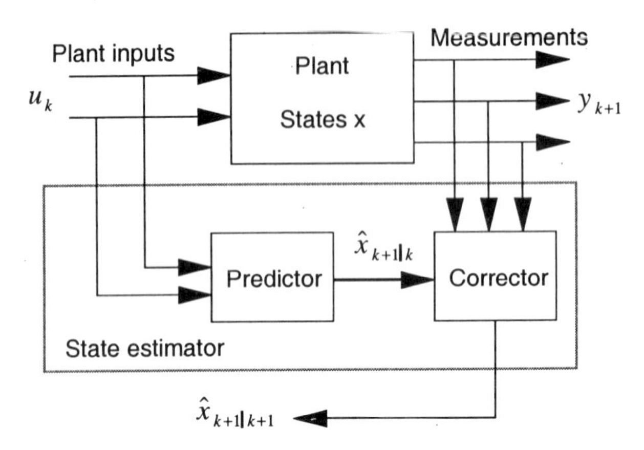

Figure 12.5: Output or state estimation.

Equation 12.17 indicates that, if the actual measurement y_{k+1} at time $k+1$ is the same as the measurement $\hat{y}_{k+1|k}$ predicted by the model $g(\hat{x}_{k+1|k})$, there will be no correction. In other words the parameters of the model are the true parameters.

As for the linear case, for a discrete nonlinear estimation filter to converge the estimates to the true values, the system must be observable. That is the measurements must be sufficient in number and in their relationship to the predicted states.

Example 12.4 (Estimation of the organism concentration) *Figure 12.6 shows the estimation of biomass concentration in a simple biological reactor (see Section 3.7.1 and Example 3.7). The nutrient concentration was the measurement on which the estimation was based using Equation 12.16 above. The value of K was tuned to get a reasonable response without excessive oscillation in the estimate. The MATLAB program* `Example_12_4.m` *was used.*

12.3.2 Filters and Observers in Noisy Systems

In previous chapters we have shown that a mathematical model is hardly ever a perfect representation of reality. There are several imperfections in the system description. In many systems the model does not manage to include all the phenomena that take place, and some states are simply neglected. For the control of systems we should always ask what is an *adequate* representation of the uncertainty.

Stochastic processes are used to model both disturbances to the process and random errors in the sensors (see Chapter 6and Section 11.2.4). A stochastic

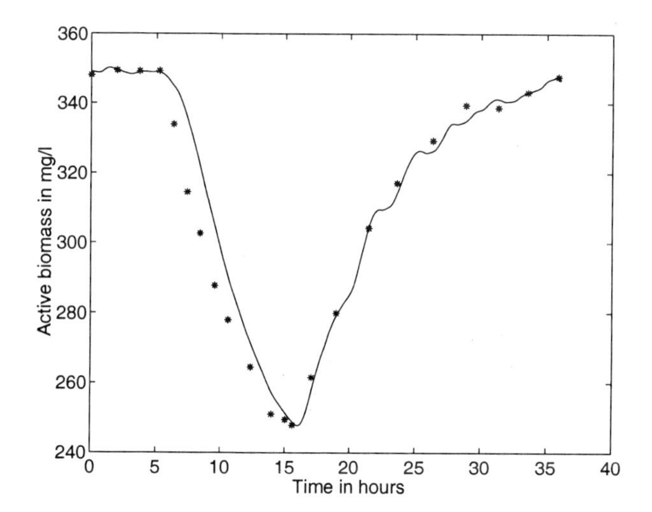

Figure 12.6: Simple biomass concentration estimator (the line is the estimation and the stars are the "true" values).

process is a sequence of stochastic variables. In principle this means that some random variable with a certain probability distribution is added to each process variable at each time instant. Similarly, measurement noise added to a sensor signal is modelled as a random variable. Other aspects of system uncertainty are discussed in Section 5.5.3.

In the previous section we assumed that the measurement information in the estimator was perfect. This is hardly ever the case since every sensor has some imperfection. For instance, the electrical noise of a sensor can be described as an additional random variable $e(t)$ in the output equation. Since each sensor obtains a random error term they can be written in compact form as a vector e added to the output Equation 12.4:

$$y(t) = C\,x(t) + e(t) \tag{12.18}$$

Each component of the noise vector $e(t)$ is modelled as a sequence of stochastic variables, in other words as random numbers. If they are independent, the amplitude of the noise at time t does not depend on the amplitudes at previous times. Often the random number amplitude can be assumed to be normally (Gaussian) distributed so the mean value and the standard deviation completely characterise the noise.

When measurement noise is present, the estimation described before has to be made more cautious. Equation 12.18 is used instead of Equation 12.4 to calculate the error. The estimator structure is changed to:

$$\frac{d\hat{x}}{dt} = A\,\hat{x}(t) + B\,u(t) + K\,[C\,x + e - C\,\hat{x}] \tag{12.19}$$

Now the choice of K has to be made to satisfy a compromise. If K is large the estimation error tends to zero fast. However, the noise term e is amplified which will cause an error. Therefore the value of K has to be sufficiently large so that $\hat{x}(t)$ approaches $x(t)$ as fast as possible, and yet sufficiently small so that the noise term does not corrupt the result. The final test for convergence is to consider the residual $\epsilon(t)$:

$$\epsilon(t) = y(t) - C\,\hat{x}(t) = C\,x(t) + e(t) - C\,\hat{x}(t) \tag{12.20}$$

Note that $\epsilon(t)$ can be calculated in real time. When we have a perfect state estimation the residual is equal to the sensor noise e. In other words, we can make a statistical test of the residual to find out if the filter has converged.

Example 12.5 (State observer with noise) *Let us reconsider the mechanical Example 12.3. Now the position measurement is corrupted with noise* e*, and the measurement Equation 12.12 is modified to:*

$$y = x_1 + e \tag{12.21}$$

From Equation 12.19 we can now see that the gain K *will also amplify the noise term* e*, and the consequence of this is demonstrated in Figure 12.7. We have to estimate the velocity based on the noisy position measurement. Again the true velocity has an initial value of 5, while the best guess for the initial value of the estimate is 0. Now the value of* K *has to be chosen with some care. First* K *is chosen to* $K^T = [3, 3]$ *and we see that the estimate converges towards the true velocity, but now the estimate is somewhat noisy. If we wish to speed up the convergence we intuitively increase* K*. However, there is a price for this, and the estimate becomes more noisy. Therefore, there is no obvious "best" estimate.*

It would seem obvious to calculate the velocity simply by taking the derivative of the position signal y*. However this is only possible for a perfect signal. It is quite apparent that by taking the derivative of the noisy signal in Figure 12.7 the result will be quite useless for any further use. Thus, by mixing the information from the model and the sensor we will get a good extraction of information.*

In order to find the best K values with noisy measurements more sophisticated methods have to be used. The best choice of K is often time varying. Typically K can be large as long as the difference between the *real* measurement $y(t)$ and the *estimated* measurement $\hat{y}(t) = C\hat{x}(t)$ is large compared to the noise e. As the error gets smaller its amplitude is comparable with the noise amplitude e and K has to be decreased accordingly.

Process Noise

The process variables themselves may contain disturbances that cannot be modelled in any simple deterministic way. Thus the model Equation 12.3 can be

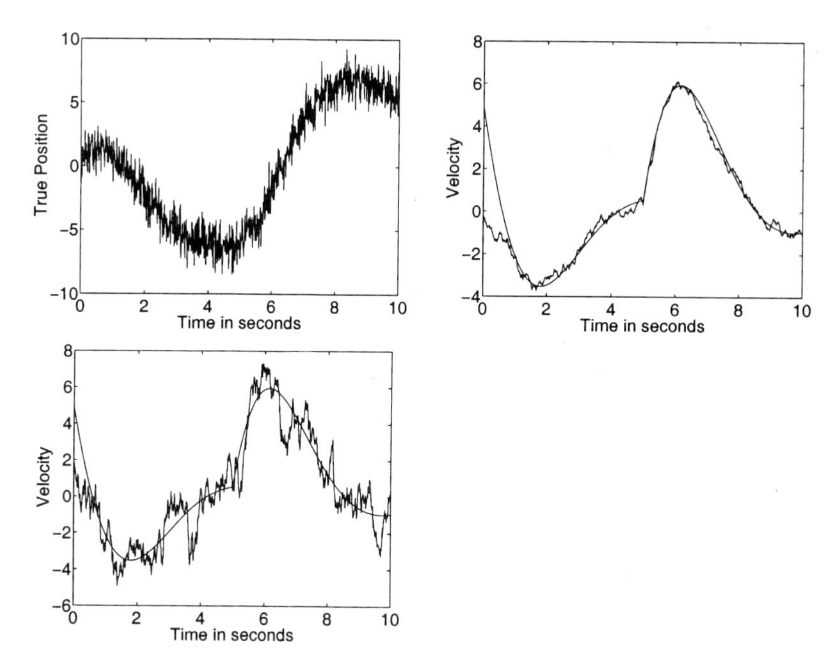

Figure 12.7: An observer for the velocity of the mechanical system with noisy position measurement: (a) the measured position, (b) estimation of the velocity, true velocity (smooth line), estimated velocity with $K^T = [3, 3]$ (noisy line), (c) estimation of the velocity, true velocity (smooth line), estimated velocity with $K^T = [6, 15]$ (noisy line).

complemented with some additional noise term that will describe either the modelling error or some real process noise. For example, the liquid surface of a large tank may not be smooth due to wind and waves causing random variations of the level. Such variations may be modelled as random numbers $v(t)$ being added to the state:

$$\frac{dx}{dt} = Ax(t) + Bu(t) + v(t) \tag{12.22}$$

The random variable v can be considered in a similar way as the measurement noise e. Given the noise description, this means that the estimator can take the uncertainty into consideration. It seems reasonable that any correction would be more cautious as soon as the state variables are distorted by noise, this means a smaller gain K. There is an optimal choice of K in a noisy situation. A **Kalman filter** has the structure given in Equation 12.19 and is based on the system description in Equation 12.22 and 12.18. The K that is obtained from the Kalman filter is time varying and represents the optimal compromise between system and sensor disturbances and the estimation error.

The degree of uncertainty (or confidence) associated with both the observations (measurements) and the model will be a decisive factor in determining how to use the model for the purpose of reconstructing the unmeasurable states. The degree of uncertainty is often embedded in the filter (linear or nonlinear).

Most estimation theory, like the theory of the Kalman filter is restricted to linear systems (linear state and output models). For nonlinear systems, it is often possible to linearise the models and use the linear observability tests and use the Kalman filter recursive design equations to calculate at least a first estimate of the filter matrix K. The MATLAB Toolbox routine `kalman.m` will perform these calculations for a linear system.

To get an intuitive feeling for the kind of reasoning in a Kalman filter we consider two distinct features of it:

- it embodies a model of the system, and
- it acknowledges that any observation and any description of the real world's behaviour is subject to uncertainty, and it embodies therefore the necessary computational effort to accommodate this uncertainty.

It is this second feature that is the most distinctive and potentially the most powerful attribute of the filter, yet also perhaps the cause of its obscurity. In the filter essentially three types of uncertainty are acknowledged:

- uncertainty in the specified relationships in the model,
- uncertainty in the input disturbances to the system, and
- uncertainty in the observed outputs from the system.

The size of the filter matrix K reflects the weighting between model uncertainty and measurement uncertainty. This weighting factor is the focal point of the computations regarding uncertainty. Expressed in words:

- if the model *is known well* relative to the accuracy of the process observations, the weighting factor K (Equation 12.19) is small. Relatively little account is taken of any mismatch between the model and reality in correcting the predicted state estimates (the mismatch is attributed to chance, spurious errors in the observations);
- if the model is *not* known well relative to the accuracy of the process observations $y(t)$, the weighting factor K is large and accordingly much account is taken of mismatches between the model and reality.

Example 12.6 (Estimation of the organism concentration) *Figure 12.8 shows the estimation of biomass concentration in a simple biological reactor (see Example 12.4). Again the nutrient concentration was the measurement on which the estimation was based. The value of K was ten times the value used in Example 12.4 and comparing Figure 12.8 with Figure 12.6 we can see that the "tracking" is much more accurate, but that there is excessive oscillation in the estimate resulting from a small amount of noise in the nutrient measurement.*

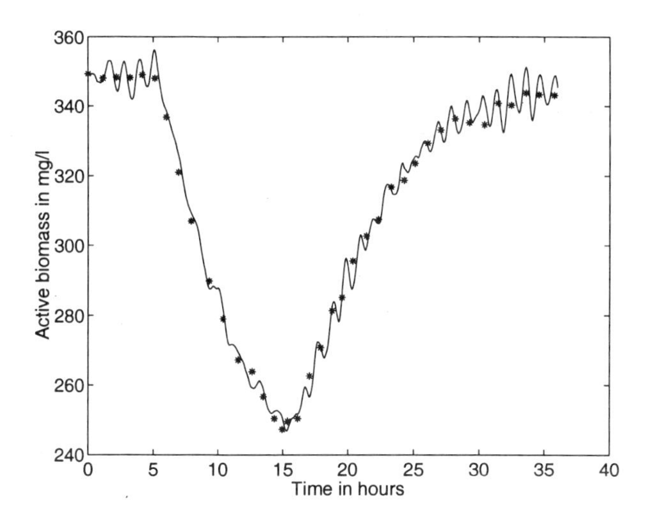

Figure 12.8: Simple biomass concentration estimator with a high gain (the line is the estimation and the stars are the "true" values).

12.3.3 Abrupt Faults

There are four types of faults that may appear in a process like a treatment plant:

- sensor (instrument) faults,
- actuator faults,
- component (unmodelled behaviour) faults, and
- parameter changes, indicating an abnormal process change.

In the model based fault diagnosis we need some mathematical model in order to make a relevant detection of the faults. A sensor or instrument fault may be seen as an additive fault to the output signal. The actuator fault can be modelled as an additive fault to some input signal of the process. Component faults are entering the system dynamics as an extra additive disturbance. If some of the system parameters are changing, we would like to be able to detect the change.

In order to incorporate various faults the model Equation 12.3 can now be described by extra sensor faults $f_s(t)$, actuator faults $f_a(t)$, process faults $f_c(t)$, as well as external disturbances $d(t)$:

$$\frac{dx}{dt} = Ax(t) + B[u(t) + f_a(t)] + Hf_c(t) + Ed(t) \tag{12.23}$$

$$y(t) = Cx(t) + f_s(t) + e(t) \tag{12.24}$$

We can see that the component fault enters the system dynamics in a similar way to the additive external disturbances $d(t)$. The idea of describing the fault term is that it is assumed that the faults will appear abruptly, more or less like a step change. Of course, a fault model like Equation 12.23 and 12.24 is simplified. The model is again compared with measurement data, as in Equation 12.5.

A model based fault diagnosis tries to identify which parameters would best fit the data, thus making the diagnosis. The fault analysis can be performed by looking at the residuals $\epsilon(t)$:

$$\epsilon(t) = y - C \cdot \hat{x} = Cx + f_s + e - C\hat{x} = \; = C \cdot (x - \hat{x}) + f_s + e \tag{12.25}$$

If no fault would appear, then $\hat{x}$ converges to x and the residual will approach the noise e. An abrupt fault will make the residual jump. The fault decision can be made in a number of ways:

- Make a statistical test of the residual, to see if it looks like the noise e or not.
- Use a set of several models like Equation 12.23 and 12.24, with and without fault terms, testing whether the system behaves like one of the assumed models with hypothesised faults included.
- Calculate a correlation function of the observed residuals with precomputed filter responses due to certain faults (fault signatures).

All these methods require accurate knowledge of the process parameters (A, B and C) and the influencing signals. This is certainly not common in environmental systems. The procedure can still be illustrated by the block diagram of Figure 12.5, where the residual is calculated in the *corrector* part according to Equation 12.25.

Example 12.7 (Faulty DO sensor) *The DO mass balance for a simple reactor was derived in Equation 3.29. An elementary test to find the oxygen transfer coefficient K_La can be performed by giving the air flowrate a step input. Neglecting the flow terms, we have the following mass balance:*

$$\frac{d}{dt}(VS_O) = r_OV + K_La(S_{O,sat} - S_O)V \tag{12.26}$$

Now we assume that the respiration rate r_O is constant and that the oxygen transfer is proportional to the air flowrate (compare Section 3.9):

$$K_La = K_a \cdot q_a \tag{12.27}$$

Consider a DO concentration change for a change in the air flowrate, and follow the procedure from Section 5.2.2:

$$\frac{d}{dt}(\Delta S_O) = -(K_aq_a)_{nom}\Delta S_O + (K_aS_{O,sat})_{nom}\Delta q_a \tag{12.28}$$

where the prefix Δ denotes the deviation from the nominal values. A broken membrane on the DO sensor will change the sensor characteristics in two ways. A *bias signal is added and the gain of the sensor will be changed, in other words the sensor signal y has a larger C value and a bias f_s added:*

$$y = CS_O + f_s \tag{12.29}$$

Figure 12.9 depicts what happens for an ideal sensor and for a faulty sensor when the air flowrate has been changed stepwise. The faulty sensor reacts not only to dissolved oxygen but also to the air flow because of a faulty membrane.

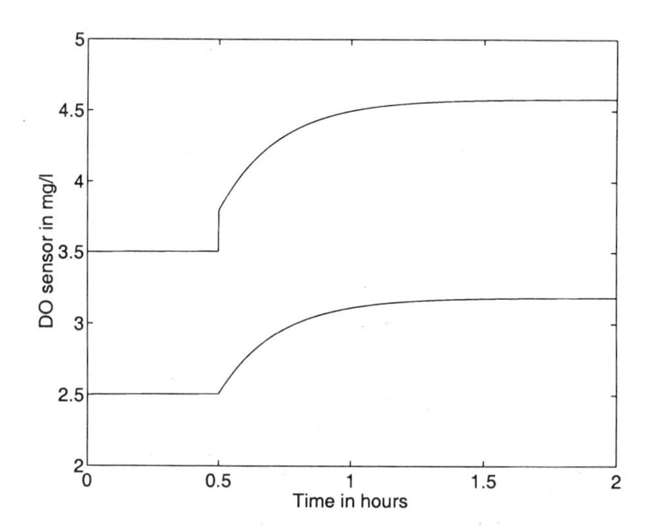

Figure 12.9: Comparison between a faulty and a perfect DO sensor step response. The air flow is increased by 10 % at time = 0.5 h. The faulty sensor has a direct reaction to the air flow change. In the figure a bias has been added only for clarity.

The simulation in Figure 12.9 is of course an idealised model of the faulty DO sensor. Figure 10.3 illustrates some practical measurements using a DO sensor with a broken membrane. Note that the response time of the faulty sensor is much faster than the true response time for the DO concentration, due to the bias error.

12.4 Recursive Parameter Estimation

In Section 7.6 time series analysis was discussed. We assumed that we knew the form of the model Equation 7.36, and that the parameters a_i and b_i were to be found. In Chapter 7 we assumed that all the measurements were available at the same time.

This section looks at the equivalent recursive estimation. Firstly we assume that the parameters a_i and b_i are constant but unknown. Then we will demonstrate how to track time varying parameters.

12.4.1 Constant but Unknown Parameters

Let us rewrite the input-output relationship Equation 7.36 in a slightly different form:

$$\begin{aligned} y_t = {} & a_1 y_{t-1} + a_2 y_{t-2} + .. + a_n y_{t-n} \\ & b_1 u_{t+m-n-1} + b_2 u_{t+m-n-2} + .. + b_m u_{t-n} \end{aligned} \tag{12.30}$$

where the subscript $t, t-1, ...$ means the sampling instants. Thus we have assumed that the sequence of inputs $(u_1, u_2, .., u_t)$ has generated an observed output sequence $(y_1, y_2, .., y_t)$. If m is smaller than n then the output is delayed in the model. The parameter vector θ is given by Equation 7.39:

$$\theta = [a_1 \; a_2 \; .. \; a_n \; b_1 \; b_2 \; .. \; b_m]^T \tag{12.31}$$

It is convenient to define another vector ϕ, the regression vector

$$\phi_{t-1} = [-y_{t-1} \; .. \; -y_{t-n} \; u_{t+m-n-1} \; u_{t-n}]^T \tag{12.32}$$

We note that the regression vector is known from the observations, while the parameter vector has to be calculated. In vector notation the autoregressive model Equation 12.30 can be written in the form:

$$y(t) \equiv y_t = \phi_{t-1}^T \theta \tag{12.33}$$

In Section 7.6 we calculated the parameter vector θ using all available input and output values. Here we will calculate θ recursively. We can assume that a certain estimate has been calculated at sampling time $t-1$. When the measurement value y_t has been observed, an updated parameter value can be calculated. The estimated parameter vector at time t is denoted $\hat{\theta}(t)$. Then $\hat{\theta}_t$ is updated according to:

$$\hat{\theta}_t = \hat{\theta}_{t-1} + K_t \epsilon_t \tag{12.34}$$

The term ϵ_t is called the residual and is defined by

$$\epsilon_t = y_t - \phi_t^T \hat{\theta}_{t-1} \tag{12.35}$$

which has an appealing interpretation. The first term on the right hand side is the true measured output, while the second term is the output $\hat{y}_t$ predicted by the model (Equation 12.30). In other words, if the residual were zero then the parameters are not adjusted, since the measured output and the model output are the same. The vector K_t is a gain vector of the same order as the parameter vector and is calculated by:

$$K_t = P_t \phi_t = P_{t-1} \phi_t (I + \phi_t^T P_{t-1} \phi_t)^{-1} \tag{12.36}$$

where I is the identity matrix and the quadratic matrix P_t is time varying and updated according to:

$$P_t = P_{t-1} - P_{t-1} \phi_t (I + \phi_t^T P_{t-1} \phi_t)^{-1} = (I - K_t \phi_t^T) P_{t-1} \tag{12.37}$$

We need some initial value for the parameter vector $\hat{\theta}_t$ and for the matrix P_t. It is common to start with the initial condition:

$$P(0) = P_o \tag{12.38}$$

where P_o is a positive definite matrix. Often it is chosen as a diagonal matrix with sufficiently large elements. The matrix P_t can be given a statistical interpretation. If the recursive estimation is started with the parameter θ_o and the matrix P_o it can be interpreted as the parameters having an initial distribution with the mean θ_o and the covariance matrix P_o. In other words, P_t tells us something about the confidence in the parameter estimates. The recursive estimator Equation 12.34 will minimise a quadratic criterion (such as Equation 7.1) and can be visualised by the block diagram in Figure 12.10.

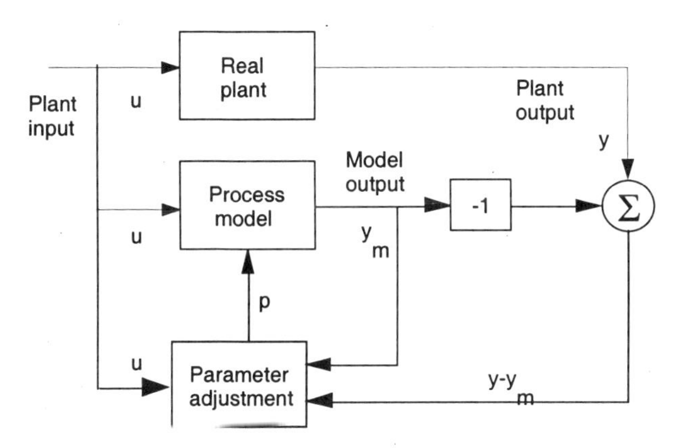

Figure 12.10: Block diagram of the least squares recursive estimator.

12.4.2 Time Varying Parameters

In many processes the parameters in Equation 12.30 are not constant, but time varying. In a wastewater treatment plant we can mention several cases. A dissolved oxygen model was derived in Section 7.6, where the parameters among other things depend on the oxygen transfer rate, which is slowly time varying. For the clarifier we derived a model (Section 8.7) with the flowrate as input and the effluent suspended solids as an output. The parameters depend on the floc settling properties, and are inherently time varying.

Here we will assume that the parameters are changing continuously, but slowly. A common way is to define the least squares criterion (like Equation 7.1) with:

$$V = \sum_{i=1}^{n} \lambda^{n-i} (y_i - y_{m,i})^2 \tag{12.39}$$

where the weighting factor is called λ, which is a number between 0 and 1, and is called the *forgetting factor*. Recent data have a weight close to 1 and old data is discarded exponentially. If the time constant for the forgetting is 10 sampling intervals, then $\lambda = 0.9$. Similarly, for time constants of the exponential forgetting of 20, 50 and 100 sampling intervals the corresponding λ values are 0.95, 0.98 and 0.99 respectively. The time constant of the exponential forgetting is a measure of the time lag of the estimator. In other words, an estimator with $\lambda = 0.9$ can track faster variations than an estimator with $\lambda = 0.99$. At the same time, the latter estimator is less sensitive to noisy data.

Using the forgetting factor, the updating of the gain vector K_t in Equation 12.36 and P_t in Equation 12.37 are slightly modified to:

$$K_t = P_t\phi_t = P_{t-1}\phi_t(\lambda I + \phi_t^T P_{t-1}\phi_t)^{-1} \tag{12.40}$$

while the quadratic matrix P_t is updated according to:

$$P_t = \frac{1}{\lambda}(I - K_t\phi_t^T)P_{t-1} \tag{12.41}$$

The parameter updating is still according to Equation 12.34. With the λ introduced the value of P_t is now slightly increased. This can be interpreted as keeping up the variance of the parameter error. This will result in a slightly larger gain K_t. In other words, the estimator takes the new measurements into consideration somewhat more than the estimator without the forgetting factor.

One has to ensure that the matrix P_t does not diverge. This can happen when new data does not contain any new information about the parameter vector.

12.4.3 Wastewater Applications

Example 12.8 (Clarifier on-line parameter tracking) *We now reconsider the settler parameter estimation from Section 8.7. The clarifier is modelled as a first order equation:*

$$y_t = a_1 y_{t-1} + b_1 u_{t-1} + e_t \tag{12.42}$$

where the output y is the effluent suspended solids concentration and the input u is the flowrate. The model error e_t is assumed to be random numbers. The parameter vector:

$$\theta = [a_1 \; b_1]^T \tag{12.43}$$

is updated recursively, according to Equation 12.34. Even if the parameters have no clear physical interpretation, their values depend on mixed liquor settleability. The updated values of the model coefficients can be monitored. Should their values deviate too much from normal, then the computer can trigger an alarm which indicates to the operator that there may be a clarification problem. The source of the problem may be a calibration problem with the effluent suspended solids sensor or a change in the settling properties of the floc. The computer does not give any exact information as to the cause of the problem but does provide an early warning so that there is more time to detect the real cause.

Note that the effluent suspended solids concentration can vary greatly in the absence of changes in floc characteristics. Variations in effluent suspended solids concentration can be caused by disturbances in the flows or the influent solids concentration.

Example 12.9 (Estimation of the oxygen transfer rate) *The DO dynamics has been described earlier, for example in Sections 3.6 and 3.7. Various attempts have been made to identify K_La during normal operation. The major problem is that the oxygen uptake rate r_O cannot be assumed to be known. The DO sensors provide the key information for the estimation. What makes the problem nontrivial is that:*

- *the respiration rate is time-varying (and cannot be assumed to be slowly time-varying or constant), and*
- *the oxygen transfer rate K_La is a nonlinear function of the air flowrate.*

Furthermore, the dynamics of the DO sensor may add to the difficulties. Note that it is assumed that the respiration rate is not measured separately. Instead the whole aerator is used as a "respirometer".

The oxygen transfer rate has been approximated in different ways. Usually the following additional assumptions are made, regarding the K_La function:

- *K_La can be sufficiently well described by a static function of the air flowrate.*
- *The parameters of the K_La function change slowly in comparison with the respiration rate.*
- *In most cases it is assumed that K_La is zero for zero air flowrate.*

In Section 3.9 we approximated K_La as a linear function of the air flowrate q_a:

$$K_La = K_a \cdot q_a \tag{12.44}$$

where the parameter K_a has to be estimated.

Other approximations being used are:

- *A square root approximation:*

$$K_La = \alpha\sqrt{q_a} \tag{12.45}$$

- *A more elaborate polynomial approximation:*

$$K_La = \alpha \cdot q_a + \beta \cdot \sqrt{q_a} + \gamma \tag{12.46}$$

- *A piecewise linear function of the air flowrate.*
- *An exponential model:*

$$K_L a(q_a) = k_1 \cdot (1 - e^{-k_2 \cdot q_a}) \tag{12.47}$$

where the two parameters k_1 and k_2 have to be estimated.

- *Spline models*

Assuming that $K_L a$ is approximated as the exponential function (Equation 12.47), the vector of unknown parameters looks like:

$$\theta = [k_1 \; k_2 \; r_O]^T \tag{12.48}$$

For a linear approximation of $K_L a$ (Equation 12.44) the corresponding parameter vector to be estimated is $\theta = [K_a \; r_O]^T$.

The various DO concentrations, the air flowrate and the wastewater flowrate can be measured.

In order to fit the time discrete formulation of the estimation problem in Section 12.4.2, the DO dynamics has to be written in a time discrete fashion. In principle the time derivative is approximated with a finite difference

$$\frac{dS_O}{dt} \approx \frac{S_O(kh + h) - S_O(kh)}{h} \tag{12.49}$$

where h is the sampling interval and k an integer. For simplicity we replace the notation kh with t and $kh + h$ with $t + 1$. The time discrete version of the DO mass balance becomes (compare Section 3.7.2 and Equation 3.35):

$$\begin{aligned} S_O(t+1) \approx S_O(t) + h \left[\frac{q_{in}(t)}{V} S_{O,in}(t) - \frac{q_{out}(t)}{V} S_{O,out}(t) \right] \\ + h \left[r_O(t) + K_L a(q_a(t))(S_{O,sat} - S_O(t)) \right] \end{aligned} \tag{12.50}$$

Equation 12.50 is now the input-output relationship 12.30 and the parameter vector θ is chosen according to the $K_L a$ approximation.

Example 12.10 (Estimating nitrification rate) *Nitrification is a two-step microbiological process transforming ammonia into nitrite and subsequently into nitrate, as described in Section 4.1.*

The dynamics of ammonia removal was defined by Equation 4.2 where the reaction rates are described by Tables 4.1 and 4.2. In this case the removal rate of ammonia is approximated by

$$r_{NH} = \frac{1}{Y_A} \hat{\mu}_A \left(\frac{S_{NH}}{K_{NH} + S_{NH}} \right) \left(\frac{S_O}{K_{OA} + S_O} \right) X_A \tag{12.51}$$

An identifiability analysis reveals that a simultaneous identification of both $\hat{\mu}_A$ *and* Y_A *is not feasible. Furthermore, no methods for measuring the concentration of autotrophic biomass exist. This makes Equation 12.51 inappropriate for the identification of the Monod kinetic expressions and of the nitrification process. Therefore the removal rate of ammonia is modified to:*

$$r_{NH} = r_{max} \left(\frac{S_{NH}}{K_{NH} + S_{NH}} \right) \left(\frac{S_O}{K_{OA} + S_O} \right) X_{SS} \tag{12.52}$$

The ammonia and oxygen concentrations as well as the suspended solids concentration X_{SS} *can be measured on-line. The parameter* r_{max} *incorporates the maximum specific growth rate of autotrophic bacteria, the observed biomass yield coefficient of ammonia and the fraction of autotrophic bacteria in the suspended solids.*

Note that Equation 12.52 includes three simultaneous processes, the ammonification, biomass assimilation and the nitrification. Still it can give an adequate estimation of the nitrification process.

Example 12.11 (Estimating filamentous biomass growth) *A Kalman filter approach has been used to estimate activated sludge bulking. Two causes that have been found to be especially important in causing sludge bulking are low organic loading and an organic overloading coupled with a low DO condition. The model of the activated sludge system is formulated according to the standard notation in Equations 12.13 and 12.14. It consists of three parts:*

- *A reactor model, essentially the IAWQ AS Model No 1, but includes both floc-forming and filamentous organisms.*
- *A thickener model similar to a flux model (Section 4.6). However, the initial settling velocity is not constant, but assumed to be a function of the floc-forming and the filamentous biomasses, so that the effects of bulking on the performance of the settler can be incorporated.*
- *An empirical clarifier model.*

The model has been simplified in order to make the identification of the unknown parameters possible. Then a so called extended Kalman filter *has been applied. The idea of such a filter is that the state* x *is extended with the unknown parameters* p, *so that an extended state* X *is created as*

$$X = \begin{bmatrix} x \\ p \end{bmatrix} \tag{12.53}$$

The parameters p *are usually updated according to*

$$p_{k+1} = p_k + v_{p,k} \tag{12.54}$$

where $v_{p,k}$ is some artificial noise that will make the unknown parameter p converge towards its "true" value. With the extended state X the system is formulated according to Equation 12.13 and 12.14. The predictor Equation 12.15 and corrector Equation 12.16 are then applied.

12.5 Summary

The idea of combining information from sensors and from models has been described in this chapter. This gives several advantages:

- The accuracy of the measurement can be increased, since a filter can extract the most from the noisy signal.
- States that are not directly measureable can be indirectly calculated.
- Non-measureable parameters can be estimated from the combination of sensor and model information.
- Faults related to sensors, actuators or parameter changes can often be detected.
- Parameters that are time-varying can be tracked on-line. This has a great impact for on-line fault detection. An early warning system can be implemented in this way.
- Controllers based on non-measureable variables can be realised.

We have shown that these ideas are applicable to wastewater treatment systems.

12.6 Further Reading

- Here we have not described computer implementation aspects of observers and filters. Some of these are described in Olsson & Piani (1992) or in Åström & Wittenmark (1990).
- A comprehensive survey of methods for the detection of abrupt faults which appear in the state variables and output variables of dynamic systems is given in Willsky (1976).
- There are many textbooks on recursive estimation. It is treated in depth in Ljung & Söderström (1983). Recursive estimators in connection with adaptive controllers are described in Åström & Wittenmark (1995).

- The applications of recursive identification of the oxygen transfer rate and the respiration rate have been published in Holmberg (1981), Holmberg (1986), Holmberg (1987), Olsson (1989), Carlsson *et al.* (1994) and Marsili-Libelli (1997).
- Estimation using a linear approximation of K_La was applied by Holmberg & Olsson (1985), Holmberg (1986), Holmberg (1987) and Bocken *et al.* (1989). Carlsson & Wigren (1993) and Carlsson (1993) applied a piecewise linear model, Lindberg (1997) an exponential function and a cubic spline function. Holmberg (1986), Holmberg (1987), Carlsson *et al.* (1994) and Lindberg (1997) have applied the estimators to real plants.
- Estimation theory has been applied to the prediction of nutrient removal kinetics by Beck (1981), Beck (1986), Holmberg & Ranta (1982), Náhlîk & Burianec (1988), Vanrolleghem *et al.* (1991), Carstensen (1994), Ayesa *et al.* (1991), Bastin & Doachain (1990), Cook & Marsili-Libelli (1981), Marsili-Libelli (1990) and Marsili-Libelli & Giovannini (1997).
- Carstensen *et al.* (1996) have applied predictive control to sewer systems.
- Jones (1989) has applied state estimation to an anaerobic process.
- Examples are cited from Carstensen (1994) and Carstensen *et al.* (1994).
- The application of the extended Kalman filter for sludge bulking estimation has been described by Chen & Beck (1993).

12.7 Exercises

1. Redundancy is often used as a means to detect faults. List some possible redundant measurements in a wastewater treatment system.
2. Consider the state estimator with a constant K. With noisy measurements we have seen that the size of K has to be determined by a compromise. Consider Figure 12.7. How would you choose K in order to achieve a fast convergence *and* a low noise level after the transient?
3. Apply the estimation filter for the mechanical system in Example 12.3 and estimate the speed given the position. Compare different choices of the initial guess of the speed and of different values of K. You can modify the MATLAB Example program `Example_12_3`.
4. Why is it often completely unsatisfactory to calculate a variable by taking the derivative of a signal?
5. Try to find variables in a wastewater treatment plant that can be related as input-output pairs, like in Equation 12.30.

Chapter 13

Knowledge Based Systems

The increasing complexity of dynamical systems in general and wastewater treatment systems in particular coupled with increasingly stringent performance criteria necessitates the use of more complex and sophisticated controllers. Current developments in knowledge based systems offer opportunities for fault tolerance systems as well as the opportunity for dealing with complex systems in an integrated manner by considering qualitative and quantitative aspects simultaneously. Knowledge based systems (KBS) are not only based on the control systems and mathematical models, that we have described in detail in the book, but on human behaviour and on computer science. Of particular importance in knowledge based systems control are direct learning schemes. The principles of KBS are outlined in Section 13.1 and we will see that the application of KBS is quite adequate for wastewater treatment systems. The representation of the data, the knowledge, is described in Section 13.2. It includes the concepts of logic, rules, trees, and graphs. In most real systems uncertainty is a fact, and the methods have to deal with it. This is discussed in Section 13.3. Part of that description is fuzzy systems. The acquisition of data, in this case called knowledge acquisition, is mentioned in Section 13.4. Finally, some software tools for KBS are discussed in Section 13.5.

13.1 Principles

Considerable research is currently being devoted to an understanding and a representation of intelligence and knowledge. In particular technological problems that can not be represented in terms of algebraic models or differential equations based on physical laws are being studied. In a wastewater treatment system there are many phenomena that are not readily quantifiable, such as smells, colours, patterns of foam, acoustical noises, and appearance of bulking or floc microscopic characterisation.

We have seen that wastewater treatment systems have a lot of special features:

- the complexity of the system is such that no single individual or small group of individuals can fully understand it;
- the models are difficult or expensive to evaluate and variables are not easily measured or causal variables may not even be known;
- the process is subject to large unpredictable disturbances;
- different operators need markedly different styles of interaction with the plant; and
- the model structure is not amenable to simple linear time invariant models. It is spatially distributed, non-linear, stochastic and time varying.

Yet wastewater treatment plants were first regulated or manually tuned by human operators before automatic controllers were installed. The human operator has few apparent problems with plant nonlinearities or adjusting to slow parametric changes in the plant or with satisfying a set of complex process constraints. The human operator is able to respond to complex sets of observations and constraints, and to satisfy multiple subjective performance criteria. However, the control actions of the human are difficult to analyse as they are variable and subjective, prone to error, inconsistent and unreliable, and in the case of safety critical situations and hazardous processes, potentially dangerous. The purpose of knowledge based control is to incorporate the positive intelligent and creative attributes of human controllers, whilst avoiding the elements of inconsistency, unreliability, temporal instability, and fatigue associated with the human conditions.

The application of traditional control methods of wastewater treatment requires:

- a valid and accurate model of the process dynamics;
- a need for reliable and complete information (measurements!);
- adjustment of flow and solids storage capabilities; and
- the ability to specify clear, precise, and unambiguous process performance objectives.

Operator control is needed to overcome the problems of inadequate models, insufficient sensors and imprecisely stated control objectives. This means that traditional control techniques, dynamic models, heuristic knowledge and human operators are all integrated within the proposed strategy.

13.1.1 Definition of a Knowledge Based System

The purpose of a knowledge based system (KBS), sometimes called an expert system [1] is to enable a non-expert user to emulate the problem solving capability of

[1] We will avoid the term *expert system* here, since the term has been the subject of so many unrealistic expectations. An expert system is not a panacea for all kinds of process problems.

an expert within some specialised domain of expertise. KBS systems utilise collections of facts, heuristics, common sense knowledge and other forms of knowledge, together with reasoning methods to make inferences or conclusions (Figure 13.1).

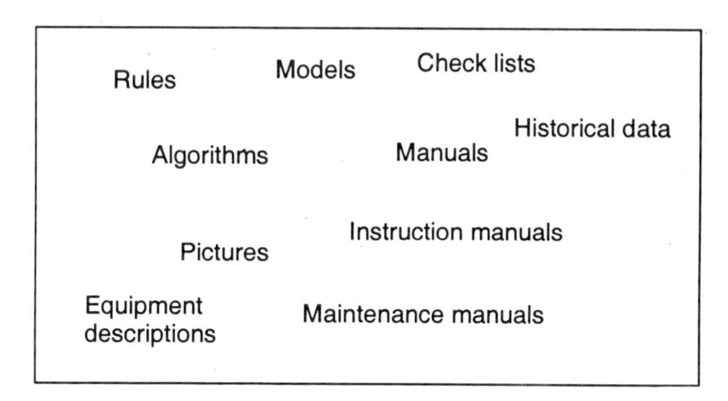

Figure 13.1: Ingredients of a knowledge base.

An advanced KBS is capable of reaching decisions on the basis of incomplete, imprecise or uncertain data or information. KBS systems are concerned with two fundamental classes of control:

- off-line applications, such as design, process planning, management and fault diagnosis, and
- on-line management and control.

Conventional computer control systems contain only the final algorithmic representation of all the knowledge involved in the design of the control system. A KBS also has other means of representing knowledge. One example is the written documentation that is today delivered together with the process and its equipment and control system. The documentation includes user manuals, installation, maintenance and operation descriptions, various checklists and instructions, component data sheets, flow schematics, mechanical assembly drawings, etc. New hypermedia techniques for presenting text and pictures will make it possible to include this in a KBS.

Another type of knowledge that should be represented in the KBS is the heuristic, experiential knowledge that different operators and engineers have about the process and which today is not made explicit and transferred to other users. The process designer bases his design on years of experience, rules of thumb, and heuristic considerations. The skilled operator knows from experience which are the important process variables to monitor or what may have caused a certain set of fault symptoms.

Models in terms of equations that describe the behaviour of the process and its components under various conditions is another important type of knowledge.

Used properly they can be the basis for training simulators, decision support simulators and model based diagnosis schemes.

The most common means of knowledge representation is by rules (like if-then-else) or protocols specifying under what condition the rule applies and how to use the rule. Most of the rules in a KBS are generated by empirical associations obtained from experts and experiences with the rules continually updated in the light of further evidence. Rules can represent shallow experiential and heuristic knowledge as well as deeper algorithmic or model based knowledge or that obtained by induction of generalised principles.

The rules can be used at a higher level of a control system to select the proper controller depending on the status of the process. Then the KBS performs the function of a consultant that previews a variety of model identification algorithms, each appropriate to a specific aspect of process behaviour, which in turn selects in an intelligent manner an appropriate controller design method that satisfies the specifications. This is called *intelligent control* or *expert control.* This will be mentioned briefly in Chapter 18 but details are outside the scope of this book. Here we look at the more limited task of a KBS, to make up a reasoning system for diagnosis where the operator is part of the reasoning process.

13.1.2 On-Line Diagnosis

At an alarm or a group of alarms a system should be able to present a more or less automatic interpretation to the operator. In other words, some of the detection and diagnosis problems discussed in Chapter 11 and illustrated in Figure 11.7 can to a certain extent be automated.

Example 13.1 (Diagnostic conversation) *When the alarm "the effluent suspended solids concentration is too high" is activated, the system should ask something like "check the sludge blanket of the settler" or "check the SVI". Depending on the answer the dialogue would continue until the system has pin-pointed the cause. The system would then present a remedy.*

13.2 Knowledge Representation

Knowledge can be represented in many ways in a computer. Here we will describe some methods to describe knowledge, such as rules, logic, equations, or trees. In Section 5.5.1 we also presented digraphs as a way to represent relationships. There is no universal way to represent knowledge. Rather, we may consider many different representations that can complement each other. The representation and the use of the knowledge are intertwined.

13.2.1 Logic

Standard logic systems are often used within KBS. The theory for different kinds of logic calculus has been firmly established. Propositions can be combined with the ordinary set of logical operators, such as **and** (&), **or** (—), **not** (_), **implies** ($\longrightarrow$), and **equivalence** ($\longleftrightarrow$).

In a standard logic system the expression $A \longrightarrow B$ simply means, that if the proposition A is *true*, then B must be *true*. The rules for the logic are not changed, so the system is called *monotonic*. Such a logic, however, can not handle common situations in complex processes: incomplete information, changing situations and generation of assumptions during the problem solving session.

13.2.2 Rules

We presented a brief discussion on rule-based models in Section 5.5.2. Here we just repeat that the basic structure of a rule is:

```
IF condition THEN conclusion
```

The *condition* part may represent process variables that are evaluated according to some scheme. The condition may also include some logical expression. The rules may be part of either a *forward* or a *backward* chaining system (compare Section 10.3). In an on-line system there has to be some time interval defined by how often the rule has to be assessed.

Rules for backward chaining can be represented as a decision tree, depicted in Figure 13.2.

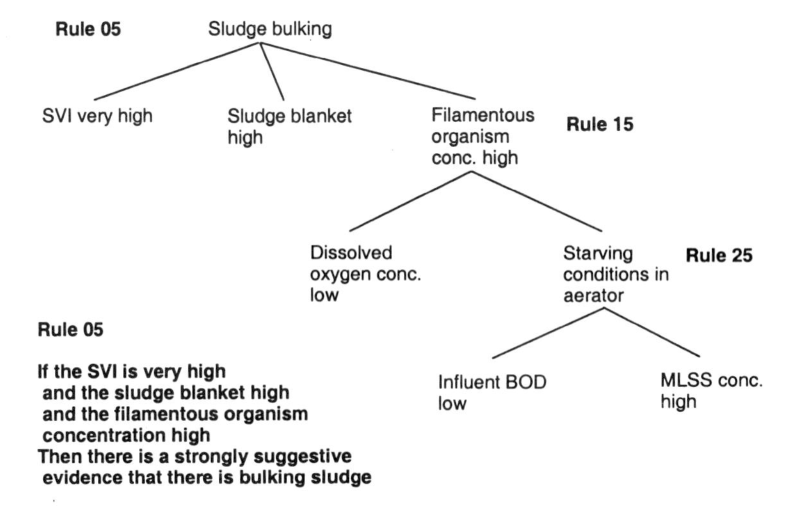

Figure 13.2: Rules and corresponding decision tree for a suggested partial diagnosis of bulking sludge.

Rules of the kind illustrated in Figure 13.2 represent associations between causes and effects. They often express heuristic knowledge. In some KBS explanations can be automatically generated from the rule chaining. For the bulking sludge case there may be rules that monitor the time history of the SVI and the sludge blanket height. When the rate of change or the magnitude of some of the values is getting too large, then the operator is notified. Another way to help the operator is alarm or fault trees, described below.

Example 13.2 (On-line KBS for operation) *An early application of a KBS for the operation of a wastewater treatment plant has been described by Ladiges & Kayser (1993). It had been noticed that a lot of plant troubles only lead to major upsets because they were recognized too late, or the operators did not react in the right way. The development of the described KBS was initiated in 1988.*

The system used the normal plant on-line data, such as water and sludge flowrates, nitrate as the signal for the aeration control, digital signals from different aggregate conditions, and laboratory values. A turbidity probe was installed at the outlet of the final clarifiers.

One example of a reasoning system is shown in Figure 13.3. The knowledge base reacts on the input of an ammonia laboratory value. If the $NH_4 - N$ *effluent concentration exceeds* $3.0 mg/l$ *the KBS starts working. The sludge retention time (SRT) is calculated (see Section 18.6.1). When the system finds that the SRT is too low, then it recommends a calculated value for the MLSS concentration.*

Sudden temperature drops and greater sludge losses can also cause nitrification problems although the SRT is sufficient. Aeration faults as well as nitrogen shock loads can result in high ammonia effluents.

Example 13.3 (Rule based diagnosis of settler problems) *In Chapter 6 and particularly Section 6.5 we have seen, that all on-line data analysis has to be preceded by data screening. The data screening in itself may contain interesting information and may lead to a diagnosis that will cause some action. The screening procedure of each individual analog signal will then reveal extreme values, steps up or down, pulses, etc. Each one of these tests can be assigned by a label for the particular signal.*

We will now describe a procedure for a rule based data screening. For each signal there are seven different tests made, shown in Table 13.1.

A pulse is recognised as a combination of a positive and negative step. At a certain time instant there may be more than one of the tests that is satisfied. For step n *a value* 2^n *is indicated. This means that, after all the tests at one time instant, we get a test value that is the sum of all the individual values. For example, if there is a* high value *and a* positive step *at a certain time, then we indicate the number* $2^3 + 2^5 = 40$. *The analog signal now has a corresponding indicator vector. This is illustrated by Figure 13.4.*

We can see that the pulse at $t = 12$ *is recognised as a combination of consecutive positive and negative steps.*

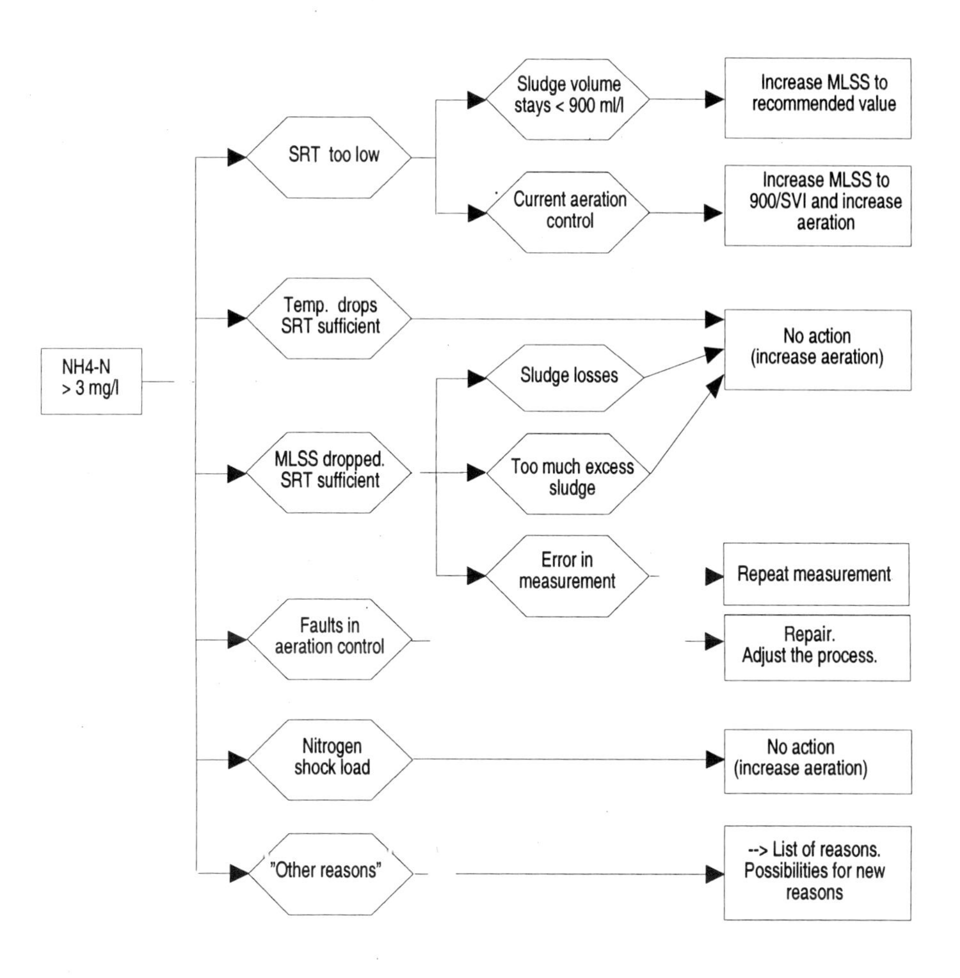

Figure 13.3: Structure of the ammonia knowledge base.

The diagnosis based on the indicator vector was tested on plant data. Looking at data records the human eye can detect a lot of strange behaviour. However, once we look at more than one variable at a time, the complexity increases quickly. Then the indicators come in handy. For each signal an indicator vector was created. These indicator vectors were combined to an indicator matrix, where each row corresponds to one analog signal and the columns to time instants. In the indicator matrix it is now possible to detect similar behaviour in several signals at the same time. The little pulse at $t \approx 1$ hour in Figure 10.5 was not considered significant by the human inspection. However, the computer diagnosis detected that the pulse appeared in both the influent flowrate (a pump was switched on for a short while) and in the settler effluent suspended solids concentration. This in turn indicated

Number	Test
0	missing data
1	extremely low value
2	low value
3	high value
4	extremely high value
5	positive step
6	negative step

Table 13.1: Data screening tests.

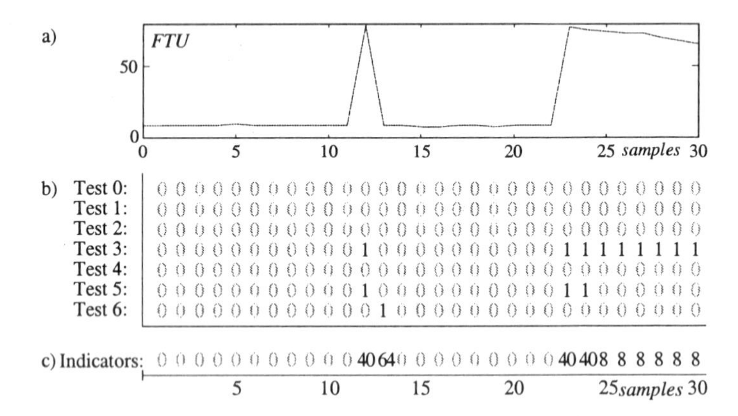

Figure 13.4: The indicator vector for the analog signal depicted: (a) a detail of the suspended solids concentration during 30 samples, (b) the composition of the indicator vector for each one of the 7 tests, (c) the condensed indicator vector.

settling problems that were later discovered. In fact, Figure 10.5 shows that a rain event appears some 25 hours later, and a washout problem is revealed. The pulse was actually the first warning. It was later shown that similar warnings had occurred earlier but were not detected in time.

13.2.3 Trees

Trees are very useful structures to represent knowledge. The natural graphical representation of a tree is a set of graphically interconnected nodes. In some cases the node in a tree may have an internal structure of other nodes, so that the tree in this sense is hierarchical.

Alarm Trees

An alarm tree is represented in Figure 13.5. The root node represents the alarm. At the bottom of the graph (the leaf nodes) physical faults are represented, and they may cause functional faults, that in turn will cause the alarm. A tree of this type may be helpful for operators or service personnel in trouble shooting off-line. Each node of the tree may include various instructions. An alarm tree then illustrates a backward chaining procedure.

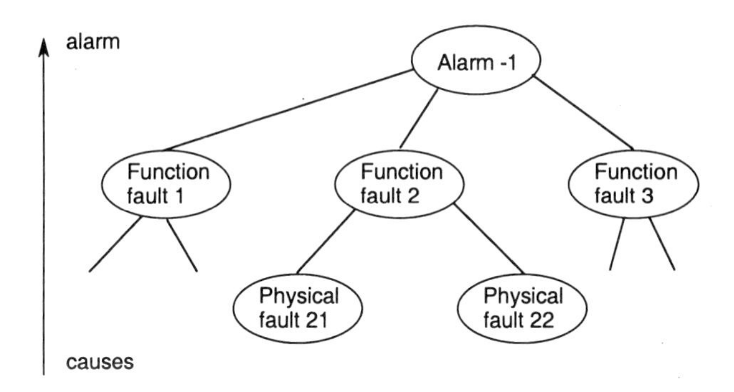

Figure 13.5: Alarm tree.

Fault Trees

A fault tree represents another kind of backward chaining description. Figure 13.6 shows a fault tree. The root node represents a possible fault, for example that the effluent concentration is too high. The other nodes represent symptoms in the form of conditions on measured process variables. If the conditions are fulfilled they will cause the fault shown.

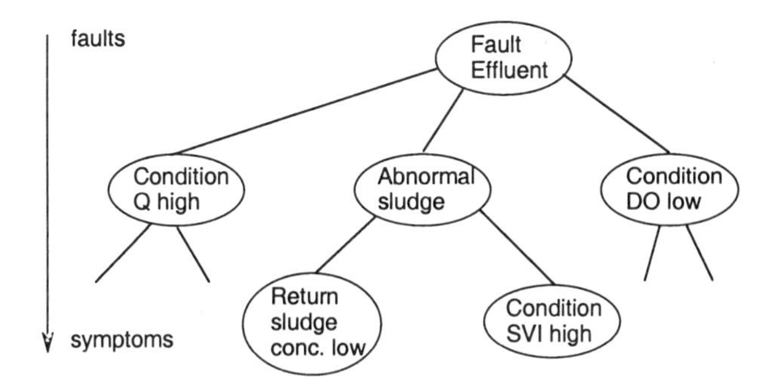

Figure 13.6: Fault tree.

In Figure 13.6, the sludge is considered abnormal if both the SVI is too high and the return sludge concentration too low. The combination of high flowrate (Q), low dissolved oxygen (DO) concentration and abnormal sludge leads to the fault condition.

Fault trees are useful for on-line diagnosis. By successively generating and testing fault hypotheses against sensor readings according to the fault trees, alarms or warnings may be generated to the operator. Backward chaining rules match the fault tree structure well and can be used to implement it. However, in the code the inherent graphical structure of the tree gets lost, and replaced with a set of rules which often can be difficult to overview.

A consultation for diagnosis begins by having the user specify how problems were identified. Initially the operator gets notified by measurements. He will add microscopic or visual observations to the information. From there different search trees will guide the operator to the final diagnosis.

Example 13.4 (Reasoning system in an activated sludge system) *As an application of fault trees let us look at the aeration basin diagnosis, as described by Gall & Patry (1989). The rules addressing the colour of the mixed-liquid, the presence of foams and scums, odours and turbulence patterns are illustrated in Figure 13.7.*

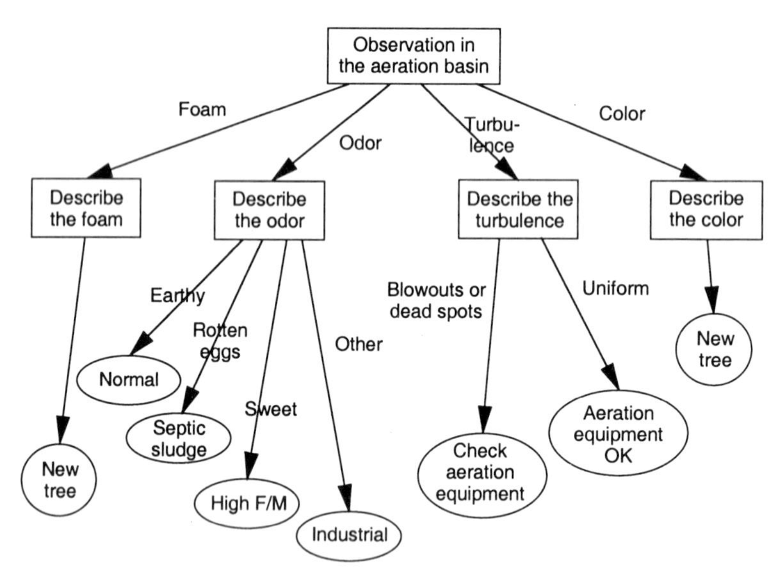

Figure 13.7: Aeration basin diagnostic logic tree. The description of foam or colour lead to new tree structures.

Having identified a foam problem there is another fault tree where the foam is identified, such as billowy white, crisp white, brown scummy, or black sudsy. This further observation will then lead to more precise conclusions and control actions. A crisp white foam should form and quickly dissipate on the surface of the aeration basin. Persistent foam is evidence of a high F/M ratio. Under such conditions microorganisms tend to release surface-active metabolic products resulting in excess foaming. A high F/M ratio can occur at start-up, during recovery from a toxic load (previous kill of organisms), during high organic load, or it may be due to excess sludge wastage.

Decision trees in themselves are limited in that they tend to be unidirectional and the intricate network can be difficult to alter when new situations arise.

13.3 Representing Uncertainty

Both the data and expertise in the real world are often uncertain. Therefore, to fully exploit the fallible but valuable judgemental knowledge offered by human experts one should introduce a measure of uncertainty into the system. There are many approaches which can be employed, and some of them are discussed below.

13.3.1 Confidence Numbers

The central notion of this is that one can associate a certainty $C(A)$ with every assertion A:

- If A is known to be true then $C(A) = 1$
- If A is known to be false then $C(A) = -1$
- If A is unknown then $-1 < C(A) < 1$

This approach was developed by the designers of MYCIN, an early well known computer program for intelligent diagnosis of infectious diseases.

13.3.2 Probability Distributions

One often wishes to find an estimate of a parameter θ from measurements of the variables $x_1, x_2, \ldots, x_n$. The classical approach is to take a sample of size n and substitute the information provided by the sample into the appropriate estimator. Now suppose that additional information is given about θ, that it is known to vary according to some probability distribution $f(\theta)$, often called a **prior distribution** with prior mean μ_0 and a prior variance σ_0^2. This means that we assume that θ is the value of a random variable Θ with the probability distribution $f(\theta)$ and we wish to estimate the particular value Θ for the population from which we selected our random sample. The probabilities associated with this prior distribution are called *subjective probabilities*, in that they measure a person's *degree of belief* in the location of the parameter. The person uses his own experience and knowledge as the basis for arriving at the subjective probabilities given by the prior distribution.

The Bayesian technique uses the prior distribution $f(\theta)$ along with the joint distribution of the sample $f(x_1, x_2, \ldots, x_n; \theta)$ to compute the **posterior distribution** $f(\theta \mid x_1, x_2, \ldots, x_n)$. This posterior distribution consists of information from both the subjective prior distribution and the objective sampling distribution and expresses our degree of belief in the location of the parameter θ *after* we have observed the samples.

We now write $f(x_1, x_2, \cdots, x_n \mid \theta)$ instead of $f(x_1, x_2, \cdots, x_n; \theta)$ for the joint probability distribution, since we indicate that θ is a random variable. The joint distribution of the sample $X_1, X_2, \ldots, X_n$ and the parameter Θ is then

$$f(x_1, x_2, \cdots, x_n; \theta) = f(x_1, x_2, \cdots, x_n \mid \theta) f(\theta) \tag{13.1}$$

The posterior distribution can now be written in the form

$$f(\theta \mid x_1, x_2, \cdots, x_n) = \frac{f(x_1, x_2, \ldots, x_n, \theta)}{g(x_1, x_2, \ldots, x_n)} \tag{13.2}$$

where

$$g(x_1, x_2, \ldots, x_n) = \int_{-\infty}^{\infty} f(x_1, x_2, \ldots, x_n, \theta) d\theta \tag{13.3}$$

The mean of the posterior distribution $f(\theta \mid x_1, x_2, \ldots, x_n)$ denoted by θ^* is called the **Bayes estimate** for θ.

Here it is assumed that the observations are exhaustive and conditionally independent. The latter is not easy to guarantee. For example, the influent flowrate and the MLSS concentration in an aerator are usually highly correlated in a short time scale, since the MLSS is simply diluted due to the flowrate changes.

In the real world, judgements are based on evidence which is subject to uncertainties, hence Bayes' estimate is often used as a basis for a plausible reasoning mechanism.

13.3.3 Fuzzy Logic

Many systems are not only nonlinear and time-variant but are generally ill defined. Measurements, process modelling and control can never be exact for real, complex processes. They can not easily be modelled by equations, nor even be represented by straightforward logic such as `if-then-else` rules. This is the background against which Lotfi Zadeh developed the **fuzzy logic** in the mid-1960s. The name fuzzy is a misnomer since the logic is firmly grounded in mathematical theory. By a short example the idea of fuzzy sets was introduced in Section 5.5.3.

Many complex processes (such as car driving) are readily controlled by humans without recourse to mathematical models, algorithms or a deep understanding of the physical processes involved. Human controllers derive adaptive experiential (empirical knowledge based on observations of the process) linguistic or qualitative models of the process. The driver (or the operator) generates qualitative based actions or responses to given situations that are loosely parametrised by the operating conditions as well as by their perceived experiential "model". The skilled human operator may not know how or why he is controlling the system, but at least he knows what to do and is able to achieve complex performance criteria.

Fuzzy logic can be regarded as a control methodology that mimics human thinking by incorporating the uncertainty inherent in all physical systems. In traditional logic and computing sets of elements are distinct, either an element is an element of a set, or it is not. This conventional (binary) logic considers only opposite states (fast/slow, open/closed, hot/cold). According to this logic a temperature of 25°C may be regarded as "hot" while 24.9°C would still be "cold", to which a temperature controller would react consequently. The SVI example in Section 5.5.3 is of the same nature, there is no sharp limit between "high" and "low" SVI.

Fuzzy logic, on the other hand, works by turning sharp binary variables (hot/-cold, high/low, fast/slow, open/closed) into "soft" grades (warm/cool, moderately fast/somewhat slow) with varying degrees of **membership**. Fuzzy logic is a logic based on truth-values in the interval [0,1] rather than just 0 or 1. An assertion can thus be more or less true corresponding to a degree between 0 and 1. A temperature of 20°C, for instance, can be both "warm" and "somewhat cool" at the same time. Such a condition is ignored by traditional logic but is a cornerstone of fuzzy logic. The "degree of membership" is defined as the confidence or certainty - expressed as a number from 0 to 1 - that a particular value belongs to a fuzzy set.

A **fuzzy set** is a set, or a class, having members with a degree of membership, rather than either a member or not a member. In a fuzzy set, the transition between membership and non-membership is gradual rather than abrupt. The word fuzzy refers to the blurred or indistinct outline of these sets in contrast to the usual sets.

Example 13.5 (A classical set) *A classical set can be defined by a membership function $\mu(x)$. For example, if we have the set of all values x with $a \leq x \leq b$ we can represent this set by a membership function as shown in Figure 13.8. For all values of the given set this function is* 1*, and for all other values it is* 0*. Values different from* 1 *or* 0 *are impossible. An element can be either a member of the set (1) or not (0).*

Example 13.6 (A fuzzy set) *The idea of a fuzzy set is to also allow membership values between one and zero as shown in Figure 13.9. All types of functions $\mu(x)$ with $0 \leq \mu(x) \leq 1$ are allowed. With this definition, it is possible to express not only the full but also the partial membership of an element to a set. For example, a is definitely a member of the set, while c is only a partial member of the set.*

Again, study the example on bulking in Section 5.5.3.

It has to be denoted that a membership value is *not* a probability. If, for example, the membership of a in Figure 13.9 is twice as large as the membership value of c, this does not mean that the probability of a being a member of A is twice as large as the probability of c being a member. A membership value is not an objective measure, but a subjective one. This subjectivity is a feature of

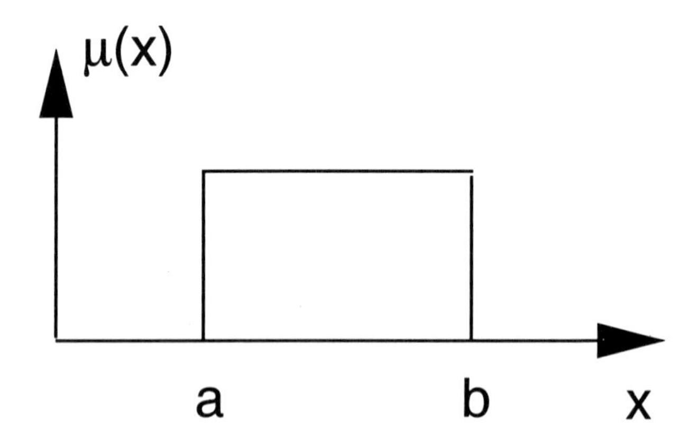

Figure 13.8: Classical set.

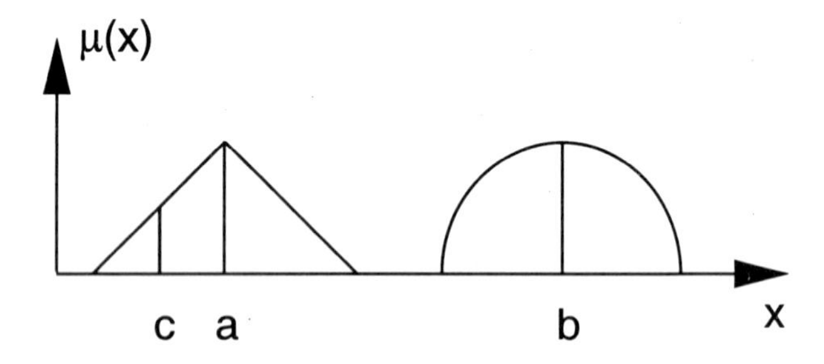

Figure 13.9: Fuzzy sets.

fuzzy sets which is sometimes an advantage and sometimes a disadvantage. We can consider a membership as a **possibility**.

What can fuzzy sets be used for? With fuzzy sets it is possible to describe vague concepts. For example, the set A in Figure 13.9 can be used to describe the set of all values that are similar to a. The membership value of c must be smaller than the membership value of a because c is less similar to a than a to itself. Obviously, with a classical set it would be difficult to describe similarity. In a similar way, the set B can describe the set of all values that are similar to b. The shape of the membership function is different from the first one. This is a result of the above mentioned subjectivity. The shape of the membership function depends on the definition of similarity given by the user.

Fuzzy systems base their decisions on inputs in the form of linguistic variables, common language terms such as "hot", "high", "slow" and "dark". The variables are tested with a small number of `if-then` rules, which produce one or more responses depending on which rules were asserted. The response of each rule is weighted according to the confidence or degree of membership of its inputs. The controller can be illustrated by Figure 13.10, and consists of a user interface, a rule base and an inference engine (analyser).

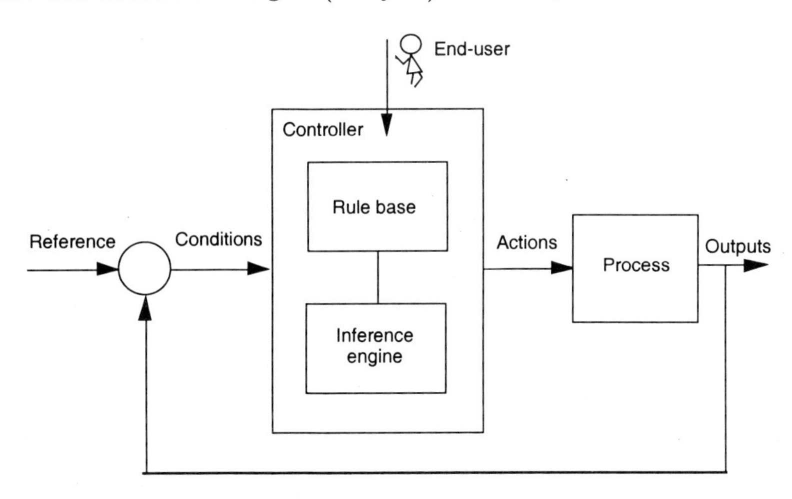

Figure 13.10: Elements of a fuzzy controller.

The user interface is primarily designed for process operators. Secondary users are either programmers or expert operators. Often the user interface is a diagram on a graphical screen showing the overall architecture of the control system. The rule base stores the control strategy and is a collection of stored control rules. Typically a rule consisting of two inputs and one output can have the form

```
IF x is A AND y is B THEN z is C
```

The inputs x and y are measured variables and z is the controller output or the control action. The variables A, B and C are linguistic terms such as *low, medium* or *high*. The *if*-part of the rule is called the condition or premise and the *then*-part is called the consequence or action. Rules can depend on each other, and chains of rules are possible.

The last part of the controller is the inference engine. This is a program that draws the actual conclusions from the actual inputs of the controller. The verb "infer" means to conclude from evidence, deduce, or to have as a logical consequence. Thus, the inference engine is a program, while the rule base is similar to a collection of data.

Example 13.7 (Low DO concentrations modelled as fuzzy sets) *The set of low DO concentrations may be characterised quite subjectively. It is obvious that* 0.1*mg/l will be a member of the set, and* 6*mg/l will not be a member of this set, but how about 1, 2 or 3 mg/l? The degree of membership for these concentrations is a number between 0 and 1 but there is no formal basis for how to determine the grade of membership. The membership for a DO concentration of 1 mg/l in the set* *of low DO concentrations depends on the conditions for the particular process. Let x be a linguistic variable with the label DO concentration. Terms of this linguistic variable, which are fuzzy sets could be* very low, low, high, *etc. and the set can be defined as*

`{very low, low, more or less high, high, very high}`

Each term is a fuzzy variable defined on the base variable x, which might be the scale from 0 to 8 mg/l, as illustrated in Figure 13.11.

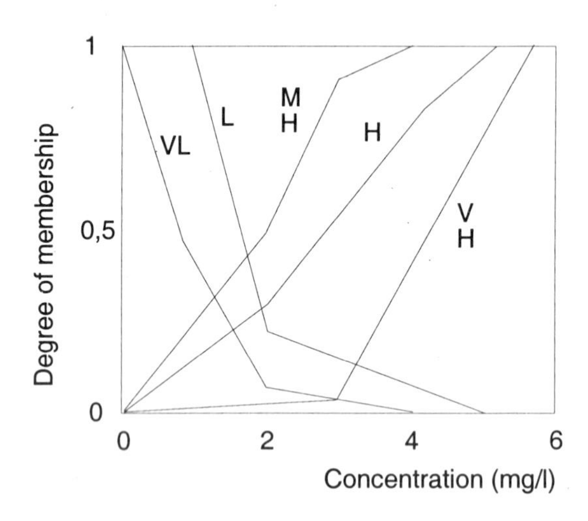

Figure 13.11: Dissolved oxygen concentrations modelled as fuzzy sets. The sets low (L) and high (H) are primary sets, while the sets more or less high (MH), very low (VL), and very high (VH) are generated from these.

Operations on Fuzzy Sets

The primary terms are the terms that have to be defined *a priori*, for example *high* and *low* in the example. The other terms are modified terms, that can be derived from the primary ones. The term *very* has an intensifying effect and is here defined by the operation:

$$very\, a = a^2 \tag{13.4}$$

while *more or less* (or *morl* for short) has the opposite effect, and is here defined by:

$$morl\, a = a^{1/2} \tag{13.5}$$

The power applies to each vector element of a in turn. The *complement* or the *negation* of a is:

$$not\, a = 1 - a \tag{13.6}$$

In our case we may define *high* as being *not low* (Figure 13.11). Accordingly *low* is the only primary term in the example above.

The membership function is obviously the crucial component of a fuzzy set. It is not surprising that operations with fuzzy sets are defined by means of their membership functions. A fuzzy set operation creates a new fuzzy set from one or several given sets. The most important ones are:

- **intersection** of a and b:

$$a \,\textbf{and}\, b = a \,\textbf{min}\, b \tag{13.7}$$

  The operation *min* is an item-by-item minimum comparison between corresponding items in a and b.

- **union** of a and b:

$$a \,\textbf{or}\, b = a \,\textbf{max}\, b \tag{13.8}$$

  The *max* operation is an item-by-item maximum operation.

These operations are illustrated in Figure 13.12.

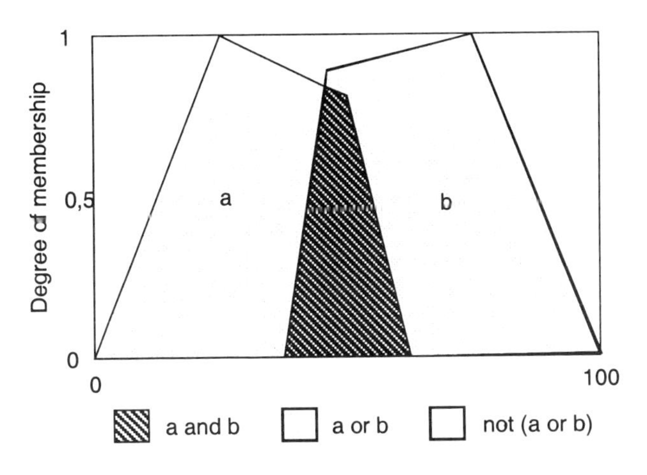

Figure 13.12: Primitive operations on two fuzzy sets a, b.

13.4 Knowledge Acquisition

There are a number of approaches and issues involved in acquiring and structuring the information required for problem monitoring and diagnosis. The knowledge structure must be chosen for:

- causes (likely operating problems)

- effects (and their observability characteristics)
- cause-and-effect relationships
- advice or solutions to the problems

There are a number of knowledge sources which could be tapped and which need to be identified:

- human experts
- experienced humans
- mathematical models
- experiments on the process

There are also a variety of techniques to elicit knowledge from the chosen sources. Techniques selected may vary with the type of knowledge and the source of the knowledge. Techniques include:

- unstructured interview
- structured interview
- questionnaire
- solution of posed problems
- completion of cause-and-effect matrices
- synthesis of cause-and-effect relationships
- diaries and casebooks

The knowledge acquisition and the evaluation of the knowledge can be very time-consuming in total, even if the time for individual sources is relatively small.

13.5 Knowledge Based System Tools

One of the more commonly used terms in KBS is the *shell*. A shell in this context is a computer program that provides a user interface, an empty knowledge base and an inference engine. The inference engine is the software module that controls the reasoning operations of the KBS. It reads from external data, from the knowledge base and deals with hypotheses, reasoning and conclusions.

Typical for KBS software is that the knowledge data base (domain of knowledge) is separated from the inference procedure. In conventional control, process knowledge (plant models) and the methodology for designing controllers are intertwined. In a KBS there is a separation of the knowledge base, the input data and the inference mechanism for applying this knowledge to the control or the diagnosis problem. Thus, a KBS can be readily altered by changing the knowledge base. The shell does not contain any knowledge of a particular application, and it has no `if ... then` rule built in. It generally represents a method of problem solving that could apply to many different areas of application. A shell can be looked at as an empty spreadsheet program without any cells filled in.

Once a result has been obtained it has to be displayed in a clean and coherent manner to the operator. This is the role of the user interface, some of which use approximations to a natural language interface.

An inference engine typically imposes a method of searching through the knowledge base. One way to check the rules is to start at the first rule and proceed sequentially through the knowledge base, trying to match conditions in the rules to the input data. This approach (albeit simplified here for explanation purposes) is known as *forward chaining* (see Section 13.2.2). *Backward chaining* is a method in which a solution is assumed by the program to be true, and then a search back through the rules is performed, attempting to prove the hypothesis.

There are many commercial shells on the market, and each one has its own constraints. Therefore it is wise to consider many options before a tool is selected. The shells are designed for ease of use, and of course some flexibility is sacrificed. For the developer a shell can be developed using a language, such as LISP or C. Having a shell the main task is to develop the knowledge base.

One of the common mistakes that KBS developers have made is to first select a tool, and then proceed to develop the application. As with all computer programs, it is crucial to first examine the problem, define the requirements for the tool, and then proceed with tool selection.

Example 13.8 (Operation support system) *An application of fuzzy reasoning for diagnosis in wastewater treatment systems is described in Example 18.11.*

13.6 Summary

The general principle of reasoning may be expressed as follows: in any inference *use all but no more information than is available.* This principle had in fact been recognized by the ancient Chinese philosopher Lao Tsu as early as the 6th century B.C. and is expressed in his book *Tao Te Ching* [2] by the following two simple statements of remarkable clarity:

> Knowing ignorance is strength.
> Ignoring knowledge is sickness.

A knowledge based system can never be better than the quality of the data included in the knowledge database. However, given proper information, KBS offer additional possibilities for both diagnosis and control.

13.7 Further Reading

There is an extensive literature in learning systems. On fuzzy logic alone there are over 9000 research papers, reports, and books written. Here we point out a few useful books and articles.

[2] Vintage Books (Random House), New York, Chapter 71, 1972

- More on statistical methods for estimation is found in standard textbooks like Walpole & Myers (1989) and Box *et al.* (1978).
- There are a lot of textbooks on KBS, and a few are mentioned here. Hayes-Roth *et al.* (1983), Alty & Coombs (1984), Negoită (1985) and Waterman (1986) explain the fundamentals of KBS.
- The MYCIN knowledge shell was developed at Stanford University during the 1970s (Shortcliffe (1978)). The application of certainty factors for analysis is found in Ristuccia & Searle (1985).
- For beginners in fuzzy systems two articles in IEEE Spectrum, Zadeh (1984) and Self (1990), are good introductions. For a more serious study Zimmermann (1991) is recommended. Jantzen (1994) has written a good introduction to fuzzy control. The book by Klir & Folger (1988) gives a lot of details of fuzzy logic. Bezdek (1981) applies fuzzy reasoning for pattern recognition.
- Several journals and scientific conferences contain a lot of articles in the areas of knowledge based systems and fuzzy control. Examples are:
  - IFAC (International Federation of Automatic Control) Symposia on AI in Real Time Control
  - Automatica Journal (IFAC)
  - Journal of Applied Artificial Intelligence
  - IEE (Institution of Electrical Engineers, UK) Journal on Intelligent Systems
  - IEEE (Institution of Electrical and Electronic Engineers, USA) - Transactions on Systems, Man and Cybernetics
  - IEEE - Transactions on Fuzzy Systems
  - IEEE - Control Systems Magazine
  - IEEE - Transactions on Control System Technology
- Årzèn (1989) and Årzèn (1990) present many examples of KBS applied to control.
- Some basic concepts for intelligent control and KBS are defined in Åström *et al.* (1986).
- A lot of software products for the development of KBS are available on the market, such as the ILOG inference engine, the EXPERT rules engine and the G2 Real-Time expert system. These are listed in the References.
- Figure 13.4 is from Bergh (1996).

Several knowledge based systems have been developed to assist in the solution of water and wastewater treatment problems:

- Rule based data screening was published by Bergh (1996).

- The Figure 13.3 is from Ladiges & Kayser (1993). Earlier experiences are reported by Kayser (1990).
- Beck *et al.* (1978) were among the first to make use of reasoning based on rules for wastewater treatment plant operation and control.
- Berthouex *et al.* (1987) extended Beck's work by integrating the knowledge based system to a database to provide plant operators with a more powerful software package.
- A structure where traditional control techniques, dynamic models, heuristic knowledge and human operators are integrated was proposed by Patry & Olsson (1987).
- Chynoweth *et al.* (1994) and Pullammanappallil *et al.* (1992) have proposed expert control systems for anaerobic digestion.
- Laukkanen & Pursiainen (1991) proposed a system for treatment plants.
- Chan & Koe (1991) proposed a knowledge based system to handle sludge bulking.
- Marsili-Libelli & Müller (1996) applied fuzzy reasoning to an anaerobic process.

13.8 Exercises

1. Define the terms fuzzy set, inference engine, `if-then` rule.
2. Isn't fuzzy logic an inherent contradiction? Why would anyone want to fuzzify logic?
3. Define some fuzzy sets in activated sludge systems.
4. Define fault trees for operational problems in a primary settler, an aeration basin, and a final clarifier. For example, it is discovered that the dissolved oxygen concentration in one tank is too low. Assume that a DO control system is in place. What would you examine?
    - air flowrate (too low?, too high?)
    - other DO concentrations
    - influent flowrate (rain storm?)
    - air flow equipment
    - return sludge flowrate
    - anaerobic digester supernatant flowrate
    - effluent ammonia
    - other measurements?

Make the fault tree based on these observations.

5. Try to define some `if-then` fuzzy rules for the control of a wastewater treatment plant.

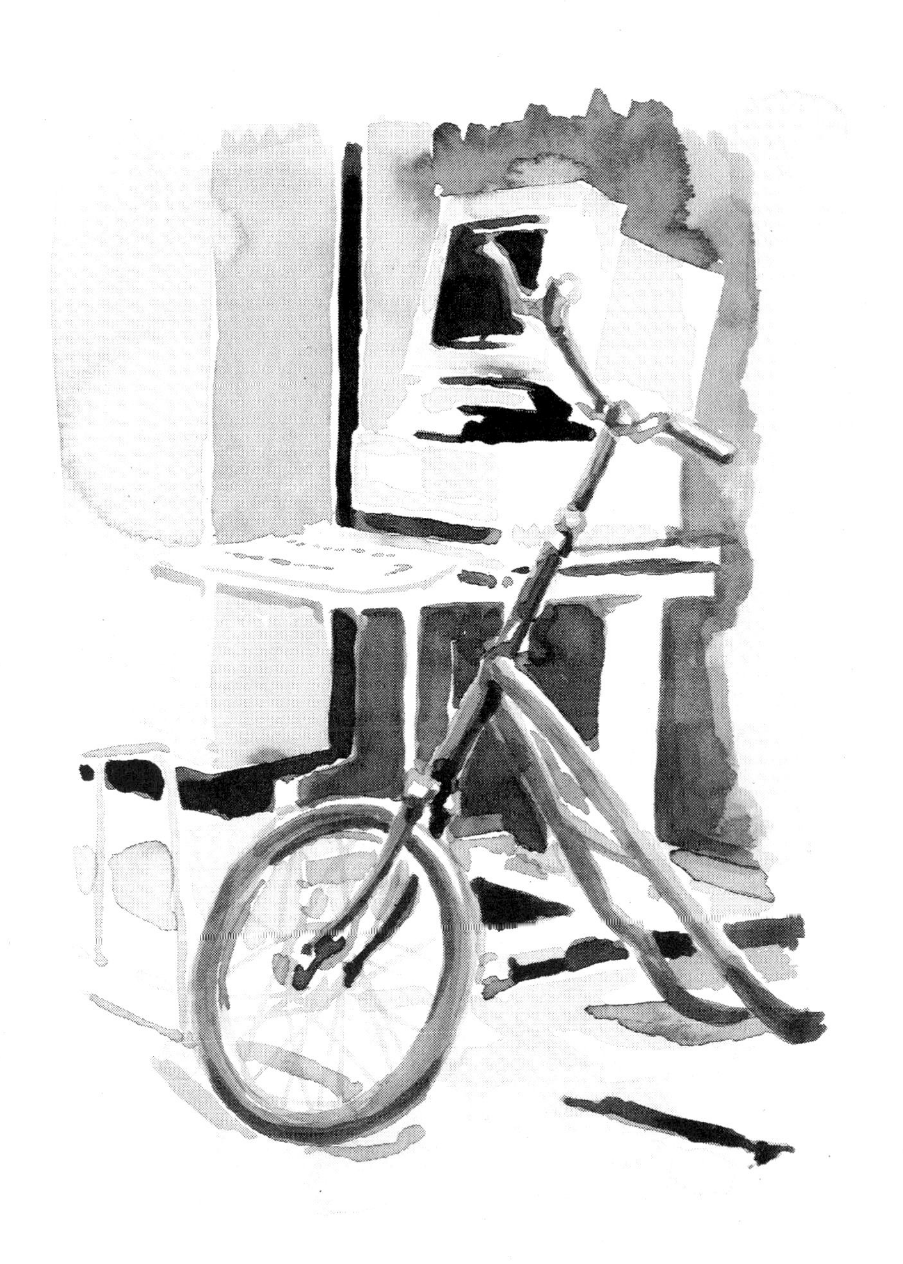

Part C

Control

Chapter 14

Goals and Strategies

The goal of this chapter is to gently introduce this part of the book on control and optimisation. The major points discussed here are why and how, and what are the major problems, Section 14.1.

So we begin with a discussion of the problem and the general operating goals of biological wastewater treatment plants, Sections 14.2 and 14.3.

Then we look at how these goals and the basic dynamics of our processes, addressed in the first part of the book, determine the general strategy that must be taken to implement control and optimisation. The general strategy is of course not unique and is generally followed for most process systems.

The particulars of implementing the strategy are of course more unique, and this is discussed in detail in the rest of the chapter. Here we address the different subsystems and propose possible strategies for control, Section 14.4 and finally we discuss that overall "helicopter view" of the whole process, Section 14.5. Some of these strategies will be taken up in more detail in the following chapters.

14.1 Defining the Problem

In the introduction we listed some unique attributes of biological wastewater treatment. Now we will add some attributes which, while not unique to our processes, do significantly affect our approach to control:

- the daily volume of wastewater treated can be huge;
- the disturbances in the influent are enormous compared to most industries; and
- the influent must be accepted and treated, no returning it to the supplier.

These attributes are the essence of our problem - major disturbances.

- the process has significant nonlinearities limiting the usefulness of simple controllers;

- there is a very wide range of response times, both a problem and a blessing;
- there are many interactions within the process with major recycles;
- there are many external couplings, from the sewer to the receiving waters;
- designers often leave too few manipulated variables and too little control authority;
- the concentrations of nutrients (pollutants) are very small, even challenging sensors;
- sensors don't exist for many states and are expensive for others; and
- the value of the product in the marketplace is remarkably low.

These attributes are why control is far from straight forward to design or to implement.

- the microorganisms change their behaviour and their population distributions;
- the separation of the effluent from the biomass is challenging and easily disturbed; and
- future effluent standards will be tight and may be based on spot checks.

The difficulties in implementing control, and these attributes, are why *disturbance rejection* must be our major goal. We must attenuate large disturbances, and we must do it as early in the process as we can. This will not only meet effluent requirements, but ease the task of controllers later in the process.

14.2 Goals and Objectives

Goals and objectives can be classified in a number of ways. The classification below is typical of the approaches taken and considers community or societal goals, process or plant goals and then operational goals. These approximately follow the control hierarchy introduced in the next section.

14.2.1 Community or Societal Goals

These should not be confused with corporate goals, which are generally a little more selfish, having to do with things such as market share and return on investment. However, any good corporate citizen will have community or societal goals as part of its own corporate goal structure.

The process industry *Responsible Care* programme is being generally accepted as a good template. It lays down the principles and suggests guidelines for corporations who wish to be responsible citizens. It covers the principal areas:

- care for the surrounding environment,
- care for its employees and contract workers, and
- care for the society in which it operates.

These are equally applicable to the wastewater treatment industry, whose very existence is to address issues of environmental and human health.

14.2.2 Process or Plant Goals

The specific goals of a wastewater treatment plant are set so that it contributes to meeting the more general corporate and community goals. It includes at least the following:

- **To meet effluent discharge requirements**. In most cases average values are specified over a period such as a week or month. This involves determining average operating conditions which will maintain average effluent quality close to but within regulatory requirements.

- **To achieve good disturbance rejection**. This involves manipulating operating conditions about their average values, in order to compensate for the effects of varying influent conditions and to maintain effluent quality constant. This ensures environmental responsibility and keeps within maximum discharge limits where these exist. It can also be very costly to allow poor effluent to be discharged, and then to try and compensate by achieving extra good quality effluent for a compensating period.

- **To optimise operation to minimise the operating cost**. The principal components of operating cost are generally interest on capital expenditure, air and chemicals, and operating and maintenance manpower. Interest can be minimised on a long term basis by maximising the utilisation of equipment, thus deferring and reducing capital expenditure.

The first goal of course includes keeping the plant running. This is the major task for many operations, at least in terms of man-power. It is not a trivial task to maintain advanced instrumentation and equipment, to make sure it runs smoothly and to detect any faults in its operation. Much of this is *traditional* automation and process control, and is not specific to running wastewater treatment systems. The environment may be very specific and harsh for much equipment, but that is also the case in other industries as well, so from a strict control point of view the problems are traditional.

However, keeping the plant running and simply recording effluent quality is too often taken as the *only* goal. This has been and is still largely due to a lack of process knowledge on the part of plant operators. Too many operators of plants rely on the consultant for expertise in both design and operation. To satisfy the effluent quality requirements, it is crucial to ask: is this task solved primarily by design or does instrumentation, control and automation play any significant role in this part? The traditional answer among designers used to be no. When only carbonaceous removal was considered the plant was considered *self-regulating* and could take care of load disturbances and still produce a satisfactory effluent.

The traditional method was large volumes, excess aeration and chemical dosage. Consequently a lot of plants have been oversized. No incentive was given for improved operation, since operational costs were not compared with investment costs. Often they were met from different financial sources.

Earlier we have seen that significant load variations and disturbances are typical for wastewater treatment systems. Thus the plant is hardly ever in steady state. Despite the wide swings of load, the plant still has to *consistently* produce a satisfactory effluent. If the volumes of the tanks are not oversized, this means that some control system has to minimise the influence of disturbances and ensure that the operation is satisfactory. Pressures to fully utilise volumes and the complexities of nutrient removal are making process control essential. This of course includes *disturbance attenuation* or *disturbance rejection.* We will look at this in more detail below and at some possible control strategies throughout this part of the book.

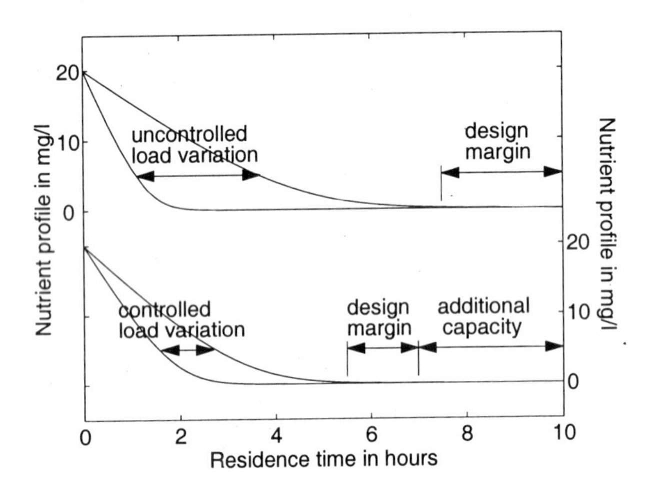

Figure 14.1: The relation between volume and disturbance elimination.

Figure 14.1 illustrates qualitatively the relationship between volume and disturbances. With a large volume most disturbances are attenuated even without advanced control. With control the volume most probably can be decreased, thus saving design costs. With a consequent disturbance rejection strategy, the volumes can be further decreased. At the same time this leaves less room for failures and poor operation. A volume minimisation depends on good and consistent operation.

All levels of government are cutting expenditure and even privatising. In both cases minimising the operating cost and maximising the use of available capacity is becoming necessary. Here control and optimisation are the key tools. Costs to be minimised include air (improved DO control and levels), chemicals (more use of the biomass), pumping and carbon addition. Of course, if control and optimisation is part of the answer, it is also part of the cost. Some control actions present little cost:

- waste sludge flowrate,
- return sludge flowrate,
- step feed, and
- recycle schemes.

Other control actions may be costly, such as chemical addition for P removal or sludge conditioning. We will look closely at the available manipulated variables and their suitability for control and the implications of this for costs and plant design.

14.2.3 Operational Objectives

Operational objectives are much more specific instructions for operation. They are developed so that a specific plant will meet the process or plant goals. Sometimes they are general in that many plants may use the same objective, but often they are very specific to a plant and its influent and its receiving waters. Examples of more general operational objectives are:

- grow the right biomass population,
- maintain good mixing where appropriate,
- adequate loading and DO concentration,
- adequate air flow,
- good settling properties,
- avoid clarifier overload, and
- avoid denitrification in clarifier

In the final sections of this chapter, we will examine the operational objectives for specific parts of wastewater plants. For example, we will look at the different zones of activated sludge bioreactors. Finally, we will stand back and look at plant-wide operation. But first we will examine why we choose to take this approach.

14.3 Hierarchical Strategies

Three factors influence how we might determine an operating and control strategy for a particular treatment plant: operating objectives (discussed above), the dynamic characteristics of the plant (discussed at length in Part A), and the structure of traditional instrumentation and management systems.

Instrumentation and management systems are organised in a hierarchical manner as illustrated by the typical pyramid shown in Figure 14.2.

On the left of the pyramid the hardware, software and manpower is classified, on the right the purpose and role is classified, and in the centre the timescale of decision making and action is classified. In principle, information in decreasing amounts flows upwards, and operating instructions in increasing amounts flow

Figure 14.2: Standard operating and control hierarchy.

downwards. As already remarked in Sections 11.2 and 11.3, humans dominate the upper levels and machines dominate the lower levels, primarily due to the human effectiveness in making assessments and decisions in the presence of incomplete and uncertain information.

In Chapter 2 the dynamics of the typical biological wastewater treatment process were classified as in Table 14.1.

Slow (days to weeks):	Biomass growth
Medium (hours to days):	Concentration dynamics Nutrient removal
Fast (minutes to hours):	Flow dynamics Dissolved oxygen

Table 14.1: Timescales of wastewater dynamics.

Comparing this classification with the hierarchy in Figure 14.2, gives us some ideas for developing the operating and control strategies, Table 14.2.

Supervisory Control:	Biomass growth
Advanced Process Control:	Concentration dynamics Nutrient removal
Basic Control:	Flow dynamics Dissolved oxygen

Table 14.2: Classification of wastewater control.

Within each category or level, we can then take the traditional engineering "divide and conquer" approach by looking at each process unit relatively independently. For example, in the nutrient removal category we can divide a BNR process into the prefermenter, the anaerobic zone, the anoxic zone, and the aerobic zone. We will now discuss operational objectives section by section, then in Chapters 17 and 18 how you could approach the design of control strategies for your process.

Finally we must take a step backwards to get the "helicopter view", and look at how the system units interact with each other. This will affect both the objectives and hence the control strategies chosen at the lower levels and also the process management strategies at the higher levels. Again, we will attempt to do this below, and in Chapter 20.

14.4 Operational Objectives of Units

Here we will look in detail at the principal subprocesses for municipal biological nutrient removal.

We will point out an important notational aspect. There are a number of block diagrams presented below. The control engineer and the process engineer often look at block diagrams quite differently. For the process engineer the arrows often indicate liquid flows or other mass flows, while the control engineer looks at the arrows as logical connections between the blocks. To avoid some confusion here, we have indicated liquid or gas flows as solid lines in the block diagrams, while the logical connections are shown as dashed lines. In later chapters on control, we hope that the interpretation of the block diagrams will be clear.

14.4.1 Activated Sludge Anaerobic Zone

The mechanisms of phosphate removal in activated sludge systems were described in Chapters 2 and 4. The key mechanism in the anaerobic zone is the uptake of volatile fatty acids (VFA) by the bio-P bacteria which store them as PHA. The energy to do this comes from the simultaneous release of phosphate into solution from stored polyphosphate.

The concentration and the type of carbon source available to the bio-P bacteria may be a limiting factor. It has been shown that enhanced biological phosphorus removal (EBPR) was favoured by low P/COD ratios. The concentration of VFA has been shown to be important in EBPR. When the influent wastewater contains only small amounts of readily biodegradable organic matter, the addition of products from the fermentation of sludge can be one way of improving the VFA concentration in the anaerobic stage.

One way to achieve this is to ferment sludge in a separate vessel or to use the primary clarifiers as a fermentation stage. By increasing the sludge level in the primary clarifiers, anaerobic conditions can be created, thus favouring the formation of VFA that can be used in the following anaerobic zone.

During high flow conditions the formation of VFA may be limited, since oxygenated water enters the clarifier, and possibly also the anaerobic tank. The retention time of the sludge in the primary clarifier should be controlled. Slow growing methanogens may be favoured, especially during the warm season, so a sufficiently short retention time is needed to guarantee that methanogenesis is not taking place, and that the methanogens are washed out.

It has been suggested to measure the amount of readily biodegradable carbon by measuring oxygen uptake rate. However, all components being readily biodegradable under aerobic conditions are not necessarily fermented to VFA in an anaerobic environment. Thus, an organic fraction determined by aerobic respirometry may be misleading if used as a measure of the amount of substrate in the wastewater for bio-P bacteria. Another way of measuring the amount of readily biodegradable carbon is to measure the COD of the filtered water.

A shortage of easily degradable matter can also be compensated for by:

- an external supply of acetic acid;
- an external supply of industrial wastewater with readily available carbon; and
- an increased hydrolysis/fermentation in the anaerobic tank by increasing the retention time.

Nitrate which is recirculated back to the anaerobic zone with the return sludge may have a detrimental effect on the EBPR process. It is usually assumed that the bio-P organisms are not able to compete successfully with other organisms for the readily degradable substrates when oxygen, nitrate or both are present.

On the other hand, there is evidence that return sludge storage to promote denitrification can also lead to phosphorus release which also affects the EBPR.

Example 14.1 (DO and nitrate competition in P release) *The MATLAB program* `Example_14_1.m` *simulates a batch reactor (equivalent to a plug flow reactor) configuration of the anaerobic zone using selected mechanisms from the IAWQ AS Model No 2. The left plot in Figure 14.3 shows a normal P release scenario, while the right plot in Figure 14.3 shows the effect of 5 mg/l of DO and 8 mg/l of nitrate. It can be seen that DO is first consumed followed by nitrate consumption at a slower rate. Nitrate consumption is initially inhibited by the DO. The acetate is consumed much faster and the P release flattens off. Perhaps more significant is the drop in PHA uptake which will result in a reduced growth of bio-P organisms in the subsequent aerobic stage.*

14.4.2 Activated Sludge Anoxic Zone

The primary purpose of this unit is denitrification (DN), the reduction of nitrate to free nitrogen by the oxidation of organic carbon. The dominating organisms in the process are the heterotrophs, a proportion of which reduce nitrate in an (ideally) oxygen free environment. Therefore, the dissolved oxygen level has to be

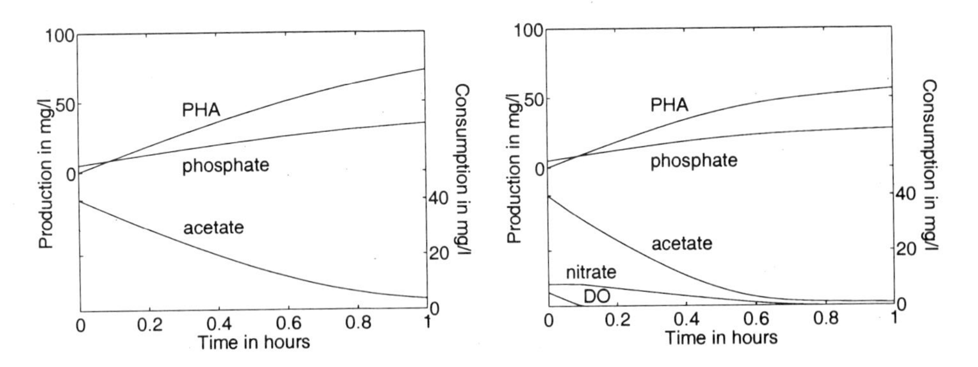

Figure 14.3: Normal P release (left) and DO and nitrate competition (right).

kept at a minimum. If not, the organisms will use oxygen instead of nitrate as the electron acceptor, and the zone will work at least partly like an aerobic reactor. The DN not only reduces the nitrate to nitrogen, it is also important for other reasons:

- carbon is consumed without using oxygen, which means an energy saving; and
- nitrification causes a pH decrease and part of this is recovered by the DN.

We will consider both post- and pre-denitrification systems, Figure 14.4. In the figure we have indicated the major liquid streams that will influence the operation of the DN zone.

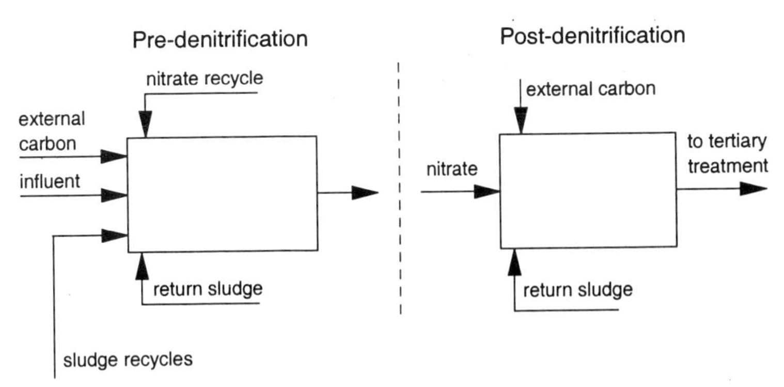

Figure 14.4: The major influent streams influencing denitrification.

From a control point of view, it is favourable to consider different time frames, as we have shown for the plant dynamics. This will make it possible to decouple the problem and make the control task manageable. We consider three time frames for DN:

- *Fractions of hours - mixing and reaction rate control*: The DN rate can be affected in minutes, and we consider two variables that can be manipulated to control the reaction rate, the dissolved oxygen concentration in the influent liquid streams, and the available carbon source for the DN.

- *Fractions of an hour to hours - hydraulic control*: The retention time of the anoxic zone can be rapidly changed by several liquid streams. In particular the nitrate recirculation in a pre-denitrification system has a very large flowrate, typically four times the influent flowrate. Other flowrates, like the influent flowrate and filter backwashing, may change significantly on an hour-to-hour basis. Furthermore the sludge recycle streams will influence the retention time of the zone, but to a lesser extent. The retention time of a continuous reactor corresponds to the phase length of a sequencing batch reactor (SBR).

- *Hours to days - reaction control*: The result of the DN reaction can be determined by the nitrate profile or the nitrate concentration at the outlet of the anoxic zone. In an SBR the nitrate concentration as a function of time can be monitored. Ideally it will approach zero at the end of the phase or at the outlet of the reactor.

Mixing and Reaction Rate Control

Even if the DN requires an oxygen free environment, there are several disturbances that may violate this condition. The various streams may have a high dissolved oxygen concentration, that will rapidly influence the DO concentration of the anoxic zone, thus the DN rate.

In a pre-DN system (Figure 14.4 left) water is recirculated from the outlet part of the aerator. It has usually high concentrations of both nitrate and DO. It is apparent that from a DN point of view, the DO concentration at the outlet of the aerator has to be kept as low as possible, so that no more oxygen than necessary is recirculated.

Another source of disturbance may be backwashing streams from filters downstream in the plant. Such water may disturb the process from two points of view. One is the hydraulic load, causing the retention time to decrease (see below). The other is the oxygen addition, since the backwash water is usually oxygen rich. The influent wastewater may be oxygen rich as well (during rain storms or during snow melting periods) and will influence the denitrification in the same way.

Example 14.2 (DO influence on the denitrification) *The MATLAB program* `Example_14_2.m` *simulates a batch reactor (equivalent to a plug flow reactor) configuration of the anoxic zone using selected mechanisms from the IAWQ AS Model No 1. Figure 14.5 shows the effect of the DO level on nitrate removal.*

Ideally the anoxic zone should be oxygen free. An increase of oxygen to 9 mg/l will slow down the denitrification reaction, and the nitrate will not disappear.

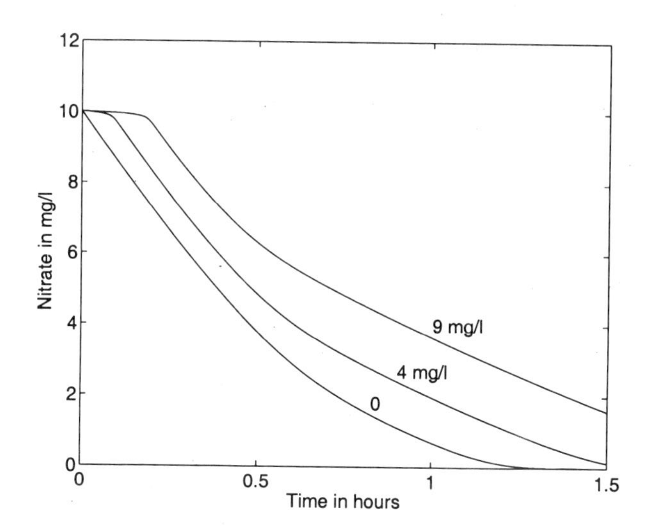

Figure 14.5: Influence of the DO concentration on nitrate removal in an anoxic zone.

The denitrification organisms need carbon as an energy source, so it is important to have a sufficiently high concentration of organic carbon in the system. This is delivered with the influent flow in a pre-DN system. During a low load period extra carbon may have to be added, like methanol or ethanol. In a post-denitrification (Figure 14.4 right) system external carbon has to be added.

Example 14.3 (Influence of carbon on the denitrification) *The MATLAB program* `Example_14_3.m` *simulates a batch reactor (equivalent to a plug flow reactor) configuration of the anoxic zone using selected mechanisms from the IAWQ AS Model No 1. Figure 14.6 shows the effect of the amount of soluble carbon on nitrate removal.*

Even with no soluble carbon nitrate is removed by the carbon formed by hydrolysis of the insoluble COD. However this is seldom sufficient, the example showing that 12 mg/l of soluble carbon is needed even without oxygen. In pre-denitrification systems, this would imply 48 mg/l in the feed for a nitrate recycle ratio of three.

Recirculation of nitrate in a pre-denitrification system is a fast process, since the flowrate is so high. Consequently the influent nitrate concentration can be changed within minutes. This in turn will rapidly change the DN rate.

The input-output relations in the fast time scale are summarised in Table 14.3. The sign of the gain is indicated with (+) or (-), while one or two signs indicate the amplitude.

The DN rate may be estimated in various ways. In a batch reactor or a plug flow reactor it can be obtained by measuring the nitrate concentration rate of change

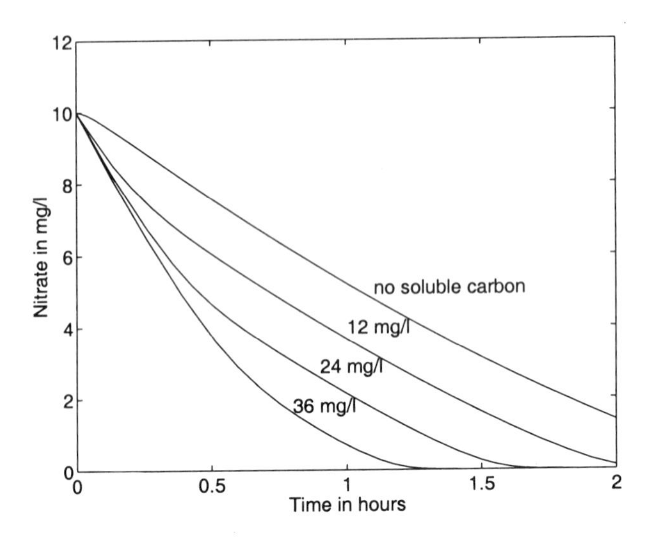

Figure 14.6: Influence of carbon addition on nitrate removal in an anoxic zone.

Output	Inputs
DO concentration	-
Nitrate recycle flowrate	+/-
Filter backwash flowrate	-
Carbon addition DN rate	++

Table 14.3: Cause-and-effect relationships in the fast time scale.

(see Chapter 23). Then the end point concentration can be used for control. This is common in sequential batch systems.

The redox potential (oxidation-reduction potential) contains certain distinctive features that can be used for DN control. Denitrifying activity has been found to decrease with an increased redox potential, that is, an increased oxygen concentration. The relationship between denitrifying activity and the redox potential has been found to be more or less linear. However, during transient conditions the relationship may be far from linear. Also, the denitrifying activity of sludges from different treatment plants shows different sensitivities to the redox potential. For some sludges, the decrease in activity with an increased redox potential is dramatic, while other sludges show only a slow decrease in denitrification when the redox potential is increased. Nevertheless, it can be concluded that oxygen is inhibitory to denitrification even in very low concentrations (when the DO can not be measured by conventional electrodes). For some treatment plants, it should be possible to at least double the denitrification rate by minimising oxygen transfer to the anoxic zone. As the nitrates disappear there is a breakpoint in the redox value, indicating the transformation from respiratory (oxygen or nitrates) to non-respiratory (fermentative) processes. Nitrogen removal control using redox measurements is described in Section 19.5.4.

Using a DN rate measurement or estimate, some control actions can be taken:

- The DO content of the nitrate recirculation has to be minimised. It has to be set as a compromise between the need for nitrification in the aeration zone and the desired limitation for the DN zone. The control signal is a setpoint command to the nitrification DO controller.
- To obtain a proper carbon dosage is a crucial operational task. Insufficient carbon causes not only limited denitrification, but influences the biomass floc formation in a negative way. Too much carbon in a pre-DN system will be oxidised in the aerated zone (and demand extra air!). The cost for external carbon is high and any overdose should be avoided. In a post-DN system there is often an extra aerator at the end to remove surplus carbon. Naturally such removal is costly and should be minimised.
- Filter backwashing can be disastrous for the operation, if it is done improperly (see Section 15.4). A smooth recycle from the filters can minimise any negative influence on the DN process.
- The DO addition from the influent water is considered negligible.

The possible control structure in the fast time scale can be summarised in Figure 14.7.

The DO set-point command has to be calculated as a compromise between the needs of the aeration tank and the anoxic tank. The carbon addition control strategy is quite simple once the DN rate can be established, while the filter backwashing strategy may be a supervisory control. Before any filter backwashing is allowed, a number of conditions have to be satisfied. For example, backwashing should never be allowed at high loads.

Example 14.4 (Batch denitrification) *Figure 14.8 shows batch reactor profiles produced by the MATLAB program* `Example_14_4.m` *using the DN reactions from the IAWQ AS Model No 1. The denitrification reaction rate (the slope of the nitrate curve) is seen to depend on the available carbon (RBCOD). Furthermore, there is almost no DN while the DO level is significant. It is apparent, that the DN phase can be terminated when the nitrate level has reached a sufficiently low level. These curves are equivalent to concentration profiles in plug flow reactor systems.*

Hydraulic Control

The retention time is influenced by all the liquid streams into the DN zone. The anoxic zone retention time has to be sufficiently large so that the DN process can go to completion. In a continuous process the volume is usually constant. Therefore the only way to ensure that the DN can be completed is to limit the

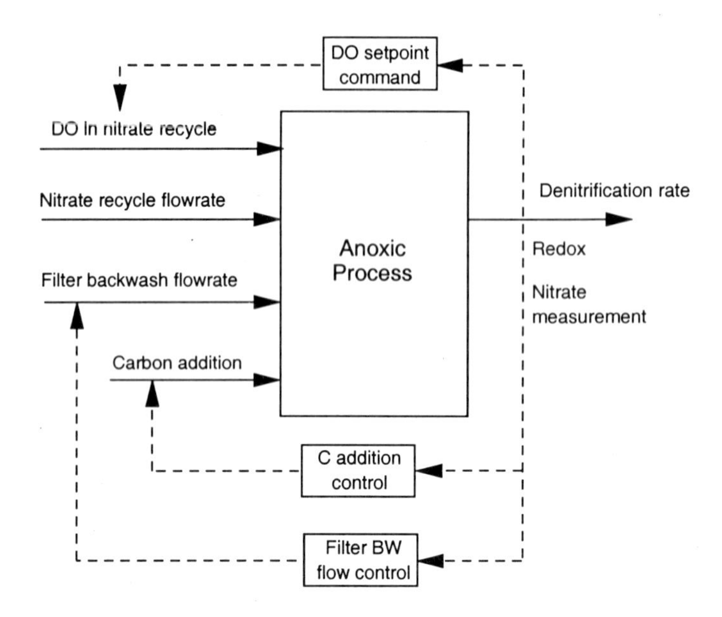

Figure 14.7: Illustration of the cause-effect relationships in the fast time scale in an anoxic reactor. The dashed lines indicate the control structures for the control of DO setpoint, carbon addition and backwash flowrate.

amount of nitrate entering the zone, in other words to decrease the recirculation flowrate.

An anoxic operation can also be part of a sequential batch reactor system or an alternating plant. Then the reaction can be continued until the nitrate concentration has reached a sufficiently low value. In a continuous flow activated sludge system there are often zones that can be alternatively aerated or just mixed. Then the volume of the anoxic zone can be changed in a slower time scale to compensate for changing nitrate loads.

The control actions for the hydraulics time-scale can be summarised in Figure 14.9. Since the retention time is influenced by all the incoming streams. In a pre-denitrification system the nitrate recirculation could be used to match other influent streams so as to keep the retention time at a desired value. Furthermore the recycle and filter backwashing streams have to be considered to keep the retention time as long as possible, especially during high loads.

Reaction Control

A measurement of the nitrate concentration at the outlet of the anoxic zone can verify if the reaction has been completed. Alkalinity is another measurement that can reflect the denitrification, since alkalinity is produced during the nitrate reduction to nitrogen gas. Measuring the pH will give an extra indication whether

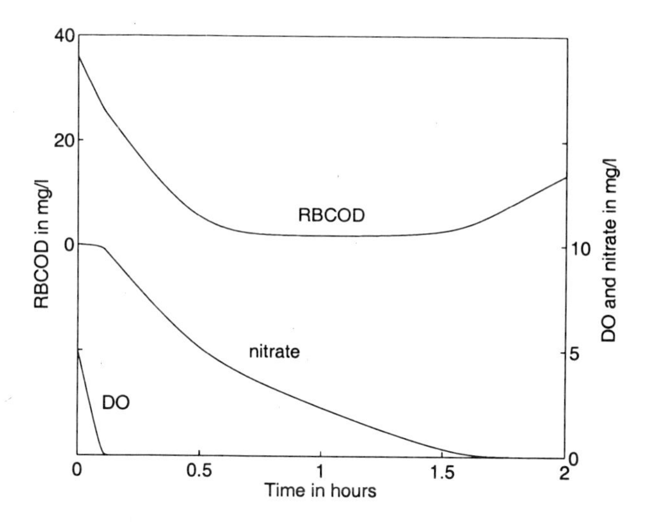

Figure 14.8: Simulation of a batch reactor denitrification.

the denitrification process is running properly. There is a pH optimum in the range pH 7-9. Like other biological processes there is different behaviour if the pH is changed rapidly or slowly. Many organisms may slowly adapt to a changing pH.

We have seen that the nitrate concentration can be influenced in many ways. A change of the DN rate will eventually lead to a change of the nitrate concentration. Likewise, a too short retention time will prevent the DN reaction from going to completion. At this time scale we still have one control variable to adjust, the nitrate recycle flowrate. Certainly, the recycle will have an influence in all the time scales, the DO concentration can be influenced in the fast time scale and the retention time in the medium time scale. Based on the outlet nitrate concentration, however, the only remaining control variable to ensure a complete reaction is the recycle flowrate of nitrate. Thus, we can consider this variable for the reaction control strategy, illustrated in Figure 14.10.

Still the long term biomass growth has not been considered. It has been observed that insufficient carbon may cause the biomass to change, and the floc properties to become unfavourable.

14.4.3 Activated Sludge Aerobic Zone

In a biological nutrient removal plant there are three competing processes in the same reactor, oxidation of organic carbon, nitrification and phosphate uptake by phosphate removing organisms. Being able to satisfy all these simultaneous reactions is a true operational challenge, and obviously there are some competing and conflicting goals.

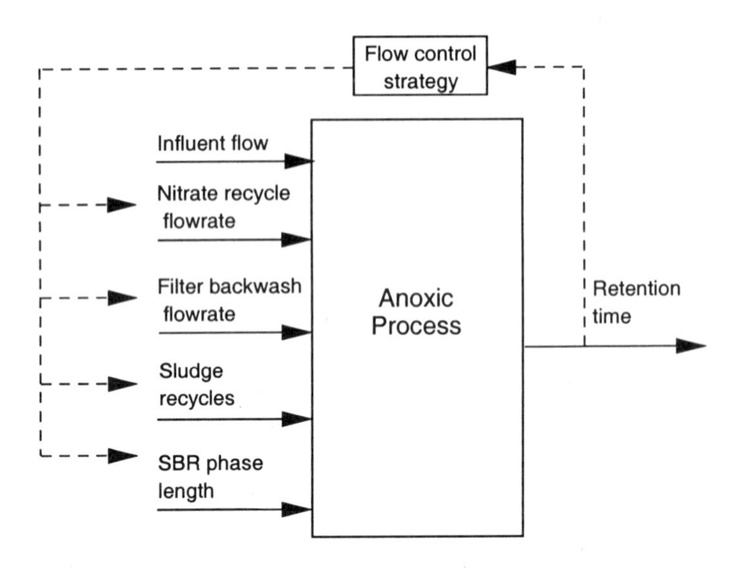

Figure 14.9: Illustration of cause-effect relationships in the hydraulic time scale in the anoxic reactor. The retention time is calculated. The dashed lines indicate the control structure for the various flowrate controls.

The operating goal of the aerobic zone is to remove as much organic carbon as possible, to oxidise all ammonia nitrogen to nitrate nitrogen, and to take up all phosphate that has been released in the previous anaerobic zone of the system. However, another equally important goal is to create an environment that supports the growth of proper floc forming organisms and to avoid the formation of filamentous growth. Furthermore foam and scum formation is highly undesirable. There remains no obvious way to uniquely define how to obtain a sludge that is readily separable. This still remains much of an art, but there are some obvious conditions that have to be recognised.

In the discussion on dynamics (Part A and Table 14.1) we have noted that mechanisms can be classified into different time scales. As for the anoxic zone, we will structure the dynamics into three time scales:

- *Fractions of an hour - mixing processes*: This is the control of the dissolved oxygen via the air flowrate. This may influence the respiration rate, that is the rate of both the carbonaceous removal and the nitrification, in minutes. Furthermore, the mixing of the system may have long term influence.

- *Hours - hydraulic and reaction processes*: The liquid flows into the aerator will affect the retention time and consequently the concentration of the sludge and the substrates. A large hydraulic load, caused by poor control of liquid flows, may cause a rapid loss of sludge. Furthermore, a change in the respiration rate will be noticed in the effluent concentration, primarily of

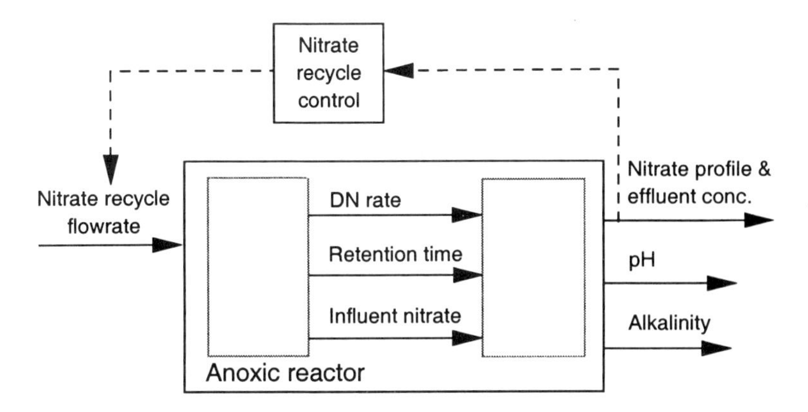

Figure 14.10: Illustration of cause-effect relationships in the reaction time scale in the anoxic reactor. Nitrate is measured and the recycle flowrate is manipulated. The dashed lines indicate the nitrate recycle control structure.

nitrate and ammonia. We will discuss the hydraulic and the reaction control separately.

- *Days to weeks - biomass growth*: The average level of the DO concentration and the respiration rates will have a significant impact on the floc formation. Filamentous growth is strongly influenced by insufficient concentrations of dissolved oxygen or substrate. The total mass of sludge will depend on the growth rate of the organisms, which in turn depends on the DO and substrate levels. By using the waste sludge flowrate, the sludge retention time will be influenced.

Dissolved Oxygen Control

Dissolved oxygen (DO) control is by far the most important control in the fast time scale. As we have seen in earlier chapters, DO is a key variable in activated sludge operation. It influences the operating costs significantly. Since air compression causes a high energy cost, and is a major operating cost of a BNR plant, there is an economic incentive to minimise the air consumption. The dynamics of the dissolved oxygen is such that the DO can be influenced within fractions of an hour. Not only the total amount of oxygen, but its distribution in space and time are essential for the operation.

From the dynamics discussion (Chapters 2 and 3) we have seen that there is an intimate relationship between DO concentration and respiration rate. The carbonaceous removal rate as well as the nitrification rate can be influenced within a fraction of an hour. The DO concentration has to be sufficiently high so that the growth of heterotrophic and autotrophic organisms is not limited due to lack of oxygen. Usually the autotrophic organisms are more sensitive to low concentrations than the heterotrophic organisms. To obtain the right DO reference concentration is always a balance between economy and the biological needs. To keep this balance is the challenge for the DO control system.

The control of the DO concentration as a physical variable does not require any in-depth knowledge of the microbial dynamics. There is extensive experience of DO control, in most cases implemented with traditional PI controllers. Despite the straightforward task of DO control there are some control difficulties involved. The DO dynamics is both nonlinear and time varying. The long time constants and random influent disturbances makes the tuning of a conventional controller tedious. For the moment we assume that it is possible to control the air flowrate to the aerator (and consequently its pressure) on a continuous basis. The total demand of air is basically determined by the total mass of viable organisms in the system. The air flow supply system may be a control system in itself, supplied with local controllers. This will be further discussed in Chapter 17.

In a suspended system mixing is crucial. Using fine bubble or coarse bubble systems combines the needs for mixing and for DO supply. From one point of view, the mixing has to be sufficiently good so that the oxygen and the substrate will have a chance to penetrate the floc. On the other hand, too much mixing towards the outlet, when flocs are being formed, may prevent a good floc formation. So we see that the air flow system plays a crucial role for the aeration basin. The mixing requirements usually define the lower limit of the air flowrate. As a consequence, the DO concentration may be kept unnecessarily high during low load conditions. Naturally, the upper limit is determined by the capacity of the air supply.

Some systems work with pure oxygen instead of air supply systems. Since the DO saturation concentration depends on the partial pressure of oxygen, the driving force for oxygen transfer can be significantly increased. Sometimes pure oxygen can be added in conventional air systems to improve the DO of the system under extreme load situations. This has been applied in some plants with extremely high organic load, such as brewery or abattoir wastes.

Example 14.5 (Carbon removal with constant air flowrate) *If there is a uniform air flow along the tank with only carbonaceous removal, a typical DO concentration profile is shown in Figure 14.11. In the inlet area the substrate concentration is high, and the corresponding oxygen uptake rate is large. Towards the outlet area most of the substrate has been consumed, and most of the activity is endogenous respiration, with decay a little larger than growth. The plot was produced by the MATLAB program* `Example_14_5.m`.

If the air flow distribution can be controlled along the aerator, then the air flow concentration is ideally constant. Instead the respiration rate has a profile, that is proportional to the substrate profile. In the inlet it is large, and will decrease towards zero at the outlet area, if the reaction is allowed to go to completion.

In a nitrifying aerator there is a combination of carbonaceous removal and nitrification. Since the growth rates of heterotrophic and autotrophic organisms

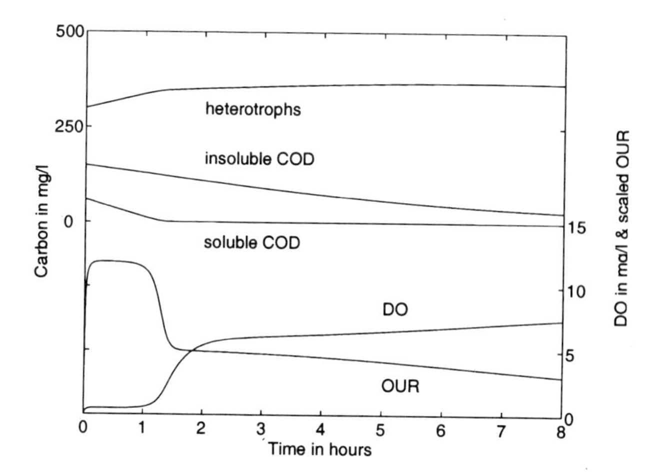

Figure 14.11: Typical profiles for uniform air flow along a plug flow aerator.

are quite different, the oxygen demand will vary significantly along an aerator. Due to the higher growth rate of heterotrophic organisms carbonaceous removal is completed prior to nitrification. The oxygen uptake rate (the respiration due to carbon removal) in the first part of the aerator will reflect the variations of the COD of the influent flow. Towards the outlet of the aerator the carbon removal has been completed, and the oxygen demand is completely dependent on the nitrification. Thus, a variation of the oxygen demand in this location will reflect variations of the ammonia load to the system. This illustrates that the oxygen consumption in at least two places has to be considered in the control.

Example 14.6 (Nitrogen removal with constant air flowrate) *The respiration rate, consisting of both carbonaceous and nitrogenous oxygen demand, can be illustrated by a sequential batch reactor transient. Such a transient is equivalent to the profile in a plug flow reactor. The COD removal is much faster than the ammonia consumption (Figure 14.12). The carbonaceous and nitrogenous oxygen demand is shown as COUR and NOUR respectively. The plots were produced by the MATLAB program* `Example_14_6.m`.

Example 14.7 (Full scale plant data) *Figure 14.13 shows data from a full scale plant. The aerator consists of a number of reactors in series, and the DO in each tank is controlled individually. Since the DO is constant, despite the load variations, the air flowrate is proportional to the load. Here we can see that the air flowrate for the inlet tank (left plot) is closely related to the TOC load variation, while the air flowrate to the tank closer to the outlet (right plot) is related to the ammonia load to the plant.*

These examples show that there is a reason to control the DO of the various

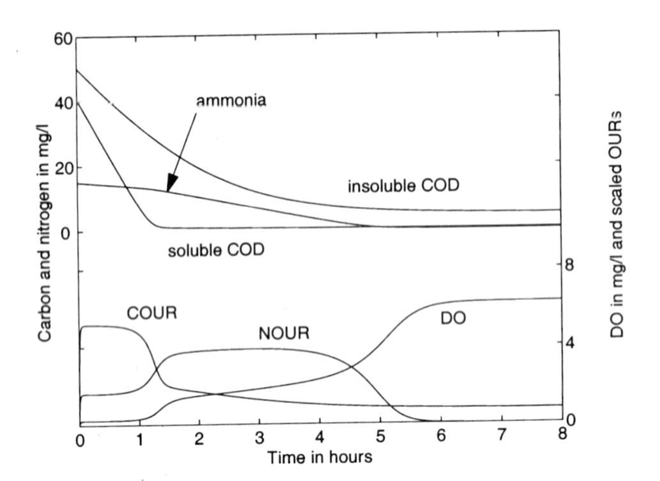

Figure 14.12: Respiration rate profiles in a nitrifying reactor with uniform air flow along the reactor.

parts of the aerator independently of each other, and more than one DO sensor is needed. This is indicated by Figure 14.14.

Choosing proper DO setpoints is not a trivial task, since there are competing biological reactions. In the previous discussion on the anoxic zone, we saw that a compromise has to be made between the nitrification need and the demand for an anoxic environment in a pre-denitrification system. The traditional assumption of a fixed DO setpoint may not be optimal. Instead, there are indications that some periodic setpoint change may influence the process favourably, so that the sludge obtains a combination of good clarification and thickening properties.

The DO control purpose can be summarised as:

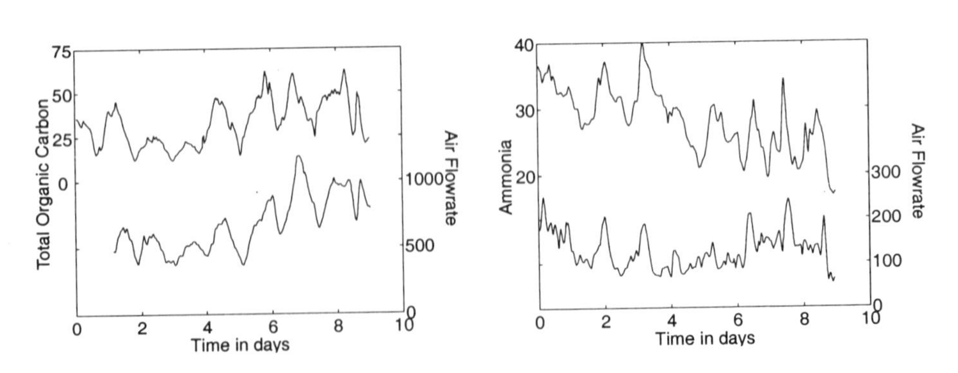

Figure 14.13: The air flowrates with TOC load and ammonia load in inlet (left) and closer to outlet (right).

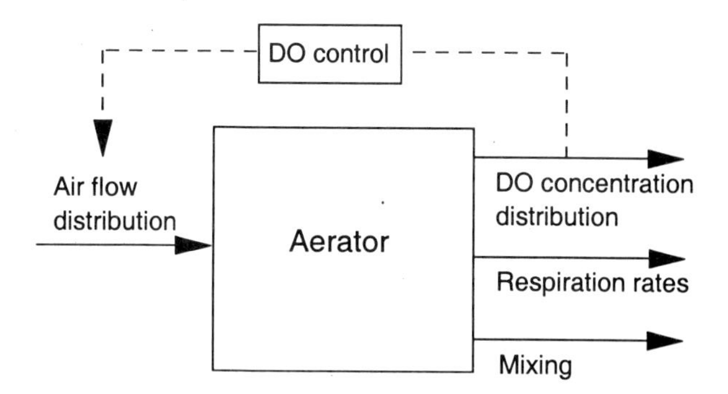

Figure 14.14: Illustration of cause-effect relationships in the fast scale in the aerator. The dashed lines indicate the dissolved oxygen control structure.

- keep the DO concentration at a sufficiently high level to support the growth of adequate organisms.
- keep the DO sufficiently low to save energy and avoid excess mixing.
- keep the DO concentration low for nitrate recirculation.

DO control will be further discussed in several sections of Chapter 17.

Reaction Control

For the reaction control we will basically ensure that the reactions have gone to completion. There are two major variables that may be manipulated, the reaction rate and the retention time. Firstly we discuss the means to control the reaction rate. In the next section we discuss how the retention time may be influenced.

In a nutrient removal process the nitrification reaction is the slowest. Therefore the outlet ammonia and nitrate concentrations are the most important indicators of completed reactions. The reaction rates depend on the DO concentration. It is obvious that any measurement of the nitrogen content at the outlet should influence the choice of the DO setpoint.

The nitrification can be improved by a high value of the DO setpoint. However, we have shown that a too high DO may adversely influence the denitrification rate in the anoxic zone. Since any over-aeration is costly, this adds another incentive to limit the aeration as much as possible. By measuring the nitrate or the ammonia concentration at the outlet, the DO setpoint can be chosen suitably.

Lowering the setpoint not only favours the DN in the anoxic zone. There are many indications that there is some simultaneous nitrification and DN (SND) in the nitrification zone. It is apparent that the DO concentration is not uniform throughout the sludge floc. Since the DO propagates by diffusion, the DO concentration in the centre of a flow may be lower than the bulk liquid DO concentration, thus favouring DN instead of aerobic reactions. Once the aeration becomes limited, the chance for SND will increase, thus saving energy in the plant. Using ammonia measurements to find the DO setpoint will be further described in Section 17.7.

The phosphorus uptake depends on heterotrophic organisms (Chapter 2 and 4) and is a faster reaction than the nitrification. The phosphorus accumulating organisms will use the available phosphorus in the liquid phase, together with available carbon and should be completed long before the nitrifiers. If the retention time is too long, there is a risk that the organisms get starved. This will cause a secondary phosphorus release from the organisms, which is naturally not desired. Therefore, a suitable retention time control is important for the bio-P organisms.

Chemical precipitation can be used for P removal, either as the only mechanism or as a complement to the bio-P removal. In any case, the P concentration towards the outlet is an indicator of whether the removal mechanism has worked or not. Thus, a chemical dosage control may be desired.

The control strategies for the reactions are illustrated in Figure 14.15. The reaction control of the aerator may be summarised as follows:

- The nitrogen concentration at the outlet (measured as ammonia or nitrate) is fed back primarily to influence the DO setpoint, to make sure that there is sufficient - but not too much - DO (see Section 17.7).
- By limiting the aeration SND may be favoured.
- Chemical dosage may be used as a complement to biological P removal, if the P concentration at the outlet is too large.

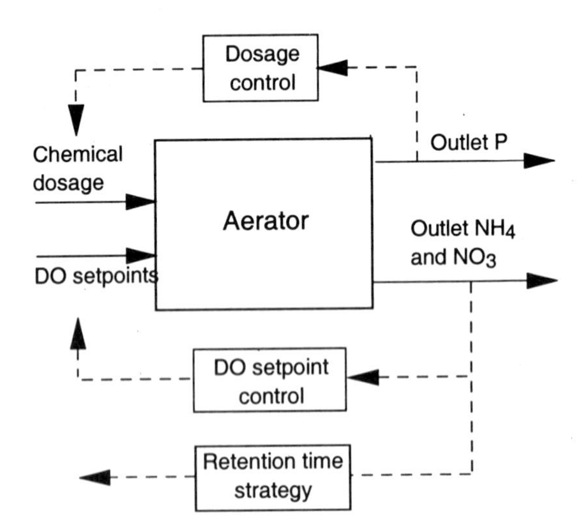

Figure 14.15: Illustration of the cause-effect relationships in the reaction time scale in an aerator. The dashed lines indicate the control structure in the reaction control.

Hydraulic Control

A change of the retention time will of course influence the completeness of the nitrification reactions. Since the variability in the retention time primarily depends on the influent flowrate, it is of interest to attenuate any peak flows by some equalisation strategy, so that the nitrification gets a better chance to be completed.

In a sequential batch reactor the retention time can be varied according to the load. The rate of change of nitrate can be monitored and the respiration rate can be calculated. It will decrease down to an endogenous level towards the end of the cycle, if the reaction goes to completion. A similar observation can be made in a plug flow reactor along the channel.

The major liquid and air streams of a pre-denitrification system are illustrated in Figure 14.16.

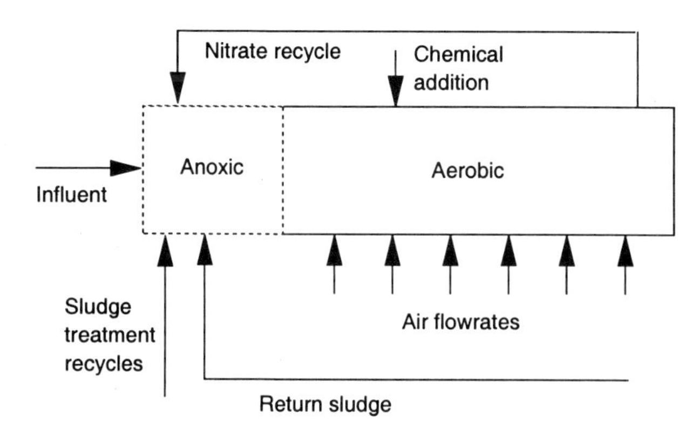

Figure 14.16: The major liquid and air streams into a nitrifying aerator with pre-denitrification.

The distribution of the sludge between the aerator and the settler is basically determined by the return sludge flowrate. The return sludge flowrate has to be balanced to meet various needs. It is used to keep the sludge blanket in the settler between its allowed levels. A large return sludge flow adds to the hydraulic loading of the aerator. Therefore the return sludge load has to be kept sufficiently low so that the hydraulic loading is not too large and that the sludge blanket does not disappear in the settler. Likewise, it has to be sufficiently large so that the sludge blanket does not rise too much. Also, it has to supply the aerator with a sufficient amount of organisms to match the incoming substrate load.

Recycle from the secondary settler is an important variable for obtaining the right operating point, but seldom useful for the control on an hour-to-hour basis. Some systems are supplied with several feeding points along the aerator for the return sludge, a so called step return sludge control. For certain loads, such as toxic loading, such a possibility can be advantageous. It may also prove to be one way of preventing bulking sludge. These topics are further discussed in Section 18.6.

The internal distribution of the sludge within the aerator can be influenced by step feeding the influent water, Figure 14.17, especially in systems for carbonaceous removal.

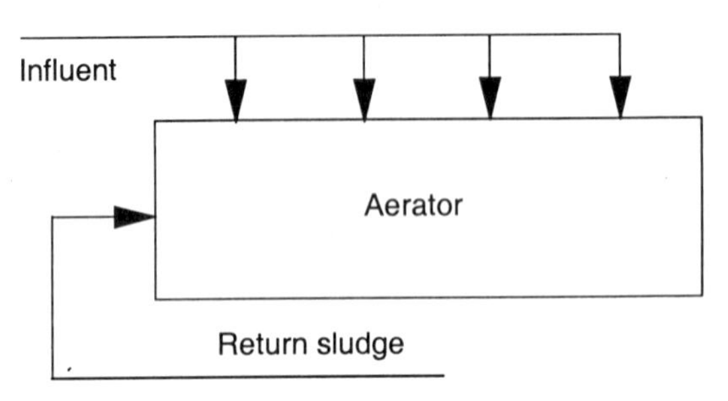

Figure 14.17: Liquid streams in an aerator with step-feed influent flowrate.

If all the influent water is fed to the outlet area of the aerator, then a contact stabilisation structure is obtained. By using step feed control the aerator can be prepared to meet various kinds of disturbances, such as a large hydraulic or organic load. It has to be emphasised that step feed control can temporarily move the sludge internally, thus locating the sludge at the right place at the right time. It is a transient condition, and should not be analysed only from a steady-state point of view. We refer to Section 18.6 for further discussion.

Figure 14.18 summarises the control structure of the hydraulic time scale operation of the aerator. The various control actions to be taken are:

- the retention time may be kept as large as possible by attenuating influent flow peaks;
- in a sequential batch reactor (SBR) the phase length can be controlled according to the effluent ammonia or nitrate concentrations;
- the return sludge flowrate is used to establish a good relation between sludge mass in the aerator and in the settler;
- step feed is used (at least in plants with only carbon removal) to temporarily control the distribution of sludge within the aerator.

Biomass Growth Control

The total amount of sludge depends on the overall growth rate and the rate of sludge wastage. This is made on purpose via the waste sludge flowrate. However, a significant portion of the sludge is wasted unpurposefully via the clarifier effluent. It is obvious, that the flow properties play a significant role here. It is obvious that the total sludge mass cannot increase faster than the biomass growth rate

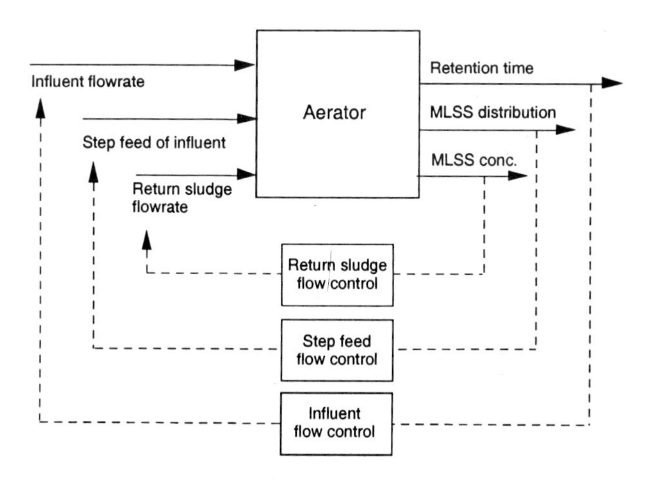

Figure 14.18: Illustration of the hydraulic cause-effect relationships in an aerator. The dashed lines illustrate the control structures in the hour-to-hour time scale in the aerator.

so it will take weeks to increase the total sludge mass. Using the waste sludge flowrate as a control variable will be further discussed in Section 18.6.1 and is illustrated in Figure 14.19.

It is apparent from the discussion of DO control that the average DO setpoint level will influence the growth of the biomass. The relationship between the floc formation and physical variables is still a subject for extensive research. It is well known that the growth of filamentous organisms is favoured by a low DO concentration. It is a common assumption that both oxygen and substrate penetration into the floc is determined by diffusion. Therefore, both a low oxygen and a low substrate level will favour organisms with a high surface-to-volume ratio, such as filamentous bacteria. *Microthrix Parvicella* growth is often observed in underaerated zones.

The term bulking is related to many different phenomena. Common to all of them is that the sludge settling properties have become so poor that the plant cannot produce an acceptable effluent. Bulking has been studied by many researchers and still remains a mystery. One important aspect of bulking is related to filamentous growth. The correlation between bulking and nutrient removal is not well established but may equally be a result of physical conditions, especially in BNR plants. Sludge age control or a fixed invariant proportional control of the recycle rate do not necessarily prevent the sludge from bulking. However, more advanced control of the recycle rate, step-feed and step-sludge have proven to be effective in controlling bulking. Although some of these strategies can cause a significant deterioration of the effluent quality, since there may be a conflict between the control of bulking and the satisfaction of the effluent standards.

Simultaneous nitrification and denitrification (SND) was discussed above and

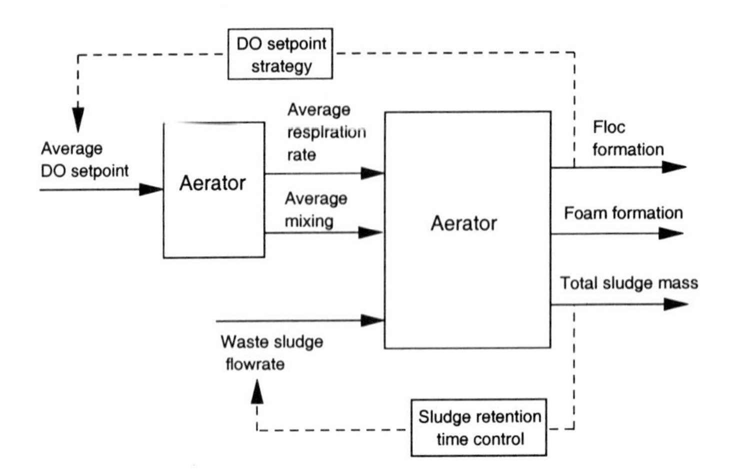

Figure 14.19: Illustration of cause-effect relationships in the aerator for biomass growth. The dashed lines indicate the control structures in this time scale.

has been known since the 1960s. The anoxic and aerobic behaviour within the floc may partly explain the SND, and emphasises the role of the floc formation for the operation. It is obvious that the aeration may influence the SND. There may be a link between underaeration to achieve SND and sludge bulking. One may add that present models of nitrogen removal, such as the IAWQ AS Models, ignore the issue of SND.

To summarise, the main control actions in a slow time scale are:

- Waste sludge flowrate control to keep an adequate sludge retention time and total sludge mass;
- Finding suitable average values of the DO concentration to support the growth of the desired species of organisms.

14.4.4 Secondary Settler

Settler performance is crucial for the whole activated sludge operation. In Chapter 4 we noted that the purpose of the settler is twofold:

- to efficiently separate the solids and liquid - the clarification;
- to thicken the sludge as much as possible - the thickening process.

The success of the settler operation depends on two factors, the hydraulic character of the various flowrates and the floc settling characteristics. We may illustrate the cause-effect relationships with Figure 14.20.

As for the aerator dynamics, the settler can be influenced in a wide spectrum of time scales,

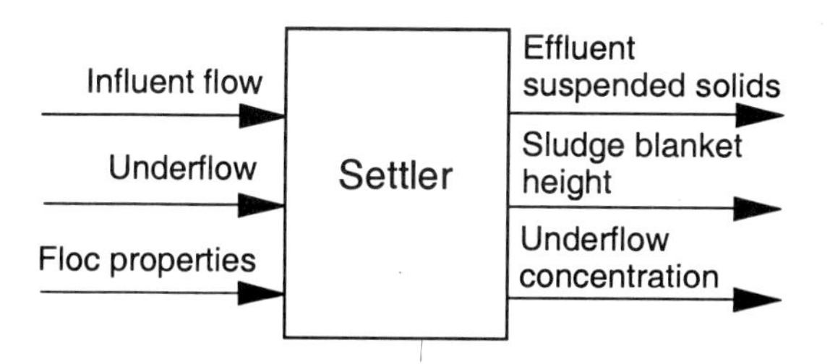

Figure 14.20: Illustration of the cause-effect relationships in a settler.

- seconds to minutes - hydraulic propagation;
- minutes to hours - hydraulic loading;
- hours - reactions in the settler;
- days - floc properties.

Hydraulic Propagation

In Chapter 4 we discussed hydraulic propagation through the treatment plant. Eventually a flowrate increase will reach the settler. It is apparent that any influent flowrate change will have a significant influence on the clarifier performance and the effluent suspended solids concentration.

Example 14.8 (Clarifier flowrates and effluent suspended solids) *Return sludge flowrate changes may have an even more direct influence on the clarifier performance, as shown in Figure 14.21. The influent flowrate Q changes as a primary pump is turned off and on. The control system couples the return sludge flowrate directly to the influent flowrate. The flowrate change causes a sudden change in the effluent suspended solids concentration.*

The actual plant has no variable speed pumps for the influent flowrate. Thus, the flowrate is changed stepwise as pumps are turned on and off. The return sludge flowrate is controlled proportionally to the influent flowrate (see Section 18.6.2), so the return sludge flowrate is changed accordingly in Figure 14.21. It was, however, not expected that the effluent suspended solids would change as quickly as in Figure 14.21. In fact, the concentration change is noticed within a minute at the effluent location of the settler.

There are two phenomena that will influence the behaviour, the hydraulic propagation along the plant, and the hydrodynamics of the settler. The graph in Figure 14.21 shows the plant influent flowrate, not the settler influent flowrate. Since the flowrate does not propagate immediately through the plant, the settler influent flowrate will be smoothed, according to Section 4.5. The return sludge flow control law, however, makes the return sludge flowrate change proportionally to the plant inflow. Therefore, there is a sudden flowrate change in the bottom of the settler. Now, the dominating fast phenomenon is due to hydrodynamics. Such a sudden change will force the liquid velocity to change, and the shock wave will

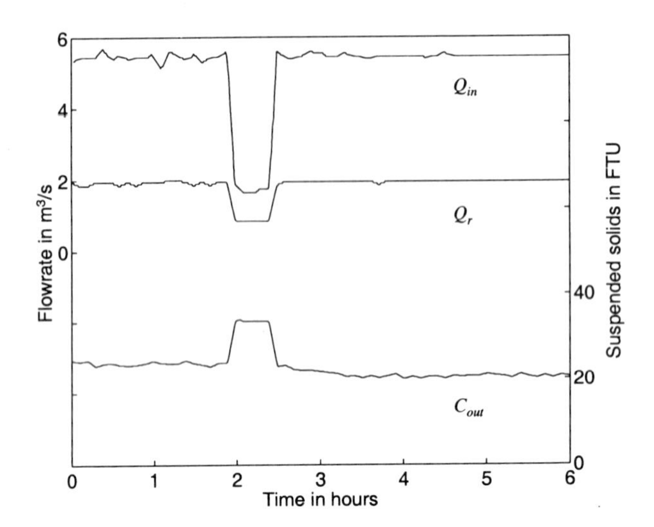

Figure 14.21: Clarifier flowrates and effluent suspended solids.

propagate within minutes to the settler outlet. If the sludge blanket is high, or the floc properties not favourable, then the effluent suspended solids will change as in Figure 14.21.

It is obvious, that any sudden flowrate change will deteriorate the settler performance. Therefore, it is crucial to be able to continuously change the plant flowrate smoothly in order to protect the settler performance.

Hydraulic Loading

According to empirical studies the effluent suspended solids concentration depends on the load to the settler, that is the flowrate times the sludge concentration. This is approximately true for almost steady state conditions. Slow flowrate changes will be reflected in increasing effluent suspended solids concentration.

In the settler both the liquid and the solids mass balances have to be satisfied. It is obvious that, if the sludge load to the settler is larger than the sludge output via the underflow or the effluent, then the sludge blanket will rise. The sludge blanket level can be allowed to vary, as long as it is kept within its maximum and minimum limits. When the blanket height becomes too high, then the clarifier performance will deteriorate. When the underflow becomes too large, then the return sludge will be dilute.

In other words, even if the return sludge flowrate has to be closely proportional to the influent flowrate at an average, it may be changed in a slower time range to compensate for other disturbances (see further Section 18.6.2).

Reactions in the Settler

The sludge in the thickener is not inert. There may be favourable conditions for biological activity. The oxygen content is of major interest.

Consider a treatment plant for only carbonaceous removal. If the plant is overloaded, then all the COD has not been removed in the aerator. Since the sludge entering the settler is oxygen rich, there may be some heterotrophic growth taking place in the thickener, removing some organic substrate until all the oxygen has been consumed. In fact, the sludge in the thickener can be purposefully used to marginally increase the organic removal capacity of the plant.

In a nutrient removal process there may be a significant denitrification in the settler, which then acts as an anoxic zone. Nitrate reduced to nitrogen gas may cause the sludge blanket to rise, so called rising sludge. If the settler holds too much sludge, then denitrification may cause large problems. This is the reason why the sludge is recycled immediately in many nitrifying plants and no sludge storage is kept in the settler.

In a bio-P removal plant still another problem may appear in the thickener, secondary release of phosphorus. If the retention time is too long in the thickener, then the conditions become anaerobic (no oxygen and nitrate available), and the bio-P bacteria get starved again. The same processes as in the anaerobic reactor take place, and phosphorus is released too early.

The return sludge flowrate is the key variable to control the sludge blanket so that the various constraints are satisfied.

Sludge floc properties

An ideal sludge should contain only well settleable flocs that will give good thickening properties. In reality the floc is very diverse. It may contain dispersed small flocs that will hardly settle at all. Such flocs will become part of the unpurposefully wasted sludge via the effluent. Filamentous organisms may have been favoured in the aerator operation. Too many filaments will create large settleability problems.

A small fraction of filaments may in fact be favourable for the clarification. They may create bridges between the flocs. Such a structure can then filter the dispersed flocs so that the clarifier effluent actually improves. At the same time it is obvious that there is a fine balance between too few and too many filaments in the sludge. Too many will obviously create sludge bulking.

Apparently the sludge properties are formed in the reactor zones. The settler operation depends crucially on the proper balance between substrate, oxygen and the microorganisms.

14.5 Plant-Wide Operational Objectives

While it is convenient to decompose the problem of operational strategies of any process by process units and by time scales, we must not lose sight of the fact that there are interactions between the process units and also between the time scales. This has been mentioned in passing several times in our discussions in the previous section.

Wastewater flow through the process propagates effects from one unit to the next, and perhaps even more seriously, recycles bring your sins back to hurt you again. We have already mentioned the hydraulic propagation of disturbances several times as well as the effects of recycled nutrients. These interactions demand that you step back and look at the global effects of your chosen disturbance rejection strategies, with a particular emphasis on recycle streams.

Also, from the point of view of minimising cost, we must remember that the optimum of the whole is not the sum of the optima of the parts. A global perspective is very important in this regard, saving a dollar in sludge production is not an economic proposition if the consequence is an extra cost of two dollars in sludge treatment.

We will examine some of the key interactions that must be considered for the plant wide operation. They involve:

- hydraulics,
- use of resources,
- recycles within the plant, and
- the operator.

Because the plant does not exist in isolation, plant wide control and optimisation must also consider influences from and influences on the world outside the boundary fence.

The whole concept of sustainable operations can only be evaluated from this perspective.

Chapter 20 will discuss several of these aspects further and propose strategies for looking at plant wide control and optimisation.

14.5.1 Hydraulics

Hydraulic disturbances are a key issue for all wastewater treatment operation. Generally speaking, we wish to avoid sudden changes and upsets of the flowrate. Since normal variations and storm events will cause the flowrate to change there is an obvious incentive to minimise such variations into the plant.

It is obvious that the interaction between the sewer network and the treatment plant is of great importance. Unfortunately, there are often different organisations responsible for the operation of the storm sewers and the treatment plant at the end of the pipe. Most often the networks for storm water and wastewater are combined, and ultimately this has a profound effect on the treatment plant. The sewer operation has to make sure that no basement floodings occur during a storm. The primary objective is to use the sewer and storage volumes as efficiently as possible, and in extreme cases by-passing into the receiving water is necessary. The treatment plant, on the other hand, has to optimise the overall treatment efficiency of the plant. As we have seen, this means that the perspective has to be longer than the time frame of a storm. If organisms are flushed out, then the recovery time may be months.

An integrated operation of sewers and treatment plants is desirable. The details of such an operation are outside the scope of this book, and are the subject of current research. However, sewers are used today as flow equalisers, and flow peaks can be reduced by such an operation. The primary pumps are often of the on/off type. As remarked in Example 14.8 this often creates problems in the settlers. Therefore the issue of variable speed pumps is important. Where are they motivated from a cost/performance point of view? Chapter 15 will consider more aspects of disturbances.

On entering the plant the flow is often split into several parallel trains. Flow splitting may seem like a trivial task, but is a major operational problem in many plants. Even if the liquid flowrates may be the same in the different plant trains, the solids loading may be significantly different.

14.5.2 Use of Resources

If the operation has not been sufficiently considered already in the design of the process, then there are fewer degrees of freedom for the operational goals. When looking at resource handling we may consider the following items:

- already invested capital costs;
- land use;
- water re-use;
- organic and nutrient use and re-use;
- overall use of energy.

Naturally, the design has a fundamental influence on the operational potential. The total capital investment in the water production and wastewater handling systems is considerable. In a country like Sweden the total investment over the years has been about 50,000 SEK (or more than USD 7000) per inhabitant. Such an investment cannot be neglected when the future development of wastewater treatment systems is planned. It makes it crucial to explore all possible improvements of the existing system. This includes the water treatment, distribution, sewer systems, wastewater treatment plants and the sludge handling in general.

The importance of land use naturally increases with the population density. Very high costs of land traditionally have motivated unconventional solutions for wastewater collection and handling, such as pure oxygen systems, two story settling tanks, or extremely deep aeration tanks and even underground plants. Wastewater treatment in urban areas has to be able to concentrate the spatial resources as much as possible. Sometimes extra energy is used to obtain a smaller land area.

Recycling of water and water saving is a reality already today in places like California, Australia and Israel. Furthermore, water savings may to some extent lead to saving of other resources, such as energy. However, water re-use raises public health issues.

In a typical wastewater treatment plant today some 50-60 % of the influent organic content is transferred to carbon dioxide and water. Some 10 % remains in the effluent water, while some 30-40 % is found in the sludge. The ultimate use of the organic material should be defined for the total system. It may be burnt, it may be used for soil improvement or it may be used for gas production in anaerobic treatment.

There is an incentive to reduce the amount of organic material, so that the sludge transportation can be decreased. The cost for the carbon removal has to be compared with other costs, like energy, transport costs, as well as the environmental impact. It is obvious that the sludge separated in the treatment plants can be utilised as a fertiliser in regions with agriculture due to its content of organic material and phosphorus. However, the sludge has to be acceptable from a food production point of view. To minimise the environmental impact it may be necessary to decrease the amount of heavy metals and unwanted organic compounds. This is one of the largest challenges for people working with the existing water systems.

Nitrogen is an almost unlimited resource in nature. Still, the removal of nitrogen from the wastewater demands a lot of resources. It may be transferred to the atmosphere (via nitrification-denitrification), it may be used as a fertiliser in one way or another, or for anoxic respiration. Some 80 % of the nitrogen in municipal wastewater comes from urine. This has motivated research to find out if urine separation is a way of taking care of this part of the nitrogen in a more economic way than today's conventional nitrogen removal systems. Again, many components of a criterion have to be compared. It is far from self-evident that the best use of nitrogen in wastewater is actually to utilise it in exactly the same way as a commercial fertiliser in agriculture. A sustainable system should be able to support structures that could improve the current environmental quality of a system. We can mention a couple:

- nitrate may be purposefully wasted from a treatment system in order to support denitrification of the bottom sediments in a lake; and
- a treatment plant may be operated in such a way that the N/P ratio is controlled to a specific value to improve the balance in receiving waters or other recycled use of the water (irrigation and fertilising).

Phosphorus is certainly a limited resource. Therefore any recycle of phosphorus will be increasingly important. It may be used directly as a fertiliser, in the manufacturing of new detergents, or it may be fractionated into new components.

The energy issue will be increasingly important in the future society. There is a significant indirect environmental impact due to the use of electrical power, heat and chemicals. Since both energy production, transmission and distribution are related to environmental consequences, there will always be an incentive to save energy. Naturally, when comparing various systems for wastewater handling, the accumulated energy consumption of the total system has to be considered. This includes transportation of the wastewater, energy demand for treatment, the use of heat content in the water, and gas production. By looking at isolated subsystems from an energy point of view it is easy to obtain false solutions, and suboptimal solutions have to be carefully avoided.

The energy use is a determining factor for both nitrogen and phosphorus removal. There is still room for better aeration systems for aerobic organisms. Other alternatives should be compared also from a total energy point of view. In the transport of water and wastewater the energy issue is critical. Many unit processes are not sufficiently energy efficient today. A significant energy price increase would almost surely force more efficient operations to be developed.

A lot of water - and warm water - is used to keep us clean and healthy. Only a minute fraction of the heat content of all this water is exploited, for example in heat pumps. If the heat content could be better utilised in cold climates in combination with the digester gas, then the wastewater treatment plants would be energy producers instead of consumers, for example to supply base heating in district heating systems.

Example 14.9 (Net production of energy) *There are various attempts to make energy evaluations of wastewater treatment systems. A detailed analysis of energy consumption and production in various treatment structures has been presented by Ødegaard (1995) and we will indicate some of his results. The energy consumption has been evaluated by relating it to a predetermined water oxygen consumption.*

Several aspects have to be considered when comparing energy consumption. As an example Ødegaard compares a compact plant with a more conventional one. The former will require less energy for its concrete design, while it may require more energy for chemicals.

Energy is consumed *as a result of:*

- *the use of chemicals in chemical plants,*
- *aeration and mixing in biological plants,*
- *transportation of sludge to disposal site.*

Energy can also be produced *as a result of the biogas production in the sludge handling.*

There are other energy consumers like heating and ventilation, but they have not been considered since they are quite site specific.

The chosen standard plant is designed for 100,000 population equivalents (pe) with the specific flow rate of 400 l/pe/day. The specific loads are given in terms of 70 $gBOD_7/pe/day$, 80 $gSS/pe/day$, 2.25 $gP/pe/day$ and 12 $gTotN/pe/day$.

The energy consumption or production is presented in terms of Wh per m^3 of wastewater flow.

The consumption due to chemicals can vary from 0 to 30 Wh/m^3, where different dosages for different configurations have been assumed.

The air consumption depends on the biodegradable loading to the plant. In the calculations it was assumed a total *energy requirement of 1.0 $kWh/kgBOD$ removed while an additional 4.0 kWh are required for each $kgNH_4 - N$ removed. Typical values for a biological treatment plant for only carbon removal is around 160 Wh/m^3 while a nitrifying plant with denitrification requires around 220 Wh/m^3.*

Energy may be produced from the sludge. Using some reasonable values of the organic content of the sludge and the degradation of the sludge the surplus energy used for electrical production is calculated to anywhere between 170 Wh/m^3 in a carbon removal plant and 120 Wh/m^3 in a nitrifying plant with denitrification. Some of the energy is also lost as heat or as usable heat for digesters.

Assuming a sludge transport of 25 km each way, the energy consumption can be computed in terms of diesel consumption, which corresponds to anywhere between 7 and 20 Wh/m^3 of wastewater flow.

It is shown that most treatment alternatives have a positive energy balance, which means that the plant can produce more useful energy than it consumes. As long as some 40 % of the biogas can be converted to electrical energy, most configurations demonstrate an energy surplus. More details and motivations are found in Ødegaard (1995).

14.5.3 Recycles within the Plant

The interactions between various unit processes make the plant wide operation far from trivial. As we have already remarked, it is not sufficient to consider one unit operation at a time. The main flow of the wastewater will influence the chain of unit processes. The recycles add considerably to the overall complexity. Some of the issues have been mentioned in the previous sections:

- recycle of oxygen and nitrate to the anaerobic zone, principally via filter backwash and via the return sludge;

- recycle or propagation of oxygen to the anoxic zone with the nitrate;

- recycle of phosphorus via the return sludge due to secondary release, and via sludge treatment wastewater streams;

- recycle of nutrients via supernatant from sludge digesters;

- effects of excess RBCOD and VFA propagated from the anaerobic to the anoxic and to the aerobic zones;
- effects of nutrient levels and residence times on biomass population distributions and floc formation; and
- reconciling sludge production goals - sludge quality for nutrient removal, sludge quantity for digestion and methane production, and minimisation of sludge disposal costs.

14.5.4 The Operator

Operators often have a good understanding of the processes. However, education of operators is important. They often get very restricted responsibilities so that they will miss the interactions. A manager would like to see the average concentrations and result, while the operator may want to see the hourly and dynamic data. This can be used to demonstrate his specific problems. Above all, he would need to know the interactions between the unit processes. The operator also needs to get cost information.

14.5.5 User Aspects

Most people expect to have safe water to drink, to cook with, bathe in, wash the car or water the garden with - at the turn of a tap. They also expect to flush the toilet, pour away wastewater from sinks and baths and have it cleaned and returned to nature without damaging the environment. Furthermore, they expect this to happen every day, round the clock.

It is obvious that most people take water for granted and see it as an almost free commodity, very much in the same way as air. This way of looking at water may imply that water should be provided by society and that it is only a matter of formalising an economic transition system in order to have a well functioning water distribution system. During the last ten years it has become apparent that this type of reasoning may have severe implications with regard to anthropogenic handling of natural resources, especially for non-depletable, replenishable ones such as water. There seems to be a growing awareness today that water can no longer be regarded as a free commodity. This is perhaps best illustrated by the fact that the "polluter pays principle" is being extended to this field. The idea to assess and value the environmental impact of water use represents a true change of paradigm.

A change of this kind is always accompanied by turmoil. There has been an intense political debate about the monetary as well environmental cost for the operation of the urban water system during the last few years. Alternative systems as well as alternative administrative systems are frequently suggested, not only to lower the actual monetary cost for the individual household, but also to improve the efficiency of the overall system in one way or another.

One example illustrating a change of the administrative system is the countries of England and Wales. A privately owned water industry serving entire river catchment areas was created on the basis of a number of previously community owned water systems during the late 1980s. The reason for creating these companies was primarily based on financial considerations. It was concluded that substantial investments were urgently needed in order to maintain a reasonable standard of the urban water system and that the only way to raise the necessary money was to create a number of privately owned water companies. Even though the different companies have slightly different images and general approaches the privatisation has mediated that a number of administrative procedures with regard to especially billing have been possible to implement, for example water meters are actually being installed in many English homes. The newly formed companies may still be said to struggle with their role in a broader context and with long-term infrastructure issues. However, it goes without doubt that they have made it apparent that water can no longer be regarded as a free commodity.

14.6 Summary

We have tried to "set the scene" for our look at control by looking at the general operating goals of wastewater treatment plants and at some general strategies for structuring their control systems. In Chapter 15 we will look in some detail at disturbances. Compensating for and mitigating the effects of disturbances are the first major task of the control system. It is therefore important to understand the disturbances, and how we might characterise them, and even model and predict them.

In Chapter 16 we will look at manipulated variables and constraints. These define what scope we have for control and for rejecting the effects of disturbances. Then in Chapters 17 and 18 we look in some detail at control strategies and implementation. Chapter 19 will introduce discrete or batch system control and look at the SBR as an example. In Chapter 20 we step back and look at the whole plant from a broader control, optimisation and management perspective. Finally in Chapter 21 we examine some methodologies for determining the benefit of investing in control systems. We will also look at some of the problems in measuring the return after the investment has been made. In Part D we have a look at instrumentation systems.

14.7 Further Reading

- The process industry Responsible Care programme is documented and discussed in ICCA (1998) and Jacob (1992).
- Discussions on the choice of DO setpoints are made in Olsson *et al.* (1985b), Hermanowisz (1987) and Lindberg & Carlsson (1996b).

- Biological activity in the settler has been studied by Sorour *et al.* (1993).
- Chen & Beck (1993) discuss various control actions to prevent sludge bulking.
- There is an increasing discussion about sustainability in urban water systems. As an indication of this, *Water Science and Technology* has devoted a special issue to sustainibility (Henze *et al.* (1997)). It is obvious that there are many factors that will influence sustainable water and wastewater treatment. Some of the issues for urban water systems are discussed in Aspegren *et al.* (1997).

14.8 Exercises

1. Consider the P release problem in Example 14.1. Use the MATLAB example program to find out how the P release would be affected by:
   (a) 20 % more or 20 % less acetate (VFA),
   (b) 20 % more or 20 % less RBCOD,
   (c) 100 % increase in initial DO concentration, and
   (d) 100 % increase in initial nitrate concentration.
2. Consider the anoxic batch reactor in Example 14.4. Use the MATLAB example program to find the nitrate reduction and how it is affected by:
   (a) A reduction of available carbon by 20 % and 40 %,
   (b) a 20 % decrease in hydraulic retention time, and
   (c) a 50 % increase in the initial DO concentration.
3. Consider the aerobic batch reactor in Example 14.5. Use the MATLAB example program to find how the DO and carbon removal profiles are influenced by the total airflow. Increase the airflow until the profiles level off towards the outlet. Notice the relationship between the DO and carbon profiles. Decrease the airflow until the DO profile has a significant slope at the outlet. Verify that the carbon removal will not be completed. Compare the OUR profiles with the carbon removal profiles.
4. List how the DO concentration can influence the operation in short and long time scales, both for carbon, nitrogen and phosphorus removal. Consider both overaeration and underaeration.
5. Assume that the influent flowrate is suddenly increased by 20 % due to a pump start. Discuss how this flowrate change will propagate in the wastewater treatment system, and how the various unit operations may be influenced. The timescale is important.

6. Give some reasons (economic and process related) to install variable speed pumping.

7. List some possible causes for bulking sludge.

8. What kind of conditions are favourable for the growth of filamentous organisms?

9. Discuss the qualitative influence of various flowrates on the settler performance. You have to consider both flowrate values and flowrate changes.

10. According to the Brundtland report a development is sustainable "when it meets the needs of the present without compromising the ability of future generations to meet their own needs". Discuss how sustainability can be formulated in more specific terms for an existing wastewater treatment system.

11. What can be done in terms of operation in order to make a wastewater treatment system more resource efficient.

12. Consider current effluent discharge limits. Can you find any other way of making them more efficient as driving forces for better operation?

13. List various interactions - in different time scales - between the wastewater treatment operation and its environment.

14. Give some reasons why it is imporant to consider plant wide operation and not only each process operation separately.

15. Simultaneous nitrification and denitrification (SND) has been discussed. How can it be favoured? What are the consequences?

16. The return activated sludge flowrate is often controlled directly proportional to the influent flowrate. Discuss some of the advantages and the drawbacks.

Chapter 15

Disturbances

The goal of this chapter is to look in detail at the major problem, disturbances. In order to design control systems, indeed to operate a wastewater treatment (WWT) plant at all, we need to understand the source and behaviour of disturbances.

Firstly, we will examine the sources of disturbances in Section 15.1. Then how we might analyse them to better understand them (Section 15.2), and even how we might model or predict some of them (Section 15.3). Finally, in Section 15.4 we will discuss disturbance rejection as the major motivation for better process control and also as a goal for designers. Without disturbances life would be simple, for us and biological waste treatment plants.

15.1 Sources of Disturbances

In the previous chapter we listed major disturbances as one of the unique attributes of biological wastewater treatment. Now we will look in more detail at where these disturbances originate.

Initially, we can categorise disturbances as *internal*, originating with the plant itself, or *external*, disturbances imposed on the plant by the surrounding environment. Figure 15.1 depicts the primary location for disturbances in a plant. The only external disturbances are related to the raw wastewater.

Table 15.1 explains the nature of the disturbances indicated with numbers in Figure 15.1.

We can also categorise disturbances as *expected*, those which we can predict with some degree of certainty, or *unexpected*, unpredictable in whether they will occur at all or at what time they may occur.

The definition of a disturbance is relative. One man's controlled output can be another man's disturbance. This is illustrated in the discussion of DO control in Section 17.6.

How do we characterise disturbances? We can not always measure them, so

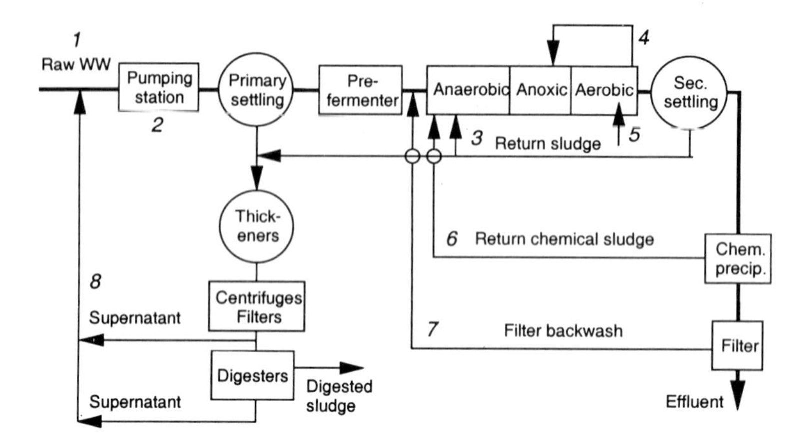

Figure 15.1: Sources of disturbances.

No.	Location	Character
1	Raw WW	Flowrate, composition, concentrations
2	Pumping	Flowrate
3	Return sludge	Flowrate, C, nitrate, P
4	Nitrate recycle	Flowrate, nitrate, DO
5	Airflow	Compressor disturbances
6	Chemical sludge	Sludge mass
7	Backwash	Flowrate, DO
8	Supernatants	Nitrate

Table 15.1: Sources of disturbances.

can we model them? Then we could predict them perhaps? We cannot control them of course, by definition, but how do we mitigate their effects (called perhaps incorrectly "disturbance rejection")?

15.2 Disturbance Characterisation

The primary disturbances to the operation of biological wastewater treatment plants come from the influent wastewater. So what do they look like and how can we describe them? Since one of our prime goals is to attenuate these disturbances, we must first get to know them and be able to describe the problem. We will discuss this issue by way of two examples, one from a small suburban plant and one from a large city plant.

Example 15.1 (Small plant disturbances) *So what do they look like? In Chapter 6, visualisation was the first technique listed for examining data. Fig-*

ure 15.2 shows four days in the life of a treatment plant. A visual inspection results in some interesting facts:

- *a very distinct daily cycle,*
- *flowrate variations from 2 to 16 Megalitres per day,*
- *total COD variations from 250 to 1100 mg/l,*
- *a high correlation between flowrate and COD (COD loading from 500 to 17,600 kg/d),*
- *total N from 40 to 150 mg/l,*
- *total P from 5 to 25 mg/l, and*
- *a less distinct but significant N and P correlation with flowrate.*

These are not atypical values for small plants. In any persons language, they are unusually severe disturbances to a process plant, and especially to a process plant that relies on biological processes and that is supposed to meet very strict effluent standards.

But what quantitative measures can help us examine disturbances, they are not always so dramatic. Of course we can look at average or mean values:

Flowrate	= 5.1 Ml/d
Total COD	= 520 mg/l
Total N	= 73 mg/l
Total P	= 12 mg/l

But that doesn't tell us a lot. We can look at the standard deviation which tells us a bit more because they are pretty big compared to the means:

Flowrate	= 5.1 ± 2.8 Ml/d
Total COD	= 520 ± 200 mg/l
Total N	= 73 ± 27 mg/l
Total P	= 12 ± 4 mg/l

We can also look at the distribution of values about the mean (Figure 15.3 produced using `resdist.m` *from the MATLAB Toolbox). The data is definitely not normally distributed or Gaussian. One might think that, because a sewer collects from a large number of small events, disturbances might be normally distributed. But of course those small events are not random, so we definitely have coloured noise rather than white noise (we discussed these types of noise in Chapter 7). In fact you can examine Figure 15.2 and clearly see many habits of people. The data comes from a small plant in a relatively new suburban setting and is for Thursday to Sunday. You can make your own interpretations, a specially interesting one to speculate upon is the phosphorus peak at about 11 am Saturday - washing the clothes, the car, the dog?*

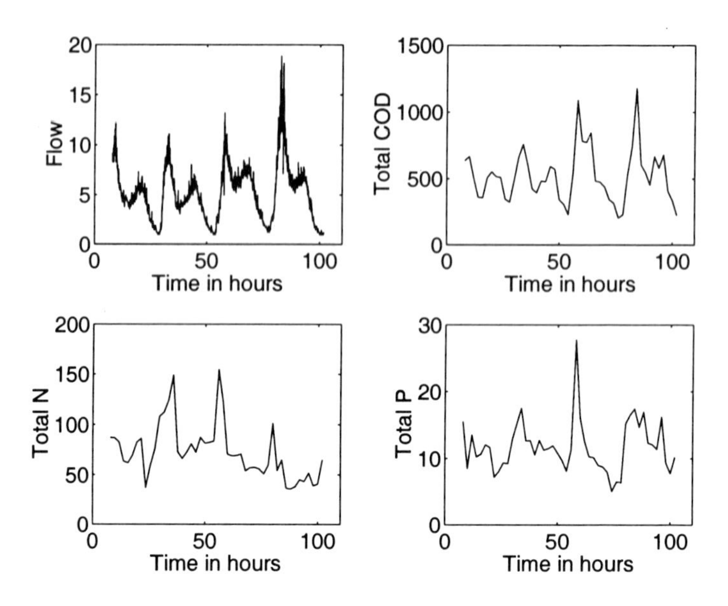

Figure 15.2: Typical influent disturbances.

More seriously, all these common quantitative tests have so far not brought out some of the obvious features from our initial heuristic examination and some of the less obvious ones. For example, using `trend.m` *from the MATLAB Toolbox, we see that Total N has a significant downward trend. You can see this from a second close look at Figure 15.2 though it is a feature of the data that could have been easily missed.*

However, now let us look at the frequency spectrum of the data shown in Figure 15.4, produced by `frspec.m` *in the MATLAB Toolbox. Here we see very clearly the periodic nature of the data. There are very distinct peaks corresponding to 24 hour (*$=2\pi/24 = 0.26$ *rad/h) and 12 hour (*$=2\pi/12 = 0.52$ *rad/h) periods. There is also less significant periodicity at 6 hours (1.05 rad/h) and Total N shows this at about 7.5 hours (0.84 rad/h) and at 75 hours (0.08 rad/h), the latter being influenced by the trend downwards in days three and four.*

We can also obtain the correlation coefficients using the MATLAB Toolbox function `shcorr.m` *shown below:*

	Flow	COD	Tot N	Tot P
Flow	1.00			
COD	0.83	1.00		
Tot N	0.33	0.36	1.00	
Tot P	0.65	0.75	0.45	1.00

This tells us the story of the correlations. Correlation between flowrate and

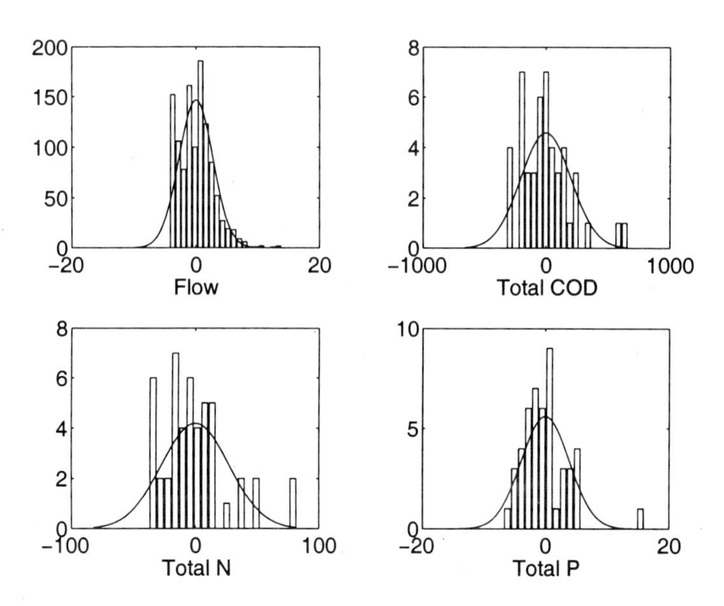

Figure 15.3: Disturbance amplitude distributions.

Total COD, flowrate and Total P and of course between Total COD and Total P. There is less of a correlation between Total N and the rest. Again, in this data we can see that, but at least we can now quantify it, and in other data sets it is not always so visually obvious.

We will now look at a second example before we make some general comments and guidelines.

Example 15.2 (Large plant disturbances) *Again, visualisation is the first step. Figure 15.5 shows the time series for influent flowrate to one of ten bioreactors, influent COD and influent ammonia (NH_4) from on-line instrumentation. Features apparent by examining the time series are:*

- *again a very distinct daily cycle,*
- *flowrate variations from 350 to 700 m^3/h ,*
- *flowrates at discrete rates due to a number of fixed speed pumps,*
- *COD variations from 120 to 210 mg/l,*
- *a correlation between flowrate and COD but shifted in time,*
- *NH_4 variations from 20 to 45 mg/l,*
- *again a weaker shifted correlation between flowrate and NH_4,*
- *several probable outliers in the COD measurement,*
- *loss of the NH_4 analyser readings after about 160 hours, and*
- *a short complete loss of influent at about 135 hours.*

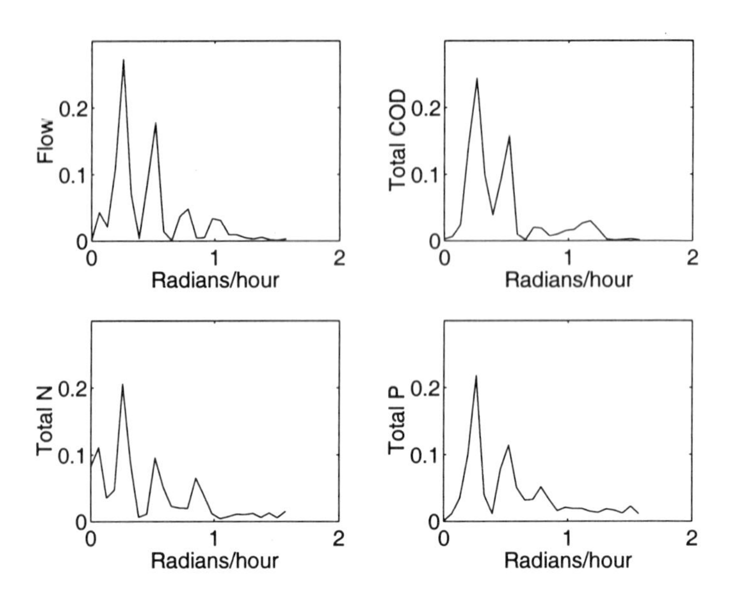

Figure 15.4: Influent frequency spectra.

Flowrate	$= 470 \pm 102$ m^3/h
COD	$= 155 \pm 26$ mg/l
NH_4	$= 26 \pm 9$ mg/l

The means and standard deviations of the time series (the first 155 hours) are:

Compared to the small plant, flowrate and COD disturbances were on average only half the size while NH_4 disturbances were on average the same amplitude. The lower amplitude variation would be at least partly due to the large plant's much larger sewer system, which itself helps dampen the disturbances. Nutrient levels were only a third of those in the small plant, likely due to different patterns of use and sewer collection policies (the data came from different countries).

Figures 15.6 to 15.8 show amplitude and frequency analysis of the flowrate, COD and NH_4 respectively. The flowrate amplitude distribution showed the effect of the pump control policy, COD amplitude is almost Gaussian, while NH_4 amplitude is distinctly bimodal. All the frequency spectra again show a very strong 24 hour ($=2\pi/24 = 0.26$ rad/h) peak with much weaker 12 hour ($=2\pi/12 = 0.52$ rad/h) peaks in flowrate and NH_4.

The much weaker 12 hour peak compared to the small plant (compared with Figure 15.4) is probably due to the mix of industrial, commercial and household effluent to the large plant while the small plant accepted predominantly household effluent.

A correlation analysis of the data using the MATLAB Toolbox function `shcorr.m` *showed the following results:*

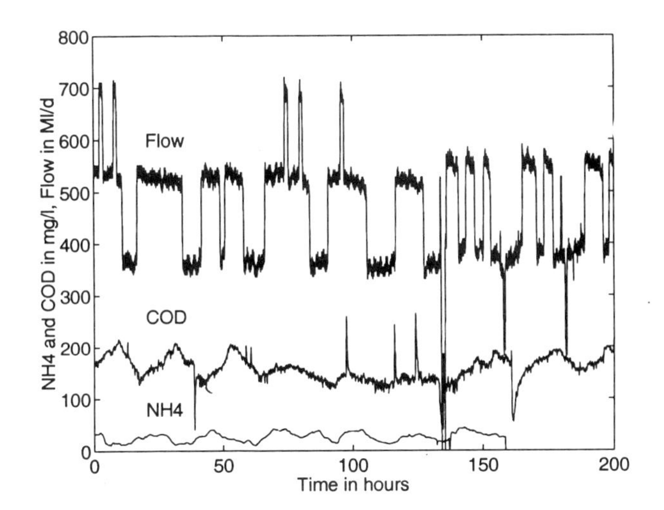

Figure 15.5: Influent time series.

	Flow	COD	NH_4
Flow	1.00		
COD	0.60	1.00	
NH_4	0.51	0.37	1.00

As in the smaller plant, the flowrate correlation with COD is better than that with NH_4. There are significant phase shifts in the time detected by the correlation. The COD lags the flowrate by 3 hours, probably due to the location of the COD analyser after the primary sedimentation. The NH_4 behaviour is more complex. It was out of phase with COD for the first few days, and then in phase for the remainder of the time. Ammonia behaviour was also the "odd man out" in the case of the smaller plant.

Hopefully we have shown you some new ways of looking at your disturbances, and quantifying some of their characteristics. Finally, a few general hints on analysing the data:

- Be aware of the guidelines and analysis possibilities for time series, for example those in Chapter 7, especially regarding sampling frequency.

- Some quantitative characteristics can be sensitive to glitches in the data, see Section 6.5. Fix up missing points, remove or filter outliers and excessive noise, remove trends, etc.

- Use all the sophisticated mathematical tools at your disposal, but temper them with a good deal of *common sense* and experience.

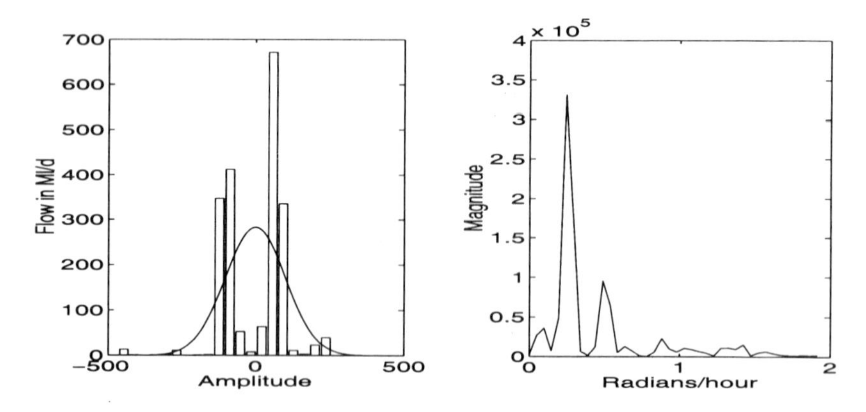

Figure 15.6: Influent flowrate characterisation.

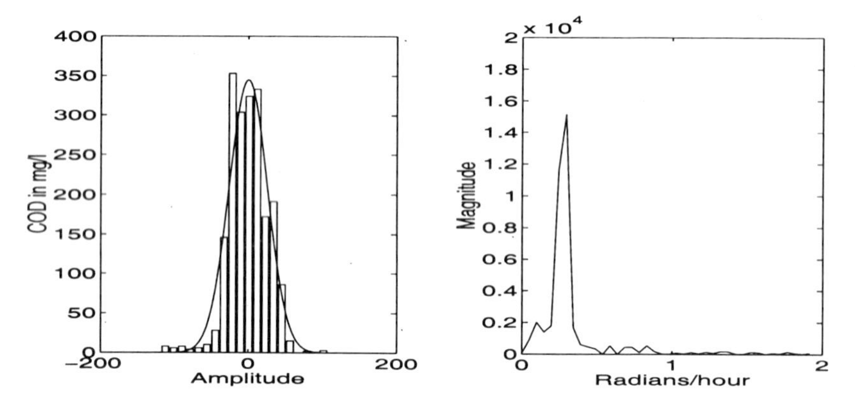

Figure 15.7: Influent COD characterisation.

15.3 Modelling Disturbances

You may remember a discussion in Section 7.6, where we said we can often derive a model for coloured noise which can give us a limited forecasting capability. This would be really nice, because obtaining all the influent time series data on a regular basis is a rather costly exercise.

Of course, in this case there is little sense in using a "black box" time series model as we did in Chapter 7. We know from our understanding of the system creating the disturbance, and from the frequency spectra, that we have very strong periodic signals. The principal periods are 12 and 24 hours. From the frequency spectra we can postulate a general disturbance model of the form:

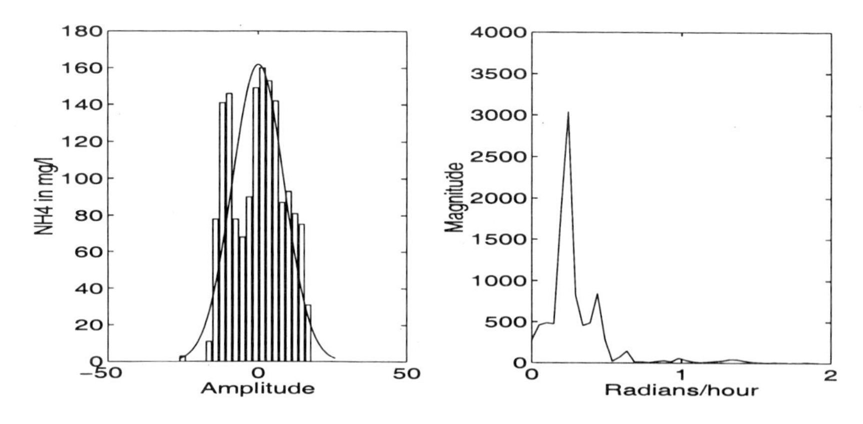

Figure 15.8: Influent NH_4 characterisation.

$$y(t) = p_1 + p_2 \left[\sin\left(\frac{2\pi}{24} t + p_4 \pi \right) + p_3 \sin\left(\frac{2\pi}{12} t + (p_4 + p_5)\pi \right) \right] \qquad (15.1)$$

where t is the time in hours. MATLAB Toolbox function `fitdist.m` fits the five parameters p_1 to p_5 or fits the parameters p_1 and p_2 (given p_3 to p_5) to a given time series of data.

Example 15.3 (Small plant disturbance models) *If we fit the five parameters, p_1 to p_5, to the flowrate data we get the values (5.1, -2.4, -0.9, 1, -0.85). Parameter p_1 is the same as the mean of the flowrate data set (as you might expect heuristically) and the parameter p_2 is 0.86 times the standard deviation of the data set (we could heuristically expect such a relationship to be reasonable). Keeping the same basic shape by fixing p_3 to p_5 and fitting the parameters p_1 and p_2 for the three nutrient concentrations, we obtain p_1 always as the mean and p_2 as 0.76, 0.33 and 0.60 times the standard deviation respectively. If we take the average of these factors and set the parameter p_2 as 0.64 times the standard deviation for all four measurements, we obtain the model predictions shown in Figure 15.9.*

The percentages of the total variation explained by this model were 63, 51, 3 and 35 respectively. These values were calculated using Equation 7.15:

$$V_{explained} = 100 \frac{\Sigma_i (y_i - \bar{y})^2 - \Sigma_i (y_i - y_{p,i})^2}{\Sigma_i (y_i - \bar{y})^2} \qquad (15.2)$$

where $y_{p,i}$ are predicted values, y_i are measured values, and $\bar{y}$ is the mean of the measured values.

If the parameters p_1 and p_2 were individually fit to the different data sets, the percentage of total variation explained was 67, 52, 11 and 35 respectively. An *arbitrary* weekend *factor could be applied to flowrate and total COD to improve their figures still further.*

Certainly this type of model is sufficiently good that it could be used to predict feedforward control of the process to compensate for at least some of the regular perturbations, almost half the total for flowrate, total COD and total P. We will show the effects of doing this in Section 18.1.

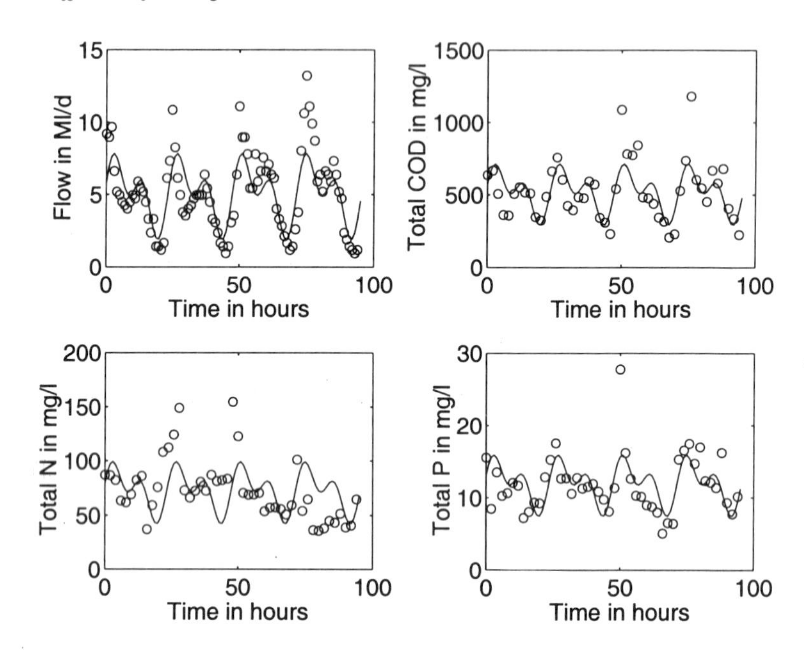

Figure 15.9: Small plant disturbance model predictions.

Example 15.4 (Large plant disturbance models) *A similar fitting exercise was performed with the data shown in Figure 15.5 after filtering to remove the discrete pumping rate effect, outliers and excess noise. The parameters of the noise models selected are shown in Table 15.2. The first and second parameters were set equal to the means and standard deviations of the filtered signals.*

Data	p_1	p_2	p_3	p_4	p_5
Flowrate	458	72	0	0.4	0
COD	153	20	0	4.0	0
NH_4	26	8	0	0.4	0

Table 15.2: Example 15.4 model parameters.

The change in the fourth parameter is the phase shift due to the different sampling locations discussed in Example 15.2 above.

These models explained 73, 46 and 33 per cent of the total variation in flowrate, COD and NH_4 respectively. Again, this type of model is sufficiently good that it could be used to predict feedforward control of the process to compensate for at least some of the regular perturbations.

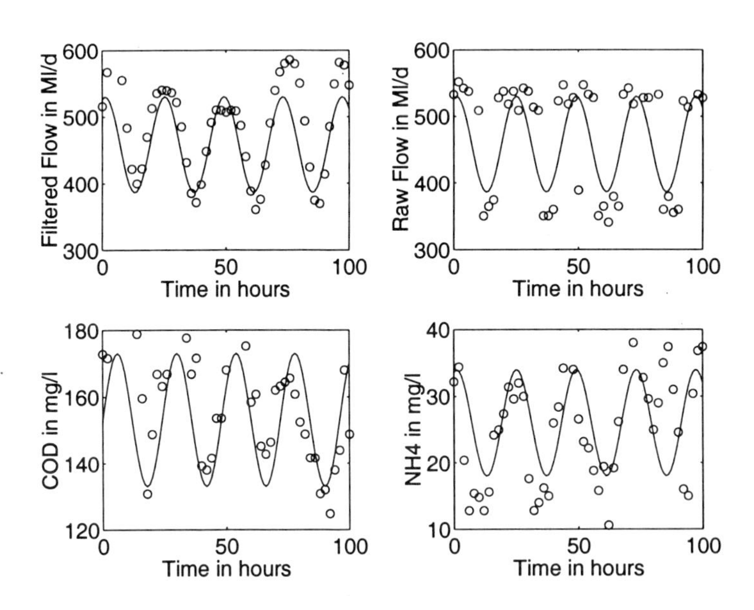

Figure 15.10: Large plant disturbance model predictions.

Finally, a few general hints on modelling with real data:

- You will not be able to model the unexpected no matter how hard you try. Choose a data set without many unexpected events.
- As in the analysis, be aware of the guidelines and analysis possibilities for time series, for example those in Section 7.6, especially regarding sampling frequency.
- Before fitting time series data, tidy it up, Section 6.5. Fix up missing points, remove or filter outliers and excessive noise, remove trends, and use a filter to *round off* effects such as the pump switching (as we did in Figure 15.10 above). Least squares fitting is sensitive to these sorts of *abnormalities.*
- Again, use all the sophisticated mathematical tools at your disposal, but temper them with *common sense* and experience.
- Remember that the goal is to predict *some*, as much as you can, of the regular pattern in the data.

- If you can afford to measure it, then do so. Measurements generally beat predictions hands down. Never-the-less, predictions could be a check on gross failures of the instruments. Never believe your measurements either until you have verified them.

15.4 Disturbance Rejection

What do we mean by disturbance rejection? Everyone will be familiar with the disturbance rejection performed by their car suspension system. Consider a car moving on a road with an *oscillating* road surface. The wheel of the car follows all the ups and downs of the road. The car body is connected to the wheels via the suspension, which can be described as a spring and a damper. For a very slow speed, the car body will follow exactly the movement of the wheel. For a very high speed the wheel will move like a sinewave, while the body does not move at all.

Fortunately most processes also act as reasonable attenuators of disturbances, and part of the task of their control systems is to improve this attenuation still further. It is important when processes are designed to keep the disturbance attenuation goal in mind and to design the process itself, and the manipulated variables used by the control system, accordingly. Of course there are other goals in design and operation that may conflict and appropriate compromises are always needed.

Here we will look at how we might characterise disturbance rejection and the basic plant design factors that affect this performance. In subsequent chapters we will look at how the control system contributes to disturbance rejection.

15.4.1 Measuring Performance

We can characterise the disturbance rejection achieved by a process with or without its control system by a simple attenuation factor.

A global figure might be the average deviation in the effluent divided by the average deviation in the influent. This raises one important factor in characterising disturbance rejection. Which effluent measurement and which influent measurement do we use? There are of course many combinations. Some of these make sense as a measure, effluent flowrate and influent flowrate, and some of these are largely nonsense, effluent temperature and influent phosphorus. *The two measurements must be significantly related.*

A second factor influencing disturbance rejection measures is that they are frequency dependent. We saw this with our car suspension example we mentioned just above. Most natural systems respond more to low frequency inputs than they do to high frequency inputs, including you and me. If you get lectured to by someone who speaks very quickly, you will understand much less than if the same ideas were presented slowly. This naturally higher attenuation of higher frequency inputs fortunately filters out most input noise to processes, but it allows low frequency operator actions to have an effect. It follows that *a disturbance rejection measure at a number of frequencies* will tell us much more about the effectiveness of the process and its control system.

The most common performance measure used is the simple gain curve plotted against frequency, part of the Bode plot for those of you who have used and remember some control. You may have one of these plots for your stereo amplifier as well. We will look at some process examples immediately below.

15.4.2 Process Design Factors

The basic design of the process is the first important step in controlling the effects of disturbances. Flow through a variable inventory is the most common attenuating factor in processes. Good use of this can reduce or even eliminate the need for some controls. It is also possible through bad design to make a process more susceptible to disturbances. An understanding of design factors is important both when the plant is built and when control systems are designed.

Example 15.5 (Hydraulic disturbance attenuation) *Consider inlet and outlet flowrates through a vessel, as we did in the introductory modelling chapter, Chapter 3. We showed that a constant volume tank had the outlet flowrate exactly equal to the inlet flowrate. This means it will not attenuate any inlet disturbance. It's gain will be equal to one at all frequencies. On the other hand, if we have an overflow weir then we have a variable volume. We saw from Exercise 3 that outlet flowrate responded more gradually to inlet flowrate disturbances. Our curves of the gain between inlet and outlet flow changes would look like Figure 15.11. The figure shows curves for the constant volume tank (horizontal line) and tanks with overflow weirs with 5 and 15 minute time constants. The vertical lines are for periodic disturbances with periods of 5 minutes and 3 hours.*

We can see that you get better attenuation for higher frequency disturbances and larger time constants. The time constant for a weir is related to the weir type (value of k), the cross-sectional area of the tank (A) and the height (h) of liquid flowing over the weir (see Section 4.5)

$$\tau \propto \frac{A\,h}{k} \tag{15.3}$$

Rectangular weirs ($k = 1.5$) are better than triangular weirs ($k = 2.5$) and a larger height of liquid over the weir (shorter or fewer weirs) or a larger tank cross-section give better disturbance rejection.

We also get attenuation of nutrient concentration disturbances due to a tank volume. The curves are the same shape with the time constant equal to the residence time. Longer residence times are better for disturbance rejection, but of course the increased volume is more costly too. It follows that the primary sedimentation tank itself is the first disturbance rejection device.

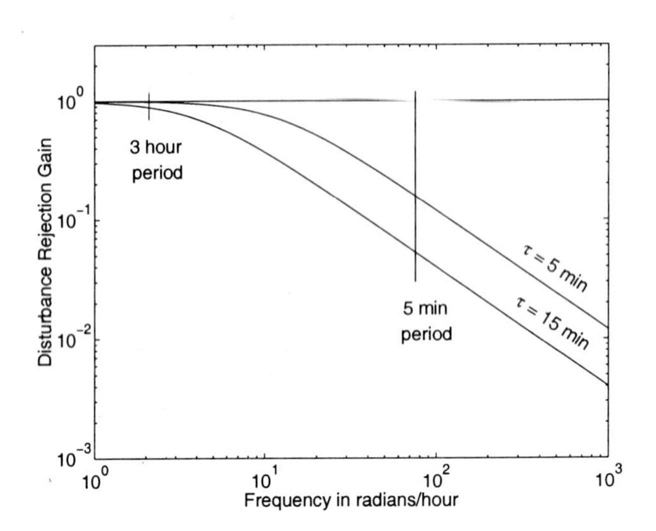

Figure 15.11: Gain curves.

Example 15.6 (Sedimentation tank disturbance rejection) *Consider the large plant data, Figure 15.5. Figure 15.12 shows Total COD and Total P nutrient data corresponding to before and after the primary sedimentation tank. The standard deviations are shown in Table 15.3.*

Variable	Raw	Settled	Settled / Raw
Total COD	190	64	0.34
Total P	4.0	2.5	0.63

Table 15.3: Attenuation by primary sedimentation tank.

The disturbance rejection is clear from Figure 15.12 and the standard deviations. The different attenuation rates probably result from the higher frequency content of the Total COD disturbance compared to the Total P disturbance. This can be seen from Figure 15.12 as well as from the frequency spectra in Figure 15.13. The even higher frequency spikes *are almost completely attenuated.*

There is also a change in mean value due to the nutrient content of the settled sludge removed.

If we took a residence time of five hours for the primary settler, and assumed periods of 12 hours and 24 hours for the two disturbances, we would predict from theory that the attenuation factors would be 0.36 and 0.61, compared to the figures of 0.34 and 0.63 from the data.

If you have tanks in series, then the attenuation factors are simply multiplied together. If we assumed the complete plant might be the equivalent of four such

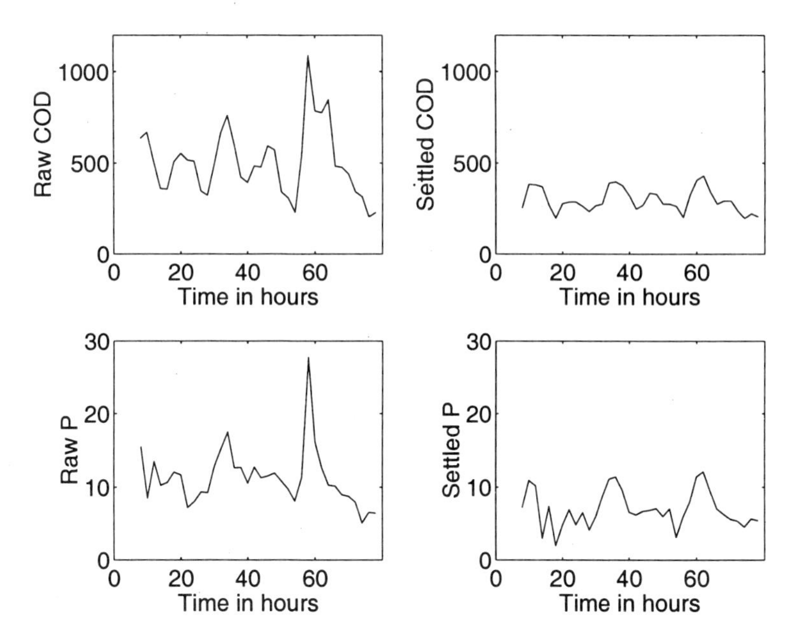

Figure 15.12: Disturbance rejection by sedimentation tank.

tanks, then the attenuation factors would be $0.344^4 = 0.017$ and $0.634^4 = 0.14$, ignoring the effects of the biological activity and the recycles which we will look at in later chapters. These attenuation factors may sound good, but are nowhere near that required by effluent regulations.

There is also an increasing understanding of the effects of hydraulic disturbances on the performance of the secondary clarifiers. Hydraulic load changes and changes in sludge withdrawal rate have a marked effect on effluent suspended solids as discussed in Section 14.4.4 and Section 18.6.

We also need to take care in our designs that we do not introduce disturbances ourselves. The design of influent pumping systems and return streams such as filter backwash can cause major hydraulic and nutrient concentration disturbances.

Example 15.7 (Backwashing disturbances) *The left plot in Figure 15.14 shows the filter backwash flow recirculated to the plant influent in a treatment plant.*

Of special note is the right plot in Figure 15.14 which shows the effects of the disturbance on both the effluent flowrate and the residual chlorine in the effluent. Without the disturbances the residual chlorine concentration should be quite constant. The correlation with the filter backwash flowrate is obvious.

You can also see the inherent disturbance rejection in the flowrate due to the plant volumes, with the effluent disturbance about four times smaller than the backwash flowrate changes.

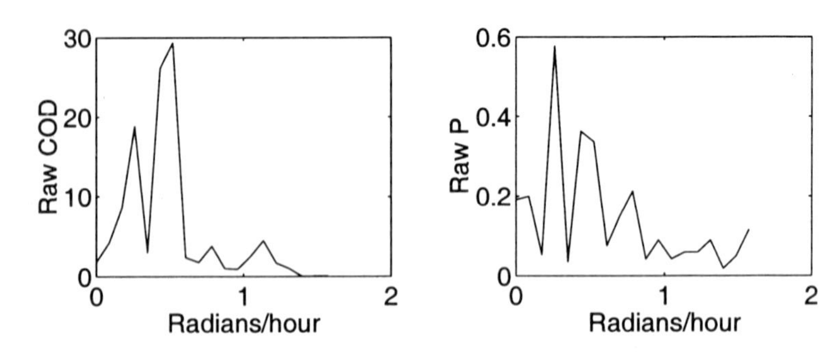

Figure 15.13: Frequency spectra of raw disturbances.

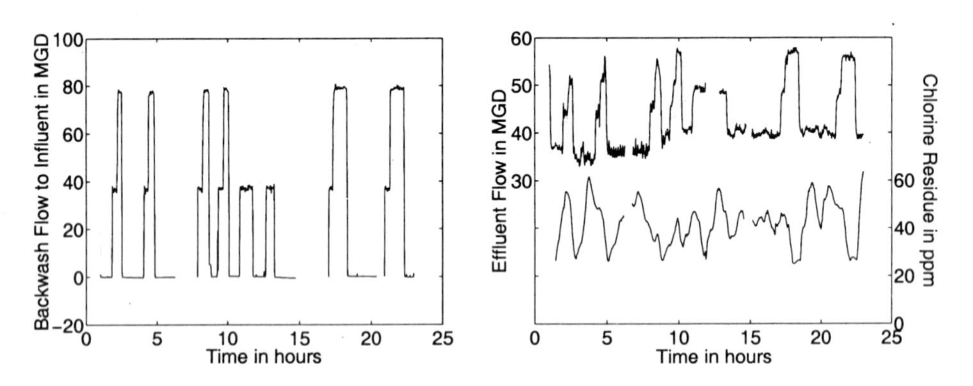

Figure 15.14: Disturbances from filter backwashing.

15.5 Summary

We have tried to *set the scene* for our look at control by looking at the general operating goals of wastewater treatment plants and at some general strategies for structuring their control systems.

Then we have looked in some detail at disturbances. Compensating for and mitigating the effects of disturbances are the first major task of the control system. It is therefore important to understand the disturbances, and how we might characterise them, and even model and predict them.

In Chapter 16 we will look at manipulated variables and constraints. These define what scope we have for control and for rejecting the effects of disturbances. Then in Chapters 17 and 18 we look in some detail at control strategies and implementation. In Chapter 19 we will describe sequencing control, that is an essential part of all process control. In Chapter 20 we step back and look at the whole plant from a broader control, optimisation and management perspective. Finally in Chapter 21 we examine some methodologies for determining the benefit of investing in control systems. We will also look at some of the problems in measuring the return after the investment has been made.

15.6 Further Reading

- Disturbance rejection due to different weir constructions has been considered by Olsson & Stephenson (1985) and in Olsson *et al.* (1986).
- Detection of unexpected disturbances has been discussed in Rosén & Olsson (1997) and in Rosén (1998).
- By making use of sewer system measurements and models the hydraulic disturbances to a treatment plant can be predicted, see for example Nielsen *et al.* (1996), Carstensen *et al.* (1996), Carstensen *et al.* (1997), Bechmann *et al.* (1998) and Ji *et al.* (1996).
- Early warning systems for toxicity loadings have been examined by Temmink *et al.* (1993).

15.7 Exercises

1. The file `Exercise_15_1.mat` in the Exercises directory contains influent flowrate, COD and ammonia data to a full-scale wastewater treatment plant with a sampling interval of six minutes. Characterise the data and attempt to construct a disturbance model for predicting nutrients from flowrate measurements.

2. The file `Exercise_15_2.mat` in the Exercises directory contains suspended solids, ammonia and nitrate data to the sand filtration unit in a full-scale wastewater treatment plant with a sampling interval of six minutes. Characterise the data and comment on the likelihood of successfully constructing a disturbance model to predict nutrients from suspended solids measurements.

3. Using the nutrient data in the file `Exercise_15_2.mat` in the Exercises directory (columns 2 and 3), and knowing that the filters have a residence time of 1.8 hours, predict the magnitude of the variations in nutrient concentrations leaving the filter. The gain curve for concentration changes passing through a mixed tank can be constructed from the equation:

$$GAIN = \frac{1}{\sqrt{\omega^2 \tau^2 + 1}} \tag{15.4}$$

where ω is the frequency in radians per hour and τ is the residence time in hours.

4. We have listed possible sources for disturbances. List in more detail possible disturbances in the:

    - aerobic zone,
    - anoxic zone,
    - aerobic zone, and
    - settler.

5. List possible differences in disturbance patterns in different plant types, such as:

    - small municipal plants with mostly domestic waste,
    - small municipal plants with a major industry connected,
    - large municipal plant with mostly domestic waste, and
    - large municipal - industrial plant.

6. List the potential for variable speed drives in various plant locations.

7. List possible disturbances generated by the operation of the plant.

Chapter 16

Manipulated Variables

The goal of this chapter is to acquaint you with the manipulated variables available in biological wastewater treatment plants to exert possible control actions. We will give a qualitative measure of their importance for manipulation of the plant and also discuss some of the constraints on their effectiveness.

There are quite a few variables which can be used to manipulate biological wastewater processes. Still, as we will see, the possibilities to control the plant in a flexible way are quite limited. However in many plants there may be potential to make a better use of the manipulated variables that are available. Here we will categorise them in the following groups:

- hydraulic, including sludge inventory variables and recirculations, Section 16.1;
- additions of chemicals or carbon sources, Section 16.2;
- air or oxygen supply, Section 16.2; and
- pre-treatment of influent wastewater, Section 16.2.

There are several other manipulated variables in a plant that are related to the equipment and to basic control loops in the process, such as flow controllers, level controllers, etc. They are not included in this discussion.

Firstly, we will systematically discuss the different manipulated variables from a qualitative point of view. Which controlled variables are influenced and the time scale of the influence will be discussed. Here we will list controlled variables that can be directly measured by or estimated from plant measurements. The results will be summarised in an *incidence matrix* that will qualitatively describe the input-output couplings and their gross time-scales.

Then there is a discussion based on operational and empirical experiences of the different constraints that are present for controlled variables, Section 16.3. Finally some of the practicalities involved are presented, Section 16.4.

16.1 Hydraulic Variables

The majority of the manipulated variables change the hydraulic flow patterns through the plant. The different flowrates will influence the retention times in the different units. Moreover, the rate of change is crucial in many parts of the treatment plant, since it influences the clarification and thickening processes. The hydraulic flows also determine the interaction between different unit processes. Here we will classify the hydraulic manipulated variables into four major groups:

- variables controlling the influent flowrate;
- variables controlling the sludge inventory and its distribution;
- internal recirculations within the biological process;
- external recycle streams, influencing the interactions between different unit processes.

In this category we will also include the control of the phase length in sequential batch reactors, since this is equivalent to controlling the retention time of a continuous unit.

16.1.1 Influent Flow Control

The influent flowrate to the activated sludge system can be manipulated in various ways:

- *Pumping of the influent flow*: In order to obtain reasonable conditions in the system the pumping has to be smooth. Continuously variable pumping is necessary. Otherwise the settler unit is easily upset by too large rates of change of the flowrate.

- *Sewer control*: By operating the sewer network as an equalisation system, the flowrate to the plant may be smoothed. This will be further discussed in Chapter 20.

- *Equalisation basins*: In smaller plants, where the load variations are often proportionally larger, it is useful and common to have equalisation basins for the influent flow. More will be said in Chapter 20.

- *Flow splitting*: Many plants are designed with two or more parallel aeration basins. The flow splitting process is crucial if the load is to be evenly distributed. This is often not the case, which results in apparent overloading in some parts of the system. In many plants the flow splitting is made by a fixed arrangement of channels, which may not at all guarantee that the real flow is split correctly. If flow splitting should be guaranteed, then the flowrates have to be measured and the individual flowrates controlled (see Chapter 17).

- *Bypassing*: This should be a manipulated variable in the sense that bypassing should never occur, unless it is ordered. It has to be compared with the alternative of not performing bypassing, and has to be based on some quantitative calculation with a suitable time horizon.

All the different modes of influent flow control are simply different ways to make the control authority larger. In other words, sewer or equalisation or bypass control all contribute to making it easier to obtain a smooth flowrate into the plant. Their goal is disturbance rejection.

A smooth variation of the flowrate is crucial for the secondary clarifier operation. It requires not only variable speed pumping at the operating level to avoid disturbances, but an adequate storage capacity in wet wells or upstream tanks to damp disturbances which cannot be avoided. Poor pumping control can deteriorate the plant performance considerably. The main reason is that usually the clarifier is quite sensitive to positive flowrate changes. Flowrate changes will propagate through the system in 15 to 30 minutes. This was modelled in Section 4.5.

From the discussion above we may consider the influent flowrate both as a disturbance and as a manipulated variable. The purpose of an advanced influent flow control is to minimise the impact of a flowrate disturbance. In the next section, we will look at the (hydraulic) impact of a flowrate change on the aeration basin and settler performance.

16.1.2 Sludge Inventory Control

The sludge inventory can be controlled primarily by three manipulated variables:

- the waste sludge flowrate;
- the return sludge flowrate; and
- step feed flowrates.

Manipulation of the waste sludge flowrate is used to control the total inventory of sludge in the process. Since the total inventory is a function of the total growth rate of organisms, it is used to control the sludge retention time or the sludge age. This manipulated variable will influence the system in a time scale of several days or weeks.

Manipulation of the return sludge flowrate is used to distribute the sludge between the aeration basins and the settler units, or between the acidogenic and methanogenic reactors in two-stage anaerobic systems. Recycle from the settling stage is an important variable for obtaining the right operating point in the reactors, but seldom useful for control on an hour-to-hour basis. Some systems are supplied with several feeding points for the return sludge. This has potential for sludge redistribution for certain loads, such as toxic loading.

A *combination* of different recycle streams may be important. In systems with chemical precipitation, sludge from the secondary settler may be combined with chemical sludge from a post-precipitation settler unit. In that way the floc properties may be influenced and the chemicals better utilised for phosphorus removal.

By controlling the step feed in an activated sludge plant, the sludge within the aeration basin can be re-distributed, given the proper amount of time. As a special case of step feed control one will obtain a *contact stabilisation* structure. Also the return sludge may be fed back not only to the inlet part of the aeration basin, but into different feeding points along the basin, a so-called *step return sludge control*. This may prove to be one efficient way of preventing bulking sludge.

16.1.3 Recirculation Streams

Internal or external recirculations provide couplings between the different units of the plant. Some of the recirculations have very large flowrates, such as the recirculations of nitrate in a pre-denitrification plant. Other streams may have extremely large concentrations, such as supernatants from the sludge treatment. They were discussed in terms of disturbances in Chapter 15. Most of them can be manipulated purposefully to achieve a better plant performance.

Some examples of recirculations within the liquid train are:

- *Recirculation of nitrate*: Having a system with pre-denitrification, it is required to recirculate the nitrate-rich water from the outlet of the nitrification reactor. The consequences of changing this flowrate were discussed in Section 14.4.2. In particular, the oxygen contained in the recirculated water may limit the denitrification rate. Even if the low rate is large, it is believed that it has a minor influence on the settler performance. The energy (pumping) cost related to this is relatively small, since the lift is small.

- *Recirculations in biological phosphorus removal*: In a bio-P system there are three types of reactors, anaerobic, anoxic and aerobic. Depending on the design, there are many recirculation patterns in such a plant (Chapters 2 and 4).

- *Recirculation in two-stage anaerobic systems*: In this case the recirculation helps to keep the methanogens washed out of the acidification stage and returns pH buffering capacity to reduce caustic usage.

Recirculations from the sludge treatment to the liquid process train can present difficulties if they are not made properly. This can occur quite quickly. If a deep bed filter is backwashed, then the hydraulic load is noted within a few minutes. The recycle streams can be considered as *controllable disturbances* to the reactor-settler system. They have to be manipulated so that their detrimental impact is minimised.

- *Supernatants*: Recycle streams of supernatants from dewatering of sludge often contribute significantly to the load of the aerators. They may appear in an hourly timescale as concentration disturbances.

- *Backwashing*: Recycles from backwashing of filters often will contribute significantly to the hydraulic load. Therefore, it is crucial to backwash filters at times when the load from other sources is low, or to use an equalisation tank.

16.1.4 Phase Length Control in Sequential Batch Reactors

The most common way to design a plant to perform biological nitrogen and phosphorus removal is to allocate a fixed volume to the anaerobic, anoxic and aerobic processes. In sequential batch reactors (SBR) the distribution of anoxic and oxic capacity is based on phase lengths or time periods. This means that a plant can change readily from a full nitrification mode to a full denitrification mode. This offers interesting potential for control to compensate for load variations.

Favourable full scale experiences have been reported. The time for denitrification and nitrification can be easily varied, based on on-line measurements of the relevant variables. The basic principle is that a nitrification phase is ended when ammonium is low and the denitrification phase is ended when nitrate is low. There are full scale commercial systems for N removal and P-N removal, and they will be further discussed in Section 19.5.

16.2 Other Manipulated Variables

In an activated sludge system selection mechanisms for controlling the viable bacteria in the system are often discussed. They are classified in the groups:

- electron acceptors (oxygen or nitrate);
- substrate;
- thickening or clarification properties;
- temperature;
- growth rate; and
- free swimming organisms.

Not all of these can be manipulated. Only the electron acceptors (mostly oxygen) and (to some extent) the thickening and clarification properties can be controlled (by chemical precipitation). The other selection mechanisms are considered as disturbances. In this section we will discuss manipulated variables that are not related to flowrates or hydraulic retention times. They are described in terms of added chemicals or nutrients and air flow supply variables. Finally, all kinds of pre-treatment possibilities are listed separately.

16.2.1 Chemical Addition

Chemicals are added for two different reasons, to achieve chemical precipitation for phosphorus removal, or to form a better settleability of the sludge.

- *Flocculant addition for phosphorus removal.* Ferrous, ferric or aluminium salts are added to obtain chemical precipitation by forming insoluble phosphates (Chapter 2). There are several ways to feed the chemicals into the system, and a great deal of experience has been collected, for example in Sweden. In order to comply with the P requirement in the 1970s post-precipitation was prescribed. Furthermore, sludge was usually stabilised in anaerobic digesters. A number of alternative process schemes were subsequently developed. Pre-precipitation was shown to be an alternative to post-precipitation. Apart from removing phosphorus, an increased portion of the organic material was also removed simultaneously in the mechanical treatment step. This in turn decreased the load on the biological treatment step and could, at least theoretically, decrease the amount of energy needed for aeration and at the same time increase the gas production in the digester.

  Simultaneous precipitation was another process alternative that gained wide acceptance. In this case ferrous sulphate, which is quite an inexpensive product, could be used instead of for example ferric chloride. In some cases two point precipitation was also introduced as a means of minimising the amount of or the cost for the precipitation chemicals. The majority of the phosphorus is removed by pre-precipitation or simultaneous precipitation, while a minor dosage is used for post-precipitation to further polish the phosphorus removal. A change in chemical dosage can have quite a fast influence on the floc formation and the settling. In Section 17.7.3 some control schemes for chemical precipitation are discussed.

  On top of the normal use of chemicals for P removal, chemicals can be added to improve sludge settling properties in the secondary settler. Sometimes chemicals are added to the primary settler to reduce the load to the aerator. However, this may sometimes lead to insufficient carbon for the nutrient removal.

- *Polymer addition*: This may be used in emergency situations to avoid major settler failures. On a routine basis it is used for sludge conditioning to improve dewatering properties. In addition, polymers could be used to further enhance the efficiency of the pre-precipitation process.

- *Caustic Addition.* This is used in two-stage anaerobic processes to control the pH which can inhibit the methanogenic microorganisms.

16.2.2 Carbon Addition

Carbon source addition is sometimes needed in denitrification to obtain an adequate carbon/nitrogen ratio in the system (Section 14.4). Too little carbon results in incomplete denitrification, while too much carbon adds a cost for the chemical and the subsequent removal of it. The time scale of such an operation is related to the retention time of the denitrification.

- *For a pre-denitrification system* carbon is usually supplied via the influent wastewater. Still this may be insufficient during low load periods, so that some carbon source has to be added.
- *In a post-denitrification system* a carbon source (such as methanol or ethanol) always has to be added. Then the problem appears how to adjust the dosage to the carbon need, without extensive measurements.

The form of the carbon is important in biological phosphorus removal. If the amount of RBCOD in the influent wastewater is insufficient, some must be added. This can be done from external sources, methanol, ethanol or glucose wastes, or by pre-fermentation of the primary sludge or the influent itself. The latter can also occur in the sewer in the right circumstances.

16.2.3 Air or Oxygen Supply

Dissolved oxygen (DO) is a key variable in activated sludge operation, as already noted in several chapters. From a biological point of view, the choice of a proper dissolved oxygen setpoint is crucial. The dynamics of the dissolved oxygen is such that the DO can be influenced within fractions of an hour (Chapters 4 and14). We just list some of the key factors related to the DO supply:

- Total air supply,
- DO setpoints,
- DO setpoint compromise in a pre-denitrification system,
- DO spatial distribution (profile): Without individual air flow measurements and feedback control over the valves it is difficult to realise an accurate air flow spatial distribution.
- Pure oxygen addition: Some systems work with pure oxygen instead of air. Sometimes pure oxygen is used in conventional air systems to add further DO to the system under extreme load situations. Since the DO saturation concentration depends on the partial pressure of oxygen, the driving force for oxygen transfer can be significantly increased.
- Controlling the relation between aerated and anoxic zones: A common design in recirculating systems is to make a couple of subreactors so that they can be either anoxic or aerobic. A more versatile way to change the relative ratio of N/DN is to use sequencing batch reactors (see Section 19.5), where the relative times for aeration and no aeration determine the N and DN result.

In a typical large scale plant, the nominal dissolved oxygen time constant is about 20-30 minutes. Given an adequate controller it is straightforward to change the DO setpoint. The time it takes can be considered instantaneous in comparison with the organism growth time constants. However, to know which setpoint is the best from a microbiological point of view is a much more intricate question. DO control will be further discussed in Chapters 17 and 18.

The air flowrate is recognised to be of major importance for the whole operation. It is reasonable to assume that a well functioning DO control system should be available. Still, since the energy cost is significant, it is of interest to minimise the air supply.

It is well known that insufficient air supply will influence the organism growth, the floc formation and the sludge settling properties. However, once non-desired organisms are formed, it is not always obvious how to get rid of them by only DO control.

The air distribution can be realised in space (different DO setpoints along the aerator) as in a continuous activated sludge system, or in time as in a sequential batch reactor. It will certainly influence the relation between denitrification and nitrification in the system. The same amount of total air can be supplied with different spatial distributions, which of course will influence the growth.

16.2.4 Pre-treatment of Influent Wastewater

Pre-treatment of influent wastewater is needed in many cases, particularly in industrial wastewater treatment systems. These manipulated variables are not considered control variables in the traditional sense, but are meant to attenuate disturbances to the plant. The most apparent operations are:

- pH adjustment of process streams;
- using equalisation basins (see above) not only for minimising hydraulic peak loads but for more sophisticated chemical pre-treatment; and
- nutrient addition (nitrogen and/or phosphorus) for certain industrial effluents.

16.3 Constraints

All real systems must deal with constraints. The most apparent constraint for a wastewater treatment system is that *all* water has to be taken care of, and bypassing should be avoided. Unsafe operating regimes have to be avoided, so the hydraulic operation has to work reliably at all times. Physical limitations impose constraints, pumps and compressors have a limited capacity and reactors have a limited volume. The traditional way to deal with such control tasks is to supplement a linear (PID type) controller with nonlinear elements such a overrides and selectors (see Chapter 17).

The plant operation present some important *soft* constraints related to environmental and economic concerns. The effluent product is soft constraint (the secondary goal) while the cost of operation is the third goal.

Manipulated variable	Lower limit	Upper limit
Influent flowrate	Capacity demand	Capacity demand Clarifier
WAS flow	Sludge age calc.	Sludge age calc.
RAS flow	Sludge blanket level Sludge retention time in settler	Hydraulic load Sludge dilution
Nitrate recirc. rate	N removal demand	DN capacity
Chem. addition	P removal demand	Economy
Carbon addition	DN capacity	Economy Excess C load
Total air flow	Respiration rate	Economy Too much mixing
Air distribution	No. of independent supply to zones	Economy

Table 16.1: List of constraints.

Earlier we have listed a number of restrictions on activated sludge control. What kind of possibilities do we have? The goal would be to operate the plant within the given constraints in such a way that the effluent is consistently satisfactory.

16.4 Practicability

In previous sections in this chapter we have discussed available manipulated variables and their use for operation and control. In this section we will discuss qualitatively their *usefulness* from a couple of perspectives. The first issue has to do with amplitude. In other words, can the control variable be manipulated enough so that it will have a significant influence on the variables that we are interested in. The second issue has to do with dynamics, that is, in which time-scale the manipulated variable be can used.

The results are condensed in an *incidence matrix*, showing available control variables and disturbances in one dimension and measured and estimated variables in the other. The couplings are qualitative and indicate if there is a notable coupling and, if so, in which time scale.

	Q_{out}	DO_x	DO_{ae}	SS_a	x_r	x_w	M_{tot}	SS_e
Q_{in}	F	F	F	M	M	M	M	F
WAS				S			S	
RAS		M	M	M	F	F		M
N_{circ}		F						
Phase	M			S			S	
Chem				M			S	M
C_{add}								
Air		F	F	S	S	S	S	S

Table 16.2: Incidence matrix, part 1.

	Temp	OUR_c	OUR_n	DN	P_{rel}	P_{upt}
Q_{in}	M	M	M	F	F	
WAS		S	S			
RAS		M	M			
N_{circ}		M	M	F		
Phase		M	M	M	M	M
Chem						M
C_{add}		M		F	M	
Air		F	F	F		F

Table 16.3: Incidence matrix, part 2.

We have not considered disturbances that may appear within the plant, such as sensor or equipment breakdowns. We also assume that the equipment control systems are available for liquid or air flow control. This means that the different flowrates are considered inputs, even if the manipulated variable itself may be a valve position or a voltage to a pump motor or compressor. For these control systems there are several internal measurement variables that are not listed in the incidence matrix, such as air pressure, compressor speed, motor currents, etc. The output variables are both directly measured and estimated variables.

The elements in the matrix (Tables 16.2 and 16.3) denote a qualitative coupling, where the letters F, M, and S indicate the fast, medium or slow range of the dynamics respectively (as denoted in Section 2.4). Since the equipment control systems are not considered here, there is no very fast dynamics noted. The abbreviations are explained in Table 16.4

The incidence matrix does not say anything about the gain between different inputs and outputs. Still an input limitation may mean that there is too little gain from a manipulated variable to make the output controllable from it.

Q_{out}	Effluent flow rate
DO_x	DO conc in anoxic zone
DO_{ae}	DO conc in aerator
SS_a	mixed liquor suspended solids
x_r	return activated sludge conc.
x_w	waste activated sludge conc.
M_{tot}	total sludge mass
SS_e	effluent susp. solids conc.
Temp	water temperature
OUR_c	carbonaceous oxygen uptake rate
OUR_n	nitrogeneous oxygen uptake rate
DN	denitrification rate
P_{rel}	P release rate
P_{upt}	P uptake rate
Q_{in}	Influent flow rate
WAS	waste sludge flow rate
RAS	return sludge flow rate
N_{circ}	nitrate recirculate flow rate
Phase	Phase length in a sequential batch reactor
Chem	chemical dosage
C_{add}	addition of external C
Air	air flow rate

Table 16.4: Abbreviations in incidence matrix table.

16.4.1 Influent Flowrate

Assume a sudden increase in the flowrate. Since the flowrate propagates fast along the plant an increase in the flowrate entering the settler is noted within a fraction of an hour (Chapter 4) (marked with an F in the incidence matrix). This in turn will influence the clarifier operation, and the effluent suspended solids concentration will increase rapidly (if the change is large).

An increasing suspended solids concentration in the effluent will cause a loss of total mass, most probably in an hourly time scale (marked by M). The flowrate increase will cause a dilution (MLSS change) in the reactor, and the solids will be moved to the settler.

Due to the dilution effect the DO concentration will be changed in the reactor. It is obvious that such a change can readily be compensated for by a DO control system.

If the flowrate increase is caused by a rain storm or melting snow, then a significant temperature change may follow (indicated by a separate disturbance variable). The effect then is a combination of dilution, decrease in growth, and loss of solids. Nitrifying organisms, in particular, are very sensitive to temperature changes. A sudden, large drop in temperature may cause the nitrification to stop. If the same temperature drop were to appear slowly (in more than weeks) the organisms could adapt to quite low temperatures, well below 10°C.

If the average flowrate increases, the resulting decrease in hydraulic retention time in combination with a larger load on the clarifier will affect the organism growth, in particular the nitrification.

In either case its concentration will be influenced by a hydraulic load change. The sensitivity of the return activated sludge (RAS) concentration depends on the RAS control. If the RAS flowrate is set proportional to the influent flowrate, the concentration may readily change with flowrate changes. Such a RAS control can in fact cause very unexpected suspended solids disturbances, as shown in Chapter 15.

16.4.2 Waste Activated Sludge Flowrate

As mentioned previously the waste activated sludge (WAS) flowrate will influence the total sludge mass, which is a slow process. This in turn influences the mean cell residence time (sludge age), which of course will have a significant impact on the formation of certain organisms, such as filaments and autotrophs. It is believed (at least to some extent) to influence sludge bulking. The WAS flow can be withdrawn either directly from the aerator or from the thickened sludge. Controlling the WAS flow will be discussed in Section 18.6.1.

16.4.3 Return Activated Sludge Flowrate

Since the return activated sludge (RAS) flowrate influences the sludge distribution between the aerator and the settler, it will (within all the constraints mentioned above) influence the MLSS concentration in a medium time scale, a few hours. A more advanced RAS control has been shown to influence the relation between floc-forming and filamentous organisms, thus preventing bulking.

The RAS flowrate has a direct impact on the sludge blanket height. This was discussed in Chapter 14. Furthermore, since the RAS flowrate is a significant part of the overall influent into the reactor system it may also have a hydraulic impact on the clarifier system. Thus, an unsuitable RAS control will lead to higher risk of sludge loss due to hydraulics.

The RAS sludge can sometimes be stepwise fed back into the aerator. This may be used to influence the relative concentration of floc-forming and filamentous organisms. Still its influence on the plant operation is somewhat unclear. Further discussion on RAS control is found in Section 18.6.2.

16.4.4 Other Control Variables

The influences in different time scales of the nitrate recycling and carbon addition are also summarised in Table 16.2 and have been discussed in Section 14.4.2. The phase length in a sequential batch reactor will be discussed further in Section 19.5. The air flowrate as an input was discussed in detail in Section 14.4.3.

16.5 Summary

We have listed quite a few manipulated variables in a treatment plant. Still many of these cannot be used in existing plants, so the flexibility is limited. Another problem is that the control authority from many of the variables is quite limited, due to constraints. Both hard constraints (upper and lower limits of the amplitude) and soft constraints have to be taken into consideration.

In the next few chapters we will illustrate how the different control variables can be used for the operation of a wastewater treatment plant.

16.6 Further Reading

- The relationship between large influent flowrate changes and the clarifier performance has been explored in Olsson *et al.* (1986).

- Hydraulic flowrate propagation was considered by Olsson & Stephenson (1985).

- A simple quantitative scheme for bypassing control was implemented in an early computer control system in 1976 (Gillblad & Olsson (1977)).

- By controlling the sewer flowrate the influent flow to the plant can be controlled. This has been discussed in Nielsen *et al.* (1996) and Carstensen *et al.* (1996). Another implementation is described by Gustafsson *et al.* (1993).

- A critical review of return sludge flowrate for control on an hour-to-hour basis was made in Olsson (1985b).

- The use of step feed as a way to redistribute the sludge within the aerator was analysed by Andrews (1974). A pilot scale implementation is described in Yust *et al.* (1981) and Yust *et al.* (1984). The potential of step feed has also been analysed by Sorour *et al.* (1993).

- One kind of SBR system is called the alternating system, commercially marketed as Bio-Denitro and Bio-Denipho systems. These will be further described in Section 19.5 (Thornberg *et al.* (1993)).

- A scheme to estimate the need for carbon addition in a post-precipitation system is described in Linde (1993). Another carbon addition control is described by Lindberg & Carlsson (1996a).

- Adding RBCOD in order to enhance the P removal has been tested at the Swedish plant in Helsingborg. The VFA potential was considered a key factor, see Jönsson *et al.* (1996).

- Selection mechanisms to encourage particular classes of bacteria in the system are discussed in Henze (1991).

- Studies of how to influence the bacterial population and in particular sludge bulking are presented in Curds (1973a), Curds (1973b), Van-Niekerk *et al.* (1988), Albertson (1991), Gabb *et al.* (1991), Switzenbaum *et al.* (1992) and Jenkins *et al.* (1993).

- State of the art control systems in full scale plants are described in Lynggaard & Nielsen (1993), Nielsen & Önnerth (1995) and Nielsen & Önnerth (1996).

- A review of the Scandinavian perspective on operation and control of wastewater treatment systems during the last 20 years is made in Olsson *et al.* (1997).

16.7 Exercises

1. The RAS flowrate has to be constrained by several factors. Discuss how the RAS flow will influence the (a) sludge blanket height, (b) return sludge concentration, (c) sludge retention time in the settler, (d) the clarifier performance.

2. Discuss in qualitative terms what happens when the WAS flowrate increases or decreases 10%. In particular consider the time scale.

3. Consider what happens in flow splitting. Assume, that the main influent stream is split into two channels, but the flowrates may deviate 20% from the nominal value. What is the consequence for the plant loading? What may be suggested by a plant designer if the flowrates are not calibrated?

4. If you were to decide when and how bypassing should be done to make the smallest environmental impact in the long run, how would you make the decision rule? Consider both a short and a long time perspective.

5. Consider primary pumping with only on-off options. How is the clarifier influenced? Compare the discussions on hydraulics in Chapter 4.

6. What is the reason to keep the sludge blanket level (a) high? (b) low? What are the risks related to a high or low sludge blanket?

7. When is the best time to allow supernatant recycling?

8. Make some comparisons between recycling plant control actions and sequential batch reactor control actions for (a) denitrification, (b) nitrification.

9. Why would you need carbon addition in a plant?

10. How would you construct an air supply system (not considering the costs!) with maximum flexibility? How would you use that flexibility?

11. Discuss some of the organisational problems in sewer control when there are separate sewer operation and treatment operation organisations. In particular, consider conflicting interests. How would you solve that if you had the political power?

Chapter 17

Feedback Control

The goals of this chapter are to introduce the concepts and algorithms for feedback controllers and to illustrate the algorithms and their advantages and disadvantages with a detailed look at the control of dissolved oxygen.

We will start our look at feedback control by reiterating some concepts and nomenclature, just to put us all on an equal footing, Section 17.1. Then we will examine the control algorithms, the automated decision making, Section 17.2. Then we look at how these are tuned for best performance, Section 17.3, and at how we handle the difficult cases using gain scheduling and auto- and self-tuning, Section 17.5. Section 17.4 describes controllers with selectors which make logical choices of the regulator configuration. Cascaded control is an important concept with many applications in flow control and drive system control, Section 17.6. Finally we look at applications in nutrient removal control, Section 17.7. These are usually the most important, after all no one complains about the easy ones. And, as Murphy would have it, dissolved oxygen control is often a "difficult one".

17.1 Feedback Control Loops

In this section we will re-introduce some fundamental control concepts, like closed and open loop control, feedback and feedforward.

17.1.1 Closed Loop Feedback

Figure 17.1 is a conceptual illustration of the key concepts of feedback and feed-forward control. We will look at feedforward control in the Chapter 18. It is clear that they get their names from the direction of the information flow in the control loop, relative to the information flow through the process. The information flow may or may not be the same direction as the flow of material.

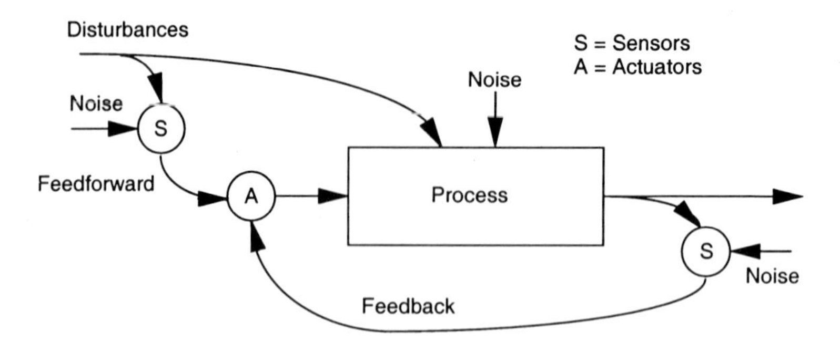

Figure 17.1: Feedback and feedforward.

Nomenclature is a tricky business because the same term often means something a little different depending on whether the person is process oriented or control oriented. You should be aware of this and always ask for clarification when it matters.

It should be pretty clear what we call disturbances, those variables that change and affect the process but over which we have no direct control. We may or may not be able or willing to measure them. Again, manipulated variables are clearly those affected by the actuators and which have an affect on the process. The internal state variables are the most important, but often we do not see them directly, only via their effect on other variables called output variables. The measurements are the signals we get from the sensors, sometimes called process variables in commercial instrumentation literature. These are the common terms to process oriented people.

How you classify a particular process variable depends on your definition of the process or system. You must always clearly draw your system boundary. This will be further discussed in Section 17.6.2.

Example 17.1 (System boundaries) *Consider the process in Figure 17.2 with three separate system boundaries drawn. In System A, the anoxic nutrient reactions, the variable* nitrate concentration *is a state variable. In System B, the oxygen dissolution, the variable* dissolved oxygen concentration *is a state variable and an output, since we are measuring it. But in System C, the aerobic nutrient reactions, the very same variables,* nitrate concentration *and* dissolved oxygen concentration, *are disturbances. Your perception of the world always differs depending on where you sit and what your goals happen to be.*

To complete our discussion of nomenclature, we now need to look in more detail at a feedback control loop, shown in Figure 17.3 in block diagram form as you will see it in the control literature.

An important thing to note here is that the *process* to most of the control community includes the actuators, the real process and the sensors. To control

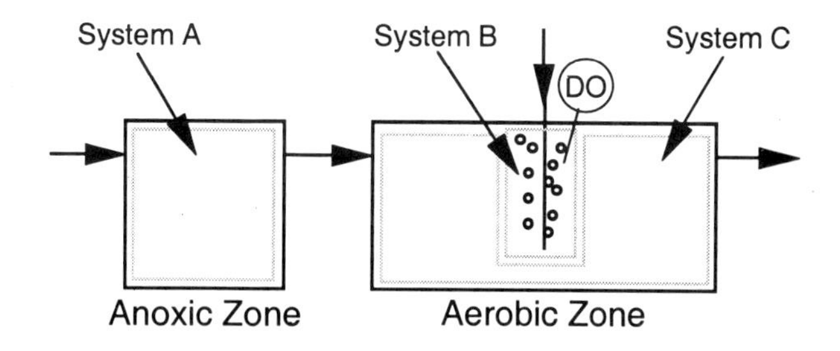

Figure 17.2: Example of system boundaries.

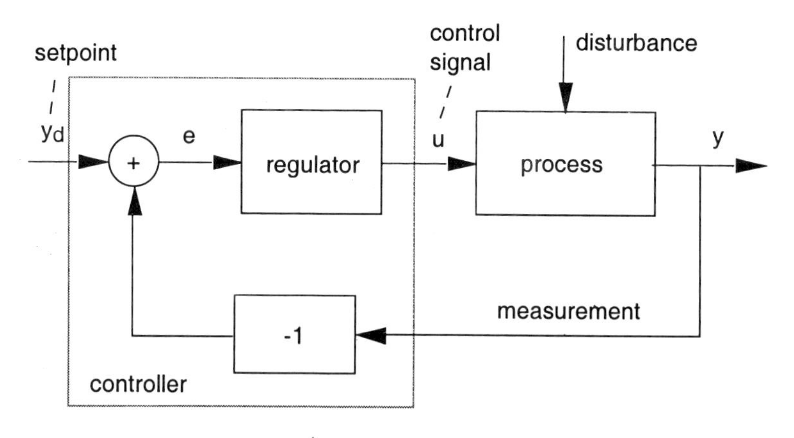

Figure 17.3: Feedback loop block diagram.

oriented people, the outputs y are the measurement signals, while to a process oriented person the process outputs are the variables being measured. The inputs u are the signals to the actuators, rather than the variables being manipulated. The desired outputs y_d are often also called **setpoints**, **reference outputs** or **reference values**. The desired outputs minus the outputs are the errors e. You will often see the "+" and "-1" blocks combined. Leaving them separated emphasises the negative feedback that is required for stability. You will all be familiar with the instability caused by positive feedback in public address systems and your stereo system.

There are many feedback loops in biochemical processes where the controller keeps just one variable at some setpoint value. This value is usually some chosen constant, but can be changed manually or by some higher level calculation. Some examples are:

- *Flow control* is applied to both liquid and gas flows. Pump control is traditionally based on flowrate measurements or some level measurement. In a wastewater treatment plant the flowrate control may be based on a wet well level.

- *Liquid flow splitting* between parallel trains is important in order to ensure good operation. A common practical problem is that weirs and splitter boxes are fixed, while the flow splitting is a dynamic process and has to be based on true flow measurements in order to work satisfactorily.

- Feedback control is used to produce a well-defined *air flowrate* to the aerators.

- *Dissolved oxygen control* in its simplest form is a conventional cascade control and is discussed below.

17.1.2 Open loop system

The other concept to make clear is that of *open loop* and *closed loop*. Closed loop is pretty clear, the connections are as they are in Figure 17.3 and the controller is functioning. Open loop means that there is a break in the feedback path which in principle could be anywhere in the path. Real control equipment has an auto-manual mode or switch which means the controller is closed or open loop respectively. The term *manual* refers to the fact that the operator can now adjust the signal to the actuator directly. In simulations, control loops are often set *open loop* by giving the controller gain a value of zero, but you have to be careful because the efficacy of this depends on the actual algorithm as we will see in the next section.

Open loop sequencing is common. In this case, there is no automatic feedback from the measurement but the control is based on a timer or a predetermined programme of action, illustrated in Figure 17.4.

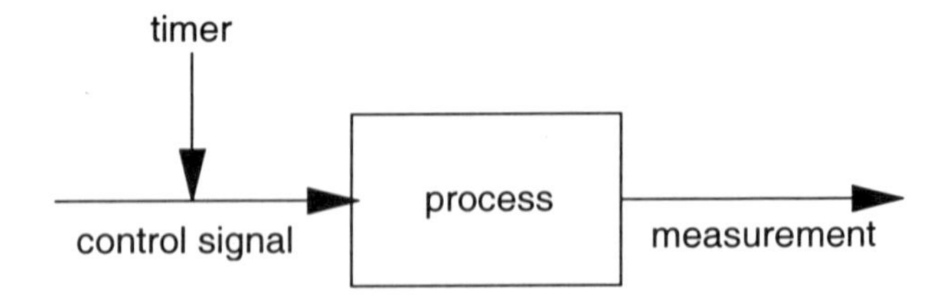

Figure 17.4: Open loop control.

Some examples of open loop sequencing are:

- *Air compressors* are often switched on and off according to timers. Sometimes a DO sensor is used to feed back information for on-off control of the compressors. Feedback DO control is further discussed below.

- *Sludge removal*, particularly in primary sedimentation, is often based on timers instead of some sludge density or sludge blanket measurements. Measurements of the sludge density can ensure that a consistent solids concentration is obtained.

- *Waste activated sludge pumping* is often based on a timer. A better action is to base the wastage rate on conventional sludge age calculations. This is further discussed in Section 18.6.1.

- *Bar screen cleaning* is sometimes based on a timer operation. However, the loop can be closed if the cleaning process is started when a high differential level over the screen occurs.

17.1.3 Loop Stability

Another concept we will examine is that of *direct* action and *reverse* action, a very important mode or switch on your controllers. Direct action means that if the input to the controller increases then the controller output will also increase. This controller action mode must be set according to the *action* of the process itself to ensure that you do indeed have negative feedback and the possibility of a stable loop.

This brings us to the final concept, that of *stability*. The most general definition is that a system is stable if its outputs are bounded when its inputs are bounded. But let us look at Figure 17.5 and think of stability in terms of our control loop.

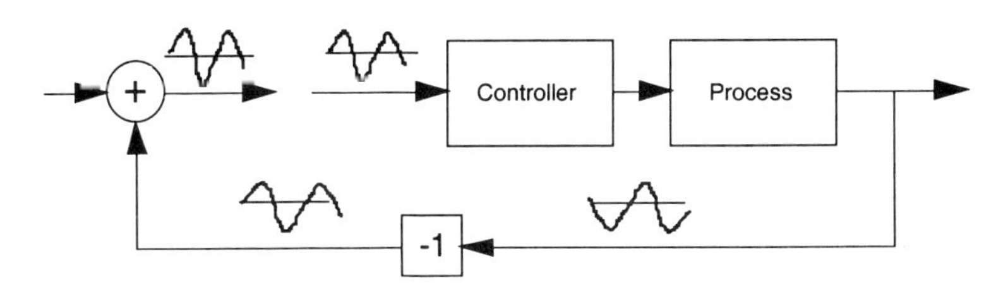

Figure 17.5: Feedback stability.

Suppose we break the loop as shown and feed in a signal of such a frequency that the controller and the process induce a 180 degree lag, that is they exactly invert the signal. Then the negative feedback will invert the signal again and it will arrive back at the break in the loop *in phase*. Now we have three scenarios which depend upon what the controller and process did to the amplitude of the signal, that is, what was the loop gain:

- return signal amplitude smaller than input signal amplitude (gain less than one),
- return signal amplitude equal to input signal amplitude (gain equals one), and

- return signal amplitude greater than input signal amplitude (gain greater than one).

What happens in these three cases if we close the loop? If you think a little about it, you will come to the conclusion that the amplitude of the signal will decrease to zero, stay the same, or increase to infinity (if practicalities allowed) respectively. If the loop gain is too big, then the latter will occur and the system is unstable. In the first two cases the system is theoretically stable. From a practical control perspective we want to keep the loop gain low enough that we always have the first case, where the oscillation decreases to zero. In theoretical terms, this is called **asymptotically stable**. The term **relative stability** is used as a qualitative measure of how fast the oscillations decrease.

Temperature control in the shower is a good place to examine stability. Good temperature control can be more difficult than you imagine, particularly if you have a very sensitive tap (high gain actuator). Large impatient adjustments are just likely to end up in wildly fluctuating temperatures. Small patient adjustments will soon achieve an asymptotically stable temperature.

Of course, if the loop gain is too low we will get too little control action and performance will also suffer. Getting the right loop gain is what controller tuning is all about. We address this in some detail shortly.

Lastly, note that we have been deliberately talking about the *loop* gain. The important thing in feedback controller performance is not the controller gain, and not the process gain, but the combined or loop gain. This will also come up again soon.

17.2 Feedback Algorithms

Two types of control algorithms predominate in wastewater treatment plants, indeed throughout the process industries in general. These are the *on-off* algorithm, and the Proportional-Integral-Derivative (PID) algorithm. We will look at both of these now.

17.2.1 On-Off Algorithm

The on-off algorithm should be familiar to us all. It predominates in our homes. Appliances such as stoves, heaters, and air conditioners use this type of control. If the oven gets too hot, turn off the power; if it gets too cold, turn on the power.

The advantage of on-off control is the simplicity. But there are disadvantages. By the very nature of the control action, the temperature in the oven will cycle up and down continuously. This is called **limit cycling**. The speed and amplitude of this cycling depends on how fast the process responds to an on or off control action. If the process responds quickly the cycling will be fast and the controller will *chatter*, and soon wear out. To avoid this problem, on-off controllers usually have a *deadband*, a range of temperature between turning on and turning off, where the controller does nothing. This slows down the cycling, but increases the amplitude.

Another aspect of on-off control is that it can create a disturbance to other parts of the process, not only *forwards* due to the limit cycling in the output, but also *backwards* due to the on-off switching of the input. We discussed this problem earlier, where the control of sewer or receiving basin level by on-off switching of influent pumps can create a major disturbance to the secondary clarifiers. The magnitude of the switching in a large treatment plant was also seen in Example 15.2 in Chapter 15.

Example 17.2 (DO on-off control) *It is not uncommon to see on-off control in use for dissolved oxygen (DO) control. Either valves in the air supply line or even the blowers are switched on and off. Figure 17.6 shows the limit cycling characteristic of on-off control produced by the MATLAB program* `Example_17_2.m`.

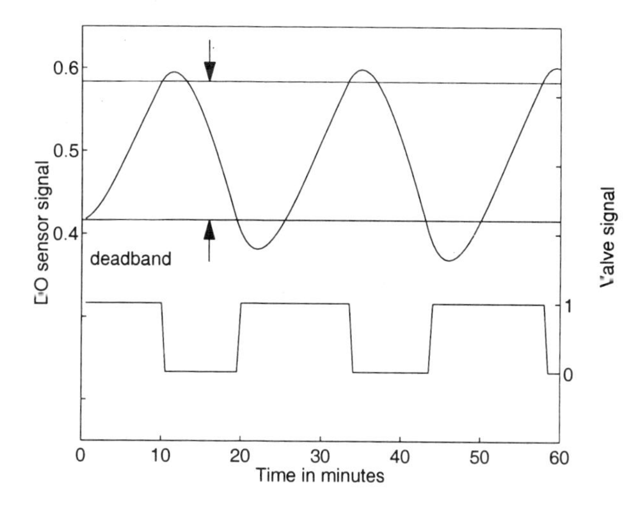

Figure 17.6: On-off control.

There is one interesting side benefit of this which was discussed in Section 8.5. When the air is turned off, the rate at which the DO level falls can be related to the oxygen uptake rate (OUR) of the biomass. When the air turns on, the rate at which the DO level increases is related to the OUR and the oxygen mass transfer rate coefficient (K_La). This can be a useful way of obtaining estimates of these two parameters on a continuing basis. OUR changes particularly are useful in gauging the activity of the biomass, which can be related to changing nutrient load or to the health of the microorganisms. Unexpected falls can indicate toxic shocks and were discussed in Chapter 10 on diagnosis.

17.2.2 PID Algorithm

The Proportional-Integral-Derivative (PID) algorithm is the most common algorithm in use in the process industries. It is a combination of three control actions, proportional (P), integral (I) and derivative (D). Which combination you use depends on the application as we will discuss later, the PI combination being perhaps the most common. Let us discuss the three actions in turn.

The P algorithm is simply that the amount of control action $(u - u_o)$ is proportional to the error e. It can be expressed by the following equation:

$$u = u_o + K_P\, e \qquad (17.1)$$

where u_o is the controller output bias and K_P is the proportional gain. The bias is set to the desired actuator signal at the steady state operating point, where the error is zero. The proportional gain is tuned to obtain the desired relative stability for the control loop.

Proportional action has the advantages that it is still conceptually simple and that the process output will reach a steady value in the absence of disturbances, unlike the limit cycling experienced with on-off control. The disadvantage is a phenomenon called *offset*, which arises when the desired output is changed or the process experiences a sustained disturbance. In these situations the controller must generate a new steady state actuator signal (u different from u_o). This can only happen if the error e takes on a nonzero value at steady state. This steady nonzero error is called the offset. You can see from the equation above that the larger the gain the smaller will be the offset. But it can only be removed by manual intervention to change the bias parameter.

It was the problem of offset that led to the development of integral action, where control action is proportional to the integral of the error. It is generally used together with P action as follows:

$$u = u_o + K_P\, e + K_I \int e\, dt \qquad (17.2)$$

The integral of the error is equivalent to, and is often implemented by, a running summation of the errors. Assume that the system is at steady state so that all the signals are constant, particularly $e(t)$ and $u(t)$. Steady state can only remain if the integral part $\int e\, dt$ is constant, otherwise $u(t)$ would change. This is only possible if $e(t)$ is zero. The integral gain parameter K_I is tuned to remove the offset quickly.

The integral gain K_I is equivalent to the inverse of the integral time T_I, $\frac{1}{T_I}$, which is used as an alternative parameter in many implementations of the algorithm. The integral time parameter appears in the denominator, which makes the dimensions of the controller terms proper and has a practical interpretation. To see this, consider a step change of the error $e(t)$ and its response in a PI controller Figure 17.7. Immediately after the step the controller output is $K_P e$. After the

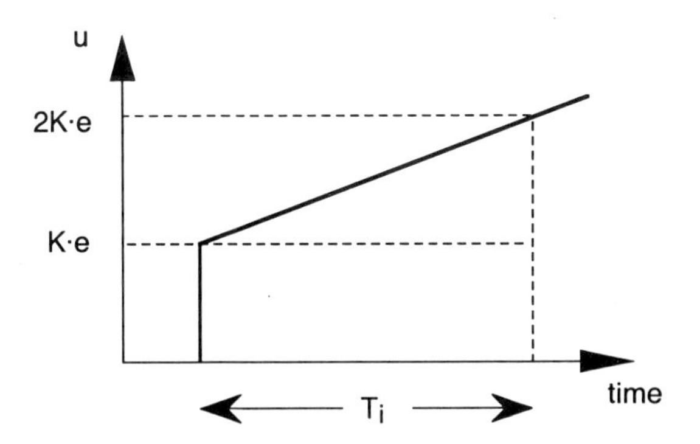

Figure 17.7: Step response of a PI controller.

time T_I the controller output has doubled. A PI controller is often symbolised by its step response.

Unfortunately, even silver clouds cast a shadow. The integral action also contributes a phase lag to the dynamics of the loop which has a destabilising effect. We can see this in Figure 17.8, where a system is responding to a change in the desired output or setpoint from 200 to a value of 205.

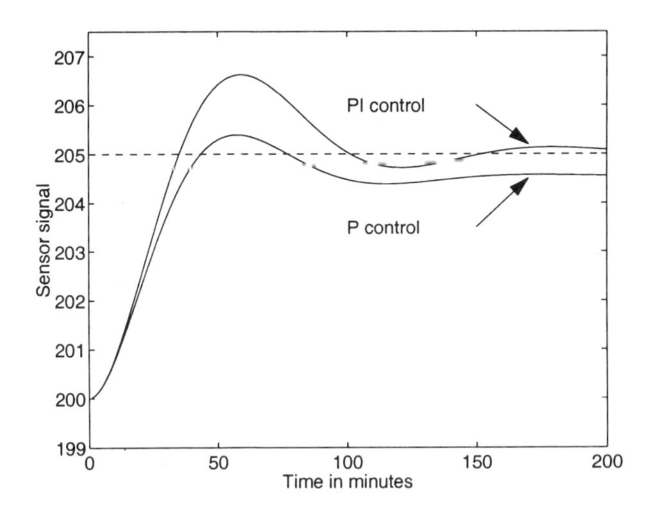

Figure 17.8: Offset and integral action.

The PI controller is eliminating the offset of about 0.5 but at the expense of a larger overshoot, showing the decrease in relative stability (the proportional gain was the same in both cases). To maintain the same relative stability the loop gain must be decreased by decreasing the proportional gain.

Another potential problem arises if the manipulated variable hits a constraint for some time, such as the valve becoming fully shut or wide open. Because of the constraint it is impossible to achieve a zero error and the error integral just grows and grows. This problem is called *integral* or *reset windup*. Commercial controllers generally solve the problem either by putting an upper limit on the integral term, or by turning off the integral action when the controller output saturates. A problem with the latter approach is that the actuator or the manipulated variable itself may saturate before the controller output.

The third control action is derivative or D action. You should be able to guess that this control action is proportional to the derivative (or slope) of the error. The full PID algorithm is therefore:

$$u \quad = \quad u_o \quad + \quad K_P\, e \quad + \quad K_I \int e\, dt \quad + \quad K_D \frac{de}{dt} \tag{17.3}$$

The derivative gain K_D is often replaced by the alternative derivative time T_D. Derivative action adds a kind of predictive capability by reacting to the rate of change of the error. In the absence of noise this adds a stabilising effect, which can counteract the destabilising effect of integral action. This can be seen in Figure 17.9, where D action was added to the same P and I action that we saw in Figure 17.8.

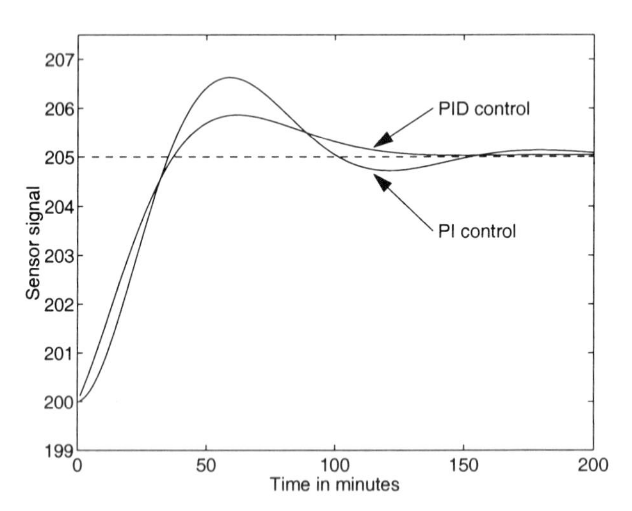

Figure 17.9: Derivative stabilisation.

If the measurement signal from the process, and hence the error, is noisy then there are problems. The slope of the noisy signal alternates rapidly between a large positive number and a large negative number so that the derivative action amplifies the noise. A similar problem arises if there is a sudden change in the desired output y_d due to operator action. The latter problem, called **derivative kick**, is solved in some controllers by taking the derivative of the measurement signal rather than the error.

You need to be a little careful about the form of the PID algorithm that your controller implements. Another form of the algorithm often implemented is:

$$u = u_o + K_P \left(e + \frac{1}{T_I} \int e\,dt + T_D \frac{de}{dt} \right) \tag{17.4}$$

Notice that in this form the proportional gain affects all three control actions, which is useful in some respects, but can confuse the tuning. One very important reason most commercial controllers use this form is that there is only one gain button that will affect the whole controller. This is important if, for some reason, the process starts to oscillate and the loop must be detuned (loop gain decreased).

17.2.3 Noise Filtering

The sensitivity of derivative action to noise is often solved by a low-pass filter on the measurement signal to reduce the noise. This is fine so long as you don't filter out the disturbances you are trying to control, so be gentle on the filter constant.

In order to examine slow variations it is necessary to remove individual spikes in the measurement data and other quick disturbances which do not contain relevant information. This can be done with a digital low pass filter, discussed in Section 10.2.5.

17.2.4 Loop Sampling Interval

Almost all controllers, even single loop controllers, are implemented on a microprocessor and therefore operate at discrete but regular times separated by the loop sampling interval.

In most commercial process control software the loops simply execute as fast as the computer allows, usually between 1 and 10 times a second. Since the dynamics of process systems are generally in the minutes to hours range, this does not create a problem. The controllers effectively operate in continuous time.

Occasionally, one or two control loops will have very fast dynamics and may require single loop controllers which generally have a smaller sampling interval.

Occasionally too, you may encounter a control package which gives you a choice of control interval for each loop that you configure. Using larger sampling intervals on loops with slower dynamics reduces the load on the computer which may allow more loops to be processed or may improve the operator interface response times. In this case you need to choose an appropriate sampling interval. It is usually recommended that the sampling rate be 6 to 10 times the bandwidth, the highest frequency disturbance you wish to control. An equivalent guideline is to sample 6 to 10 times within the rise time of a step response.

You need to take particular care with sampling interval when nutrient analysers are to be used for closed-loop control. Because the analysis time can be of the order of 15 to 30 minutes, and because the cost of the analyser encourages analysing multiple streams with the one analyser, you need to do your sums carefully or your sampling interval may be too large for effective or stable control.

17.3 Controller Tuning

The first task in tuning a control loop, that is determining values for its parameters, is to be clear on your objectives. What is the purpose of the control loop? There are three main types of loop:

- *servo loops* need to be tuned to follow frequent and quite rapid changes in the desired output or setpoint as quickly and as accurately as possible. Examples are the inner loops of cascade controllers (see Section 17.6) and valve positioners.

- *regulator loops* are for disturbance rejection, so they try and keep as close as possible to a desired output which seldom changes. This is the majority of control loops in a process plant.

- *averaging loops* are used to damp out disturbances, usually controlling the level or pressure of buffer tanks. The controller tries to maintain a constant outflow at the average of the quite variable inflow. Examples are equalisation tank level control in two stage anaerobic digestors and level control in tanks for buffering filter backwash.

The second task is to decide which parameters will be tuned, usually what combination of P, I and D action should be used. General guidelines for choosing control actions include:

- use as few as possible (there are fewer parameters to tune).

- use only P action for liquid levels in vessels, unless it is an averaging loop or the actual level is important.

- use only P action on the inner loop of cascade loops. This is further discussed below in Example17.10.

- use only I action for averaging loops.

- add I action where adherence to the exact desired output setting is important.

- add D action for high order process dynamics where the initial reaction is slow.

- use extra care with D action when the measurement is noisy.

No matter what tuning technique you choose, and there are nearly as may techniques as there are control engineers, the general procedure involves three steps:

- identify a simple model of the process (the form of the model).
- estimate parameters for the model form chosen, usually by some type of stimulus-response experiment on the process.
- design controller parameters according to some procedure.

Among the many techniques, we recommend IMC tuning which will be discussed below. It has at least some theoretical basis, and some theoretical drawbacks, but in general gives good stable responses particularly suited to the process industries. You can also use one of the many auto-tuners currently on the market, but evaluate them first because different auto-tuners work best on particular classes of loops. You may even need several different tuners to cover your needs.

The IMC tuning rules for PID controllers are summarised in Table 17.1. The parameter τ_m is a tuning parameter. The other parameters are defined in Section 7.5.

Model	K_P	K_I	K_D	T_I	T_D
first order	$\frac{\tau}{K\tau_m}$	$\frac{1}{K\tau_m}$	-	τ	-
second order overdamped	$\frac{\tau_1+\tau_2}{K\tau_m}$	$\frac{1}{K\tau_m}$	$\frac{\tau_1\tau_2}{K\tau_m}$	$\tau_1+\tau_2$	$\frac{\tau_1\tau_2}{\tau_1+\tau_2}$
second order underdamped	$\frac{2\zeta\tau}{K\tau_m}$	$\frac{1}{K\tau_m}$	$\frac{\tau^2}{K\tau_m}$	$2\zeta\tau$	$\frac{\tau}{2\zeta}$
integrator	$\frac{1}{K\tau_m}$	0	0	-	-
integrator plus first order	$\frac{1}{K\tau_m}$	0	$\frac{\tau}{K\tau_m}$	-	τ

Table 17.1: IMC tuning rules.

As with other tuning techniques, the first step in IMC tuning is the identification of a simple input-output model of the process from amongst the options in Table 17.1. The input being the control signal to the actuator and the output being the sensor signal into the controller. This can be done by one of the following techniques:

- A simple open-loop step test with parameters estimated by one of the techniques discussed in Chapter 7.
- A closed-loop step test with proportional only control and a proportional gain high enough to give a decay ratio of about one third. The parameters can be estimated by identifying an under-damped second-order "closed loop model" and then back-calculating a second-order "open loop model" knowing that you have a proportional controller.
- Fitting the chosen model to time series data, using standard least squares regression as discussed in Chapter 7.

Once the model has been identified the controller parameters can then be calculated from the formula in Table 17.1. The parameter τ_m is a single tuning parameter. Small values will give fast oscillatory responses and large values will give slow overdamped responses (see Figure 17.10). A value about equal to the average of the model time constants is usually a safe starting point.

Example 17.3 (PID tuning techniques) *An open-loop step test on the DO control loop simulated by the MATLAB program* `Example_17_3.m` *gave an overdamped second-order response, with second-order model parameters estimated as gain $K = 1.47$, time constants = 5 and 20 minutes. Force fitting a first order model gave the same gain and a time constant of 25 minutes. The left plot in Figure 17.10 shows IMC tuned responses with tuning parameter values of 1, 5 and 12 minutes for a PID controller using the second order model.*

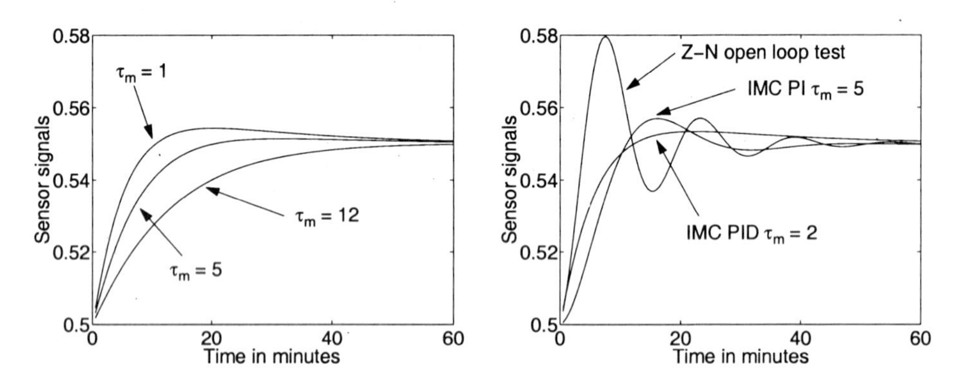

Figure 17.10: IMC tuned responses.

The right plot in Figure 17.11 compares IMC tuned PI (first order model) and PID (second order model) responses compared to the traditional Ziegler-Nichols tuned response. Detuning Ziegler-Nichols settings in the same way as IMC settings slows the response but marginally increases the oscillations. The two plots in Figure 17.10 show the usually more conservative responses possible with IMC tuning.

The disadvantage of IMC tuning can be seen in Figure 17.11 where the responses are the result of a ten percent decrease in the substrate load. The Ziegler-Nichols tuned controller achieves better disturbance rejection but the response is more oscillatory.

Comparing Figures 17.10 and 17.11 also shows the trade-off between tuning for setpoint changes and tuning for disturbance rejection.

The tuning of *servo* controllers differs slightly from that for *regulatory* controllers. The latter are generally tuned a little tighter (higher gains) with the emphasis on tracking changes in setpoint as quickly as possible. Slave controllers in cascade loops (see Section 17.6) are often tuned for a servo response.

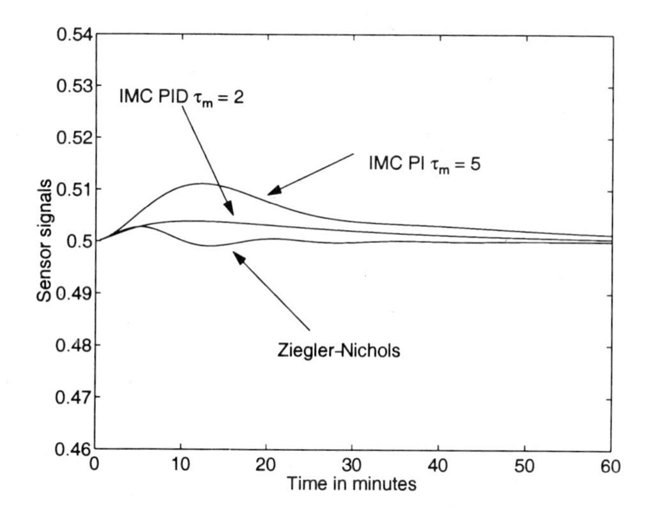

Figure 17.11: Disturbance rejection tuning.

Averaging level control of buffer vessels is a very special application. Here the emphasis is on maintaining a constant outflow despite changing inflows, with the constraints that the vessel should neither overflow nor empty. The level controller generally has proportional action with a large deadband to satisfy the constraints and a very slow integral action. The integral action allows the outflow to track the inflows, maintaining the long-term average level at half full.

Example 17.4 (Averaging level control) *The MATLAB program called* `Example_17_4.m` *simulates a buffer tank with an inlet flowrate disturbance similar to what you might expect from a backwashing filter. Figure 17.12 shows the disturbance and the outlet flowrate in the top plot, and the tank level in the bottom plot. Responses are shown in the plots for level controllers tuned both conventionally and detuned by a factor of ten to show averaging performance. The disturbance rejection performance of the averaging controller is clearly superior. Care must be taken that such loops are not accidentally tuned conventionally.*

17.4 Controllers with Selectors

In Section 17.1 we have considered simple regulators. In many industrial systems, including wastewater treatment systems, there are most often many other elements to complete the regulator structure. A very common element is a **selector**. It works like a *max* or *min* function selecting the input signal with the largest or smallest value and producing this signal as its output. A selector is a nonlinear function in the control loop since the selector function can suddenly

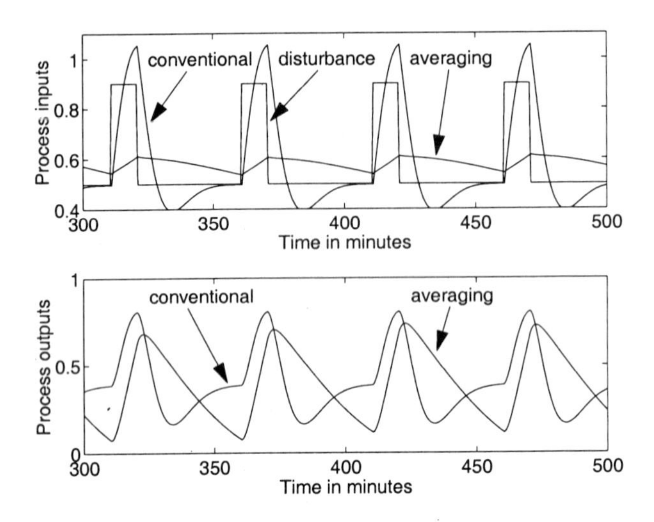

Figure 17.12: Averaging level control.

disconnect a signal causing a discontinuity. It is used to handle logical choices in the configuration.

Example 17.5 (Control of DO in parallel basins) *Assume that the DO is controlled in two parallel aerators. The DO controller will produce an air flow such that the aerator DO concentration is kept at the setpoint value. We assume that a compressor will keep a prescribed pressure in the air system. For a given pressure it is required to keep a certain air valve position in order to produce the required air flowrate. From an energy point of view there is a power loss across the valve, if it is not fully open. Therefore, a good control system would act such that the most open valve of the parallel channels should be almost fully open. This can be achieved by controlling the pressure in the air system.*

Figure 17.13 illustrates the idea. The valve openings are measured and the most open valve is chosen and compared to a setpoint, say 95%. If the valve is less than 95% open, then the pressure can be decreased by controlling the compressor. This can in turn be realised either by control of the compressor speed or of guide vane control, which can limit the air flow out of the compressor. As the pressure is decreased the valves have to open more in order to produce the same DO concentration in the aerator.

The pressure controller is part of the complete DO control system, as depicted in Figure 17.20. It is obvious, that the pressure regulation has to take place in a much slower time scale than the DO concentration change. Otherwise slow oscillations may be excited by the control system itself.

MAX and MIN selectors are nonlinearities that are introduced into the system and will create certain difficulties. Some of them resemble the kind of prob-

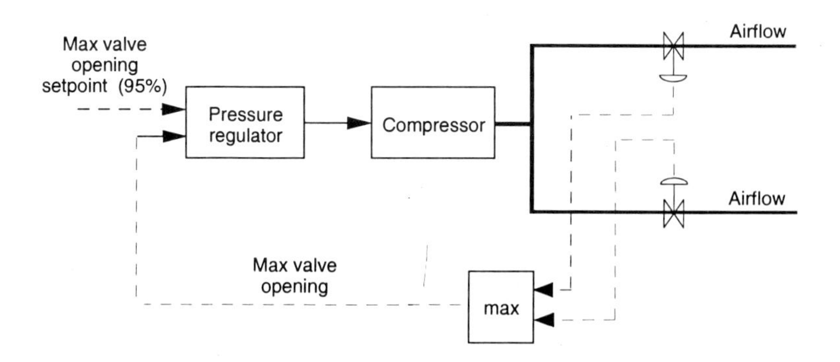

Figure 17.13: Controlling air flowrate with a selector.

lems that are introduced by constraints on control signals. We have seen in Section 17.2.2 that a limited control signal can cause integrator windup that has to be compensated. A similar problem can appear due to selectors. This can be illustrated by the DO control example above.

Assume that the second instead of the first aerator suddenly becomes the most loaded channel and demands more air than the other aerator. This means that the selector now has chosen the other valve opening and sends this signal to the pressure controller. If the control error was not eliminated, then there is an integrator value in the pressure controller that may be too large, since a new control action will now start.

The solution, as in the regulator windup case, is to tell the pressure controller that the nonlinearity (the selector) exists. It is not as simple as for control signal limitations since we are talking about two different control loops. The problem is solved by **tracking**.

Tracking is readily understood if the pressure controller is written in an incremental form. Normally a control signal can be written in the form

$$u(t) = u(t-1) + \delta u(t) \tag{17.5}$$

In other words, the control output at time t is equal to the previous control output plus an additional value $\delta u(t)$. When the pressure controller is changed to the other aerator then the control signal can be written as

$$u(t) = v(t-1) + \delta u(t) \tag{17.6}$$

where $v(t-1)$ is the real control signal, which is the output from the other regulator. At each new calculation we act as if the output from the selector agrees with the control signal from the previous measurement. In that sense the controller tracks the real control signal.

17.5 Loops with Variable Dynamics

We have seen that, to achieve a desirable relative stability and performance, we tune our controller parameters to match the process characteristics. But what happens if the process characteristics change? The obvious answer is that we need to re-tune our controllers, or suffer a loss in performance at best, or have an unstable process at worst.

Manual re-tuning is only practical if the process changes are quite infrequent, monthly or less often than that. Auto-tuners are a solution if changes occur not too frequently. When performance deteriorates or you change operating conditions, simply initiate a re-tune by the auto-tuner as a part of the normal operating procedure.

Example 17.6 (Variable process characteristics) *In biological wastewater treatment plants, the well studied example of changing process characteristics is that of DO control. The simulated results we will discuss were generated by the MATLAB program* `Example_17_6.m`.

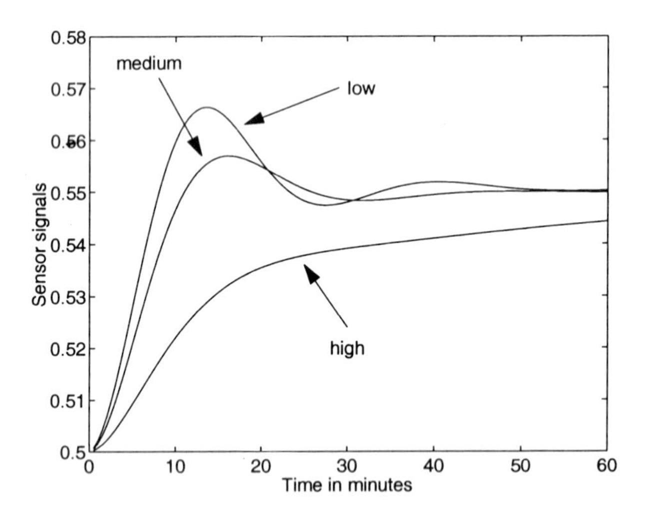

Figure 17.14: DO change at different operating points.

Figure 17.14 shows the controller response to a small setpoint change at three different operating points (different OUR values and hence different air flowrates). The PI controller parameters remained the same. A controller that is quite well tuned for medium substrate levels is less stable but just acceptable for low substrate levels, but pretty well unacceptable at high substrate levels.

We showed in Chapter 15 that wide swings in load occur in 12 and 24 hour cycles, much too often to rely on the auto-tuning approach. Two other approaches are possible, *gain scheduling* and *self-tuning*. Both of these retune the loops continuously.

17.5.1 Gain Scheduling

The variation in the performance of DO control shown in Figure 17.14 is a result of changing process gain and dynamics at different operating points, due to nonlinearities in the process. The changing process characteristics change the loop gain characteristics and therefore the relative stability.

What is the answer? As we have implied above, the objective is to maintain a constant loop gain to maintain a constant relative stability. If the process gain changes because we move to a different operating point, or because the process is a time-varying process, then the controller parameters must move in the opposite sense to maintain the status quo.

If we knew how the loop gain changed with changing conditions, then we could make a correction to the controller gains. This is demonstrated in the following example.

Example 17.7 (Gain scheduling DO control) *The dissolved oxygen balance used in the simulation in Example 17.6 giving the results in Figure 17.14 is given in Equation 17.7 below.*

$$\frac{dS_O}{dt} = \frac{q_F}{V} S_{OF} - \frac{q_F + q_R}{V} S_O + \frac{Y_H - 1}{Y_H} \hat{\mu}_H \left(\frac{S_S}{K_S + S_S}\right) \left(\frac{S_O}{K_{OH} + S_O}\right) X_H + k_1 \left(1 - e^{-k_2 \cdot q_a}\right) \left(S_{O,sat} - S_O\right) \tag{17.7}$$

where the variables are defined in Table B.1 in Appendix B.

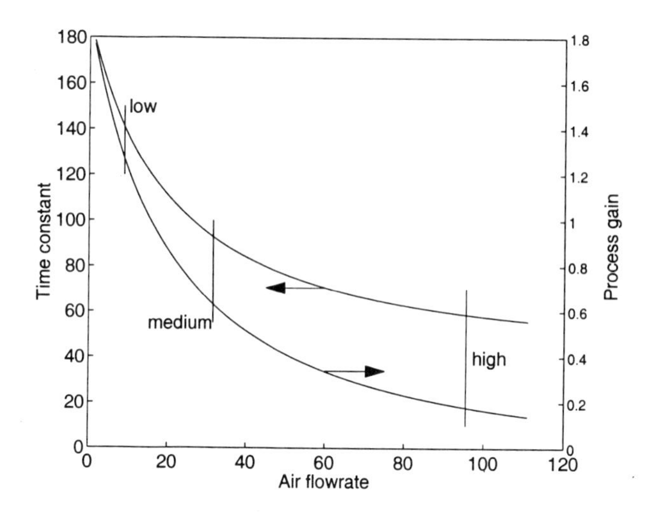

Figure 17.15: Changing gain and dynamics.

If you linearise the above equation, following the procedure in Section 5.2.2, you can determine the approximate process gain and time constant for different air flowrates (Figure 17.15). This figure shows the positions at which the three responses in Figure 17.14 were obtained, the increasing gain and time constant as the air flowrate decreases combine to decrease the relative stability and slow down the response time.

For example, if we read the process gains and time constants from Figure 17.15 and use them to scale the proportional gain and integral time inversely, then we get the results in Figure 17.16 which all have a relatively constant relative stability.

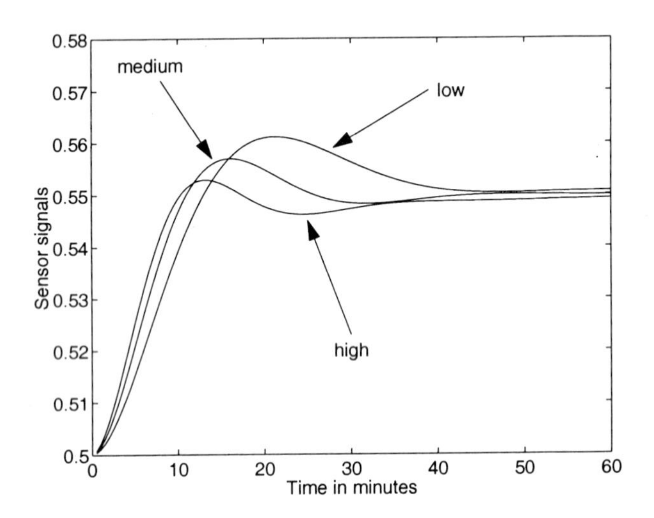

Figure 17.16: Retuned DO responses.

Of course the speed of response remains slower as the substrate level decreases, but unfortunately we can not change the fundamental dynamics of the process.

This sort of scaling of controller parameters using some measure of the operating point, in this case as a function of an averaged air flowrate, is called *gain scheduling*. It is the simplest of the solutions for processes with continuously varying characteristics. There are several commercial controllers with the capability to construct gain schedules of various levels of sophistication.

In many cases gain scheduling will not work because the schedule is a function of process conditions that change and that are not measurable. For example, the DO gain scheduling done above depends on knowing the residence time and on using the steady state version of the oxygen balance above to estimate the term:

$$\left(\frac{S_S}{K_S + S_S}\right) X_H \tag{17.8}$$

If the residence time changed rapidly, and we had no measurement of flowrate, we would have no way of calculating the schedule.

The only solution in this case is to use a self-tuning controller. This is a much more sophisticated controller.

17.5.2 Self-Tuning Control

If the controller parameters must be continuously updated, the controller may be called a *self-tuning* or *adaptive* controller. Such controllers have three parts which perform quite distinct tasks:

- an estimation part to identify the process characteristics (if we could measure them directly we would use gain scheduling),
- a design part which calculates the controller parameters, and
- a controller part with parameters calculated from the estimation results.

This arrangement is shown graphically in Figure 17.17.

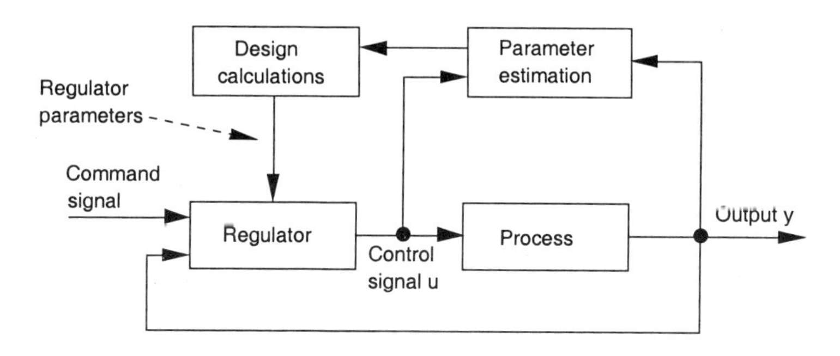

Figure 17.17: The principal parts of an adaptive controller.

The estimation part measures the process output and input signals to recursively update the process parameters, in general based on a time series model like those we discussed in Sections 7.6 and 12.4. Typically a second order plus time delay model is used:

$$y_t + a_1\, y_{t-1} + a_2\, y_{t-2} = b_0\, u_{t-d-1} \tag{17.9}$$

The design algorithm then updates the controller parameters a_i and b_i. The dead time d must be specified beforehand. The controller part can be a minimum-variance discrete controller of the form:

$$u_t = -\frac{1}{\hat{b}_0}\left(\hat{r}_1\, u_{t-1} + \ldots + \hat{s}_o\, y_t + \hat{s}_1\, y_{t-1} + \ldots \right) \tag{17.10}$$

or it can be a standard PID controller.

There are several variants of this general scheme. Instead of updating the process parameters, the controller parameters can be updated directly.

Self-tuning control was implemented in full-scale at the Käppala Sewage works close to Stockholm, Sweden, in 1984 to examine the potential of self-tuning control in activated sludge systems. The Käppala controllers (six parallel trains) have been in use for several years and perform satisfactorily. These self-tuning controllers do not give an explicit estimate of K_La or the oxygen uptake rate OUR, but calculate the controller parameters directly.

From the DO control point of view, the oxygen uptake rate is considered a disturbance (see Section 17.6.2).

While self-tuning control can be made to work, it suffers from some major drawbacks. One of these is the dilemma faced by all adaptive controllers. On the one hand they need to identify process parameters and need excitation of the process in order to do this. On the other hand they are attempting to control the process or to remove the excitation. If the excitation is insufficient the parameter estimation can very quickly become unstable, which will often also make the control unstable. Careful *tuning* of the self-tuner parameters is required together with several *safety nets* to try and catch instabilities in the estimation before they affect the controller. An external excitation signal applied to the control signal may be required.

17.5.3 Exact Linearisation

An alternative to empirical gain scheduling or self-tuning is so called *exact linearisation* where an invertible model of the nonlinearity is used.

Example 17.8 (A nonlinear DO controller) *We have seen that the dissolved oxygen dynamics is nonlinear, Equations 3.35 and 17.7. One solution to this problem is to let the regulator compensate for these nonlinearities. The dissolved oxygen dynamics in a well mixed aerator was presented in Equation 3.35, repeated here:*

$$\frac{d}{dt}(VS_O) = q_{in}S_{O,in} - q_{out}S_{O,out} + r_O V + K_La(S_{O,sat} - S_O)V \qquad (17.11)$$

The oxygen transfer rate K_La is a nonlinear function of the air flowrate, see Example 12.9:

$$K_La = f(q_a) \qquad (17.12)$$

The multiplication of K_La with $(S_{O,sat} - S_O)$ further adds to the nonlinearities. By introducing a variable:

$$x(t) = K_La \cdot (S_{O,sat} - S_O) = f(q_a) \cdot (S_{O,sat} - S_O) \qquad (17.13)$$

the dynamic description becomes linear with $x(t)$. This approach is sometimes called exact linearisation. The manipulated variable, the air flowrate q_a (compare Equation 3.38), is then calculated from the controller output $x(t)$ according to:

$$q_a(t) = f^{-1}\left(\frac{x(t)}{S_{O,sat} - S_O}\right) \tag{17.14}$$

where f^{-1} is the inverse of the K_La function. If the K_La function is approximated by an exponential function (see Equation 12.47 and 17.7):

$$K_La(q_a) = k_1 \cdot (1 - e^{-k_2 \cdot q_a}) \tag{17.15}$$

then the air flowrate is given by:

$$q_a(t) = -\frac{1}{\hat{k_2}} \cdot \ln\left[1 - \frac{x(t)}{\hat{k_1} \cdot (S_{O,sat} - S_O)}\right] \tag{17.16}$$

where $\hat{k_1}$ and $\hat{k_2}$ are the estimated parameters in the K_La function (see Chapters 12 and 18). The regulator structure is illustrated in Figure 17.18. The variable x can be considered an intermediate control variable, calculated by a conventional PID controller. The true air flowrate is then calculated from Equation 17.16.

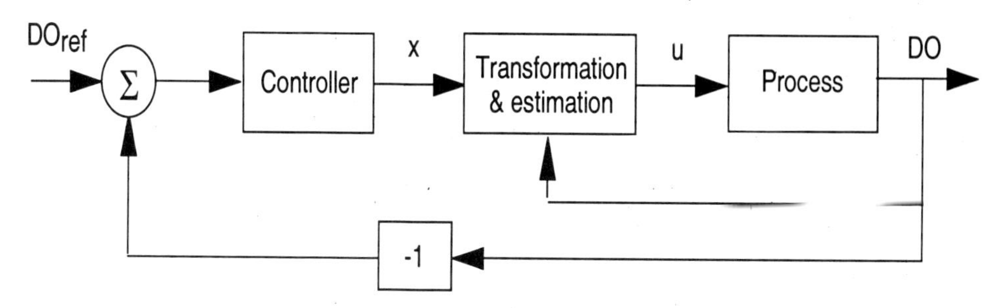

Figure 17.18: Structure of the nonlinear DO controller.

A conventional PID controller was compared with the nonlinear controller in pilot scale, Figure 17.19. The controller was tuned for large air flowrates. At low air flowrates the process has a larger gain (the slope of K_La is larger), then the linear regulator makes the control oscillatory, which is clearly shown in the control signal. The nonlinear regulator will adjust the controller gain according to the valve opening and has a satisfactory behaviour for both small and large loads.

17.6 Cascade Control Loops

Cascade control is a variation of the feedback control topology which is common in practise though not always obvious. It consists of two (or more) feedback controllers with the outputs of the higher-level controllers cascading to the setpoints of those below as shown in Figure 17.20 below.

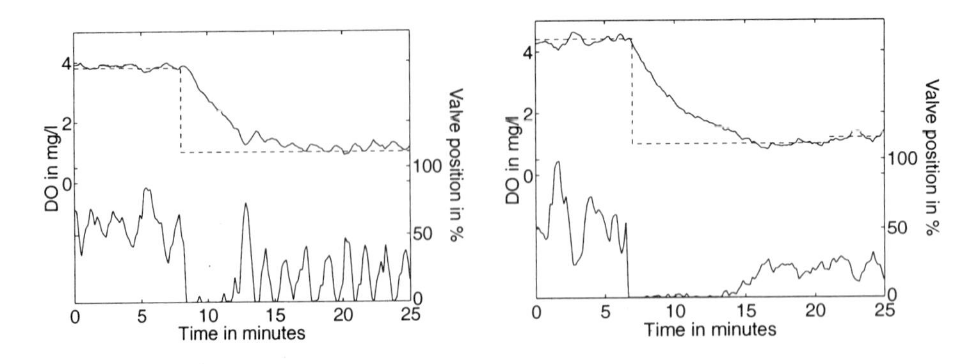

Figure 17.19: Linear (left) and nonlinear (right) dissolved oxygen control in a pilot scale plant

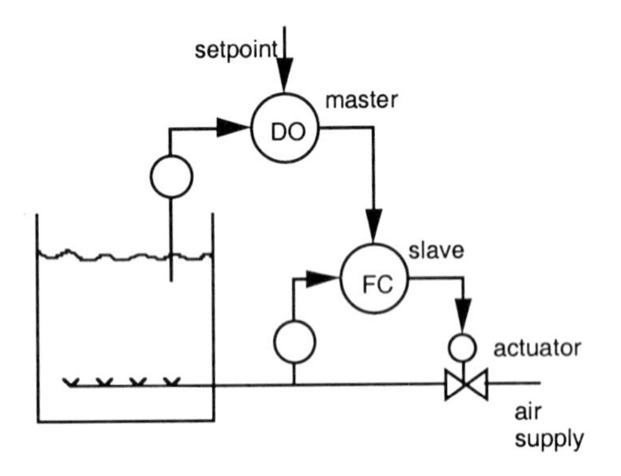

Figure 17.20: Cascade controller configuration.

The lower-level controller is called the secondary or slave controller. The upper-level controller is called the primary or master controller. Just as in the equivalent human situation, the slave must be capable of reacting faster than the master to ensure stability. In fact, if a slave controller is added to an existing feedback loop, the original (now master) controller may need to be detuned to ensure that this is the case.

The slave controller is always tuned first with the master controller open-loop, and then the master controller is tuned. Unless the master controller has a very slow response, the slave controller will generally be a simple proportional controller. The master controller will remove any offset which may occur.

17.6.1 Applications of Cascade Control

Cascade control is used for a number of reasons:

- *Disturbance rejection.* It helps to reject disturbances which affect the process via the manipulated variable. DO controllers will often be cascaded onto a flow or pressure slave loop. This will filter out disturbances caused by changes in the air supply pressure before the DO concentration is affected.

- *Gain scheduling.* When a slave loop is introduced, as far as the master loop is concerned the gain characteristics of the original actuator are replaced by the gain characteristics of the slave sensor. This is another way of linearising or otherwise shaping the loop gain of the master loop.

- *Hysteresis removal.* Actuators often require a strong control signal to overcome mechanical friction or opposing forces such as those caused by the fluid pressure drop across control valves. This can cause hysteresis and oscillations in slow responding loops. A slave loop can be added to supply the strong control signal and remove the oscillations. Valve positioners on control valves are the most common example.

Example 17.9 (Cascaded DO control) *A DO control loop at a first glance seems to be straight-forward. In Example 3.4 we looked at a simulated and simplified example. The DO is measured and compared with the setpoint value, as illustrated in Figure 17.20. The DO controller calculates the desired air flowrate that will cause the DO to approach the setpoint. However, the oxygen transfer rate is not proportional to the air flowrate, as discussed in Section 8.5 and Example 17.1.*

In a single loop arrangement the DO controller would directly change the air flow valve, which in turn would create the desired air flowrate. The problem is that such a valve is usually quite nonlinear, as discussed in Section 24.1.2. For example, a throttle valve has a high gain at low air flowrate and a low gain at large air flowrates. Such a nonlinearity adds to the overall nonlinearity of the DO controller.

The valve nonlinearity is elegantly solved with a separate flowrate controller. It receives its setpoint value from the DO controller and a measurement of the air flowrate. In order to illustrate its implication we show a block diagram of the flow controller.

We assume that the controller is a simple proportional (P) controller, which is usually adequate, represented by its gain K_c. The valve behaviour is represented by a simple gain K_v which is variable, depending on the valve opening. The gain from the flowrate setpoint u_f to the flowrate y_f is readily calculated. We have:

$$e = u_f - y_f \tag{17.17}$$

$$u = K_c \cdot e \tag{17.18}$$

$$y_f = K_v \cdot u \tag{17.19}$$

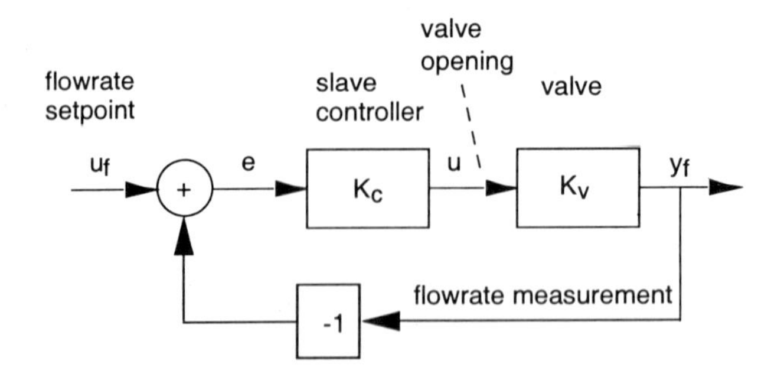

Figure 17.21: Block diagram of the slave controller in Figure 17.20.

Eliminating e and u we get the closed loop slave controller gain

$$\frac{y_f}{u_f} = \frac{K_c \cdot K_v}{1 + K_c \cdot K_v} \tag{17.20}$$

We can tune the gain K_c, and if it is made large then the control system gain approaches 1, independent of the gain K_v. In this way we have *eliminated the nonlinearity of the valve.* The price is a measurement of the air flowrate. This means that the DO controller commands an air flowrate, and the slave controller produces a real air flowrate that is close to the desired one. If there is any bias then an integrator in the DO controller adjusts the air flowrate setpoint until the DO has reached its DO setpoint.

Example 17.10 (Electric drive system control) *Several times we have discussed variable speed control of pumps. Let us look at the speed control of electric motors that will drive pumps or fans. The control structure is a cascade control as depicted in Figure 17.22, which is the common standard for electric drive systems.*

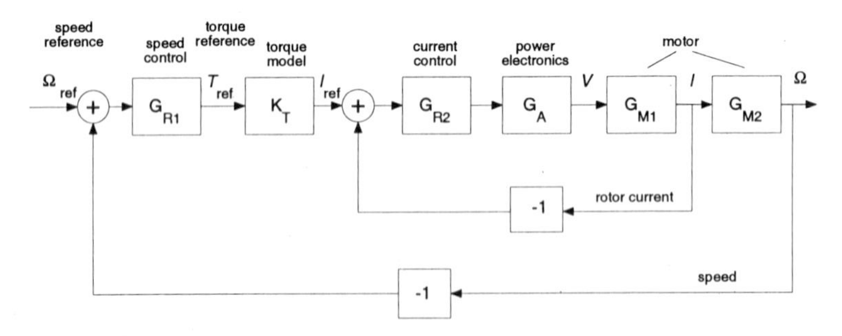

Figure 17.22: Block diagram of the speed controller of an electric drive system.

In principle the speed could be controlled by a simple controller, measuring the speed error and then giving the motor such a voltage that the speed is corrected.

Such a controller would be impractical due to the complex characteristics of the system. Primarily there are several nonlinearities in the system.

From the block diagram in Figure 17.22 we find that the speed controller G_{R1} *computes an output signal that is the torque needed to accelerate the motor to the desired speed. The desired current,* I_{ref}*, to produce the torque is calculated from a mathematical model of the motor. Here we denote this with a simple gain* K_T*, which is adequate for DC motors.*

The inner loop controls the current that is needed to produce the torque. The output signal of the torque controller G_{R2} *is a command to the power electronics unit that produces the necessary voltage input to the motor (see Section 24.1.1).*

Let us calculate the transfer function from the rotor current setpoint I_{ref} *to the rotor current* I*. In this book we have not defined the concept of a transfer function, but you may superficially think of it as a gain, the output signal divided by the input signal. The power electronics and the electrical part of the motor are represented by the transfer functions* G_{PE} *and* G_{M1} *respectively. The real system is not linear, but we can still illustrate the point by this approximation. Now the transfer function* G_I *of the inner loop:*

$$G_I = \frac{I}{I_{ref}} \tag{17.21}$$

From the block diagram we calculate:

$$(I_{ref} - I) \cdot G_{R2} \cdot G_{PE} \cdot G_{M1} = I \tag{17.22}$$

which is simplified to

$$G_I = \frac{I}{I_{ref}} = \frac{G_{R2} G_{PE} G_{M1}}{1 + G_{R2} G_{PE} G_{M1}} \tag{17.23}$$

Let us now make the gain of the controller G_{R2} *large. Then the transfer function* G_I *will approach 1 and will also be quite insensitive to variations in the power electronics and motor transfer functions. Nonlinear behaviour of the power electronics and the motor is consequently eliminated due to the feedback from the current* I*. As a result the speed control system can be simplified as in Figure 17.23 where the current control loop is represented by* G_I *that is close to 1.*

The mechanical part of the motor G_{M2} *is usually quite well behaved. Thus we can see that the cascade structure eliminates many of the inherent complexities in the power electronics and the motor dynamics.*

The rotor current feedback serves yet another purpose. Since the rotor current has to be limited, the inner loop serves as a current limiter.

17.6.2 System Boundaries Revisited

In order to find a controller structure the system boundaries have do be defined, as described in Section 17.1 and in Example 17.1. Due to the strong interaction

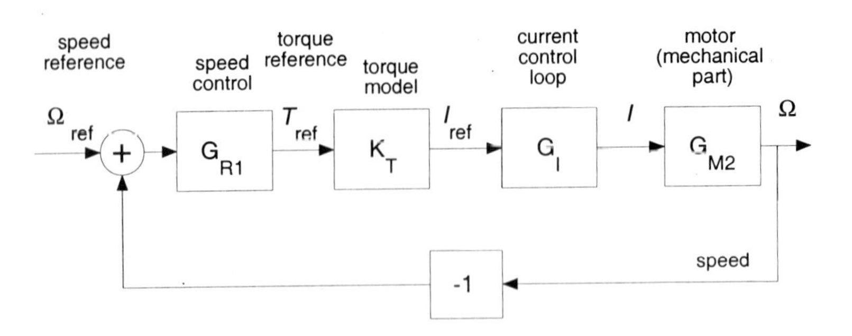

Figure 17.23: Simplified block diagram of the speed controller of an electric drive system.

between DO concentration and respiration rate the definition of system boundaries in DO control and respiration rate control are not obvious. Before we look into the control configurations we have to understand the various cause-effect relationships. Figure 17.24 depicts three system variables, the substrate concentration (here we only consider one substrate), the DO concentration and the respiration rate r. The ovals indicate the system variables and the arrows the cause-effect relationships.

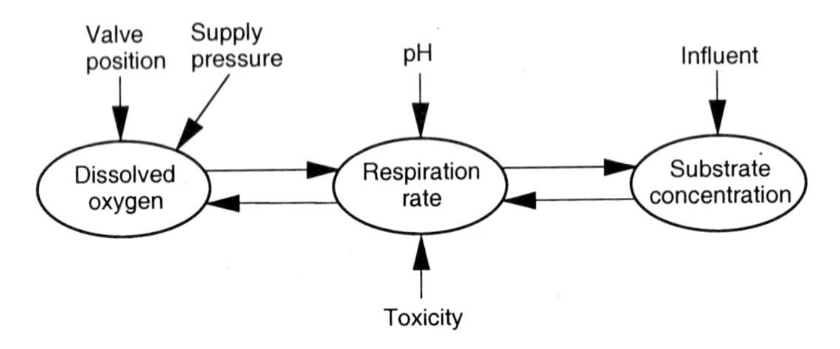

Figure 17.24: The relationship between substrate concentration, dissolved oxygen and respiration rate.

It is apparent that r will be influenced by external inputs like pH and toxicity, so they are considered disturbances. The influent substrate concentration will affect the aerator substrate concentration, that in turn will influence r. Likewise the DO concentration is manipulated by the air flowrate and will affect r. However, the coupling between r, substrate and DO concentrations is bi-directional. An increasing respiration rate in turn will influence the substrate utilisation and the oxygen uptake rate and thus the DO concentration. The actual direction of the cause-effect relationship depends on the view defined by the control system structure.

Example 17.11 (System boundaries for DO control) *Conventional DO control was discussed in Chapter 3. The DO measurement is compared with the*

DO set-point and the air flow is subsequently manipulated to keep the DO close to the set-point despite disturbances, Figure 17.25. As remarked, the DO controller is cascaded with a slave controller to control the valve position.

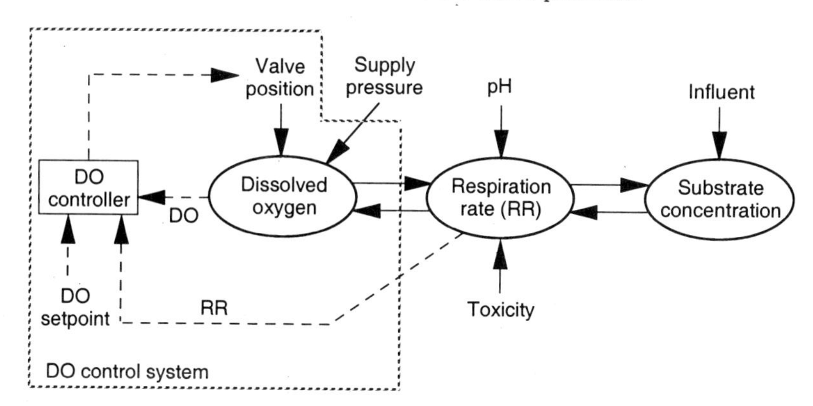

Figure 17.25: Conventional dissolved oxygen control. The respiration rate is considered a disturbance.

A simple feedback DO controller does not explicitly consider the respiration rate. It is only noted implicitly via the DO concentration. For example, assume that a toxic substance enters the plant. As a result the respiration rate will drop. The DO controller will notice that less air will be needed to reach the DO setpoint value, but does not explicitly recognise the toxic disturbance. A decreasing substrate concentration would have caused a similar control action.

From the controller's point of view the respiration rate is a disturbance in the same way as the influent substrate concentration, pH or toxicity. If the respiration rate is measured, either in the influent flow or early on in the aerator, the signal can be fed forward to the DO controller, thus improving the control performance (see further Chapter 18). The fact, that there is a coupling back from the DO concentration to the respiration rate, as shown in Figure 17.25, is not considered by this control structure.

The DO control with feedforward can also be depicted as a conventional block diagram, more easily recognised by control people, Figure 17.26.

Example 17.12 (Respiration rate control) *Consider respiration rate control, Figure 17.27. The purpose of this controller is to keep the respiration rate close to its setpoint value. Consequently the respiration rate is measured and compared to the setpoint. The controller manipulates the air flowrate to achieve this goal.*

For the DO controller the respiration rate was considered a disturbance. Here it is the controlled variable. Note, that the DO concentration is considered an intermediate variable, that is influenced so that the respiration rate will be changed adequately. The controller does not specifically measure the DO concentration.

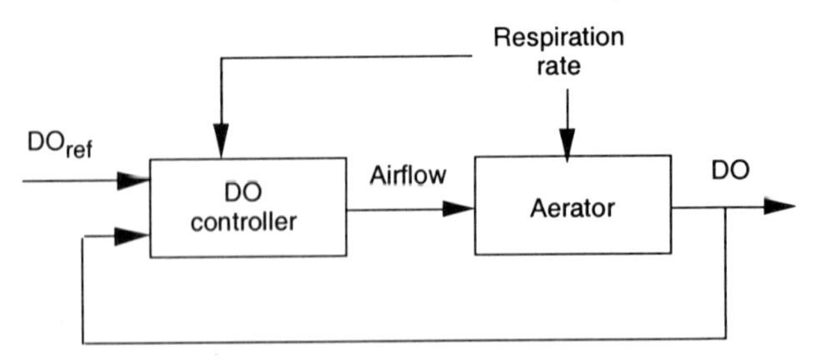

Figure 17.26: Conventional block diagram of DO control with feedforward from the respiration rate.

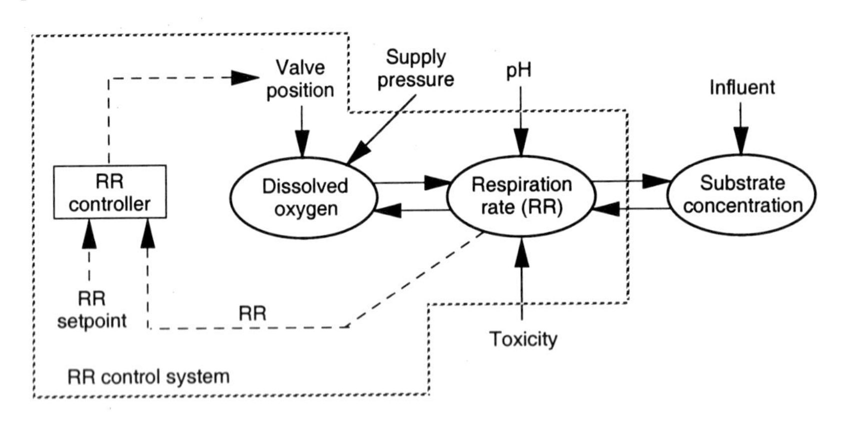

Figure 17.27: Structure of respiration rate control.

The respiration rate is changed due to variations in some of the external parameters like the influent substrate, pH or toxicity. If the influent substrate is measured, then this signal can be fed forward to the controller. Ideally, the feedforward controller should command an air flowrate change before the disturbance has appeared in the respiration rate. To do this a good model of the dynamics of the respiration as a function of the influent substrate is needed. Note that the controller structure defines the directions of the cause-effect relationships between the substrate, respiration and the DO concentration.

Example 17.13 (Cascaded respiration rate control) *The scheme from Figure 17.27 can be further refined, as depicted in Figure 17.28. Instead of letting the respiration rate controller influence the air flowrate directly it will only change the DO setpoint, so we obtain a cascaded control.*

The DO controller is now a slave controller that will make the real DO concentration change until the respiration rate has reached the desired set-point. Notice that the DO controller can contain another slave controller, to control the air flowrate. The real advantage is that the outer loop, the respiration rate controller, does not have to consider the DO dynamics. This is taken care of by the slave

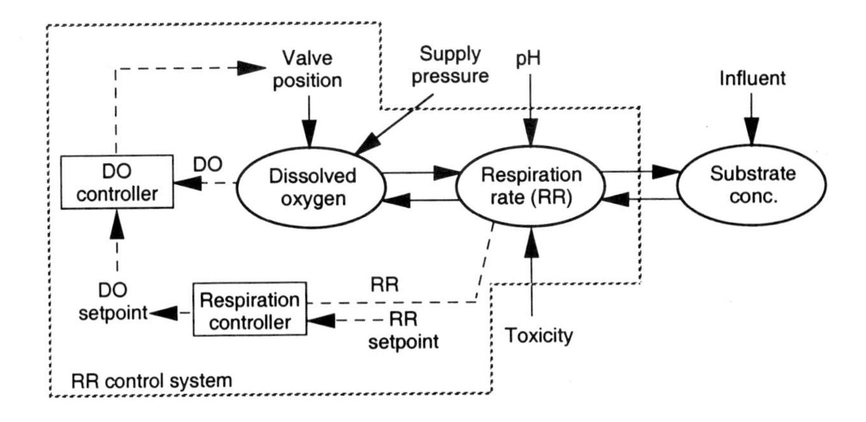

Figure 17.28: Cascaded respiration rate control.

controller. Such a control structure makes commissioning and tuning easier.

Naturally, the cascaded control can be combined with a feedforward compensator. For example, the influent TOC concentration can be measured and fed forward to the respiration rate controller.

17.7 Nutrient Removal Control

The development of advanced nutrient analysers has made it possible to introduce better control (see Chapter 23). In biological nitrogen and phosphorus removal there are many factors that influence the reaction rates, such as the amount of microorganisms, temperature, substrate composition and concentration. There are only a few ways to influence the nitrification/denitrification rates in practice. One is to adjust the dissolved oxygen set-point in the aeration zone, another to control the dosage of external carbon. Below we will describe some control schemes that make use of sensors or instruments for ammonium, nitrate and phosphorus. In Chapter 19 we will consider similar control problems in sequential batch systems.

17.7.1 Dissolved Oxygen Variable Setpoint Control

In a nitrifying aerator, oxygen is consumed as a result of both carbonaceous and nitrogenous removal. Since the nitrification is the slowest reaction it will be the determining factor. It is essential to control the DO in such a system. Once the DO setpoint is sufficiently high the nitrification is guaranteed. Here we will look at a *time varying* DO setpoint. The setpoint value will be determined by the ammonium concentration at the end of the aerator. The control structure is a cascaded control system and is shown in Figure 17.29.

The reason for implementing the variable DO setpoint controller is to save

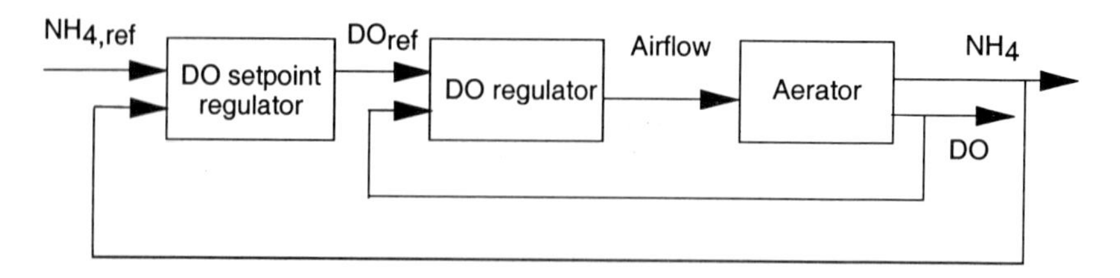

Figure 17.29: Block diagram of the variable DO setpoint control.

energy by minimising the air flow requirement. The idea is to reach only the required, not the minimum, ammonium concentration at the end of the aerator. If the ammonium concentration is too high, then the DO setpoint is increased and vice versa. The setpoint value is sent to the DO controller that keeps the DO at the required value. Note that the DO controller itself is usually a cascaded controller, as described in the previous section. For safety reasons it is natural to define upper and lower limits of the DO set-point.

Example 17.14 (Test of variable DO setpoint) *The strategy was tested in a pilot plant in Uppsala, Sweden. Some results are shown in Figure 17.30. The control is turned on at about 1.5 days. Two parallel liquid lines in the Uppsala pilot plant are compared, a time varying setpoint control (P1) and constant DO set-point control (P2). The same nitrate analyser is used for the two lines.*

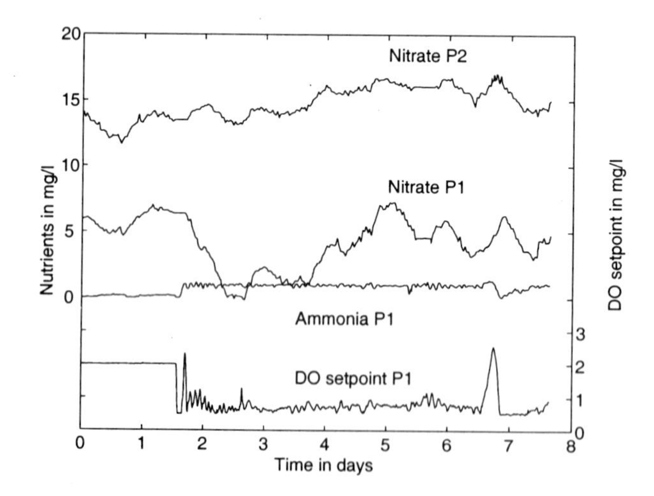

Figure 17.30: Dissolved oxygen control with variable setpoint.

The controller was turned on at about 1.5 days, using an ammonium set-point of 1 mg/l. Since the current ammonium value was lower than 1, the DO set-point was decreased from 2 mg/l to less than 1 mg/l. Due to the lower DO concentration the effluent nitrate concentration was also decreased. The reason is most probably that there is a simultaneous denitrification in the aerated zone. With a lower DO level the simultaneous denitrification is favoured, and the available carbon can be used.

It has to be established if the floc properties will be negatively affected by this control. Furthermore, the influence on the nitrate change may be different if there is a larger anoxic zone preceding the aerator.

17.7.2 Carbon Dosage Control in Denitrification

In Chapter 14 we described that the denitrification rate depends on adequate carbon concentration. The nature of the carbon source will not only influence the denitrification rate. It may also induce selected microorganisms, and the adaptation of the organisms may take weeks. The adaptation time is quite different for different carbon sources. Carbon sources like ethanol can be switched off for several days without losing the denitrification capacity. It has been argued that the flowrate of external carbon should not be controlled automatically, due to the coupling between the carbon source and the composition of the organisms. It also depends on reliable nitrate analysers.

A reasonable control strategy for the carbon dosage is to keep the nitrate level at the outlet of the anoxic zone at a prescribed low level. Later in Chapter 19, we will describe control in sequential batch reactors. There a model of the denitrification rate can be used to determine the carbon dosage, so that the denitrification is completed within the current operation cycle.

A PID controller may be adequate. Feedforward has been suggested to improve the capability of the controller, Figure 17.31.

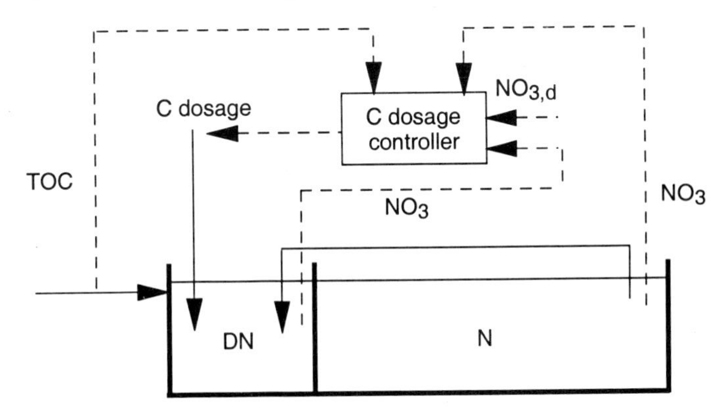

Figure 17.31: External carbon control with feedback and feedforward.

The mass flow of influent TOC and mass flow of nitrate in the recirculated water can be measured and used as feedforward signals. To the authors' knowledge such feedforward has not been implemented, but simulations have shown favourable behaviour. Sensors for nitrate and TOC are further discussed in Chapter 23.

17.7.3 Control of Chemical Dosing for Phosphorus Removal

As the precipitation chemicals are added to the process many approaches to control the dosage based on different parameters have been tried over the years. A proper dosage of chemicals for phosphorus precipitation should of course rely on on-line P analysis. Wet chemical phosphate analysers were tried unsuccessfully during the late 1970s and early 1980s. Basically these instruments were too labour intensive in comparison to what could be gained from an improved operation. In addition in certain areas with hard water it could be discussed whether phosphate was the appropriate parameter to measure. Other approaches have explored less complicated instruments which measure turbidity or particles in the water. In most cases parameters such as turbidity and phosphorus are correlated in time and it should consequently be possible to devise a control structure based on other parameters than phosphate.

The approach to use a parameter which is correlated to the phosphorus concentration has been further extended and today the dosage is in most cases based on the flowrate to the plant. A successful approach was made in the early 1980s at the SYVAB Himmerfjärden Plant in Sweden, where information on historic records of normal daily or hourly variations of the phosphorus concentration was added to the flowrate measurements. The dosage was also adjusted during large flowrates due to rain. Furthermore, the dosage rate was adjusted to the water quality obtained in off-line laboratory tests.

Since the reliability of phosphorus analysers has increased during the last few years, chemical dosage based on phosphorus analysis has been implemented in some plants. The dosing is typically a combined calculation of a basic dosing adjusted once a day and a more sensitive dosing based on the actual phosphorus measurement, effectively feedforward and feedback control.

17.8 Summary

We have introduced the concepts and algorithms for feedback controllers and illustrated them by looking primarily at aspects of the control of dissolved oxygen concentration. Some of the more complex issues in dissolved oxygen control will be addressed in Chapter 18. The use of oxygen uptake rate (OUR) as a diagnostic tool was addressed in Section 12.4.3. The issue of selecting appropriate setpoints for DO controllers will be further addressed in Chapter 18.

The major limitation of feedback control, the fact that the disturbance has already affected the process before the controller sees an error, is addressed by various forms of model-based predictive algorithms configured as *feedforward* controllers. We will look at this in detail in the next chapter, so keep reading, the best is yet to come.

17.9 Further Reading

- There is a whole literature on *process control* but Luyben (1990), Franklin *et al.* (1986), Seborg *et al.* (1989), Stephanopoulos (1984) and Lee *et al.* (1998) give a comprehensive view of the subject. A control perspective of the process industry is presented by Morari (1993).

- Both theoretical and practical aspects of *PID controllers* are discussed in Åström & Hägglund (1988). They also describe controllers with nonlinear elements such a overrides and selectors.

- *IMC tuning* is discussed in Rivera *et al.* (1986) and closed-loop model identification is summarised in Taiwo (1993).

- For a light hearted look at *controller tuning* and other control issues, try McMillan & Weiner (1994).

- *Auto-tuning* is described in detail in Åström & Hägglund (1988).

- *Self-tuning control* as well as a comprehensive treatment of other adaptive controllers is found in Åström & Wittenmark (1995). Cameron & Seborg (1983) also discuss self-tuning control using a PID controller. Self-tuning control was implemented in 1984 at the Käppala plant in Sweden (Olsson *et al.* (1985b)) and has been in use for several years and performs satisfactorily (Rundqwist (1988)).

- Early applications of dissolved oxygen control are presented in Flanagan *et al.* (1977), Olsson (1977), Gillblad & Olsson (1977) and Wells (1979).

- Other early applications of wastewater treatment control are found in Flanagan (1979) and Manning & Dobs (1980).

- The *nonlinear DO controller* and the *variable setpoint DO controller*, Figures 17.19 and 17.30, are published by Lindberg & Carlsson (1996c).

- *Carbon dosage* in recirculating plants has been studied by Aspegren *et al.* (1992) and Hellström & Bosander (1990). The control structure in Figure 17.31 is published in Lindberg & Carlsson (1996a) and in Hallin *et al.* (1996). Nyberg *et al.* (1993) and Nyberg *et al.* (1996a) warned against automatic control of carbon due to the adaptive sensitivity of microbial systems. Hasselblad & Hallin (1996) have shown that ethanol dosage can be switched off for several days without losing denitrification capacity. Experiences of methanol and ethanol in post-denitrification has been demonstrated by Andersson *et al.* (1995). A strategy where the carbon dosage is based on presence or no presence of nitrate in the effluent has been suggested by Vanrolleghem *et al.* (1993). The carbon dosage in a post-denitrification pilot plant has been controlled by considering the oxygen demand (Linde (1993)).

- More on *electrical drives* can be found in Leonhard (1997).
- Early approaches to *chemical dosing control* were made at the Himmerfjärden Plant in Sweden (Hellström *et al.* (1984)).

17.10 Exercises

1. The MATLAB program `Example_17_3.m` implements PID control of the dissolved oxygen in a simple aerated carbon removal tank. Modify the program to implement respiration rate control (substitute `our` in place of `so` in the "sensor dynamics" part of the process model and choose suitable sensor characteristics, gain `sgain` and time constant `stau`). Retune the PID controller using the IMC tuning rules.

2. The MATLAB program `Exercise_2.m` simulates the control of nitrification by adjusting the dissolved oxygen level using a PID controller. Tune the PID controller using the IMC tuning rules.

3. The MATLAB program `Exercise_3.m` simulates the control of denitrification by adjusting the carbon dosage rate using a PID controller. Biomass adaptation is not included. Tune the PID controller using the IMC tuning rules.

4. Using the respiration rate control problem in Exercise 1, determine the "process gain" $(\partial y/\partial u)$ as a function of influent substrate loading for the two cases of a linear valve and of an equal-percentage valve (these two options are already coded into the MATLAB program `Example_17_3.m`). You can determine the gain curves either algebraically or numerically. For one or both cases design and simulate a suitable nonlinear gain controller. You can use either gain scheduling or an exact linearisation approach (if the latter is invertible).

5. Choose one of the previous exercises and investigate the effect of increasing the control interval (by increasing the MATLAB variable `conint`). Then examine the interaction between the control interval and the IMC tuning parameter τ_m.

6. The MATLAB program `Example_17_3.m` implements PID control of the dissolved oxygen in a simple aerated carbon removal tank. Modify the program to implement a proportional only slave controller measuring air flowrate `va` and adjusting the valve signal `u`. The existing PID controller will now control the setpoint of the slave controller. Investigate the disturbance rejection

performance with and without the slave controller to disturbances in the influent substrate and the air pressure.

Chapter 18

Model Based Control

The goals of this chapter are to introduce the concepts and algorithms for commonly used model based controllers, and to illustrate the algorithms and their advantages and disadvantages with examples appropriate to wastewater treatment.

We will begin by looking at feedforward compensation, perhaps the simplest and most widely applied model-based control, Section 18.1. Then we will look at the principles of predictive control and some techniques for implementation, Section 18.2. We will also look at one of the nonlinear model based controllers that has also been implemented in the process industries, Section 18.3.

One of the inherent problems of more advanced model based controllers is that they require more information about the process. The increasing availability of analysers, for nutrients and oxygen uptake rate analysers, certainly helps in this regard. In fact, utilising model based control is one way of exploiting the potential of and paying for the investment in such analysers. State feedback, Section 18.4 utilises information from all the states. Despite the new analysers, there are often measurements of states needed for which analysers are not available. Here we can sometimes make use of state estimation, as discussed in Chapter 12. We will apply this in Section 18.5.

Sludge inventory control makes use of relatively simple models for an adequate control, Section 18.6. A couple of examples of nutrient removal control are presented in Section 18.7.

Fuzzy sets and fuzzy logic were introduced in Sections 5.5.3 and 13.3.3. Here we will complete the picture by control using fuzzy logic, Section 18.8.

Finally we will look at some of the considerations that you have to take into account for the design of model based controllers, Section 18.9.

18.1 Feedforward Compensation

Figure 18.1 was presented in the previous chapter on feedback control. It is a conceptual illustration of the key concepts of feedback and feedforward control. We looked at feedback control and its most popular application in biological wastewater treatment plants, DO control.

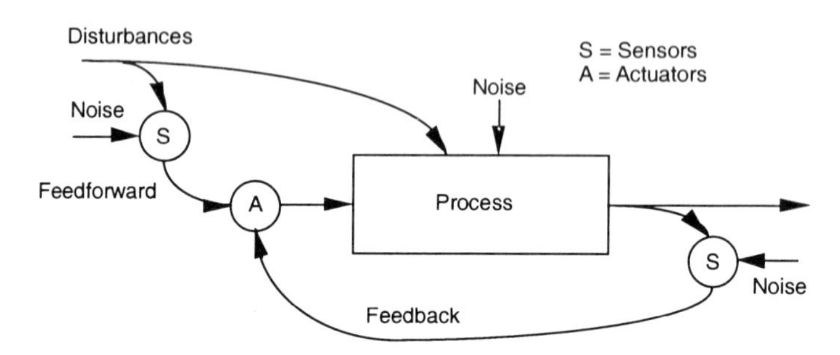

Figure 18.1: Feedback and feedforward.

The whole principle of feedforward control is very simple, measure the disturbance and adjust the manipulated variable to compensate for the effect of the disturbance. The ideal is that the effects of the disturbance and the feedforward compensation will exactly cancel out, so that there will be no disturbance of the output.

We can draw a more detailed block diagram of this as shown in Figure 18.2. It is clear that the net effect of the sensor plus the compensator plus the manipulated variable dynamics should be equal and opposite to the effect of the disturbance variable dynamics. This is the basic design principle of a feedforward compensator.

It is also clear that, to design the compensator quantitatively, we must have models of the sensor, the disturbance variable dynamics and the manipulated variable dynamics. That is why feedforward is model based.

Our block diagram in Figure 18.2 and the discussion above imply some conditions for the physical realisation of a feedforward compensator:

- We must be able to measure the disturbance, either directly or by estimation from other measurements.

- The dynamics via the sensor, compensator and manipulated variable must be at least as fast as those via the disturbance dynamics.

The process output can be considered a sum of two terms, the disturbance dynamics output y_t^d and the manipulated variable dynamics output y_t^u.

Suppose we model the disturbance dynamics by a simple first-order time series model:

$$y_t^d = a_1^d \, y_{t-1}^d + b_0^d \, d_{t-d^d} \tag{18.1}$$

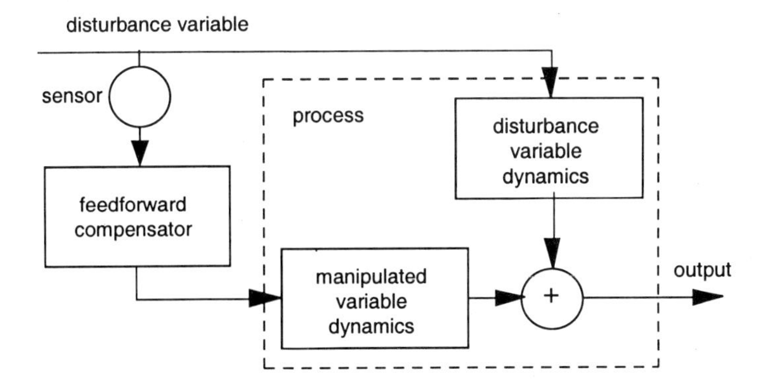

Figure 18.2: Feedforward block diagram.

where d_t is the disturbance variable and d^d the dead time.

We have seen these types of models several times in earlier chapters. We can write this model in a different form:

$$y_t^d = \frac{b_0^d \, q^{-d^d}}{1 - a_1^d \, q^{-1}} \, d_t = G_d \, d_t \tag{18.2}$$

where q^{-1} is known as a *shift operator* since it implies a shift backwards in time by one sampling interval:

$$y_{t-1} = q^{-1} y_t \tag{18.3}$$

We could also combine the sensor and the manipulated variable dynamics and model them similarly:

$$y_t^u = \frac{b_0^u \, q^{-d^u}}{1 - a_1^u \, q^{-1}} \, u_t = G_u \, u_t \tag{18.4}$$

where u_t is the manipulated variable and d^u the dead time for the manipulated variable dynamics.

Our design was based on the principle that the net effect on the output is zero. Assume that the feedforward compensator is modelled by G_{ff}. Then we get:

$$y_t = y_t^u + y_t^d = G_u \, u_t + G_d \, d_t = G_u \, G_{ff} \, d_t + G_d \, d_t = 0 \tag{18.5}$$

so that it follows from Equations 18.2, 18.4 and 18.5 that the compensator must be:

$$G_{ff} = -\frac{G_d}{G_u} = -\frac{b_0^d \left(1 - a_1^u \, q^{-1}\right)}{b_0^u \left(1 - a_1^d \, q^{-1}\right)} q^{-(d^d - d^u)} \tag{18.6}$$

We have three terms to our feedforward compensator:

- a gain term, $-\frac{b_0^d}{b_0^u}$

- a dynamics term, $\frac{(1-a_1^u\, q^{-1})}{(1-a_1^d\, q^{-1})}$ (this is the so-called *lead-lag* compensator)
- and a dead time term, $q^{-(d^d-d^u)}$

The dead time term is generally small and is frequently neglected. However, if $d^u > d^d$ then it says that the term is a pure predictor that is unrealisable. In this case, a perfect compensation is not possible, and even a partial feedforward compensation (neglecting the term) may not be successful if the difference is large.

If the major time constants for the manipulated variable and disturbance dynamics are similar, $a_1^u \approx a_1^d$, then the dynamic term is often also neglected, leaving us with a static compensator with just the gain term, $-\ b_0^d/b_0^u$.

Example 18.1 (Feedforward control) *Let us look at a simple example that has been implemented in the MATLAB program* `Example_18_1.m` *for you to play with. The block diagram for our example is shown in Figure 18.3.*

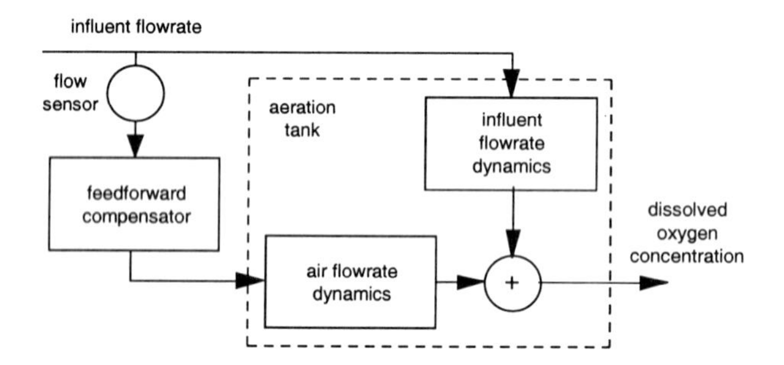

Figure 18.3: Feedforward example.

The process is simulated using the nonlinear mass balances for substrate and oxygen, the first two equations of our simple carbon removal plant detailed in Appendix B. The variables and standard values are listed there in Table B.1.

The first task is to obtain even simpler models for designing the feedforward compensator. This design was achieved in three ways:

- *A dynamic design based on open-loop step tests (marked STD in the plots in Figure 18.4). Up and down steps were performed on the process and the responses were fit by first-order models (see Chapter 7). This gave a lead-lag compensator design as discussed above:*

$$G_{ff} \quad = \quad 8.7\,\frac{(1\ -\ 0.9785\, q^{-1})}{(1\ -\ 0.9840\, q^{-1})} \tag{18.7}$$

- *A dynamic design based on a linearised model (marked MBD in the plots in Figure 18.4). The nonlinear equations were linearised and the design performed by equating the output to zero. This resulted in a gain plus lag compensator:*

$$G_{ff} \quad = \quad \frac{0.646}{(1 \; - \; 0.942 \, q^{-1})} \tag{18.8}$$

- *A static design which gives a compensator with a gain of 11.7.*

These compensators were simulated using the `Example_18_1.m` *program. The left plot in Figure 18.4 shows DO concentration responses after a ten per cent influent flowrate disturbance. Clearly the model-based design produced a more accurate dynamic compensator, which in this case turned out to be an even simpler compensator.*

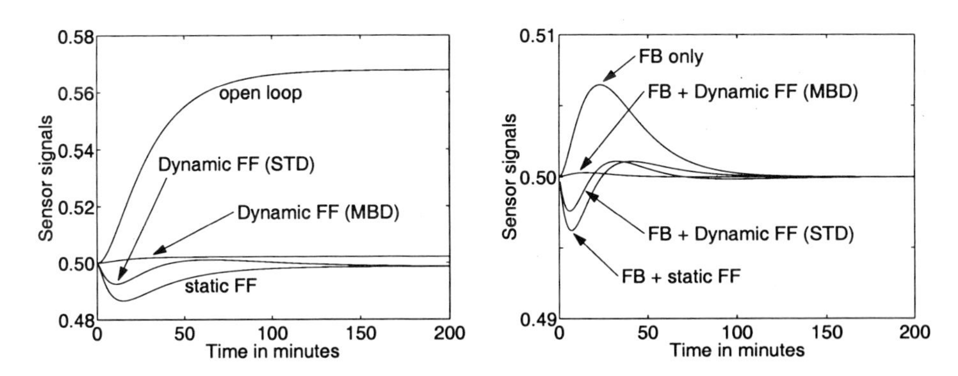

Figure 18.4: Feedforward control (left) and feedforward plus PID (right) (FB = feedback, FF = feedforward, STD = step test design, MBD = model based design).

The plant-model mismatch evident particularly with the static and step-test (STD) designs is common with feedforward designs. It is usually reduced by combining a feedforward compensator with a feedback controller as shown Figure 18.1 above. The simulated results of doing this are shown in in the right plot in Figure 18.4 where a feedback only response is shown for comparison.

Clearly we get a considerable improvement from very simple feedforward compensators. Of course you must measure the true disturbance. Now we will consider a second example, by combining the disturbance modelling we did in Chapter 15 with our simple carbon removal process simulation used in the previous example.

Example 18.2 (Feedforward from modelled disturbance) *Again we consider the control of DO with influent disturbances in flowrate and COD of the same magnitude and form as those we modelled and displayed in Figure 15.9. Figure 18.5 shows the results of feedback, feedback and feedforward from measured influent flowrate, and feedback and feedforward from both measured influent flowrate* *and influent COD predicted from the disturbance models (MATLAB program* `Example_18_2.m`*). The mean disturbance rejection factors are shown in Table 18.1.*

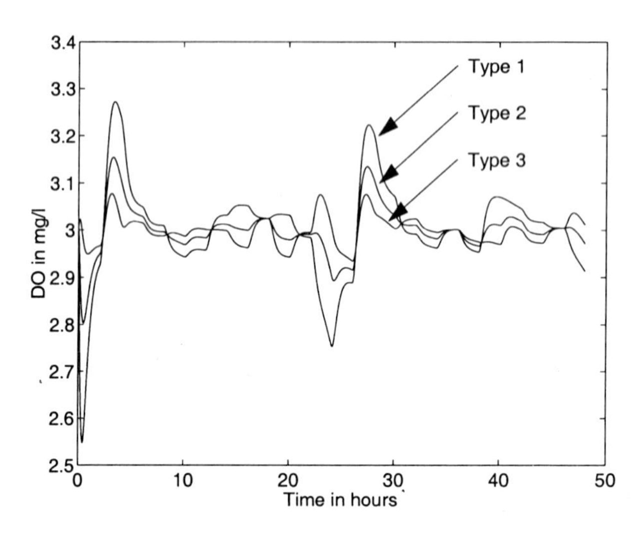

Figure 18.5: Influent feedforward (Table 18.1 defines control system types).

Type of control (DO setpoint at 3 mg/l)	Range of DO	Standard deviation	Rejection factor
0 - No control	0.2 - 6.9	2.00	1
1 - Feedback only	2.5 - 3.3	0.10	20
2 - FB plus FF from measured influent flow	2.8 - 3.2	0.05	40
3 - FB plus FF from measured flow and modelled COD	2.9 - 3.1	0.03	70

Table 18.1: Feedforward disturbance rejection factors.

While feedforward from only part of the disturbance (flowrate) gives an improvement, full feedforward is preferable. In this case we could measure the influent flowrate and calculate the influent COD from the flowrate using our disturbance models. Regular updating of the disturbance models could be performed on the basis of laboratory COD analyses.

Similar results could be obtained using effluent COD as the output variable of interest instead of DO, in which case feedback would require a COD analyser.

Three applications of feedforward control implemented in a biological wastewater treatment plant were:

- MLSS and temperature to air flowrate also with feedback from DO sensors.
- Influent flowrate and COD to DO setpoints. Here it is feedforward only with effluent COD as the output variable.
- Influent flowrate and COD and aeration tank temperature to MLSS setpoint. Here it is feedforward only with F/M ratio as the output variable.

Other common applications of feedforward control in biological wastewater treatment plants include:

- Using respirometry as an early warning system for toxicity.
- Using respirometry to determine the influent biodegradable organic substrate concentration.
- Using *in-stream sensors* to analyse the influent stream, to be used as feedforward information.
- Various proposals have been made of feedforward/feedback control strategies based on respirometry. The respiration rate is determined in various parts of the activated sludge system, and several manipulated variables have been suggested including air flowrate, return sludge, waste sludge, internal nitrate recycle, sludge treatment recycle flowrates as well as chemical dosage. Most of these schemes are only tested in simulation.

18.2 Predictive Controllers

Predictive controllers are the most successful of the advanced multivariable controllers in use in the process industries. There are several commercial implementations which can be purchased if you have a deep enough pocket. Despite their commercial price tag, they are relatively simple to implement for any control engineer who is familiar with modern matrix manipulation techniques.

Their popularity is due to the simple form of a linear process model which they use, and to their ability to incorporate practical constraints. They are also simple to understand in concept. The basic concept is to choose a number of future adjustments of the manipulated variables to minimise the errors between the controlled (output) variables and their setpoints, based on the predicted response of the process. The model is used to calculate the predicted response. The first one or two control moves are implemented and then a new set of predictions are made.

Because we are looking ahead in time it is possible to predict our approach to constraints and to avoid control moves which will violate them.

The model used is the *dynamic matrix* or convolution model which is usually derived from a number of step tests on the process under *test run* conditions. It could also be derived from a validated mechanistic model, though if you had one of these you might use different control algorithms, such as the generic model controller discussed in the next section. Figure 18.6 shows a step response and the step response weights on which the model is based (similar in concept to the impulse response models we discussed previously in Chapter 7).

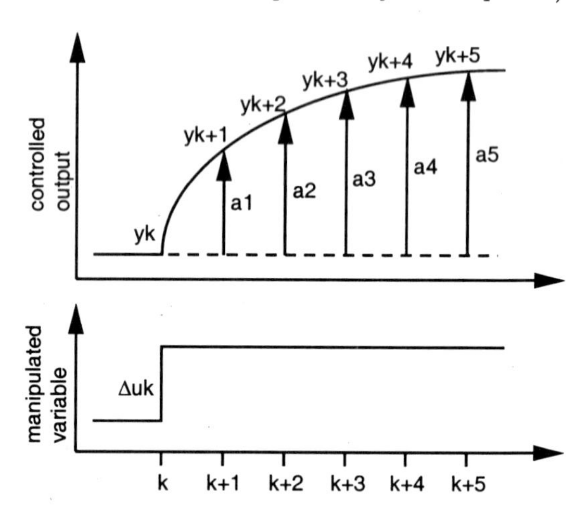

Figure 18.6: Step response weights.

The controlled variable at different sampling times after the step can be expressed in the form:

$$\begin{aligned} y_1 &= a_1 \, \Delta u_1 \\ y_2 &= a_2 \, \Delta u_1 + a_1 \, \Delta u_2 \\ y_3 &= a_3 \, \Delta u_1 + a_2 \, \Delta u_2 + a_1 \, \Delta u_3 \end{aligned} \tag{18.9}$$

or in matrix notation as:

$$y = A \, \Delta u \tag{18.10}$$

where y is the output deviation from steady state, Δu is the change in the input manipulated variable and the *dynamic matrix* is:

$$A = \begin{bmatrix} a_1 & 0 & 0 \\ a_2 & a_1 & 0 \\ a_3 & a_2 & a_1 \end{bmatrix} \tag{18.11}$$

A is constructed to the appropriate dimension to encompass the step response of the slowest variable of interest. Multiple manipulated variables and controlled outputs are accommodated by simply extending the vectors y and Δu and the dynamic matrix A.

$$A = \begin{bmatrix} A_{11} & A_{12} & .. \\ A_{21} & A_{22} & .. \\ .. & .. & .. \end{bmatrix} \tag{18.12}$$

This then is the model that we use to predict the effect of past control moves into the future. From the prediction of the controlled variables we can calculate the *predicted error*, the difference between the predicted and desired values. On the basis of this error prediction we must design our controller to calculate some manipulated variable moves:

$$\Delta u \quad = \quad f(e) \tag{18.13}$$

The literature gives a number of algorithms for $f(e)$ with various checks and techniques for handling constraints. The theoretically ideal controller is the *inverse model*:

$$\Delta u \quad = \quad A^{-1}\, e \tag{18.14}$$

which selects N_P manipulated variable moves based on N_P error predictions. This is unlikely to be robust since our model is not perfect. The simplest practical design is to select N_C manipulated variable moves, where $N_C < N_P$, using a least squares solution of the inverse:

$$\Delta u \quad = \quad \left(A^T Q A + R\right)^{-1} A^T Q\, e \tag{18.15}$$

This algorithm also includes diagonal weighting matrices for the error, Q, and for the manipulated variable moves, R. Since matrix inverses are numerically intensive, there are improved algorithms based on singular values and principal component analysis (see Section 11.4.1) to avoid calculating the inverse.

Example 18.3 (Predictive control of denitrification) *As an example application of predictive control we will examine the control of anoxic zone denitrification. The output is the signal from a nitrate sensor and the manipulated variable is the rate of external carbon addition. The left plot in Figure 18.7 shows open loop step tests of 20% in the carbon addition and the predictions of the dynamic matrix constructed from an average of the two step responses. The control interval is 10 minutes and the prediction horizon N_P is 30.*

The MATLAB program `Example_18_3.m` *produced the right plot in Figure 18.7 which shows responses of a PI controller (dashed line) and a predictive controller (solid line). A 50% increase in nitrate loading occurred at time zero and a setpoint change occurred at time 250 minutes.*

Both the IMC tuned PI controller and the DMC (dynamic matrix control) predictive controller were tuned for zero setpoint overshoot using the filter constant τ_m and the control action weighting R respectively. Comparable load rejection was possible with the PI controller by tuning it more tightly, but this resulted in overshoot and oscillation for setpoint changes.

One drawback of predictive control is its reliance on a linear model. If the process is nonlinear, and since you are making predictions over a significant time horizon, the model will need to be designed to give worst case predictions. This will inevitably lead to conservative control action and reduced performance. The following section describes one of the few nonlinear model-based controllers.

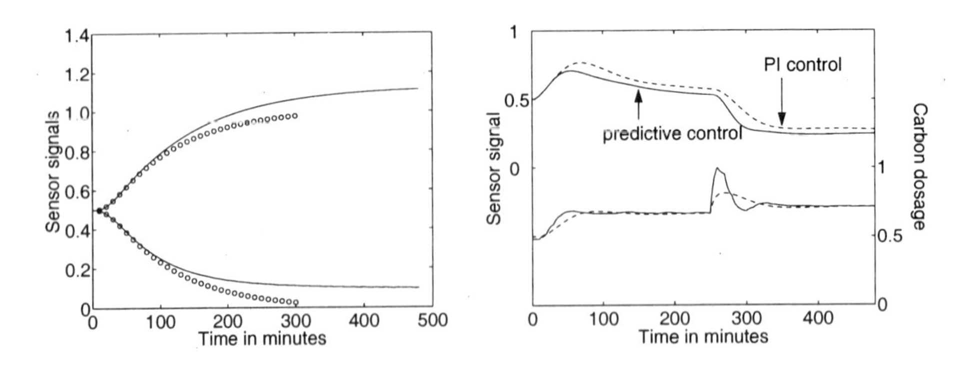

Figure 18.7: Step tests and controller responses.

18.3 Generic Model Controllers

Generic Model Control (GMC) is a control technique which can utilise nonlinear process models. We will examine the simplest form of the algorithm as it was originally published in 1988. There has been considerable development of the concept and several industrial applications.

The design of GMC controllers is based on the concept that, if the process is away from its desired operating state (there is an error), then we would like the rate of change of the state to be a function of that error. We would like the desired rate of change to be proportional to the error (the further we are away, the faster we are returning) and we would like the error to go to zero (implying integral type action). This leads to a control algorithm of the form:

$$\left(\frac{dy}{dt}\right)_d = K_1 \, (y_d - y) \, + \, K_2 \int (y_d - y) \, dt \tag{18.16}$$

Now suppose that we have a process model of the general nonlinear form:

$$\frac{dy}{dt} = f\,(\,x,\;u,\;d\,) \tag{18.17}$$

Ideally the actual rate of change is the real rate of change which implies that:

$$f\,(\,x,\;u,\;d\,) \;-\; K_1 \, (y_d - y) \;-\; K_2 \int (y_d - y) \, dt \;=\; 0 \tag{18.18}$$

The control law is obtained by solving this equation for the manipulated variable u, where y are the measured outputs, d the measured disturbances (feedforward action is included) and x the measured and/or estimated states.

One other advantage of GMC is that the developers also give us a systematic procedure for selecting the *controller parameters* K_1 and K_2 based on some relatively intuitive performance specifications. The expressions:

$$K_1 = \frac{2\tau}{\xi} \qquad\qquad K_2 = \frac{1}{\tau^2} \tag{18.19}$$

relate the parameters to a rise time τ, generally set to the major open-loop time constant of the system, and a damping factor ξ, generally set to a value between zero (better performance) and one (more robust).

Of course there is no such thing as a free lunch. Disadvantages include the requirement that we measure or estimate the state x, and that the nonlinear function $f(x, u, d)$ should preferably be explicitly solvable for the manipulated variable u. We could solve an implicit $f(x, u, d)$ numerically, but that would increase our computational load, which may or may not be a concern.

Example 18.4 (GMC control of dissolved oxygen) *As a simple example, we will consider the control of dissolved oxygen on the Simple Carbon Removal process described in detail in Appendix B.*

The control law is derived from the oxygen mass balance (Equation B.5):

$$\begin{aligned}\frac{dS_O}{dt} &= \tfrac{q_F}{V} S_{OF} - \tfrac{q_F+q_R}{V} S_O \\ &\quad + \tfrac{Y_H-1}{Y_H}\hat{\mu}_H \left(\tfrac{S_S}{K_S+S_S}\right)\left(\tfrac{S_O}{K_{OH}+S_O}\right) X_H \\ &\quad + k_1 \cdot \left(1 - e^{-k_2} \cdot q_A\right)\left(S_{O,sat} - S_O\right)\end{aligned} \tag{18.20}$$

and the GMC control algorithm:

$$\left(\frac{dS_O}{dt}\right)_d = K_1 \left(S_{O,d} - S_O\right) + K_2 \int \left(S_{O,d} - S_O\right) dt \tag{18.21}$$

where $S_{O,d}$ is the desired dissolved oxygen value. By combining these equations, and rearranging terms, we can obtain an expression for the air flowrate:

$$q_A = \frac{1}{k_2} \cdot \ln\left[k_1 \cdot \left(S_{O,sat} - S_O\right)\right] - \frac{1}{k_2} \cdot \ln \left(\begin{array}{l} \tfrac{q_F}{V} S_{OF} - \tfrac{q_F+q_R}{V} S_O \\ + k_1 \left(S_{O,sat} - S_O\right) \\ + \tfrac{Y_H-1}{Y_H}\hat{\mu}_H \left(\tfrac{S_S}{K_S+S_S}\right)\left(\tfrac{S_O}{K_{OH}+S_O}\right) X_H \\ -K_1 \left(S_{O,d} - S_O\right) \\ -K_2 \int \left(S_{O,d} - S_O\right) dt \end{array}\right) \tag{18.22}$$

This control law requires values for the variables S_O, S_S and X_H. The dissolved oxygen S_O is a measurement. The term in Equation 18.22 containing S_S and X_H

is the oxygen uptake rate or OUR which can also be measured. Alternatively it would be possible to use the process model consisting of Equations B.4 to B.6 in Appendix B to construct a state estimator to get estimates of S_S and X_H (see Chapter 12 and Example 12.4). This leaves us with a nonlinear feedback control law from DO, and nonlinear feedforward compensation from OUR and the flowrates q_F and q_R.

The GMC controller in Equation 18.21 used tuning parameters K_1 and K_2 based on a rise time of 0.3 hrs and damping factor of 0.5.

The MATLAB program `Example_18_4.m` *simulates PI or GMC control of dissolved oxygen, using Equations B.4 to B.6 in Appendix B as the* process. *The plots in Figure 18.8 (PI control on the left and GMC control on the right) show DO and substrate responses to influent flowrate and substrate concentration disturbances of +50% and +33% respectively at 1 hour, -50% and -33% at 3.5 hours, and returning to the initial values at 6 hours. The improvement of the inherent feedforward and nonlinear compensation is clear despite the 5 minute lag applied to the OUR measurement to allow for analyser dynamics.*

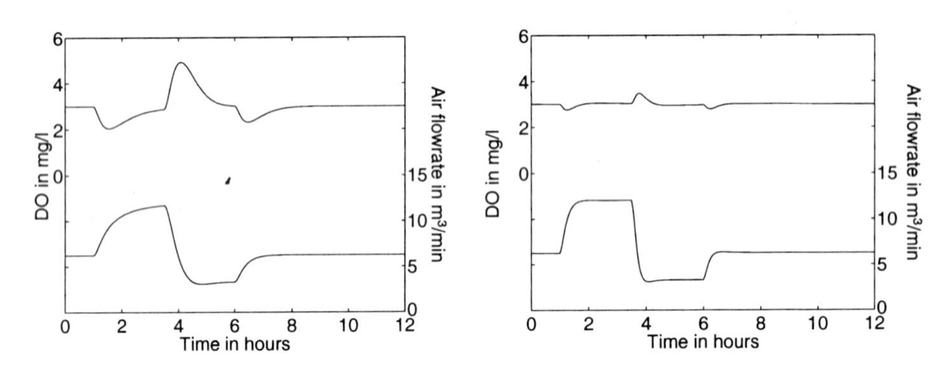

Figure 18.8: PI control (left) and GMC control (right) of the dissolved oxygen concentration.

18.4 State Feedback Controllers

In many process environments, including wastewater treatment, it is not possible to measure all the variables of interest. In other cases it may be possible to measure a variable but the measurement is heavily contaminated with noise. Since the variables of interest are always functions of the state variables, the problem becomes one of estimating and/or filtering state variables. This has been described in Section 12.3. Here we will now describe how the information of the estimated states can be used for a controller.

In Chapter 3 we found that the activated sludge system is characterised by a large span of time constants that range from seconds (pressure in air pipes) to weeks (sludge retention time). To cope with the control we have divided the plant system into smaller systems. The phenomena within the smaller model have approximately the same time constant. The set-point of a fast process can then be controlled by a slower process (cascade control, Section 17.6). In that way the fast dynamics may be neglected when considering the slower models.

Example 18.5 (Multivariable control) *Consider the incidence matrix in Table 16.2. We find that the external carbon dosage, internal nitrate recirculation, and the DO setpoint affect the ammonia and nitrate concentrations and the response time is of the same order of magnitude (hours). Now we may consider a system with three inputs and two outputs depicted in Figure 18.9. This example is taken from Lindberg (1997).*

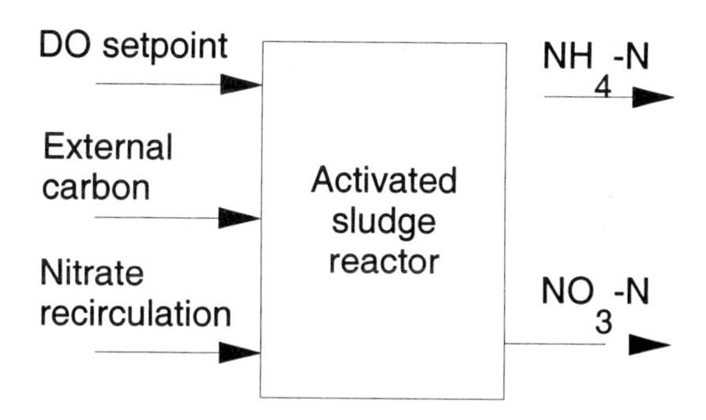

Figure 18.9: The activated sludge system as a multivariable system.

The model indicates that there are six possible feedback loops in the system, since each one of the inputs can influence both the outputs. Let u represent an input vector and the two concentrations as a state vector x. Then we will design a controller of the form

$$u = K \cdot x \tag{18.23}$$

where K is a matrix with 6 elements. In other words, we have to determine 6 controller gains:

$$\begin{aligned} u_1 &= k_{11} \cdot x_1 + k_{12} \cdot x_2 \\ u_2 &= k_{21} \cdot x_1 + k_{22} \cdot x_2 \\ u_3 &= k_{31} \cdot x_1 + k_{32} \cdot x_2 \end{aligned} \tag{18.24}$$

It seems obvious, that we need some systematic method to tune the regulator. It is not always true that the state variables can be measured. Then we may replace x in Equation 18.23 with the estimated values $\hat{x}$ obtained from an estimator or observer (see Section 12.3).

We have seen in Chapter 5 that a nonlinear system can be linearised and written in the form Equation 5.20. We have noted that the measurement vector y usually has a lower dimension than x, and that x can be estimated from a model, given y (Section 12.3).

We now have to define what is meant by a good control. This is expressed as a criterion:

$$\min_u J = \frac{1}{2} \int_0^t \left(x^T Q x + u^T R u\right) dt \tag{18.25}$$

where Q and R are positive definite matrices. The control objective is a quadratic expression in x and u. Assuming that the matrices Q and R are diagonal and the system is of third order with two control variables, then Equation 18.25 looks like:

$$\min_u J = \frac{1}{2} \int_0^t \left(q_{11}x_1^2 + q_{22}x_2^2 + q_{33}x_3^2 + r_{11}u_1^2 + r_{22}u_2^2\right) dt \tag{18.26}$$

This really says that we want to minimise the weighted sum of squares of the deviations in the state variables and the amount of control action taken. Obviously the choice of the weighting matrices Q and R is an important issue.

This *linear quadratic control* problem is an optimization problem with a linear model and a quadratic criterion. It formed the back-bone of the "modern" control theory development during the 1960s, when state space theory was developing.

The problem can be stated as an optimisation problem, where Equation 18.25 has to be minimised under the constraint

$$\frac{dx}{dt} = A\,x + B\,u \tag{18.27}$$

It appears that the solution looks very attractive, and has the form of Equation 18.23, where

$$K = -R^{-1} \cdot B^T \cdot P \tag{18.28}$$

The matrix P is a quadratic symmetric matrix that satisfies the matrix differential equation

$$\frac{dP}{dt} = -PA - A^T P + PBR^{-1}B^T P - Q \tag{18.29}$$

This equation is known as the *matrix Riccati equation*, a famous equation in the control literature. Usually we consider the quadratic criterion Equation 18.25 with $t = \infty$. Then the solution of the Riccati equation becomes a steady state solution with P constant.

Many computer packages, including MATLAB, possess functions to evaluate the feedback gain matrix K, given the system matrices A and B and the weighting matrices Q and R.

The weighting matrices Q and R place importance on different variables. The ratio between Q and R determines the trade-off between tight control and a lot of control action. If Q is much bigger than R then the control action will be more vigorous.

18.5 Model Based Dissolved Oxygen Control

It is of great interest from a diagnostic viewpoint (see Section 12.4.3) to continuously know the value of the oxygen uptake rate OUR while controlling the DO concentration. Since K_La is generally not known, the parameters are not identifiable under normal circumstances. In Example 12.9 we studied the estimation of K_La and OUR in open loop. In other words, the air flowrate could be manipulated freely so that the estimation would succeed. We will now consider the problem of estimating the OUR under *closed loop*. This means that the air flowrate can not be excited arbitrarily, but has to keep the DO at a desired value during the estimation phase.

Example 18.6 (Estimation of respiration under DO control) *We will now consider the problem of simultaneous estimation of K_La and the respiration rate under closed loop DO control. In Example 12.9 the oxygen transfer rate and the respiration were estimated during open loop, and the DO was not automatically controlled. The problem of estimating the respiration under closed loop control becomes more elaborate.*

The estimation part of this scheme looks similar to the previous example, and both the K_La parameters and the respiration are assumed to be unknown. In the open loop case it is all the time assumed that the air flowrate is changed in such a way that the DO concentration will be sufficiently excited. The estimation can not be performed under steady state conditions.

To estimate both K_La and r_O under closed loop DO control a special DO controller can be constructed. The controller has to make a compromise between probing *and* control. *Such a controller is called a* dual controller. *This means that the air flow is used partially to excite the system in order to get relevant information about the unknown parameters and partially to control the DO concentration around its setpoint. The probing part of the controller simply means that it forces the DO concentration to oscillate slightly around the setpoint. The excitation caused by this "relay" function of the controller will perturb the system every sampling instant, typically every 10-15 minutes. In a practical implementation, at the Malmö Sewage Works in Sweden, the DO concentration was allowed to vary 2 $\pm$ 0.2mg/l. For a larger variation the accuracy of the estimation could be increased further.*

To understand how the controller works, we define the error e between the DO setpoint $S_{O,sp}$ and the true DO value S_O,

$$e = S_{O,sp} - S_O \tag{18.30}$$

The controller should work so that the error will approach zero. Therefore we prescribe that the error should obey the differential equation

$$\frac{de}{dt} = -e \tag{18.31}$$

Take the derivative of Equation 18.30:

$$\frac{de}{dt} = \frac{d}{dt}\left(S_{O,sp} - S_O\right) = -\frac{dS_O}{dt} \tag{18.32}$$

since the setpoint is assumed to be constant.

Combining Equation 18.31 and the DO dynamics, Equation 3.35, we obtain:

$$e = \frac{dS_O}{dt} = \frac{q_{in}}{V} S_{O,in} - \frac{q_{out}}{V} S_{O,out} + r_O + K_L a(q_a)(S_{O,sat} - S_O) \tag{18.33}$$

We assume that $K_L a = K_a \cdot q_a$ *(Equation 12.44). The control variable* q_a *can be solved from Equation 18.33:*

$$q_a = \frac{-\left[\frac{q_{in}}{V} S_{O,in} - \frac{q_{out}}{V} S_{O,out}\right] - r_O + e}{K_a \cdot (S_{O,sat} - S_O)} \tag{18.34}$$

This control would be perfect if the estimated values are equal to the true ones, that is $\widehat{K_L a} = K_L a$ *and* $\hat{r_O} = r_O$.

Now we make a little modification of the desired controller behaviour, Equation 18.31:

$$\frac{de}{dt} = -a_c \cdot e - d \cdot sign(e) \tag{18.35}$$

The term a_c *is used to tune the controller. A large value of* a_c *makes the controller more ambitious and causes a larger gain and larger amplitude of the air flowrate changes. The last term in Equation 18.35 is the relay part of the controller that causes the probing. The function sign(e) is +1 if e is positive and -1 if e is negative. The relay gain d is to be tuned to obtain the proper compromise between probing and control. The relay gain is tuned such that the amplitude of the DO concentration oscillations becomes* e_{sp}, *which is a chosen parameter. Typically the value 0.2 mg/l has been chosen.*

With Equation 18.35 replacing Equation 18.31 the control signal is changed accordingly and Equation 18.34 is slightly modified to:

$$q_a = \frac{-\left[\frac{q_{in}}{V} S_{O,in} - \frac{q_{out}}{V} S_{O,out}\right] - r_O + a_c \cdot e + d \cdot sign(e)}{K_a \cdot (S_{O,sat} - S_O)} \tag{18.36}$$

The relay tuning has been chosen according to a recursive scheme:

$$d_{t+1} = d_t + k_d(e_{sp} - | e |) \tag{18.37}$$

where:

$$k_d = \begin{cases} k_{d,0} \ when \ \ | e_t + e_{t-1} | < e_{sp} \\ 0 \ otherwise \end{cases}$$

Now the true values of K_a and r_O are replaced by the estimated values. Thus the final version of the dual DO controller becomes:

$$q_a = \frac{-\left[\frac{q_{in}}{V} S_{O,in} - \frac{q_{out}}{V} S_{O,out}\right] - \hat{r_O} + a_c \cdot e + d \cdot sign(e)}{\widehat{K_a} \cdot (S_{O,sat} - S_O)} \tag{18.38}$$

As shown in Equation 18.38 the controller requires information on the wastewater flowrates, the influent DO concentration, and the controlled DO concentration. However, if the corresponding signals are measured in other parts of a long aerator, then not only the parameter variation in time at one point is possible but also the parameter variation along the aerator can be estimated.

Experiments with the dual controller have been performed in a full scale plant. It is proven that the dual controller performs as expected. The DO concentration is kept around $S_{O,sp} \pm e_{sp}$ which in this case was $2 \pm 0.2 mg/l$. The only tuning of the controller has been the choice of e_{sp} and the controller gain a_c.

18.6 Sludge Inventory Control

Sludge inventory control was introduced in Section 16.1.2. Here we will analyse the three types of control using the waste activated sludge, return sludge or the step feed distribution as control variables.

18.6.1 Waste Activated Sludge Control

To understand the role of the waste sludge control flow, consider a simple solids mass balance of the aerator-settler system according to Figure 18.10.

Let us assume that the sludge mass is represented by only one concentration x. The mass balance for the aerator-settler is then:

$$\frac{d}{dt}(V_a x_a + V_s x_s) = r_x \cdot V_a \cdot x_a - q_e \cdot x_e - q_w \cdot x_w \tag{18.39}$$

where r_x is the net specific growth rate of the sludge, V_a and V_s the volumes of the aerator and the settler respectively, and x_a and x_s the average sludge concentrations in the respective volumes.

We assume that the sludge in the settler has no growth. The second term on the right hand side is the amount of sludge unintentionally wasted from the system

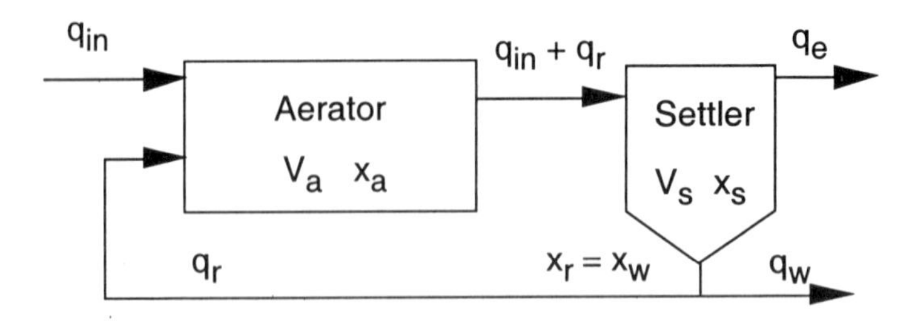

Figure 18.10: Basic configuration of activated sludge system with flowrates and sludge concentrations indicated.

while the third term accounts for purposeful wastage. The second term cannot be neglected since it can be of the same order of magnitude as the third term.

It is common (but of course not always accurate) to neglect the sludge mass in the settler, so we obtain:

$$\frac{d}{dt}x_a = \frac{1}{V_a}\left(r_x \cdot V_a \cdot x_a - q_e \cdot x_e - q_w \cdot x_w\right) \tag{18.40}$$

The common definition of sludge age (or sludge retention time SRT) is a steady state definition that is assuming all flowrates and concentrations constant:

$$SRT = \frac{V_a \cdot \bar{x}_a}{\bar{q}_e \cdot \bar{x}_e + \bar{q}_w \cdot \bar{x}_w} \tag{18.41}$$

where the bar over the variables denote averaging over a sufficiently long time. Assuming steady state, Equation 18.40 simplifies to:

$$\bar{r}_x \cdot V_a \cdot \bar{x}_a - \bar{q}_e \cdot \bar{x}_e - \bar{q}_w \cdot \bar{x}_w = 0 \tag{18.42}$$

which leads to the well-known expression:

$$SRT = \bar{r}_x^{-1} \tag{18.43}$$

that says that the sludge retention time is inversely proportional to the net growth rate. *This of course is valid only under steady state conditions.*

Under *dynamic* conditions we see clearly from Equation 18.39 that the flowrate q_w basically can be used to change the total mass content (there is no difference from this point of view to withdrawing the sludge directly from the aerator). A decreasing q_w can make the total mass grow, but the increasing rate is always limited by the difference between the growth rate and the unpurposefully wasted sludge $q_e \cdot x_e$. An increasing q_w may decrease the total sludge mass quicker if the pump rate is sufficiently large.

This simple calculus demonstrates that it is relatively easy to decrease the total sludge mass, and consequently the (average) SRT, while it takes longer to increase the (average) SRT. As an example assume that the SRT is 10 days. Setting the wastage rate to zero the growth rate of the total would be limited to 0.1 day^{-1} (and probably quite smaller due to the unintentional wastage via the clarifier).

The waste sludge flowrate is used to manipulate the sludge age. Since this is a quasi-steady state concept we can only talk about slow changes where the flowrates and concentrations are averaged over a long time, typically at least as long as the SRT. Equation 18.41 illustrates that a proper sludge age calculation involves the estimation of the average total sludge content of the system as well as the average total sludge withdrawal. This requires an accurate sludge concentration measurement as well as a realistic model of the sludge distribution within the aerator and the settler. If there is a significant amount of sludge in the settler we require a good estimate of the sludge concentration distribution in the thickener, which is far from trivial.

With lack of instrumentation various approximate methods for SRT control have been applied. The unpurposefully wasted sludge term $q_e \cdot x_e$ is often neglected. This will lead to an overestimated value of the SRT. The effluent concentration is certainly much smaller than the mixed liquor suspended solids concentration, but since the effluent flowrate is typically two orders of magnitude larger than the average waste sludge flowrate the value of $q_e \cdot x_e$ may be as high as 10-25 % of the purposefully waste sludge $q_w x_w$. The simplified definition of the SRT then becomes:

$$SRT = \frac{V_a \cdot \bar{x}_a}{\bar{q}_w \cdot \bar{x}_w} \tag{18.44}$$

If the wasted sludge is withdrawn directly from the aerator instead of the settler underflow then the two concentrations are identical, and the SRT is calculated simply by the average waste flowrate:

$$SRT = \frac{V_a}{\bar{q}_w} \tag{18.45}$$

From Equation 18.45 it is straight-forward to calculate the required waste sludge flowrate to produce the required SRT value. Again, we remind you that all the values are average values. This means that an instantaneous change from the current flowrate to the new required flowrate causes the SRT to gradually approach the desired value.

18.6.2 Return Sludge Flow Rate Control

The return sludge flowrate can seldom be used for control on an hour-to-hour basis. Basically there are two common practices for the return sludge flowrate,

- constant flowrate;
- ratio control, that is a constant ratio between the return sludge flowrate and the influent flowrate.

The traditional control schemes do not calculate how the aerator concentration varies or how the settling conditions change. Therefore let us consider some of the constraints. They are related to:

- hydraulic loading to the clarifier;
- dilution effects on the return sludge;
- sludge blanket level;
- sludge retention time in the settler.

An increasing return sludge flowrate will always cause a dilution of the return sludge. The purpose of an increasing return sludge flowrate can be to increase the sludge content of the aerator. The result may be a decrease of dissolved BOD due to more activity. However, the increased hydraulic load may influence the clarifier efficiency, so that what is gained in dissolved BOD removal is lost by an increase in effluent particulate BOD. One example of this was demonstrated in Example 14.8.

The return sludge line has a maximum dry mass transport capacity. For a small flow the dry mass transport may increase linearly with the flowrate. For a large flow, however, one may encounter a dilution of the sludge, and the dry mass reaches a maximum as a function of the flowrate.

The sludge blanket has to be kept within certain limits. This implies a lower and an upper limit for the return sludge flowrate. In nitrifying systems it is common to have a minimum sludge inventory which leads to a relatively large return sludge flowrate.

Sometimes the settler (or a separate tank between the settler and the aerator) is used for sludge storage. The retention time has to be kept limited in order to keep the organisms viable, that is the return sludge flowrate has a lower limit.

All these considerations taken into account, it is not very practical to use the return sludge flowrate as a control variable on an hour-to-hour basis.

In summary the lower limit of the instantaneous value of the return sludge flowrate is determined by the sludge blanket height and the sludge retention time in the thickener. The upper limit is determined by the hydraulic load to the clarifier and the maximum dry mass flowrate in the return sludge.

Steady State Constraints

The return sludge is used to keep the sludge within the system. As demonstrated in Section 18.6.1 the waste sludge is used to balance the growth of organisms, so that the average sludge mass is constant. Actually this means that the recycle ratio has to be kept within quite narrow limits in order to keep the overall sludge mass balance.

Consider the steady state mass balance in the settler. Based on the nomenclature defined in Figure 18.10 we calculate the steady state total mass balance over the settler. The left hand side of the equation is the mass inflow and the right hand side the mass outflow.

$$(q_{in} + q_r)x_a = q_e x_e + q_w x_w + q_r x_r \tag{18.46}$$

At steady state we can assume an algebraic relationship between the MLSS concentration and the effluent and waste sludge concentrations, as indicated in Figure 18.10.

$$x_w = x_r = g \cdot x_a \tag{18.47}$$

$$x_e = e \cdot x_a \tag{18.48}$$

The mass balance is now rewritten as:

$$q_{in} + q_r = q_e e + q_w g + q_r g \tag{18.49}$$

The flowrates at steady state can be related to the influent flowrate:

$$q_r = r \cdot q_{in} \tag{18.50}$$

$$q_w = w \cdot q_{in} \tag{18.51}$$

where r and w are constants. Then the mass balance is simplified to:

$$1 + r = (1 - w)e + (w + r)g \tag{18.52}$$

Let us now reconsider the SRT definition (Equation 18.41). Using Equation 18.46 the SRT can be defined by:

$$SRT = \frac{V_a \cdot \bar{x}_a}{\bar{q}_e \cdot \bar{x}_e + \bar{q}_w \cdot \bar{x}_w} = \frac{V_a \cdot \bar{x}_a}{q_{in}\left[(1 - w)e + wg\right]\bar{x}_a} = \frac{\theta_H}{1 + r - rg} \tag{18.53}$$

where θ_H is the hydraulic retention time of the aerator. The SRT has to be kept above a certain value:

$$SRT = \frac{\theta_H}{1 + r - rg} > \theta_x \tag{18.54}$$

which leads to:

$$1 + r - rg < \frac{\theta_H}{\theta_x} = \delta \tag{18.55}$$

This gives a lower steady state limit for the return sludge flowrate ratio:

$$r > \frac{1 - \delta}{g - 1} \tag{18.56}$$

The second limitation on r is given from the definition of the SRT. The denominator in Equation 18.54 has to be positive, that is:

$$1 + r - rg > 0 \tag{18.57}$$

This gives an upper bound for r of:

$$r < \frac{1}{g - 1} \tag{18.58}$$

Example 18.7 (SRT control bounds) *To get a better feeling for the numbers involved we consider some typical plant data:*

1. $\theta_H = 6$ *hrs;* $\theta_x = 200$ *hrs* (≈ 8 *days) gives* $\delta = 0.03$. *Assume the thickening ratio at steady state is* $g = 2$. *This requires* r *to be kept within* $0.97 < r < 1$, *where* $q_r = r \cdot q_{in}$.
2. $\theta_H = 8$ *hrs;* $\theta_x = 360$ *hrs* (≈ 15 *days) gives* $\delta = 0.022$. *Assume the thickening ratio at steady state is* $g = 2.5$. *This requires* r *to be kept within* $0.64 < r < 0.67$.

Again note that the analysis has assumed steady state. Dynamically the return sludge flowrate may vary, but the average ratio r has to be within very narrow bands in order to keep the mass balance of the settler. If the return sludge flowrate is not close to proportional to the influent flowrate, then the sludge blanket will move up or down.

Dynamic Control of the Return Sludge

We have shown that the steady state return sludge flowrate has to satisfy very narrow limits. In a shorter time scale there are some possibilities to control the return sludge flow. However the different constraints have to be considered.

During some disturbances the buffer capacity of the settler should be taken into account. Consider a large hydraulic disturbance. If the disturbance can be predicted then the plant can be prepared for the increasing flow by lowering the sludge blanket to a minimum. As the large flow enters the plants the return sludge flowrate can temporarily be decreased and the buffer can be filled. Thus a decreasing return sludge flowrate can temporarily dampen the load to the clarifier.

18.6.3 Step Feed Sludge Control

Step feed control has been advocated for a long time as a means to redistribute the sludge within the aerator for a carbonaceous removal system. Using step feed control the system can better meet organic, hydraulic or toxic loads. It has to be emphasised that step feed should be used under transient conditions and it is only dynamic changes that motivate the changing feed pattern.

Consider Figure 18.11, consisting of only two reactors in series. In order to quantify the effects of step feed control we will make a simplified analysis of the sludge mass balance of the system.

In the time scale we consider (less than a day) organism growth is neglected so the time constants become functions only of the hydraulics of the system. We assume a perfect settler, that is all sludge leaving tank 2 is returned back to tank 1 via the settler and the return sludge. It is assumed that no sludge enters the system via the influent flow. The sludge concentrations are denoted x_1 and x_2 respectively so the mass balances become:

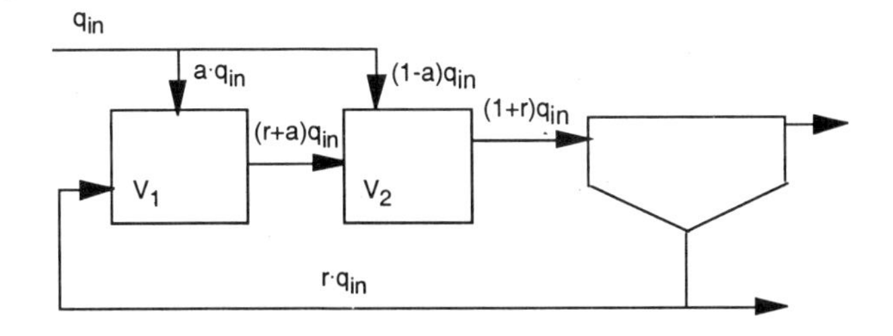

Figure 18.11: A simple two reactor system with step feed.

$$\frac{d}{dt}(V_1 x_1) = -(a+r)q_{in} \cdot x_1 + (1+r)q_{in} \cdot x_2$$
$$\frac{d}{dt}(V_2 x_2) = (a+r)q_{in} \cdot x_1 - (1+r)q_{in} \cdot x_2 \tag{18.59}$$

where $0 \leq \mathrm{a} \leq 1$.

Let us calculate the steady state sludge concentrations. Then we find that the two equations are linearly dependent, so only one relation can be derived:

$$-(a+r) \cdot x_1 + (1+r) \cdot x_2 = 0 \tag{18.60}$$

or

$$x_1 = \frac{1+r}{a+r} \cdot x_2 \tag{18.61}$$

There is no unique steady state solution. This is natural since the model does not include any source or growth term. It simply describes the relative concentrations of the sludge between the two reactors.

For the case $a = 0$ we notice from Equation 18.61 that the concentration in the first reactor is larger than that of the second reactor since r is usually smaller than 1. This means that most of the aerator sludge has been concentrated into reactor 1. In fact, the concentration there is the same as that of the settled sludge.

This configuration is a so called *contact stabilisation plant* where the return sludge is kept aerated without any further substrate addition before it is contacted with the influent flow. The contact time in tank 2 is so short that the organisms will only absorb the substrate having too little time for metabolism. When the sludge returns to tank 1 there is time for organism growth and endogenous respiration.

Step feed control can be used favourably during the following general conditions:

- *large hydraulic influent flow*: To simplify the reasoning let us assume that the step feed control $a = 0$ has been applied for some time (typically a few hours) before the disturbance enters the system. Most of the sludge will then be located in the first reactor. The large hydraulic load will cause an increasing load to the clarifier. In the worst case there will be a washout. However, for a limited duration the sludge can be kept in the first reactor so the damage can be limited. After the hydraulic shock has passed the influent water can enter the first reactor, $a = 1$. Due to the step feed control most of the sludge has been kept in the system.

- *toxic load to the system*: Assume that a toxic load is discovered. If a is set to 0 the toxicity is entering only tank 2. If the return sludge can be kept to a minimum while the toxicity passes the plant, then the damage may be limited to the organisms from tank 2. After the disturbance has passed through the system a is set to 1.
- *large organic load*: Again assume that a large organic load is expected within a few hours. As a preparation for the load we set $a = 0$. This will make sludge organisms in tank 1 more starving. When the load enters the plant a is switched to 1, and the load will meet the starving organisms causing a more efficient carbon removal.

In the second and third cases the load to the system is indicated by the oxygen uptake rate (OUR) in tank 2. When the OUR is expected to rise a should be set to 0 before the disturbance and to 1 during and after the disturbance. Conversely, when a decreasing OUR is expected ideally a is set to 0 before and during the disturbance, and to 1 after the disturbance.

Typical for step feed is that it takes some time to change the configuration from one flow pattern to the other. We have seen that the system has to be able to predict the disturbance if the step feed control should be efficient. The time to change from one configuration to another depends on the hydraulic retention times. For simplicity we assume that the volumes are equal (V). The time it takes to change the concentrations can be calculated from Equation 18.59 and is determined by the eigenvalues of the system. They can be calculated analytically:

$$\lambda_1 = -\frac{F}{V}(a + r); \quad \lambda_2 = -\frac{F}{V}(1 + r) \tag{18.62}$$

The time constants of the system are simply the inverses of the eigenvalues:

$$T_1 = \frac{V}{F(a + r)}; \quad T_2 = \frac{V}{F(1 + r)} \tag{18.63}$$

This corresponds to the retention times of the tanks. Assume that $a = 1$ initially. With all the influent flow coming into the first reactor the concentrations in the two tanks will be the same at steady state. Switching to $a = 0$ the tank 1 will be filled gradually with concentrated sludge at a flowrate of $r \cdot F$ with the corresponding retention time $V/(r \cdot F)$. We can see that it takes several hours to change from one pattern to another. In other words one needs to be able to predict a large load change several hours ahead if the step feed control should be efficient.

18.7 Nutrient Removal Examples

In Chapter 14 we discussed the decomposition of the control strategies for biological wastewater treatment processes. We decomposed the problem in a vertical sense based on the time scale of the response time or the basic process dynamics, and in a horizontal sense based on the equipment units. We then discussed the characteristics of each subproblem and possible operating objectives and strategies.

The biological nutrient removal reactions all fall into the *hours-to-minutes* time scale and can be decomposed horizontally by considering each process section, piece of equipment or unit process. Control strategies can be developed for each subproblem on the basis of the operating goals and on the process cause-and-effect relationships. We will present simulation examples of developing model based control strategies for some of the principal unit processes below.

Of course there will be interactions, as there are interactions in the vertical direction between different time scales. Chapter 20 on plant wide control will consider process interactions and how they might be handled.

Example 18.8 (Activated sludge anaerobic zone) *The primary purpose of this zone is substrate uptake and storage by phosphate removing organisms with an accompanying phosphate release. The substrate stored is used by the microorganisms for growth and phosphorous uptake in the subsequent anoxic and aerobic zones. A secondary purpose may be the degradation of fermentable organic compounds to produce volatile fatty acids (VFA) for the phosphate removing organisms. This latter process can also occur incidentally in the sewer and in the primary settler, or deliberately in a primary clarifier with sludge recycle or in a prefermenter.*

We saw in Chapter 14 that the most likely operating goal is maximisation of the substrate storage by phosphate removing organisms. This is equivalent to maximisation of the use of the available VFA. VFA are most likely to be the limiting or most expensive substrate. A small but nonzero level of VFA leaving the anaerobic zone would be the ideal situation (see Figure 18.12). This would indicate that there was sufficient VFA to fully utilise the substrate uptake capacity of the anaerobic zone, but not so much that it would cause unwanted downstream effects. Excess VFA will carry over to the anoxic and aerobic zones where they will enhance nitrification, but will also consume more oxygen and could trigger phosphate release if anaerobic conditions are found in poorly aerated spots or within large flocs.

The plots in Figure 18.12 are batch reactor profiles produced by the MATLAB program `vfa_utilisation.m` *using the hydrolysis, fermentation and P release equations from the IAWQ AS Model No 2. They are equivalent to concentration profiles in a plug flow bioreactor.*

The curves shown on the left in Figure 18.12, starting from the bottom at time zero in order, are phosphate, VFA, fermentables, polyphosphate and hydrolysable insoluble carbon. The vertical dashed line at 1.5 hours shows the exit concentrations of an equivalent plug flow reactor with a residence time of 1.5 hours. It

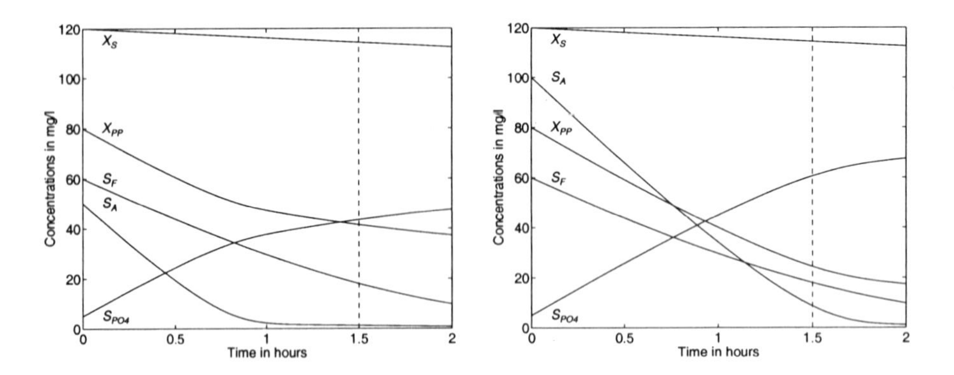

Figure 18.12: Insufficient VFA (left plot) and excess VFA (right plot).

is clear from the profiles that hydrolysis is slow relative to fermentation and P release and could be omitted from the model with little consequence. The curves shown on the right in Figure 18.12 are the same conditions with twice the initial VFA concentration.

Possible constraints on achieving the desired VFA uptake are the presence of influent oxygen and nitrate and too little volume (residence time) in continuous bioreactors. In an SBR the volume (reaction time) is an additional degree of freedom.

Disturbances in the same dynamic range include changes in influent flowrate and in influent concentrations of phosphate and of VFA (those not deliberately added as the manipulated variable) and also changes in the polyphosphate level in the microorganisms. Factors which may disturb the system through competition for VFA or through reaction rate inhibition include oxygen in the influent water and nitrates in the influent water or sludge. Changes via the hydrolysis reactions are much slower (as seen in Figure 18.12) and via microbial population dynamics are very much slower.

We also saw in Chapter 14 that the most likely control strategy is adjustment of the initial concentration of VFA or of the supplemental rate of addition of VFA.

There are two issues from a control perspective. A manipulated variable is required in order to manipulate the influent concentration of VFA. This could be done using a separate stream from a prefermenter or from an external VFA source. If an activated *primary clarifier is used, then the sludge recycle could be manipulated affecting the elutriation of VFA from the sludge.*

The second possibly more difficult issue is that of measurement. The possibilities include:

- *an acetate or VFA measurement. This is the ideal measurement, being directly related to the operating goal.*

- *a phosphate in solution measurement. There is a somewhat indistinct change in slope but this would be very difficult to detect, especially in a continuous flow reactor.*

- *a phosphate in solution and a model-based estimator.*
- *a REDOX measurement. There can be a quite distinct change in slope when VFA have been consumed, but there is much debate on the efficacy of this measurement and especially its stability during transient conditions.*

A possible control strategy is to manipulate VFA addition to maintain a small excess of VFA in the effluent, say 5 mg/l.

Example 18.9 (Anaerobic equalisation tank) *Two-stage anaerobic processes are primarily used to pretreat high strength wastewater from food processing plants such as canneries, breweries and abbatoirs. The first stage, the equalisation tank, presents an interesting control problem where dynamic models are useful to determine a good control strategy. The equalisation tank has a number of competing operating goals:*

- *The reduction of the influent COD to levels acceptable for traditional treatment in a municipal activated sludge plant.*
- *The generation of volatile fatty acids (VFA) by hydrolysis of influent COD for digestion to methane in the second-stage reactor.*
- *Keeping methane production in the first-stage to a minimum.*
- *Maintenance of a moderate pH that will not inhibit the methanogens in the second-stage.*
- *Buffering influent flowrate disturbances, the so-called "equalisation".*
- *Buffering influent concentration disturbances.*

Some of the competing strategies are obvious. A larger volume will better absorb concentration variations and flowrate reductions, but a moderate volume is needed to absorb flowrate increases. A higher production of VFA is desirable but this will lower the pH which could inhibit second-stage digestion.

Other interactions are more clearly seen if a model is available. For example, Figure 18.13 shows some concentration profiles when the HRT in an equalisation tank is decreased by 33 per cent (MATLAB program `equalisation_tank.m`*).*

The decrease in the HRT was achieved by a 50 per cent increase in flowrate so that an increase in insoluble COD and a decrease in VFA could be expected without any reactions being affected. The decrease in the HRT also reduces the hydrolysis time which should lead to a further increase in insoluble COD and decrease in VFA. This initial response is seen over the first few days.

In this reactor there was initially significant methane production so that over the longer term a washout of methanogens occurred. This is clearly seen in the

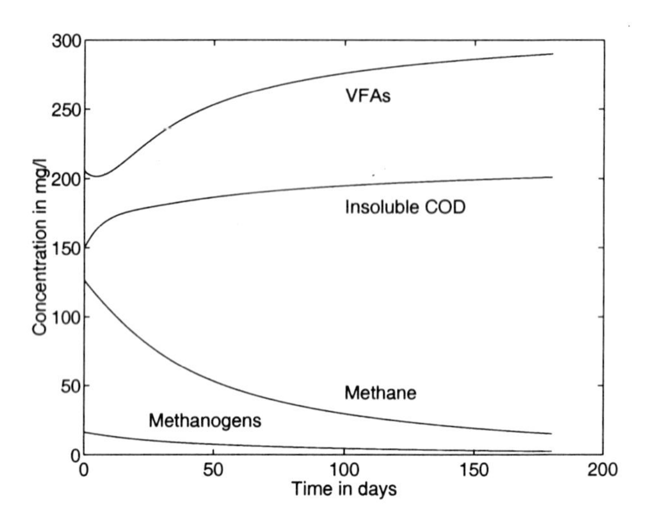

Figure 18.13: Some concentration profiles in an anaerobic equalisation tank after a HRT reduction.

decrease in methane production and in methanogen population. This also results in a marked increase in VFA production, but this is only seen after almost a week and it would take months to find a new "steady state".

It is clearly a complex task to optimise the operation of the equalisation tank and the difference in timescales between often competing hydraulic, reaction and biomass effects would make it a difficult task without a process model.

Common control strategies include:

- *Buffering pH and controlling HRT by stage-two effluent recycle.*
- *Controlling pH by caustic addition to the equalisation tank or to the stream from stage-one to stage-two.*
- *Controlling volume to absorb predictable disturbances.*
- *Maintaining a maximum HRT to reduce methane production.*

Measurements monitored include volume, flowrates, pH, gas production, and methane and VFA concentrations.

18.8 Fuzzy Controllers

The structure of a fuzzy controller is generally to follow some reference value of the system. The measurements can either be made by sensors or be based on manual observations. The measurements that are crisp (non-fuzzy) have to be characterised in terms of *low*, *high*, etc. As described in Section 13.3.3 they have to be given some degree of membership in the reference sets. Then the control is made in terms of some rule base that performs operations on the fuzzy sets and inference, such as:

```
if input e is small then output u is large
if input e is large then output u is small
```

The resulting rule then has to be translated from a fuzzy control action (which is still a membership function) into a crisp control signal (a scalar). This is called *de-fuzzification.* The various steps are illustrated in Figure 18.14.

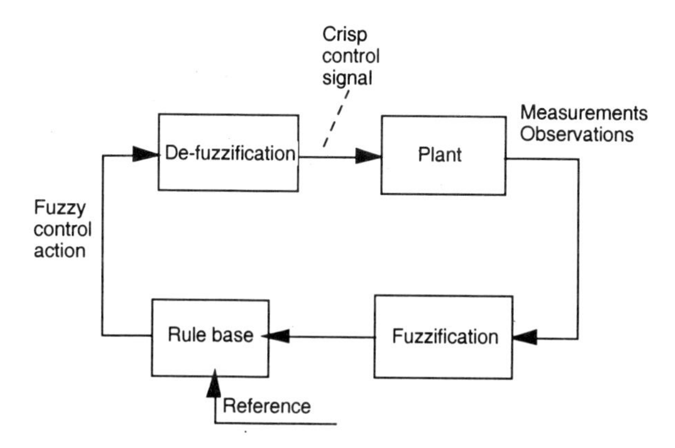

Figure 18.14: A closed loop configuration for fuzzy control. The control action may be automatic or manual.

18.8.1 Fuzzification

In order to apply the fuzzy control law we have to define what *small* and *large* means. A classical static controller is simply defined by a relation $u = f(e)$ that connects any input value to an output value. For example, a proportional controller is defined by $u = K \cdot e$ which is a straight line in the $e - u$ plane. Even a classical controller can be represented by a membership function. Every point on the line $u = f(e)$ has a membership $\mu = 1$ while all other points have membership $\mu = 0$ (Figure 18.15).

The vague concepts *small* and *large* are now defined by fuzzy sets as shown on the e-axis and the u-axis of Figure 18.16. The problem in defining sets is the subjectivity of *small* and *large.* Different people have different ideas of small, and in addition to that a small value of e means something different than a small value of u. Therefore this step may consume more time than defining the rules.

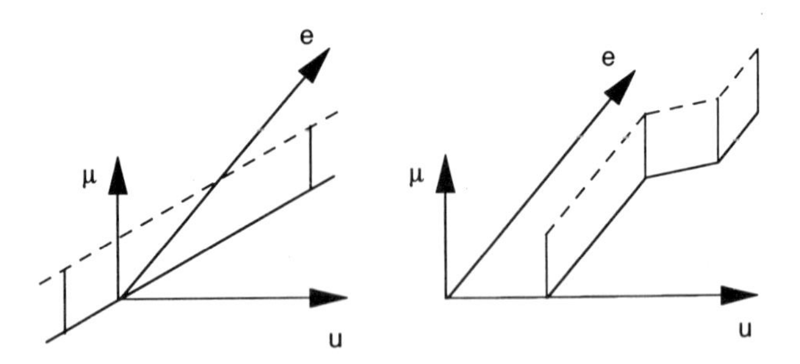

Figure 18.15: Representation of a classical non-dynamical controller by a membership function, a proportional controller (left) and another function $u = f(e)$ (right).

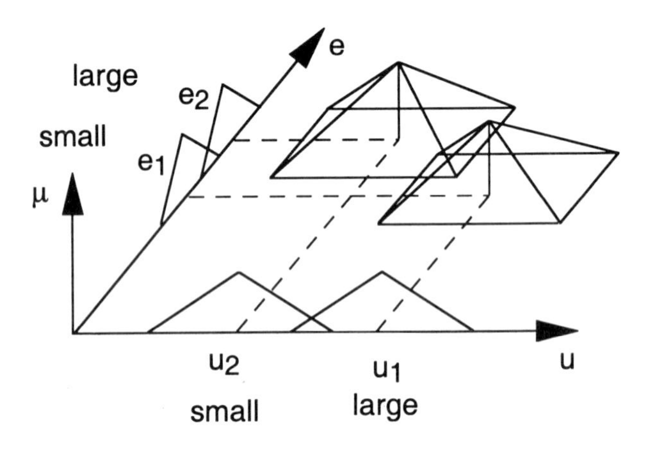

Figure 18.16: Representation of fuzzy rules by fuzzy relations.

18.8.2 Fuzzy controller

The fuzzy controller is defined by fuzzy sets for the input and output variable and by linguistic rules that connect any input fuzzy set to an output fuzzy set. Note that this has to be formulated differently than the classical relationship.

The fuzzy rule can be represented by a two-dimensional fuzzy set (the pyramids in Figure 18.16) in the $e-u$ plane, called a fuzzy relation. An algorithm to compute this relation can be derived. Here we renounce large formulas and just want to give a graphical description of this algorithm. For example, for the first rule we take the input fuzzy set small and make a projection onto the $e-u$ plane. Then we do the same with the output fuzzy set large. The intersection of the two projections is a pyramid.

This pyramid is the fuzzy relation R_1 that represents the first fuzzy rule. After computing all the fuzzy relations we have to connect disjunctively all relations R_i.

That means we have to overlay all the pyramids. The resulting ridge in the $e-u$ plane is the complete set-theoretical representation of the input-output behaviour of the fuzzy controller. This should be compared with the function in Figure 18.15.

Comparing one of the single fuzzy relations R_i with the fuzzy set A in Figure 13.9 leads to an interesting interpretation of a fuzzy controller. As A is the set of all values similar to a, R_i is the set of all points (e, u) similar to (e_i, u_i). The fuzzy relation hence defines the output u_i for the input e_i and for all inputs that are similar to e_i there will be an output similar to u_i, while similarity is defined by the shape of the fuzzy relation.

We now have to recognise how the output is computed for a given input. Like the computation of a fuzzy relation this step is well-defined by classical set theory. Again we just give a graphical description of the algorithm. We obtain the output fuzzy set U_m for a given input e_m by looking at the cut at $e = e_m$ in the e-plane, as depicted in Figure 18.17.

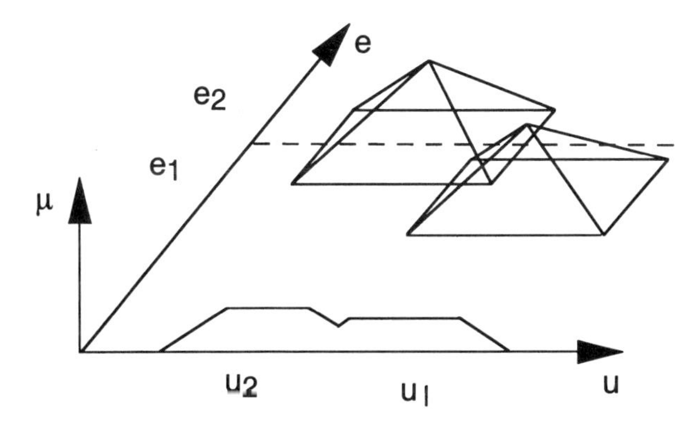

Figure 18.17: Output computation of a fuzzy controller.

18.8.3 Defuzzification

The fuzzy set of the output now has to be *defuzzified*, meaning that one value of the set U_m must be chosen to be the output value u_m . Usually the u-coordinate of the centre of the area given by U_m is chosen.

It is easy to see that for any i, the closer the input value is to e_i the closer the output value is to u_i. Therefore the complete procedure is just a kind of interpolation (Figure 18.18). For any input value e with $e1 \leq e \leq e2$ we will get an output value u with $u1 \leq u \leq u2$. The only difference to a characteristic curve controller lies in how the interpolation is done.

We have shown one method for defuzzification and there are various methods used. We mention the most popular ones (compare Figure 13.12):

- *centre of area method*: this picks the x-coordinate of the vertical line that

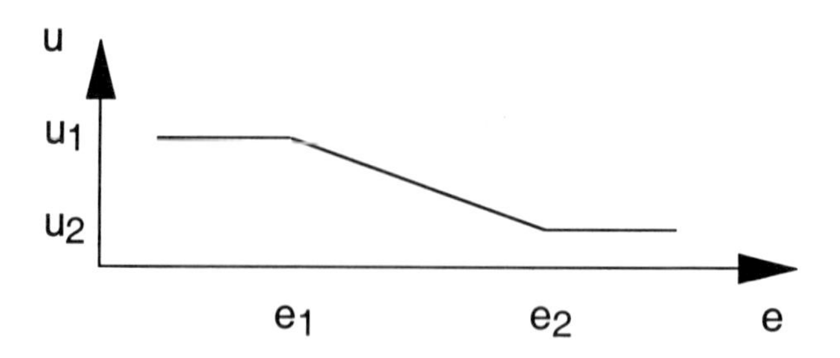

Figure 18.18: Output behaviour of a fuzzy controller.

splits the area under the membership curve into two equal halves;

- choose the point with the *maximal membership*, that is the strongest possibility. If there are several such points then take the mean of the maxima;

- if the fuzzy set has several peaks both the methods above will choose a point between the peaks. The controller sometimes has to pick one of them. Depending on the application it may choose the first or the last maximum.

The rule base has to be consistent. If the inferred fuzzy set has several peaks the rule base is inconsistent and is pointing to several different control signals at the same time. It is useful if contradicting rules can be discovered. Some methods are available. Finally, a rule is redundant if the information is otherwise present in the rule base.

18.8.4 Some Remarks on Robustness

A fuzzy controller can be combined with dynamic elements like integrators and indeed this is usually done. But even with dynamic elements a fuzzy controller is nothing magic. Any response behaviour that we can get with a fuzzy controller can also be obtained with classical methods. The difference between a fuzzy controller and a classical nonlinear controller is not the resulting behaviour but the way this behaviour is achieved. It is often believed that fuzzy control brings some special robustness in the control. The discussion above demonstrates that this is just a fairy tale. A characteristic curve like the one in Figure 18.18 is as robust as any other characteristic curve derived from classical methods.

Most people writing about the special robustness property of fuzzy controllers do not know what robustness really means. A controller derived for a nominal plant is called robust if the closed-loop system is stable not only for the nominal plant but also for a plant that differs slightly from the nominal plant. Using this definition nearly any controller in the world is robust because few controllers are designed to drive the closed-loop system at the border of stability.

Therefore, the interesting question is not if a controller is robust or not. It is to define a *measure of robustness*, that means how large may the difference be between the nominal and the real plant so that the controller is able to stabilise not only the nominal but also the real plant. There is a special theory, H_∞ theory, that addresses this problem precisely. For a H_∞ controller the measure of robustness can be given, but not for a fuzzy controller.

18.8.5 Fields of Application

Where does it make sense to derive a controller by fuzzy rules rather than by classical methods? The use of a fuzzy controller only makes sense if

- there is no analytical model of the plant available, or
- the model is too complicated to compute a controller by classical methods, or
- the control goals are not defined precisely.

Consequently, for the control of a servo system, an electrical motor, a liquid level, or a pressure, the use of fuzzy control will not give any improvement. For such systems there exist precise models from which a controller can be derived, and also algorithms to simplify these models so that the controller can be well designed.

However, at a higher level in the hierarchy where the local controllers are integrated in large systems the conditions are somewhat different. Here it is difficult to derive an analytical process model and there might also be several different control goals to be taken into account (compare Chapter 14). It is obvious that these vague goals cannot lead to precise analytical solutions. Therefore it could make sense to use fuzzy control, because for this solution neither precise control goals nor an analytical process model is needed. The control rules can be derived from a heuristic description of operating experience.

When deriving a controller in this way the question of stability arises. A controller based on fuzzy logic is in practice an estimator of the system not based on a particular model. It is very difficult to prove stability of such a controller.

There are some methods available to test if the rule base is complete, that is that all possible input combinations generate some fuzzy set that is not full of zeros. However, recently some algorithms have come up dealing with this problem. With these algorithms it is possible to investigate the stability of a system containing a plant which can be described only qualitatively or numerically. Certainly these investigations cannot give an analytical proof of stability, but they can give a qualitative overview of the closed loop system behaviour. However, this lack of analytical exactness is not caused by fuzzy control. If we want to handle plants that cannot be described analytically we have to accept that we cannot analytically prove stability.

18.8.6 Applications of Fuzzy Control to Wastewater Treatment

An activated sludge plant is characterised by a lack of relevant instrumentation, control goals that are not always unambiguously stated, the use of qualitative information in decision making, and poorly understood biological behaviour mechanisms. Therefore it appears to be a good candidate for fuzzy control.

Example 18.10 (Fuzzy control of an activated sludge process) *An early attempt to apply fuzzy control in wastewater treatment was made by Tong* et al. *(1980). Here we relate some of their findings. The control problem they state is to manipulate the return sludge flowrate, the dissolved oxygen setpoint and the waste sludge flowrate so that the effluent quality is maintained despite large variations in the influent. Furthermore, the controller was to prevent or quickly recover from process failures like sludge bulking. Here sludge bulking is assumed to be caused by the presence of filamentous organisms that prevent settling of the sludge in the clarifier. The growth of the filamentous organisms is partly governed by the dissolved oxygen concentration.*

A rising sludge is caused by denitrification in the clarifier whereby nitrogen gas is formed and then rises to the surface of the clarifier bringing sludge with it.

The controller has to obtain measurements and observations. Measurements may be non-fuzzy to begin with but have to be fuzzified, and may be of the form "MLSS is very high" or "total effluent BOD is more or less normal". The key terms are "low", "high", "large", "more or less" and correspond to imprecise but useful information about the process. In general many rules are required to specify a controller and may have the form:

```
WHEN "effluent suspended solids concentration is medium"
 AND "denitrification is indicated"
DO "make a small positive change in sludge wastage rate"
```

Another rule can be:

```
WHEN "MLSS is very small"
 AND "sludge wastage rate is small"
DO "make a small positive change in return sludge set-point"
```

In this case incremental changes are made in the return sludge or the waste sludge until desirable performance has been achieved.

Example 18.11 (Operation support system) *An operation support system was presented by Watanabe* et al. *(1993). Data was received on-line from for example flowrates, MLSS concentration, and air flowrate. An image processing apparatus received data from a high resolution submerged microscope. Furthermore data was obtained off-line, such as manually analysed water qualities and human observations.*

The knowledge database was divided into two parts based on whether it was general or specific. Knowledge was further divided into smaller classes such as "sensor trouble diagnosis". Data was checked for contradictions.

Operator guidance was defined through membership functions presented to the operator. The rules were in principle arranged as if-then *rules in a fuzzy reasoning framwork. Messages concerning bulking sludge were obtained from the microscopic information and presented on the operator screen (see Chapter 22.4.3).*

Sensor troubles were diagnosed based on the average values, standard values, upper and lower limit values and rates of change obtained by statistical analysis using historical data. Furthermore periodical change patterns during a day and the concurrent values in the same type sensors on other lines were considered for the diagnosis.

18.9 Design of Model Based Control

We have tried to be consistent in offering you a taste of the *theory* and listing references and further reading, but also in presenting approaches you can take to trialling it in the real world. For model based control we offer you the following possible approach to the design and implementation of such controllers.

While we believe such advanced controllers are valuable in certain applications and should be trialled where appropriate, we must say that it is not a trivial task. There is a lot of solid engineering and experimentation involved which should not be underestimated.

On the other hand we are also sure you will find it an enjoyable and very rewarding task, and hopefully so will your employer.

At least the following design tasks will be involved:

- *Gaining Process Knowledge.* This is a key factor in the success of the exercise. If you are not familiar with the theory, design and operation of the process, then do some homework. Preferably involve other people with first-hand knowledge. Refer to Chapters 2 and 4.

- *Defining the Goals and Objectives.* Again a key factor. Know where you are going and why you are going there. Refer to Chapter 14.

- *Determine Available Measurements.* This is a particular issue for WWT plants, even if it is not unique to WWT. Model based control relies not only on the model but also on extra process information for its increased performance. Refer to Chapters 22 to 24.

- *Evaluate Prior Plant Performance.* You must benchmark plant performance before and after the exercise if you are to demonstrate its success (and get funds for more projects in the future).

- *Specify a Model.* Knowing the process, the objectives and the available measurements enables you to set the goals for the model development and validation. Refer to Chapter 2.

- *Developing an Appropriate Model.* The trick is to include the minimum number of mechanisms to ensure success, something that is easier said than done. However it can generally be achieved with two or three iterations. Refer to Chapters 3 and 4.

- *Model Verification and Validation.* Preliminary verification simulations, good experiment planning, good data analysis and your specifications are all you need. Refer to Chapters 5 to 8.

- *Specify Estimators and Controllers.* By this stage you will know much more about the process than you ever thought possible. Together with your goals and objectives, available measurements and a knowledge of the different model based controllers available (don't hesitate to consult with the "experts" on this) you can select and specify the requirements for the estimation and control tasks.

- *Design and Simulate the Estimators.* Procedures for this can be found above or in the literature. The principal challenge lies in defining (modelling) the relationships between states and inferential measurements and in tuning the various design parameters available for peak performance.

- *Trial the Estimators.* It is important to implement and run the estimators on the plant to get first hand performance data. You will have some plant-model mismatch, so some retuning of design parameters may be necessary. Once you have confidence and have defined the accuracy of your estimators you can proceed with controller design.

- *Design and Simulate the Controllers.* Again, procedures for this can be found above or in the literature. The primary task will be tuning the various design parameters available for peak performance given the now known uncertainty in your real and estimated measurements.

- *Trial the Controllers.* Again, you have some plant-model mismatch so some retuning of design parameters may be necessary for peak performance.

- *Evaluate Post Plant Performance.* This is the moment you have been waiting for, how big will your salary bonus be? It will also ensure you can do it all again with least hassle with proving benefits. Techniques and problems for defining and comparing performances can be found in Chapter 20.

- *Publish the Results.* Don't be shy, tell us all about it so we can spread the gospel more easily too.

- *After Sales Service Manual.* This is a good idea (unless you are leaving the company). Write a manual to document what has been done, the successes and problems and obstacles and especially on what is needed to retune the system to maintain it at peak performance and how peak performance is measured. It helps reduce those nuisance telephone calls.

Like any design procedure, while it is presented as a sequence of tasks, there is much that can be done in parallel and many iterative "feedback" paths to be taken. Hopefully the convergence will be fast.

18.10 Summary

We have introduced the concepts and algorithms for three model based controllers and illustrated them by examples. Feedforward controllers are already applied in some wastewater treatment plants. Predictive control (a linear controller) and generic model control (a nonlinear controller) have been applied in the process industries. Pilot scale trials are known to be planned on nutrient removal bioreactors, so watch the literature or try it yourself now that you know how. State feedback and linear quadratic control is a more elaborate control, but we hope to have demystified it a little. Likewise, fuzzy control is useful but not magic. Still process knowledge is crucial for a good control result. Sludge inventory control demonstrates that the various interactions of the process units have to be considered.

18.11 Further Reading

- For more information on *feedforward control* see Seborg *et al.* (1989) and Lee *et al.* (1998).

- Some of the *predictive controllers* are best described by some of the earlier developers, such as Richalet *et al.* (1978), Cutler & Ramaker (1980) and Maurath *et al.* (1985), before they got studied to death by the theorists. For more information on predictive control see Lee *et al.* (1998).

- Application of predictive control for sewer operation is found in Carstensen *et al.* (1996).

- The original publication of the *GMC algorithms* is found in Lee & Sullivan (1988). A collection of papers on GMC and on some industrial applications can be found in the book edited by Lee (1993). For more information on generic model control see Lee (1993) and Lee *et al.* (1998).

- For more information on state estimation see Luenberger (1971) and Lee *et al.* (1998).

- The idea of the simplified SRT was first proposed by Garrett (1958). The sludge age calculation (Section 18.6.2) was first proposed by Olsson & Andrews (1978).

- The problem of simultaneous estimation and control was solved and tested at the Malmö Sewage works in southern Sweden (Holmberg (1987) and Holmberg *et al.* (1989)).

- The IAWQ AS Model No 2 is described in Gujer *et al.* (1994) and Henze *et al.* (1994).

- An early attempt to use reasoning for the control of a medium sized plant in Sweden was made in 1976 (Gillblad & Olsson (1977)). Logical reasoning was used to determine if bypassing was necessary.

- Some results of fuzzy control stability are found in Kiendl & Rüger (1995) and Klawonn *et al.* (1997). Kruse *et al.* (1994) contains a detailed discussions on fuzzy control. A good introduction into fuzzy control is Jantzen (1994). Further references on fuzzy control are found in Chapter 13.

- Applications of fuzzy control in wastewater systems are described in Marsili-Libelli (1992), Marsili-Libelli & Müller (1996) and Chang *et al.* (1998).

- Several Japanese experiences on fuzzy reasoning applied to wastewater systems are reported by Tsai *et al.* (1993), Tsai *et al.* (1994), Watanabe *et al.* (1993) and Enbutsu *et al.* (1993).

- *Respirometry* used for early warning of toxicity has been described by Temmink *et al.* (1993). Control based on respirometry has been reported by Klapwijk *et al.* (1993). In Witteborg *et al.* (1996) it has been used to estimate the influent biodegradable organic substrate concentration. A large number of feedforward/feedback schemes based on respirometry have been reported in Spanjers *et al.* (1996). More details are published in a scientific and technical report from IAWQ (Spanjers *et al.* (1998)).

- An early description of feedforward in activated sludge systems is found in Davis *et al.* (1973). Applications of feedforward in activated sludge are found in Nam *et al.* (1996).

- The dual controller in Example 18.6 has been presented by Holmberg *et al.* (1989).

18.12 Exercises

1. If you could measure influent TOC, influent ammonia concentration and OUR, how would you use the information in a feedforward structure?

2. Why is it important to combine feedforward control with feedback control?

3. Use the MATLAB program `Example_18_1.m` applied in Example 18.1 and try different structures and gains for the feedforward controller.

4. Use the MATLAB program `Example_18_3.m` applied in Example 18.3 to study the effects of:

   (a) the relative weighting between control variables and manipulated variables, using Q and R.

   (b) the prediction horizon, N_P (Use the MATLAB program `dmc.m` to generate new dynamic matrices).

5. Use the MATLAB program `Example_18_4.m` applied in Example 18.4 to examine the effects of different tuning parameters τ and ξ.

6. Compare the dual controller with a proportional controller. Assume that S_0 is almost constant and that the influence from the flowrate to S_0 can be neglected.

7. Often the waste sludge is pumped intermittently, a few times a day. In calculating the sludge age, how do you consider such a pumping?

8. You wish to increase the sludge retention time from 10 days to 11 days. How long would it take to establish the new sludge retention time?

9. There are two common ways to control the return sludge flowrate, either proportional to the influent flowrate or constant. What are the consequences of the two control methods? In particular consider the sludge blanket level and the clarifier behaviour. Compare the experience described in Example 14.8.

10. Step feed is considered a *dynamic* and not a *static* control action. Explain! Why do we need to predict a disturbance in order to use step feed control successfully?

11. Formulate fuzzy rules to control the carbon addition to the anoxic zone.

12. Formulate fuzzy rules to set the DO setpoint in the aerator according to the effluent ammonia measurements (Compare Section 17.7.1).

Chapter 19

Batch Plant Control

There are a number of applications involving sequenced operations in any wastewater treatment plant. Operations which proceed through a number of distinct steps in a set order depending on process conditions. Influent pump control and sand filter backwashing are typical examples.

19.1 Introduction

Sequencing equipment operation has traditionally been realised with relay techniques. Until the beginning of the 1970's, electromechanical relays and pneumatic couplings were dominating in the market. During the 1970's microcomputer-based programmable logical controllers became commonplace, and today sequencing is normally implemented in software.

Programmable logical controllers (PLCs) are microcomputers developed to sense on/off signals, handle Boolean operations and activate on/off actuators. The early PLCs were designed only for logic-based sequencing jobs involving digital on/off signals. More sophisticated PLCs will now also handle analog input and output signals and a range of analog operations, from simple comparators to PID control algorithms. Today there are hundreds of different models on the market. They differ in their memory size (from 256 bytes to several kilobytes) and I/O capacity (from a few to many thousands). They also offer different optional programming operations, from timer and counter capabilities to quite complex arithmetic calculations.

The logical decisions and calculations may be simple in detail, but the decision chains in large plants are very complex. This naturally raises the demand for structuring the programming problem, its implementation and documentation. Ladder diagrams, which are simply relay system circuit diagrams, are being phased out of many automation systems but they are still commonly used to describe and document sequencing control implemented in software. GRAFCET, now an

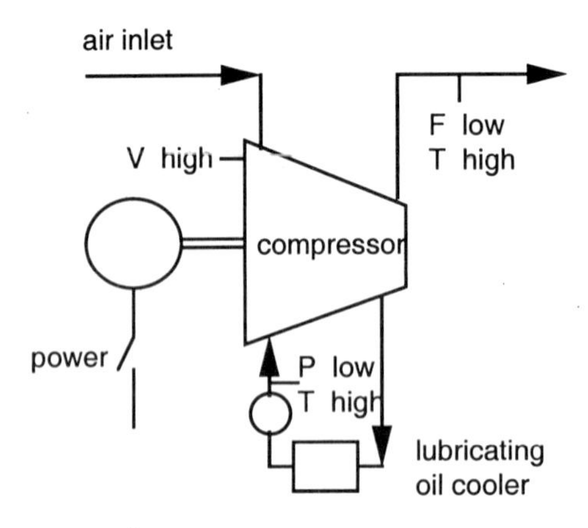

Figure 19.1: Simple compressor.

international standard IEC 848, is an important and more sophisticated notation to describe binary sequences, including concurrent processes. It is being more commonly used both as a documentation tool and as a programming language.

The design of correct and robust sequences of operations remains an *art* although there is an increasing research effort in this area. Even design principles and guidelines are few and far between.

In Section 19.2 we discuss some typical applications in wastewater treatment plants. Section 19.3 discusses the issue of representation mentioning ladder, logic and GRAFCET diagrams, as well as specialist programming languages. In Section 19.4 we discuss the design of sequences, and finally the sequenced batch reactor (SBR) is used as an example in Section 19.5.

19.2 Applications

The most common purely digital applications of sequencing involve the start-up and shutdown of more complex equipment. In a wastewater treatment plant, the larger air compressors, energy recovery equipment, grit and sludge handling equipment, and pumps can be applications in this category. Start-up systems can involve staged activation of equipment and safety interlocking from lubrication and cooling systems. Once the equipment is running safety shutdown systems monitor a range of inputs such as pressures, temperatures, vibrations, and excessive loads and trip in relief systems or trip out the operation when dangerous conditions are detected.

Example 19.1 (Simple compressor shutdown I) *Consider the simple air compressor (Figure 19.1) with an electric drive and a lubricating oil cooler and recirculating pump.*

Power will normally be turned on or off by the plant operator. However, power may be automatically turned off (tripped) if one of the following conditions occurs:

- *low air flowrate for more than a set time,*
- *high air temperature,*
- *low oil pressure,*
- *high oil temperature, and*
- *high vibration for more than a set time.*

This is a typical equipment shutdown system designed to protect the well-being of the equipment and in extreme cases the people nearby. Inputs to the shutdown controller will be digital on/off signals from sensors with preset limits which are not normally adjustable. There will be one digital on/off output signal to the power to the drive motor. In order to prevent tripping directly at the start, the air flowrate and the vibration are allowed to be outside the limits for a short time after startup.

Combined analog and sequential applications generally involve the implementation of control strategies. Examples of feedback control applications include the control of influent pumps from sewer or receiving well levels. Examples of batchwise operations are sequencing batch reactors.

Example 19.2 (Control of influent pumps) *Consider a process with an influent wastewater well whose level is controlled by the rate at which water is pumped into the treatment plant. The three pumps have variable speed electric drives. One possible control system structure is shown in Figure 19.2.*

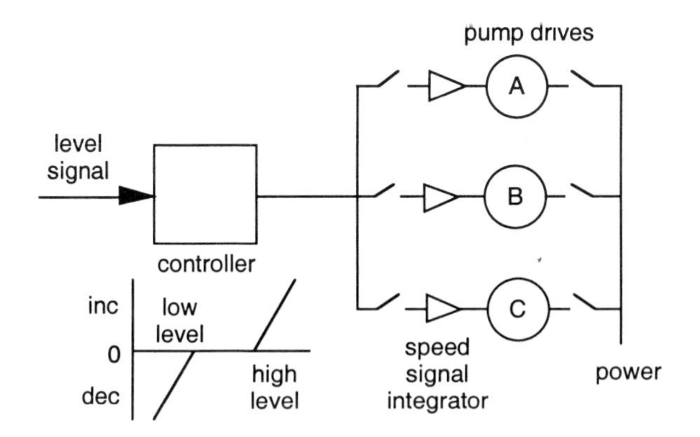

Figure 19.2: Influent pump control.

The analog level signal drives a proportional analog controller with a deadband which produces an analog rate-of-change signal to increase speed, decrease speed or take no action. This is switched to an integrator on the pump, currently varying the flowrate. If the speed signal out of the integrator remains at the maximum for more than a set time, then the next pump is turned on and the control signal is switched to it. If the speed signal remains at the minimum for more than a set time, then that pump is turned off and the control signal is switched to the previous pump.

19.3 Representation

The representation of sequenced operations is a key issue in the design, implementation, programming, commissioning and subsequent maintenance phases. Particularly in the design and commissioning phases, it is important to facilitate communication and understanding between process people and instrumentation people. During implementation and the programming phases there are often computer people and instrumentation people involved. For subsequent maintenance, the ability for people hitherto not involved to readily understand the systems is of course important.

19.3.1 Low Level Representation

Traditionally sequenced operations have been described in words by process people and converted into a low-level representation by instrumentation people. Such low level representations as ladder logic diagrams, Boolean logic diagrams and even Boolean algebra have been used. The following example illustrates the usual gulf between the process design representation and the representation from which systems are typically implemented.

Example 19.3 (Simple compressor shutdown II) *The ladder diagram in Figure 19.3 was produced by the instrumentation engineer for the compressor trip system described in Example 19.1. He expected comments on and approval by the process engineer who thought ladder diagrams were for building fire escapes. The communication gulf should be clear.*

19.3.2 Ladder Logic Diagrams

While statistics show that the share of electromechanical relays is decreasing, they will remain a part of sequential control systems for some time to come. Their influence is also evident in the number of PLCs whose programming and displays use relay terminology.

Relay circuits are usually drawn in the form of **ladder (logic) diagrams**. Even if the relays are replaced by solid-state switches or programmable logic they are still quite popular for describing combinatorial circuits or sequencing networks. They also serve as a basis for writing programs for many programmable controllers.

A ladder diagram reflects a conventional wiring diagram (Figure 19.4). A wiring diagram shows the physical arrangement of the various components (switches, relays, motors, etc.) and their interconnections and is used by electricians to

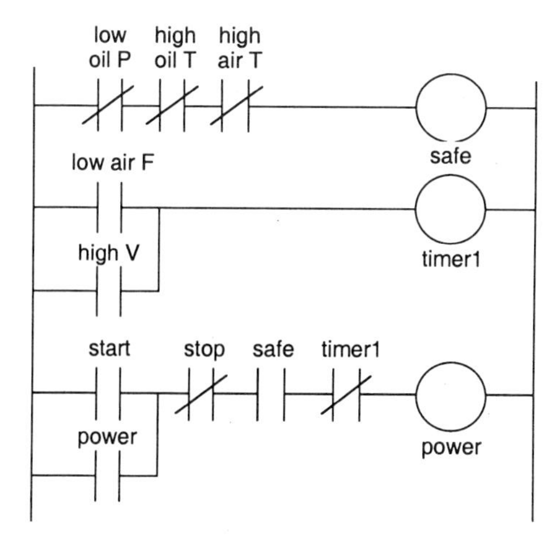

Figure 19.3: Ladder logic diagram.

do the actual wiring of a control panel. The ladder diagrams are more schematic and show each branch of the control circuit on a separate horizontal row (the rungs of the ladder). They are meant to emphasise the purpose of each branch and the resulting sequence of operations. The base of the diagram is two vertical lines, one connected to a voltage source and the other to ground.

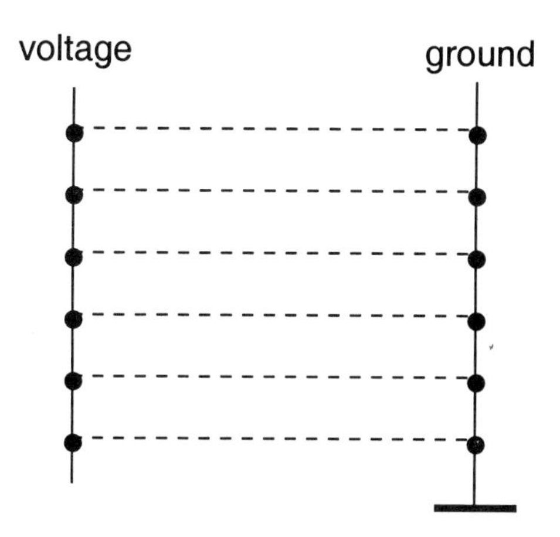

Figure 19.4: Framework of a ladder diagram.

Relay contacts are either normally open (n.o.) or normally closed (n.c.) where *normally* refers to the state when the coil is not energised. The relay symbols are shown in Figure 19.5.

A control sequence progresses through a number of **states** or **steps**, where

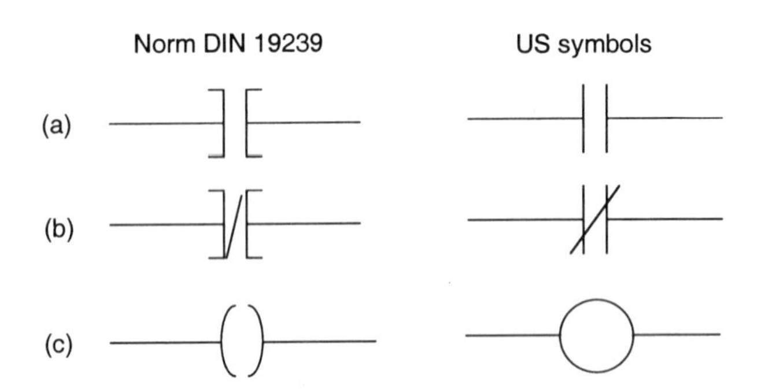

Figure 19.5: Relay symbols (a) n.o. contact, (b) n.c. contact, (c) relay coil.

each state or step is associated with a certain control **action**. Only *one state at a time* can be active and the progression from one state to the next depends on one or more **conditions**. The conditions normally depend upon the process (for example a certain temperature is reached after a heater was turned on) or the control system (for example a timer times out after being activated) responding to the previous action. Note that the *state* here is different than the state of a dynamic system, that is defined in Chapter 3.1.

The structure of a sequence is shown as a ladder diagram in Figure 19.6. The execution jumps one step at a time and returns to step 1 after the last step. Step N in Figure 19.6 shows the basic ingredients of a step. The Step N relay coil can only be activated if:

- Step N+1 is inactive. If all steps have this n.c. contact belonging to the next step then at most one step or state can be active at any one time.
- Step N-1 is active. If all steps have this n.o. contact belonging to the previous step then it ensures that the sequence 1, 2, ..., N-1, N, N+1, ... is followed.
- The "Condition" is satisfied. This will be the process or the control system response to a previous action.

When the Step N relay activates it will:

- Deactivate the previous step through its Step N+1 n.c. contact now opened.
- Latch into an active state through the Step N (self-holding) contact.
- Activate process or control system actions via one or more Step N action contacts.

Step 1 has one difference, an extra parallel contact from a start button or some other initiating device.

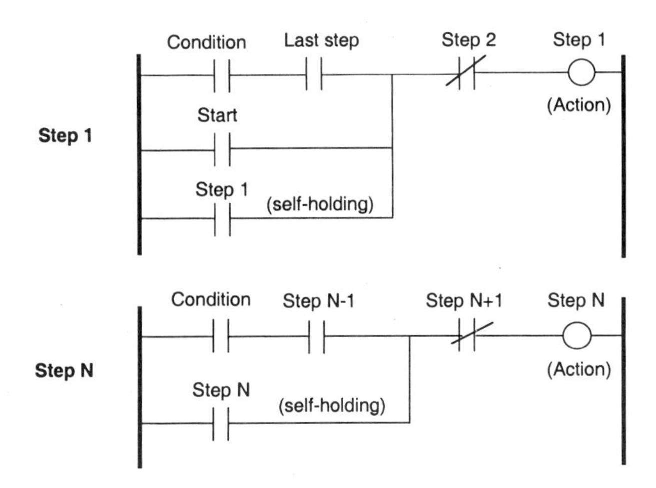

Figure 19.6: A sequence described by a ladder diagram.

Example 19.4 (Simple batch process I) *By way of an example, we will illustrate the use of a ladder logic diagram for a batch process. A tank is to be filled with a liquid. When it is full the liquid is heated until a certain temperature has been reached. After a specified time the tank is emptied and the process starts all over again. The ladder logic diagram of the sequence is shown in Figure 19.7.*

An indicator Empty, *signals that there is no liquid left. A* Start *signal together with the* Empty *indication initiates the sequence by closing Step 2 which in turn closes Valve and starts Pump. An indicator* Full *causes a jump to Step 3 when the pump is switched off and a heater is switched on. The heater remains on until the final temperature has been reached (*Temp*) and there is a jump to Step 4. The heater is switched off and a timer is started. When the waiting time has elapsed the outlet valve is opened. The sequence then starts again when the* Empty *signal occurs.*

19.3.3 High Level Representation

A solution to the common communication problem between process and instrumentation people has been the emergence in recent years of high-level representations. Typical of these high-level representations is the GRAFCET graphical representation and computer-like high-level languages.

The need for structuring a sequential process problem is not apparent in small systems. But as a control system becomes more complex the need for better functional description increases. Each block also has to be able to include more and more complex functions. This means that logical expressions in terms of ladder logic diagrams or logical circuit diagrams are not sufficiently powerful to allow a structured description. In order to facilitate a more rational top-down analysis the

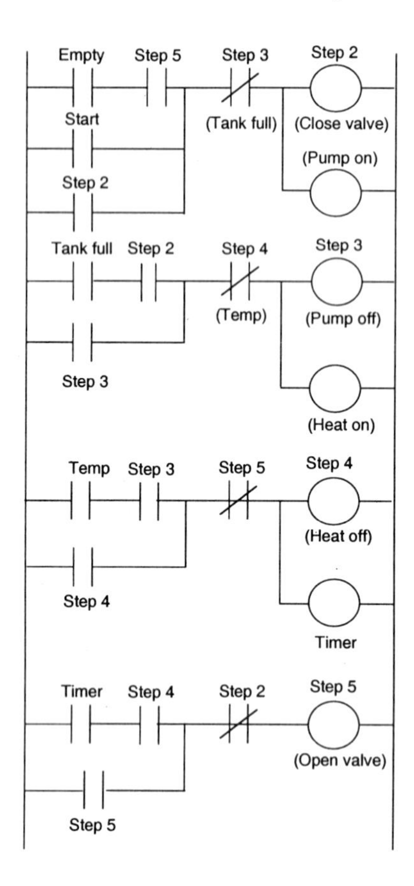

Figure 19.7: Ladder logic diagram for batch tank.

functional diagram **GRAFCET** (GRAphe de Commande Etape-Transition) was developed by a French commission in the late 1970s and was adopted as the French national standard. Since 1988 GRAFCET has been specified in an international standard, the current version being IEC 1131-3 where it is called the Sequential Function Chart.

19.3.4 The GRAFCET Diagram

GRAFCET is a method that was developed for specifying industrial control sequences by means of a diagram. A similar method originated in Germany, called FUP (FUnction Plan). The basic ideas behind the two methods are the same and the differences are of minor importance.

Example 19.5 (Simple batch process II) *The GRAFCET diagram of the sequence for the batch tank is shown in Figure 19.8. The description of the process* *can be found in Example 19.4 together with the equivalent ladder logic diagram (Figure 19.7).*

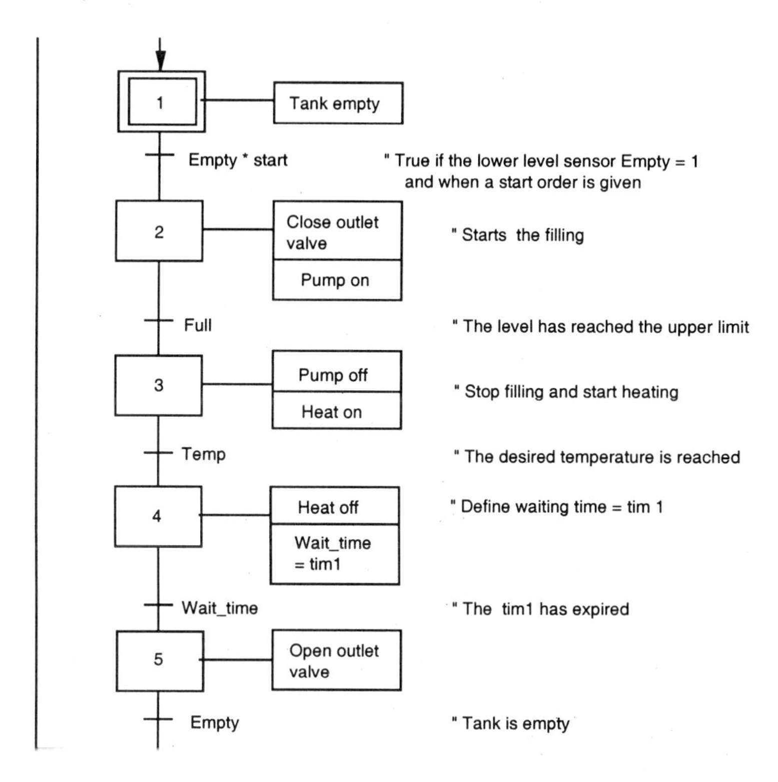

Figure 19.8: GRAFCET diagram for the batch tank.

An indicator Empty *signals that there is no liquid left. A* Start *signal together with the* Empty *indication initiates the sequence. In Step 2 the bottom valve is closed and the pump is started. An indicator* Full *tells when to stop the pumping and causes a jump (called a* **transition***) to Step 3 when the pump is switched off and a heater is switched on. The heater remains on until the final temperature has been reached (*Temp*) and there is a jump to Step 4. The heater is switched off and a timer is started. When the waiting time has elapsed (*time_out*) there is a transition to Step 5 when the outlet valve is opened. The sequence then returns to the start.*

Figure 19.8 illustrates that the GRAFCET representation consists of a column of numbered **blocks** each one representing a **step** or a state. The vertical lines joining adjacent blocks represent **transitions**. Each transition is associated with a logical condition called **receptivity**, which is defined by a Boolean expression written next to a short horizontal line crossing the transition line. If the receptivity or condition is logical 1, the transition is executed and the system passes to the next step.

A GRAFCET diagram basically describes two things according to specific rules:

- which order to execute actions, and
- what to execute.

The function diagram is split into these two parts. The part describing the order between the steps is called the **sequence part**. Graphically this is shown as the left part of Figure 19.8 including the five boxes. The sequence part does not describe the actions to execute. This is described by the **object part** of the diagram that consists of the boxes to the right of the steps.

A *step* can be either active or passive, whether it is being executed or not. The *initial* step is the first step of the diagram. It is described by a box with a double frame. An **action** is a description of what is performed at a step. Every action has to be connected to a step and can be described either by a ladder logic diagram, by a logical circuit diagram or by Boolean algebra. When a step becomes active its action is executed. However, a logical condition can be connected to the action so that it is not executed until both the step is active and the logical condition is fulfilled. This feature is useful as a safety precaution.

Several actions can be connected to one step. They can be of the type *outputs*, *timers* or *counters* but can also be controller algorithms, filtering calculations or routines for serial communication. A *transition* is an "obstacle" between two steps and can only originate from an active step. Once a transition has taken place the next step becomes active and the previous one inactive. The transition consists of a logical condition that has to be true in order to make the transition between two steps possible.

By combining the three building blocks *initial step*, *steps* and *transitions* it is possible to describe quite a large number of functions. The steps can be connected in:

- simple sequences,
- alternative parallel branches, and
- simultaneous parallel branches.

In a **simple sequence** there is only one transition after a step and after a transition there is only one step. No branching is possible. In an **alternative parallel sequence** (Figure 19.9) there are two or more transitions after one step. This means that the flow of the diagram can take alternative paths. Typically this is an *if-then-else* condition and is useful to describe alarm situations for example.

It is very important to make sure that the transition condition located immediately before the alternative branching is consistent, in other words the alternative branches are not allowed to start simultaneously. A branch of an alternative sequence always has to start and end with transition conditions.

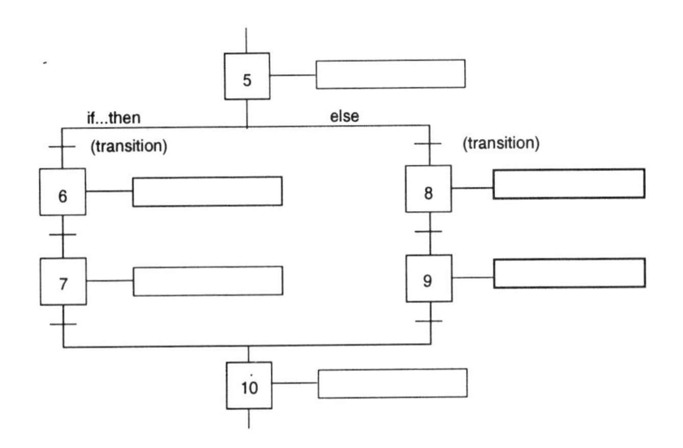

Figure 19.9: GRAFCET for alternative parallel paths.

In a **simultaneous parallel sequence** (Figure 19.10) there are two or more possible steps after a transition that are active simultaneously. In other words this is a concurrent (parallel) execution of several actions.

The double horizontal lines define the parallel processes. When the transition condition is true both the branches become active *simultaneously* and are executed *individually* and *concurrently*. A transition to the step below the lower parallel line cannot be executed until both the concurrent processes are completed. In translating the GRAFCET notation into a real-time program, synchronisation tools such as semaphores have to be used. The branches become concurrent processes that have to be synchronised so that racing conditions are avoided.

The three types of sequences can be mixed but have to be done in a correct way. For example if two alternative branches are terminated with a parallel ending (two horizontal bars) then the sequence is locked since the parallel end waits for both branches to finish, while the alternative start has started only one branch. Also if simultaneous parallel branches are finished with an alternative ending (one horizontal bar) then there may be several active steps in the code and it is not executed in a controlled manner.

In GRAFCET there are several inherent real-time features that have to be observed in an implementation. The realisation of real-time systems requires intensive efforts with considerable investments in time and personnel. The implementation of the GRAFCET function diagram into computer code is not part of the definition and of course varies in the different systems. Obviously any implementation makes use of real-time programming tools. GRAFCET compilers are available for many different industrial control computers. Typically, the block programming and compilation are performed on a PC. After compilation the code is transferred to the PLC for execution. The PC is then removed in the real-time execution phase. More advanced PLC systems have GRAFCET compilers built into the system.

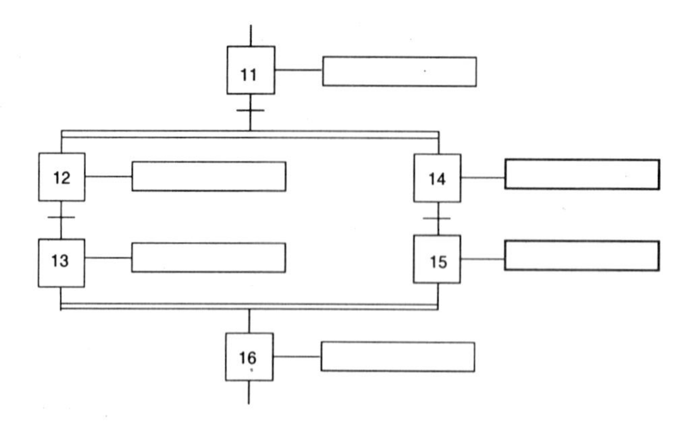

Figure 19.10: GRAFCET for simultaneous parallel paths.

The obvious advantage of GRAFCET and similar types of abstract descriptions is their independence of specific hardware and their orientation to the task to be performed. Unfortunately it has to be said that high level languages such as GRAFCET do not yet enjoy the success they deserve. It seems odd that so many programmers start anew with programming in C or Assembler while control tasks of the type we have seen are much more easily solved with a functional block description.

As in any complex system description the diagram or the code has to be structured suitably. A GRAFCET implementation should allow the division of the system into smaller parts and of the GRAFCET diagram into several sub-graphs. For example each machine may have its own graph. Such hierarchical structuring is of fundamental importance in large systems. Of course GRAFCET is useful also for less complex tasks. It is quite easy for the non-specialist to understand the function as compared to the function of a ladder logic diagram. By having a recognised standard for the description of automation systems the chances for re-utilisation of computer code are increased.

The translation of GRAFCET to computer code depends on the specific PLC. Even if there is no automatic translator from GRAFCET to machine code the functional diagram is still very useful, since it allows the user to structure the problem. Many machine manufacturers today use GRAFCET to describe the intended use and function of the machinery. Of course it makes subsequent programming much simpler if GRAFCET can be used all the way. An implementation tool also may simulate the control code of a GRAFCET diagram on the screen. During the simulation the actual active state is shown.

19.3.5 High Level Languages

An example of a high-level language representation is **Structured Text**, also defined in the International Standard IEC 1131-3 (1993). There are also various proprietary languages defined by manufacturers for their own sequence or batch control software products. The following example illustrates a high level language representation.

Example 19.6 (Simple batch process III) *Table 19.1 shows the Structured Text representation of Example 19.4 whose GRAFCET representation is shown in Figure 19.8.*

```
FUNCTION_BLOCK BATCH_TANK
 VAR_INPUT
 START, (* start button *)
 EMPTY, (* low level sensor *)
 FULL, (* high level sensor *)
 TEMP, (* high temperature sensor *)
 WAIT_TIME : BOOL ; (* timeout of timer *)
 END_VAR
 VAR_OUTPUT
 VALVE, (* tank outlet valve *)
 PUMP, (* filling pump *)
 TIMER, (* timer *)
 HEAT : BOOL ; (* tank heater *)
 END_VAR
INITIAL_STEP STEP1 : END_STEP
TRANSITION FROM STEP1 TO STEP2 := EMPTY & START ;
END_TRANSITION
STEP STEP2 : VALVE(S) ; PUMP(N) ; END_STEP
TRANSITION FROM STEP2 TO STEP3 := FULL ;
END_TRANSITION
STEP STEP3 : PUMP(N) ; HEAT(S) ; END_STEP
TRANSITION FROM STEP3 TO STEP4 := TEMP ;
END_TRANSITION
STEP STEP4 : HEAT(R) ; TIMER(N) ; END_STEP
TRANSITION FROM STEP4 TO STEP5 := WAIT_TIME ;
END_TRANSITION
STEP STEP5 : VALVE(R) ; END_STEP
TRANSITION FROM STEP5 TO STEP1 := EMPTY ;
END_TRANSITION
END_FUNCTION_BLOCK
```

Table 19.1: Structured text for batch tank.

19.4 Sequence Design

As you may have inferred from the previous section a discrete control sequence consists of a number of **steps**, sometimes called states. Each step has an associated set of **actions** that it performs, which may affect the process (close a valve, start a pump) or they may be actions affecting the control system itself (start a timer, switch a continuous controller to manual). The steps will usually be executed in sequence (hence the name sequence control) although sometimes several steps may operate in parallel. The progression from one step to the next is termed a **transition** and it occurs only if specified **conditions** are met. Like actions these conditions may be process related (valve has closed, level is high) or they may be control system related (timer times out, previous steps completed).

So where do these control sequences come from and what tools are there to assist us in coming up with a working sequence? There are several tasks involved in designing or developing a discrete control sequence:

- setting the process goals for the task the sequence controller is to perform.
- the identification of actuators and sensors available for implementing the goals and checking that actions taken have indeed happened (feedback).
- the synthesis of the control sequence consisting of the steps, actions and conditions.
- documenting the sequence developed.
- testing or proving the correctness of the sequence, that is checking that it actually achieves the initial process goals.
- implementation of the control sequence for a particular hardware platform such as a PLC.

In Section 19.3 we discussed the issue of representation or documentation of sequences where there is a degree of standardisation and some software tools. There are also established techniques and quite a bit of software for the final implementation task of translating a documented sequence into an equipment specific program or circuit.

The other perhaps more important and difficult tasks are much less formalised. It is probably no exaggeration to say that current practise relies almost solely on experienced practitioners who have learnt "on the job". Several authors have proposed hierarchical models of the design process but formal procedures remain elusive. We will discuss these areas in more detail below.

19.4.1 Process Goals

By process goals we mean a detailed specification of the desired behaviour of the process that is to be driven through a sequence of steps or phases by the discrete controller. Such a specification is equally necessary for all types of applications, start-up, shutdown, or batchwise operation.

Not only is a detailed specification necessary for design but it is also the basis of testing that the designed sequence actually performs the intended purpose. Unfortunately there are few appropriate methodologies or even heuristics or guidelines. This makes it difficult to write about, but we will mention a few factors we consider important.

The ultimate starting point has to be a general operating goal and a detailed description of the process. The general top-down design and successive refinement methodologies developed for software engineering can be applied to gradually define more and more specific and detailed statements of goals. This is done down to the point that one goal is one step in the sequence and all actions and conditions are specified.

Here again the issue of representation is encountered. How do we document our goals unambiguously, especially bearing in mind that this documentation is often the communication between process people (who may or may not understand instruments) and instrument people (who may or may not understand processes). In practise this is invariably done in words occasionally supported with some sort of flowchart borrowed from the world of computer programming. There has been some research into goal modelling and representation for processes and Multilevel Flow Modelling is perhaps the most advanced in terms of concepts and definition. The process is modelled at three **functional** abstraction levels, goals, functions, and components, and a graphical language has been defined for the representation.

Sequence control is primarily open loop in nature, however there are elements of feedback as well. The transition between steps is frequently based on a feedback signal from the process telling the controller that it has responded to the previous actions. The importance of this feedback cannot be overestimated (compare Section 10.5). The purpose is to confirm that the action requested by the sequence controller really did happen. The best feedback is a measurement close to the specified goal for the step, for example if we are turning on a pump to fill a vessel the feedback should be, not "did the switch close" or "is pump the running" or "is there delivery pressure" but, "is there flow". The general policy should be to avoid sensors such as limit switches on valves (notoriously unreliable anyway) and to use process variables (such as flow, level, temperature, and concentration) which tell whether the goal has been achieved.

19.4.2 Control Sequence Synthesis

Given that we can develop a detailed specification of the objectives, the synthesis of the control sequence itself, duly documented in GRAFCET for example, is a less daunting task. The less detailed the specification, the bigger the synthesis problem.

Again there are no commonly used systematic techniques for synthesising sequences. There is promising research into systematic techniques but it is some distance from commercial implementation.

The conclusion can only be that good and detailed specifications must be written. At least we have some general methodologies for that.

19.4.3 Proving Correctness

Even leaving aside the issue of equipment or process failures during operation, how do we ensure that a given control sequence will meet the goals, and which goals? This is again a vexing issue. There are three possibilities that are under development:

- **Proof by theory**. There is certainly a body of research into Discrete Event Systems (DES) but this is relatively recent. Current less-than-elegant analysis techniques are plagued with the curse of combinatorial explosion, especially with systems of a realistic size from an industrial perspective. There are also tentative steps towards applying some of the work in computer science on proving the correctness of computer code.

- **Modified HAZOP techniques**. HAZOP is a structured analysis procedure developed by ICI for evaluating process designs. A modified version for analysing PLC and other control software has been trialled in industry with some success reported.

- **Simulation**. Of course you can argue that this is a demonstration but hardly a proof. But then again the saying says that "beggars cannot be chosers" (that hasn't been proved either). The difficulty with simulation is that we must simulate a hybrid system, partly discrete event and partly continuous. Few simulators handle hybrid systems robustly though some prototypes exist (see Section 9.6).

19.4.4 Handling Exceptions

Of course we cannot leave aside the issue of equipment failures (sensors, actuators, even the controller itself) or process failures (pumps, valves, utilities) during operation. How do we identify likely failures, detect when they occur, diagnose what happened and handle subsequent processing of the sequence?

Identification is facilitated by systematic techniques such as HAZOP studies. Detection involves artefacts such as "watch dog" timers and autonomous sequences. Even if a "stuck sequence" is detected, diagnosing the cause is a nontrivial exercise. The huge number of possible permutations and combinations of failures generally means that only very hazardous situations warrant the expense of automatic detection and correction. Generally we will simply generate an alarm and leave the problem to the operators.

19.5 SBR Operating Sequence

Batch processes were introduced in Section 4.7. As we will see, batch operation is often very attractive from a process control point of view. We may measure the concentration of oxygen or of nutrients like ammonia and nitrate nitrogen as functions of time. This gives us the possibility of estimating reaction rates. In a recirculating plant we will measure at one point in space at a time but this will not give the same amount of information. The batch reactor is all the time in a transient state (even at a constant load). A control system will always get more information from transient conditions than from almost stationary conditions.

Example 19.7 (Carbon removal in a batch reactor) *The simple batch carbon removal modelled in Example 4.9 shows the nutrient profiles of Figure 19.11.*

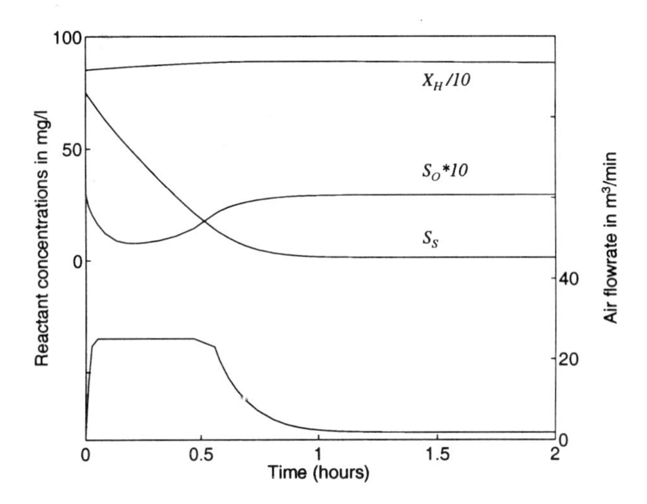

Figure 19.11: Nutrient profiles for simple batch carbon removal.

Monitoring the slope of the S_S and S_O transients can give valuable information on the $(\hat{\mu}_H X_H)$ and K_La parameter values in the reaction rate and oxygen transfer terms of the component balance (Equation 4.50). This information provides parameter estimates for modelling as well as diagnostic information on the state of the biomass and the aeration equipment.

The initial value of S_S and the length of time the air flowrate is saturated give load related information.

19.5.1 Sequential Batch Reactors

The Sequential Batch Reactor (SBR) was described in Section 4.7. It sequentially cycles through fill, react, settle and decant phases of a total cycle time of typically 4 to 6 hours.

The react phase for carbon removal was shown above in Example 19.7. Subphases within the react phase can supply anoxic conditions for denitrification and even anaerobic conditions for VFA uptake.

Various solutions have been implemented to overcome the intermittent filling problem including multiple SBRs operating out-of-phase, equalisation tanks and reactors with continuous feed.

While there are both capital and operating cost advantages and the monitoring advantages mentioned above, the disadvantage is increased complexity in operation and control.

19.5.2 Alternating Plants

In Denmark low loaded alternating activated sludge plants designed for nitrification and simultaneous precipitation had to be upgraded for nutrient removal. The nitrogen removal process Bio-Denitro was developed in the early 1970s from coincidal observations of an unintentional N removal in alternating plants during the late 1960s. The Bio-Denitro process was extended to the Bio-Denipho process with the capability for both N and P removal. These two processes are patented processes developed by Krüger Systems in cooperation with the Department of Environmental Engineering at the Technical University of Denmark.

The principle of the Bio-Denipho operation is illustrated in Figure 19.12. The figure shows only the biological part of the plant. The two subprocesses involved with biological nitrogen removal, nitrification and denitrification, are performed in a semi-batch manner by periodically changing the path of flow through two parallel aeration tanks which are aerated according to a fixed or controlled strategy. An anaerobic zone at the front of the process promotes P release.

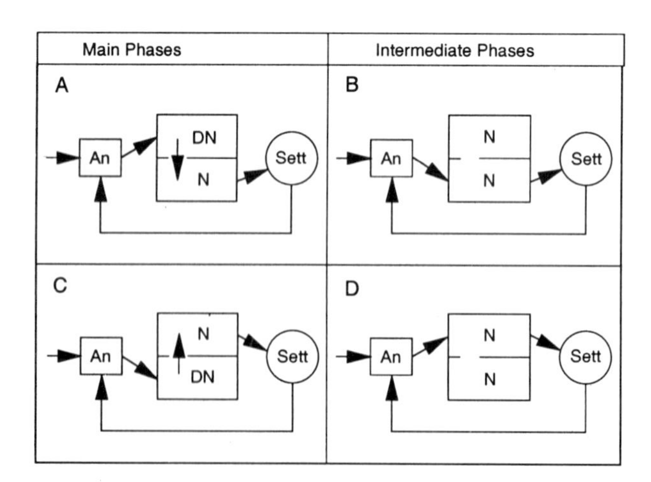

Figure 19.12: The Bio-Denipho cycle.

The phases A to D execute in sequence making up the one operating cycle. The intermediate phases B and D last from 15 to 30 minutes and the main phases A and C from 60 to 90 minutes. The complete cycle typically takes about 3 hours. The Bio-Denitro process is in principle just the Bio-Denipho but without the anaerobic tank. In the alternating process there is no recirculation of nitrate, instead the carbon source is directed to the nitrate.

19.5.3 Phase Length Control of Nitrogen Removal

In an intermittent aeration control (phase length control) the division between the nitrification phase and the denitrification phase is given by the treatment time. Such control can be implemented both in a typical sequential batch reactor as well as in a traditionally recirculating plant with given volumes for denitrification and nitrification.

The distribution of anoxic and oxic capacity in an SBR or alternating plant is based on phase lengths. Normally the phase lengths are controlled by a time scheme driven by computer software with variations on workdays/weekends and during summer/winter. These time schemes are open loop control schemes and do not take into account variations in the load. The time schemes are often planned with safety margins inducing "waste of process time" in some periods. From a control point of view the alternating processes can change from 100 % nitrification to 100 % denitrification in 5 minutes. This of course makes it possible to perform on-line feedback control.

The N removal can be controlled by installing on-line measurements of ammonium and nitrate from the aeration tanks. The basic principle is that a nitrification phase is ended when ammonium is low (and nitrate is sufficiently high) and the denitrification is ended when the nitrate is low.

Both the nitrification and denitrification are sensitive to the DO concentration (see Chapters 4 and 14). By changing the setpoint in the DO control it is possible to affect the reaction rates thereby following the dynamic variations in the nitrogen load. If the ammonium load is increasing then the DO setpoint can increase thereby increasing the nitrification rate. This is completely analogous to the DO setpoint control in Section 17.7.1 and will be further described in Section 20.3.3.

19.5.4 Nitrification Control using Redox Measurements

The oxidation-reduction potential (ORP), or Redox, has been used for the control of nitrification-denitrification in sequential batch systems. The ORP can give information that is related to the biological state of the system and can control the "nitrate breakpoint". Consider a sequential batch reactor with intermittent aeration. During the anoxic part when no air is added, Figure 19.13, the nitrate is decreasing towards zero. The ORP then displays a breakpoint associated with the disappearance of nitrates. From a bacterial point of view this corresponds to a transformation from respiratory (oxygen or nitrates) to non-respiratory (fermentative) process.

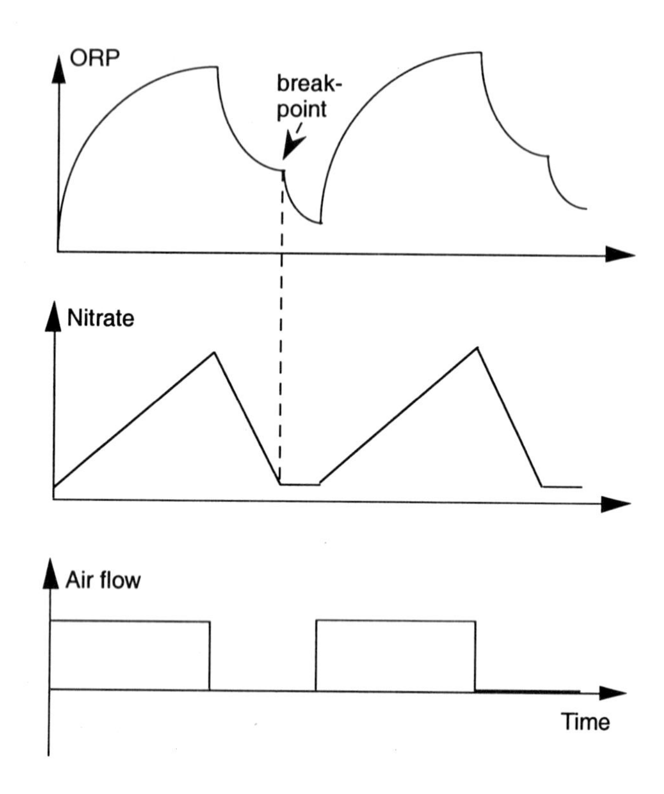

Figure 19.13: Principal behaviour of the ORP, nitrate concentration and aeration during intermittent aeration with fixed phase length operation. The nitrate "breakpoint" in the ORP is noticed when the nitrates disappear.

The obvious control strategy is then to turn on the air supply once the ORP breakpoint has been noticed.

19.5.5 Carbon Dosage Control

The carbon dosage control in an alternating denitrification-nitrification plant is different compared to a continuous recirculating plant, Section 17.7.2. A model for the denitrification is used to predict the nitrate concentration at the end of the current operation cycle. That prediction is then used to find the required carbon dosage.

19.6 Summary

We have introduced aspects of batch or discrete event sequencing and control. Representation and design of control sequences are key problem areas where recent advances and guidelines should be implemented.

We also discussed wastewater treatment examples of sequencing equipment startup and of control strategies for biological nutrient removal in the sequenced batch reactor (SBR) and in alternating bioreactors.

19.7 Further Reading

- A proposed hierarchical model of the design process is described in Rosenof & Ghosh (1987)
- DeMarco (1978) has developed general top-down design and successive refinement methodologies for software engineering.
- The software standard IEC1131-3 is documented in IEC1131 (1993).
- A good introduction to GRAFCET is found in David & Alla (1992).
- High level programming, including GRAFCET, for sequencing control is further described in Lewis (1997).
- Multilevel Flow Modelling is described in Rasmussen & Lind (1981), Lind *et al.* (1987) and Lind (1987).
- The HAZOP analysis is described in Kletz (1986), Ozog & Bendixen (1987) and Burk (1992).
- HAZOP analysis of PLC programs is presented in Nimmo (1994) and Lear (1995).
- A good overview of the various high-level representations of batch processes can be found in Fisher (1990). Batch reactors for wastewater treatment are described in Irvine & Ketchum Jr. (1989), Nakazawa & Tanaka (1991) and Oles & Wilderer (1991).
- An overview of the problems in hybrid systems control is found in Fabian & Lennartson (1995), Lennartson *et al.* (1994) and Lennartsson *et al.* (1996).
- Control of the phase length of aeration based on ammonium and nitrate measurements has been extensively applied in Denmark. This includes the Bio-Denitro and Bio-Denipho processes (Lynggaard & Nielsen (1993), Thornberg *et al.* (1993), Önnerth *et al.* (1996), Nielsen & Önnerth (1996), Zhao *et al.* (1994b), Zhao *et al.* (1994a), Bundgaard (1988), Thornberg (1988), Sørensen *et al.* (1994) and Nielsen *et al.* (1981)).

- Applications of ORP control are found in Wareham *et al.* (1993), Fröse & Köhler (1995), Plission-Saune *et al.* (1996) and Sasaki *et al.* (1993).
- The application of respirometry is described in Klapwijk *et al.* (1998).
- Using a model for estimation of the denitrification for carbon dosage control has been applied by Isaacs *et al.* (1992) and Isaacs *et al.* (1994).

19.8 Exercises

1. Compare the definition of a state in a sequencing operation and in a dynamic system.
2. Consider the ladder diagram in Figure 19.6. The sequence is started by pushing the *start* button. Why is it possible for the process to continue after you have released the *start* button?
3. We have shown that high level programming like GRAFCET can simplify the structuring of complex sequential operations. Still ladder diagrams are very popular and the change from low level to high level programming is a slow one. What can be the reasons?
4. Compare the operation of a sequential batch system and a conventional recirculating system. Make a list of arguments for and against each type of system.
5. A hybrid operation consists of both sequencing operations and continuous control of dynamic variables. Make a simple structure of the hybrid nature of the control of an alternating plant.

Chapter 20

Plant Wide Control

In Chapter 14 we examined some control strategies for the different subprocesses that make up a wastewater treatment plant. This decomposition of the process into subprocesses is a convenient way of studying and designing control systems. But eventually we must step back and take a *helicopter* view of the whole plant. In any complex system the best policy for the whole is seldom the sum of the best policies for the individual parts. There are a number of reasons why this is so:

- some plant objectives are achieved by utilising more than one subprocess, for example phosphorus and nitrogen removal.
- even subprocesses in simple sequences are frequently affected to some extent by what is done in previous subprocesses, for example the utilisation of COD and biomass in BNR processes.
- recycle streams complicate the interactions because they mean that a subprocess can be downstream of itself so that the effects of changes can be reduced or exaggerated after they have been transmitted around the circle. Biomass and nitrate recycles are examples.

Because of these interactions between the parts of the plant, we need some plant wide coordination to decide how the parts should operate to achieve the overall goals. This chapter will introduce **optimisation** which is the tool available to assist in these decisions (Section 20.1), and plant wide control techniques which can help to implement the decisions (Section 20.3).

While we will look at some aspects of the *broader picture* we will mainly remain within the plant. You could stand back further and include sewers and receiving waters, or stand much further back and include wastewater producers and clean effluent consumers. The latter system definition will include more recycle *streams* involving water, nutrient and financial resources. That is the topic of another book.

20.1 Optimisation

The optimisation of a system is selecting conditions to achieve the best possible result within some given limits. There are several issues in this definition that we need to address:

- As we have discussed so often before we must define our system boundaries and the interfaces to the "outside".
- There must be some selectable conditions usually called degrees of freedom.
- We must define what "best possible" means, that is we need an objective function.
- The limits or constraints must be determined.
- We must have a way of making the selection, an optimisation technique.

The first four issues will be discussed in this section and the bigger issue of optimisation techniques will be addressed in the following section.

20.1.1 Defining the System

We have discussed the importance of explicitly defining your system in Chapter 3 with respect to modelling and in Chapters 14 and 17 with respect to control. We are now going to discuss this topic again, this time with respect to optimisation. Unfortunately, or perhaps fortunately, it is not possible to optimise the universe. We can only select some small part of the universe and do our best with that, realising on the one hand that we must accept what goes on *outside* as disturbances, and realising on the other hand that what we do will in turn disturb the *outside* (Figure 20.1).

The inputs are again of two types, those we can manipulate which we call *degrees of freedom* in optimisation terms, and those we cannot affect directly which again we term disturbances. The best we can do is to manipulate the degrees of freedom to optimise the system for given disturbances. If the disturbances change then the optimum may or may not change too. Disturbances also dictate how close we can operate to constraints.

The outputs will in turn disturb the outside world. This means that if the outside was optimal before then it may no longer be optimal after we have optimised our system. A tricky but very important concept to recognise. In addition there will frequently be relationships in the outside world between our outputs and our disturbances, so that our decisions will definitely come back to haunt us.

Yet again we must talk of time scales. When we optimise from a plant wide control perspective we must also define the time scale of interest. We can ignore the dynamics of disturbances in much faster time scales, and in much slower time scales simply consider average values.

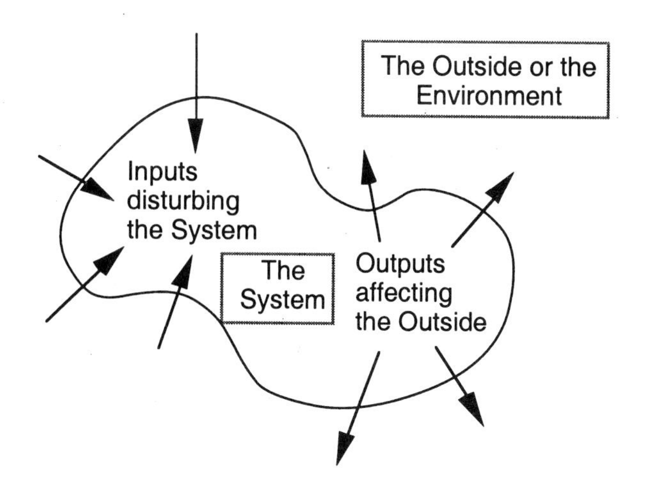

Figure 20.1: The system and the outside.

As in our previous system definitions there will also be system state variables. 'hese in turn determine the output variables and are also important in the defiition of the objective function.

xample 20.1 (A larger and longer term problem) *We could be interested ı defining the parameters for a treatment plant to be built for a small commuity which discharges wastewater into an already partly polluted lake with heavy ediments (Figure 20.2 left plot). Interactions within the system will occur in the mescale of years to tens of years. This tells us that our disturbance inputs will variables such as population growth, patterns of water usage and possible indusialisation. Outputs will be variables such as lake recreational use, downstream ater users and sludge disposal. Because of our system boundary definition we are noring external relationships such as those between recreational use and populaon. System state variables will include the treatment plant effluent composition, e lake water quality and the amount of sediment.*

xample 20.2 (A smaller and shorter term problem) *Here we may define ır system as optimising nitrogen removal in the anoxic and aerobic zones in a e-denitrification treatment process (Figure 20.2 right plot). Disturbance inputs ould be variables such as influent water composition, COD supplement and air age. Outputs would be variables such as effluent composition, sludge composition ıd settleability. The time scale will be of the order of tens of minutes to a few ıys. System states will include carbon, oxygen, ammonia, nitrate and biomass ncentrations in each zone.*

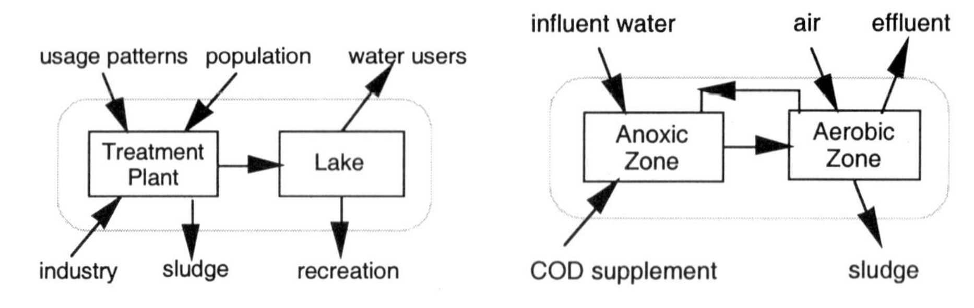

Figure 20.2: Systems for Example 20.1 (left plot) and Example 20.2 (right plot)

20.1.2 Degrees of Freedom

Of the variables that we are free to manipulate, it must be clear which will b utilised for system optimisation. These are the degrees of freedom.

Degrees of freedom are frequently control system setpoints but there may als be some variables that were not utilised for control purposes. Of course the degree of freedom must have an impact on our objective function which will be discusse below.

The degrees of freedom define the dimensions or axes of the operating space The purpose of the optimisation is to locate where in this operating space th system should operate.

20.1.3 Objective Functions

The objective function defines our concept of what is *best* (compare Sections 7. and 7.2). An objective function is a summation of one or more terms, each terr being made up of some measure of performance multiplied by a correspondin weighting.

$$J = \sum_i w_i M_i(u, x, y) \quad (20.1$$

where w_i is the weighting factor, M_i the measure which is a function of inputs u states x and outputs y.

The weighting factor is very important and serves two functions. Firstly th weighting factor brings all the measures to a common basis, for example to an oper ating cost in dollars. It is possible to have more than one basis and hence objectiv function, although this requires the use of so-called *multi-objective optimisatio* which involves some heuristic trade-off type decision making. Secondly the weight ing factor determines the relative importance between the different measures. Th relative importance of different measures will frequently change over time, anothe of those external disturbances, which will of course affect the ultimate choice c the optimum.

In the operation of wastewater treatment plants we can define two general categories of objective measures:

- Those determining *environmental impact*, typically effluent water quality, sludge disposal, odours and noise.
- Those related to *resources*, typically energy (pumping water and recycles and compressing air), chemical additives and manpower, but also those relating to nutrient reuse. Nutrient reuse could include phosphorus recovery, methane production and the use of sludge as a fertiliser or soil conditioner.

These will usually be combined using weighting factors to produce an operating cost.

In the case of design it is the usual balance between capital and operating cost. They can be combined by including the cost of the capital in the operating cost.

Example 20.3 (A larger and longer term problem) *Degrees of freedom would include:*

- *size of wastewater treatment plant (volumes that can be treated),*
- *degree of nutrient removal in the design,*
- *sludge production, and*
- *sludge treatment design.*

The objective function would include measures for:

- *treatment plant capital and operating costs,*
- *recreational quality of the lake, and*
- *quality of water leaving the lake.*

Example 20.4 (A smaller and shorter term problem) *Degrees of freedom would include:*

- *nitrate recycle flowrate,*
- *COD supplement flowrate,*
- *DO profile, and*
- *sludge age.*

The objective function would include measures for:

- *nitrogen removal,*
- *cost of COD supplement,*
- *cost of air, and*
- *sludge production.*

Example 20.5 (Sustainable urban water systems) *The issue of sustainability has been raised in Section 14.5. Any future wastewater treatment system has to be evaluated according to a quantitative criterion which has to consider:*

- hygienic aspects: *We believe that nobody will accept a lower hygienic standard than today.*
- environmental impact: *The wastewater impact on the environment has steadily increased the need for better treatment. We need to be prepared for changes in society by having a continuing dynamic research environment in this area. Too often the knowledge build-up has been reactive instead of proactive.*
- economising resources: *wastewater treatment looks and probably should look quite different in densely populated urban areas and in rural areas. Considering resource handling one has to consider already invested capital costs, land use, water re-use, organic and nutrient use and re-use, as well as overall use of energy.*
- user aspects: *Technical functionality, economy and liability in different aspects require profound consideration.*

A sustainable wastewater treatment has to adapt to a local environment and the total resource demand has to be calculated, including a direct environmental impact on receiving waters, air and soil as well as energy consumption and nutrient recycling. Thus there is a need for a quantitative performance index. The quest for sustainable development has to be based on objective as well as subjective reasons. An integrated performance index is part of a necessary decision making model for the design and operation of wastewater treatment systems. In this index it is evident that a trade-off should be made between the pursued quality of the process outputs (liquid, solids, gas) and the associated efforts (investments, operation) required to achieve this considering the inputs (wastewater). In order to make this trade-off a common framework is needed to quantitatively compare the different objectives.

A metric to judge the sustainability of different options will facilitate a fruitful dialogue between politicians, ecologists, engineers and economists. Only a truly interdisciplinary approach can help to solve the challenges ahead. It is important that an objective function has to be able to reflect the consequences of different changes. For example what would happen if:

- *the energy price is doubled?*
- *the water consumption is reduced by 30%?*

20.1.4 Constraints

A system cannot operate just anywhere in its operating space. There will be many constraints on its operation which must be defined in terms of the degrees of freedom. The constraints will define a feasible region for the process and the optimisation. There are three general types of constraints: external, process, and equipment constraints. Examples of these constraints are:

- *external constraints*: effluent discharge limits and other environmental standards, safety (environmental, human, and equipment), political policies.
- *process constraints*: mass and energy balances, other functional relationships such as rate expressions and solubilities, flowrates and concentrations must be positive.
- *equipment constraints*: volumes, maximum and minimum flowrates, some equipment can only be on or off.

Process constraints and some of the equipment constraints may be defined in two ways:

- They may be mathematically defined so that we optimise a mathematical model and then apply the results to the system. This approach is called off-line optimisation. It is generally the preferable approach since the optimisation is faster which means the process will be more often at its optimum. It also means that faster more aggressive optimisation techniques can be used since we can do things with a mathematical model we would never do with the process, for example we can violate some constraints. This is discussed in Section 20.2 on optimisation techniques. The off-line approach also allows an evaluation of the results before implementation.
- Alternatively, the constraints may be defined by the process itself so that we optimise the process on the run. This approach is called on-line optimisation. It is generally not the preferred approach for the same reasons given above. However it does avoid the problem of plant-model mismatch which can lead the off-line approach to recommend an infeasible optimum. It also means we can optimise without developing and validating a mathematical model.

Example 20.6 (A larger and longer term problem) *Constraints could include:*

- *local, regional and national discharge regulations,*
- *treatment plant sites,*
- *sludge disposal options, and*
- *capital availability.*

Example 20.7 (A smaller and shorter term problem) *Constraints could include:*

- *local, regional and national discharge regulations,*
- *existing volumes,*
- *pumping and compression rates,*
- *sludge disposal facilities, and*
- *COD supplement availability.*

20.1.5 Response Surfaces

We saw the importance of the shape of the objective function in Chapter 7 on parameter fitting. Parameter fitting is simply an optimisation with the parameters as degrees of freedom and for example a least squares measure as the objective function. The shape of the objective function is equally important in system or process optimisation.

It is wise to perform some response surface mapping before attempting an optimisation. In particular it will enable you to assess the sensitivity of the objective function to the degrees of freedom and to detect correlation between degrees of freedom. *A priori* response surface mapping requires a mathematical model of the system. Where the process itself is optimised it is wise to build up a response surface mapping as data is collected at different operating points.

Three dimensional contour or surface plots of the objective function in the feasible region should be examined. These can be plotted against degrees of freedom two at a time using a fairly coarse grid of operating points. A very flat surface relative to a particular axis indicates a lack of sensitivity (Figure 20.3 left plot) and the corresponding degree of freedom is best removed. Valleys at an angle indicate correlation between the degrees of freedom (Figure 20.3 right plot). It may be possible to derive a mechanistic or empirical relation between the parameters and therefore remove one as a degree of freedom, for example by fitting the line along the bottom of the valley. You must take care that this relationship does not change with time or you must update it when system or objective function parameters change.

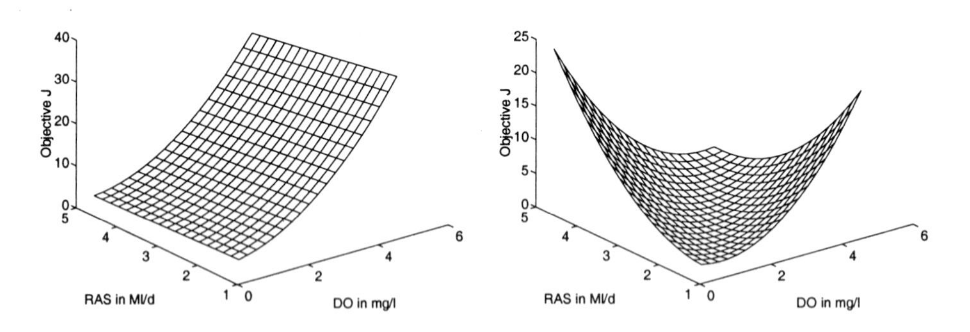

Figure 20.3: Insensitivity to RAS (left plot) and RAS - DO correlation (right plot).

The biggest problem with complex systems is the existence of multiple minima (Figure 20.4). In such instances it is preferable to redefine your objective function, but you can also use the response surface plots to guide the selection of starting points for the optimisation.

We will now examine the definitions of optimisation problems for some real wastewater examples.

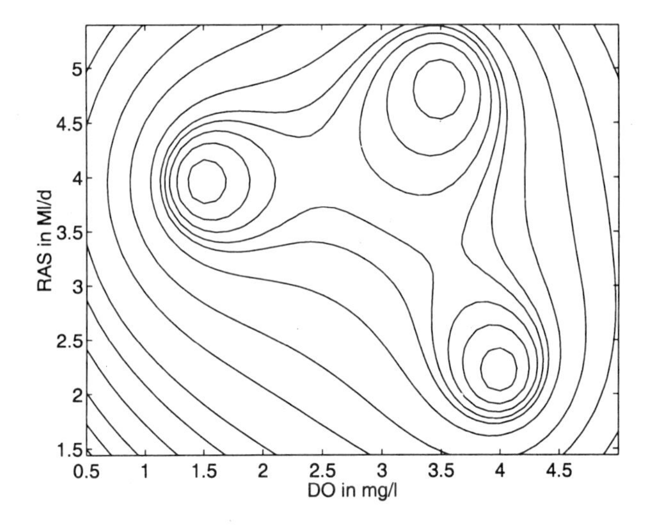

Figure 20.4: Multiple minima.

Example 20.8 (Two-stage anaerobic digestion) *This process shown in Figure 20.5 involves the breakdown of organic carbon to volatile fatty acids (VFA) in the first stage and conversion of VFA to methane in the second stage. The acidogenic microorganisms in the first stage grow much faster than the methanogenic microorganisms in the second stage. A solid packing is often used in the second stage to help prevent washout of the organisms. Furthermore the methanogens are inhibited by low pH, so caustic soda is used in the first stage to keep the pH at acceptable values and a recycle is often employed to buffer the influent and reduce caustic consumption.*

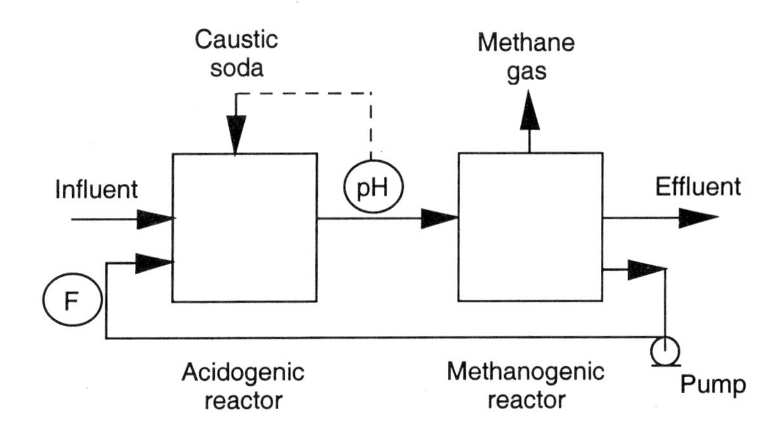

Figure 20.5: Two stage anaerobic digestion process.

To illustrate the technique for defining an optimisation problem we will make a simplified qualitative analysis of this process.

- System Definition. *The adjustable inputs are the setpoint on the pH controller and the recycle flowrate. The disturbances are the flowrate and COD of the influent wastewater. Outputs of major interest are the COD of the effluent, the flowrate of methane gas, the flowrate of caustic soda and the power consumption of the pump.*
- Degrees of Freedom. *There are two degrees of freedom, the setpoint of the pH controller and the recycle flowrate.*
- Constraints. *There is an external constraint on the allowable COD of the effluent stream. There is a minimum value of the pH before the second stage fails by washout of the organisms and a maximum value of the pH where all microbial activity stops. These are process constraints. There are maximum flowrate equipment constraints on the recycle and caustic addition. These define an operating space something like the left plot in Figure 20.6.*
- Objective Function. *This is defined in terms of the caustic soda consumption, municipal charges on the effluent COD load, and power consumption for the recycle pump. The response surface might be something like the right plot in Figure 20.6 which shows the minimum cost at maximum recycle with a flat optimum about a pH of around 6.5. This would probably imply operating in the constraint corner shown by the star in the left plot of Figure 20.6.*

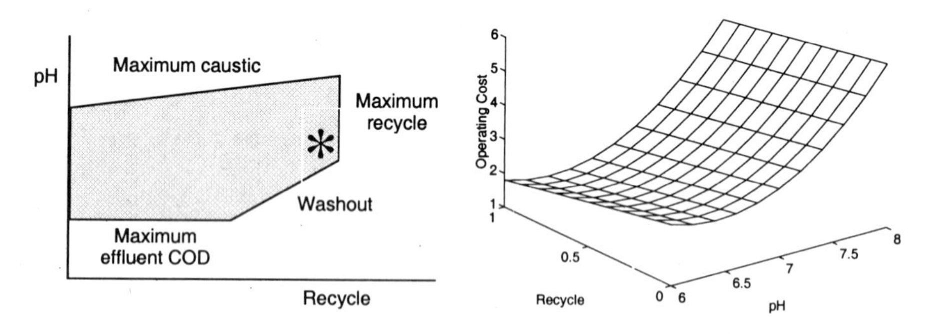

Figure 20.6: Operating space (left) and response surface (right) for the anaerobic digestion process.

Example 20.9 (Activated sludge control) *One of the important aspects of an activated sludge process which affects most of the secondary treatment units is sludge control. While we can consider the different bioreactor zones and the* *clarifier to a certain extent separately from the point of view of nutrient reactions, the recycle and wastage of sludge (Figure 20.7) affects all of the units. Again as an example of posing an optimisation problem we will take a somewhat simplified look at sludge control. It might be a starting point you could use in reviewing your sludge control strategies.*

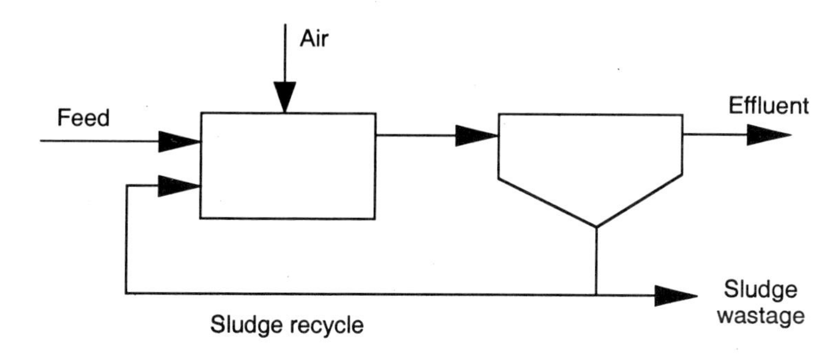

Figure 20.7: Activated sludge process.

- System Definition. *The primary adjustable inputs are the sludge recycle flowrate and the sludge wastage flowrate. In a more detailed study you might also include the DO setpoints in the aeration tanks or even the setpoints or targets for phosphate, nitrate and ammonia in the relevant zones (see the goals and strategy discussions in Chapter 14). The disturbances are the bioreactor influent flowrate and nutrient concentrations on a short time scale and ambient temperature on a longer time scale. Operational changes to DO setpoints could also be considered disturbances.*

- Degrees of Freedom. *There are two degrees of freedom, the setpoints of the sludge recycle flowrate and the sludge wastage flowrate.*

- Constraints. *Here we could write a thesis or two. The external constraints are simple, they are the effluent quality specifications and possibly the capacity of the sludge treatment facilities. Equipment constraints are the pumping capacity for the recycle and wastage streams and the volume of the clarifier. Process constraints are many and complex. There will be an optimal range of sludge age, too small causing washout of autotrophs or too low a value of the MLSS, too large causing sludge settling problems (filament growth).*

  The sludge cannot remain in the clarifier or return sludge tank for too long or there will be a loss of sludge viability. On the other hand we may have a minimum inventory constraint to facilitate faster recovery from toxic shocks. Too large a recycle can cause a hydraulic overloading of the clarifier (high effluent suspended solids) and/or short residence times in the bioreactors (less nutrient removal) especially when combined with high influent flow disturbances. These define an operating space something like the left plot in Figure 20.8.

- Objective Function. *Again we must define the operating cost. The power consumption for the sludge recycle is important. High nutrient concentrations in the effluent could attract regulatory penalties and/or higher costs for tertiary chemical treatment. High effluent suspended solids could incur higher costs for tertiary filtration. The cost of sludge treatment is involved.*

  The response surface might be something like the right plot in Figure 20.8, which shows that minimum cost is minimum recycle rate and say one third of the available wastage rate. This would probably imply operating along the sludge viability constraint as shown by the star in the left plot in Figure 20.8. Different formulations of the objective function will of course imply different optimal operating points.

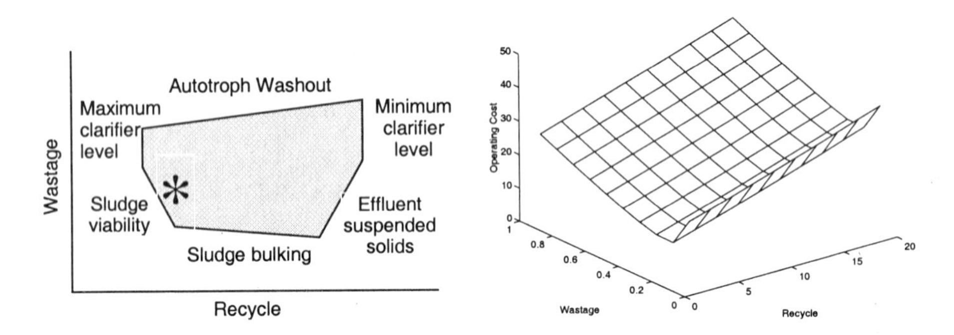

Figure 20.8: Operating space (left) and response surface (right) for the activated sludge process.

Having formulated our optimisation problem the next issue is finding the optimum which may move from time to time due to disturbances, changing costs, constraints and process characteristics. This was not such a difficult problem in our two examples where there are just two degrees of freedom. Where there are more degrees of freedom it becomes less easy to visualise and we must resort to computational techniques. We will now discuss some of these.

20.2 Optimisation Techniques

Optimisation is simply a searching procedure to find the minimum value of some function within the space defined by the degrees of freedom. We can use the analogy of wandering about a landscape of hollows trying to find the lowest point. For example consider the landscape represented by a contour plot in Figure 20.9 where the straight lines are fences.

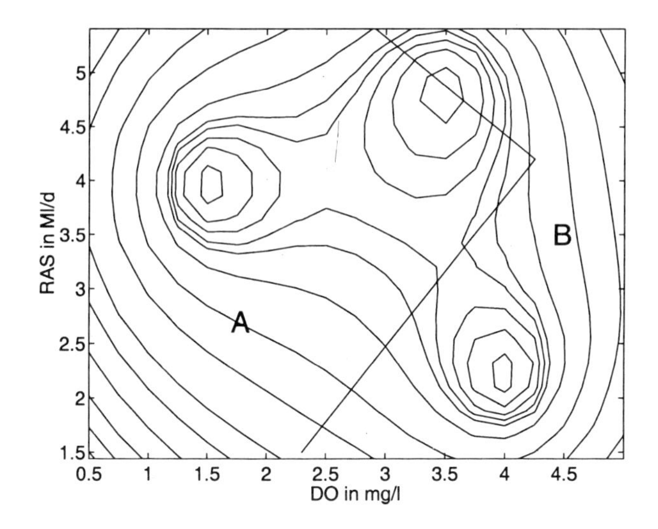

Figure 20.9: Landscape of hollows.

We will discuss the principles involved in the commonly used techniques for optimising small nonlinear problems of the type most likely to be encountered in wastewater treatment plant operation. Very large scale optimisations using such techniques as linear programming or sequential quadratic programming for nonlinear problems will not be considered here.

20.2.1 Unconstrained Optimisation

The simplest optimisation or search techniques are for unconstrained problems. That is, we are free to wander over the landscape and cross the fences at will. There are a great many search techniques possible. We will look at just three strategies, a line search, a gradient search and a simplex search. The line search is a simple but important strategy since it is the basis of many of the more complex techniques:

- The direction in which we search is restricted to the axis directions. In our analogy we only walk on a straight line in either a north-south or an east-west direction.
- We start walking in the favourable direction, in this case downhill. If we were looking for a hill instead of a hollow, it would be uphill.
- We stop walking when the next step is unfavourable, that is uphill.

- We walk each axis direction in turn. This means we would go north-south, then east-west, then north-south, and so on.
- We quit when we cannot find a favourable direction to walk. How close we are to the bottom of the hollow will depend on the size of our steps. The baby will be closer than its father.

The left plot in Figure 20.10 shows three line searches over our landscape of hollows performed by the MATLAB Toolbox program `linesrch.m`. For nice round holes you can see that we get to the bottom after changing direction two or three times with an average of 107 function evaluations. But if we had the angled hollow in Figure 20.10 we would be busy. It made 20 changes of direction, 577 function evaluations and still didn't quite get there.

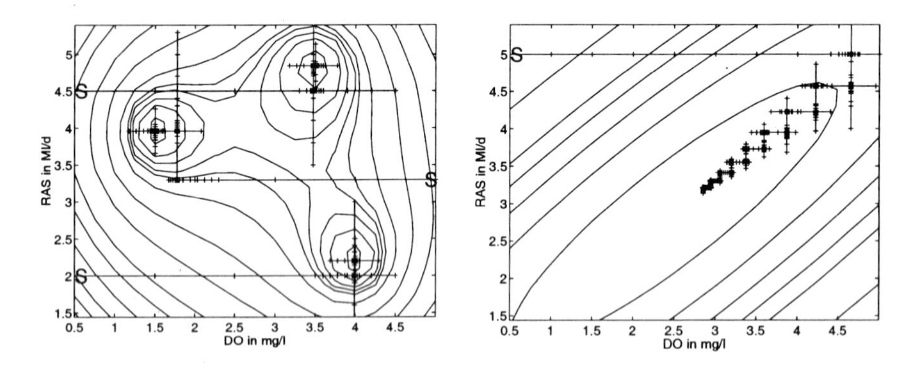

Figure 20.10: Easy (left) and difficult (right) line searches. **S** indicates the starting point.

Two very important lessons are illustrated in Figure 20.10:

- If there is more than one hollow the one we find depends on where we start.
- The hollow we find may not be the deepest hollow. We could get stuck in a big pothole.

The gradient search is a variation of the line search. Instead of walking only in the axis directions we can walk in any direction but still in a straight line. When we start we head off in the most favourable direction at that point. Starting is a little more complex (determining the most favourable direction) but we search fewer times for more complex surfaces (see the straight line searches in Figure 20.11 performed by the MATLAB Toolbox program `gradsrch.m`).

In the left plot in Figure 20.11 the gradient algorithm made 3 to 5 linear searches and averaged 92 function evaluations finding different minima from the same starting points used in Figure 20.10. In the right plot in Figure 20.11 the superiority of the gradient search is clear compared to the right plot in Figure 20.10.

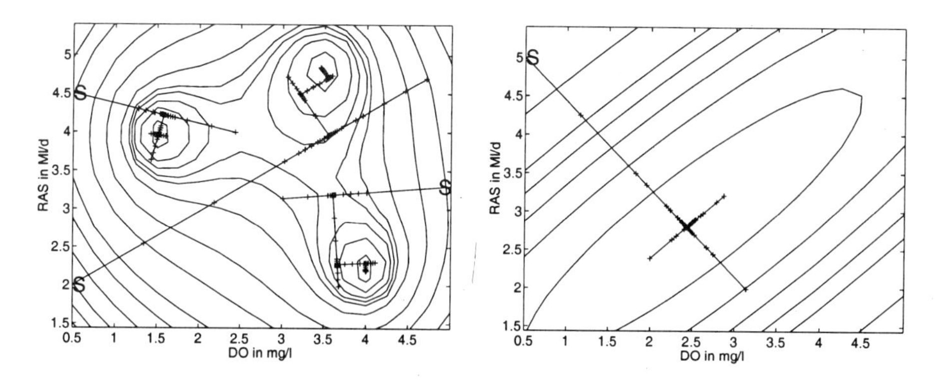

Figure 20.11: Easy gradient searches (left) and a not so difficult gradient search (right). **S** indicates the starting point.

It converged to the minimum with two linear searches and 81 function evaluations from the same starting point.

Examining the plots in Figure 20.11 you can imagine that the natural thing that you or I would do is to follow a curved path from the start to the minimum. We would effectively evaluate the gradient after each step taking the next step in the most favourable direction. Each time we evaluate the gradient we must assess or calculate the objective function $N + 1$ times, where N is the number of axes or degrees of freedom. In the case of walking downhill on your Sunday stroll the very efficient human sensors and the huge computing power of the brain has no trouble, you don't even notice the effort. But if we have a mathematical model of a wastewater treatment plant and a desktop computer, we try to optimise our optimisation and perform the fewest function evaluations by taking more simple steps and evaluating fewer gradients. A strategy that tries to emulate the human strategy in an efficient manner is called the simplex search.

- We choose $N+1$ points in a regular shape at our starting point, an equilateral triangle for two dimensions.
- We evaluate the objective function at each point.
- We discard the least favourable point and replace it by taking its reflection in the surface of the remaining points.
- We evaluate the objective function at this new point and go back to the second step.

This strategy illustrated in Figure 20.12 is performed by the MATLAB Toolbox program `simplex.m`. It is a very simple strategy and fairly efficient. In the left plot of Figure 20.12 the algorithm averaged 15 size reductions and 45 function evaluations, better than line or gradient searches. In the right plot in Figure 20.12

the algorithm stopped at 50 size reductions having made 81 function evaluations and not quite reaching the minimum. There are potential problems when the algorithm runs into valleys where it can get stuck simply flipping back and forth or shrinks excessively resulting in premature halts. There are various strategies for dealing with these situations and for judging when it has found the end point.

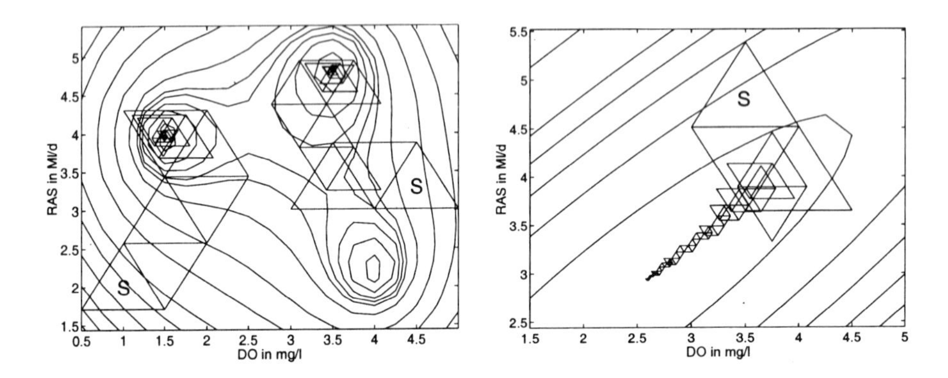

Figure 20.12: Simple simplex searches. **S** indicates the starting point.

You may have noticed that all of these search techniques will find different end points when there are multiple minima and the one they find will depend on where they start. Indeed this is so for all optimisation techniques which is why we advise examining your response surface before you start. If you have no idea what the surface looks like then only an infinite amount of computing will give you a certain answer.

The other obvious conclusion is that different problems are most efficiently solved by different algorithms. The *art* of optimisation is to be able to match problem and technique with a minimum of trial and error. Again it is important in making such a decision to have a good knowledge of the shape of the response surface.

All these search methods involve some form of step size which determines how fast they proceed with a search, how accurately they track the response surface and how close they approach to the true answer. Unfortunately there is no free lunch here either. The faster they proceed using a larger step size, the less accurate they track, and vice versa. Most search algorithms include some scheme for starting with a large step size, perhaps even increasing it on long straight searches, and then reducing it as searches become smaller closer to the optimum. Unfortunately it is often difficult to distinguish between sharp curves in the response surface and the final optimum. Most techniques will stop at the bottom of steep narrow valleys even if the optimum is some way off. Parameters such as step size, acceleration factor and deceleration factor can be used to *tune* the algorithm for best performance for a particular problem.

20.2.2 Constrained Optimisation

What happens when we have constraints, when we cannot cross those fences shown in Figure 20.9? In real life there are always constraints. Unconstrained optimisation techniques have no way of handling either type of constraint, hard constraints or soft constraints. You must not cross a hard constraint, for example if you run into a sheer cliff face, but you can cross a soft constraint a little way, for example if you run into a very steep hillside.

There are basically two approaches to constrained problems.

- Firstly by modifying the objective function. Penalties are added to the objective when you cross a constraint. This is a good approach for soft constraints but less efficient in the case of hard constraints.
- Secondly by modifying the search strategies within the optimisation algorithm. For example by changing direction and following the fence. This is the best approach for hard constraints.

The addition of penalty functions to the objective function is a good simple solution for problems with soft constraints. You can then use any of the standard unconstrained optimisation strategies. Penalty functions should be chosen so that they discourage straying too far across the constraint but so that they do not introduce discontinuities into the response surface (see left plot in Figure 20.13).

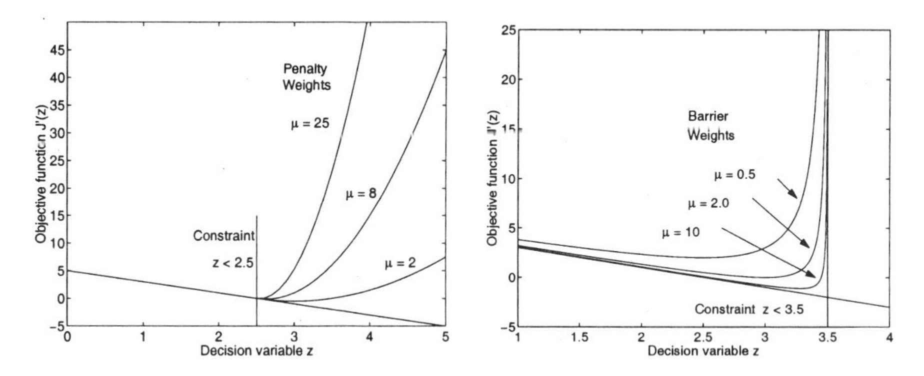

Figure 20.13: Penalty functions (left) and barrier functions (right).

Most search strategies do not handle discontinuities very well. Consider the optimisation problem:

$$\textit{minimise} \quad J(z) \quad \textit{subject to} \quad g_i(z) \leq 0, \; i = 1, ..., p \tag{20.2}$$

The following equation shows a typical penalty function.

$$J'(z) = J(z) + \mu \sum_{i=1}^{p} (\max(0, g_i(z)))^2 \tag{20.3}$$

where μ is a penalty weighting factor. As μ becomes large the penalty becomes more severe and a constrained value will be closer to the constraint. The weighting factor can be progressively increased as the optimiser step size is reduced. Barrier functions (see right plot in Figure 20.13) are a form of the penalty function which can be used for hard constraints and ensures that the constraint is never crossed. The following equation shows a typical barrier function.

$$J'(z) = J(z) - \frac{1}{\mu} \sum_{i=1}^{p} \frac{1}{g_i(z)} \tag{20.4}$$

where μ is a barrier weighting factor in this case. As μ becomes large the barrier becomes less severe and a constrained value will be closer to the constraint. Again the weighting factor can be progressively increased as the optimiser step size is reduced. One of the most common modifications of optimisation techniques for constraint handling is the implementation of gradient projection. When a constraint is active the true gradient direction is determined and it is then projected onto the constraint to obtain a search direction (see Figure 20.14). In the case of our simple analogy, gradient projection is running into the fence and following it downhill.

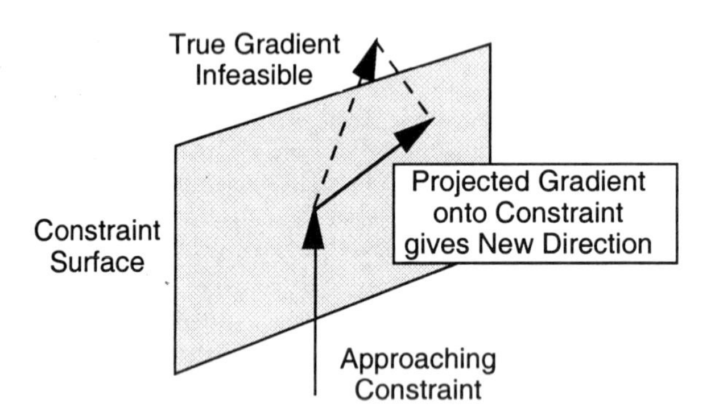

Figure 20.14: Gradient projection.

Another common modified technique is the complex technique, a variation on the popular simplex technique mentioned above. In this case more than $N + 1$ points are used and when a reflected point crosses a constraint it is reflected onto the constraint. The increased number of points makes the technique more robust and allows it to move along and around constraint corners. Animated versions look very much like some sea slug crawling over the constraint surfaces changing shape as it goes.

20.2.3 Solution Strategies

Small optimisations involving up to five degrees of freedom are best solved as a single problem. Larger optimisations can be solved either as a single integrated problem or they can be subdivided into smaller problems. These small problems can be solved individually in a coordinated fashion using hierarchical multi-level optimisation strategies as illustrated by Figure 20.15. Multi-level optimisation problems can be solved at least as efficiently as integrated problems and are often much more easily understood.

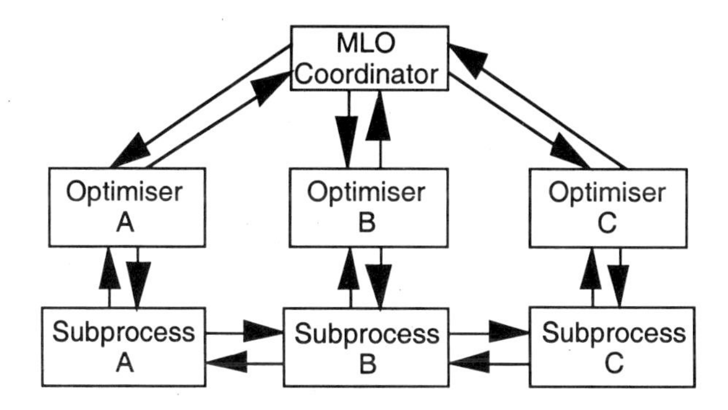

Figure 20.15: Multi-level optimisation.

20.3 Plant Wide Control

Having determined the best operating conditions by optimisation we are then faced with two further problems, getting there from where we are now, and once we get there staying there. This is where control enters the plant wide perspective.

20.3.1 Optimal Control

Optimal control is concerned with driving a process between operating points in some optimal fashion. Traditional objective functions for optimal control problems are minimum time and minimum energy (in the sense of minimum effort by the manipulated variables). Other objective functions can also be formulated and constraints can also be applied. A dynamic mathematical model is required to solve the problem as well as efficient large scale optimisers such as `MISER` (Jennings *et al.* (1990)).

Processes are more typically moved between operating points *heuristically* and on the basis of past experience. It has been shown in numerous examples that improvements of two to ten in transfer time can be achieved using optimal control. Often an optimal control study is used to improve the heuristic operating strategies without actually implementing the results automatically on-line.

20.3.2 Constraint Control

Constraint control is a control strategy to keep processes at their optimal operating point. The concept behind it is based upon the observation that optimal operating points are generally fully constrained, that is they lie in one corner of the feasible operating region (Figure 20.16).

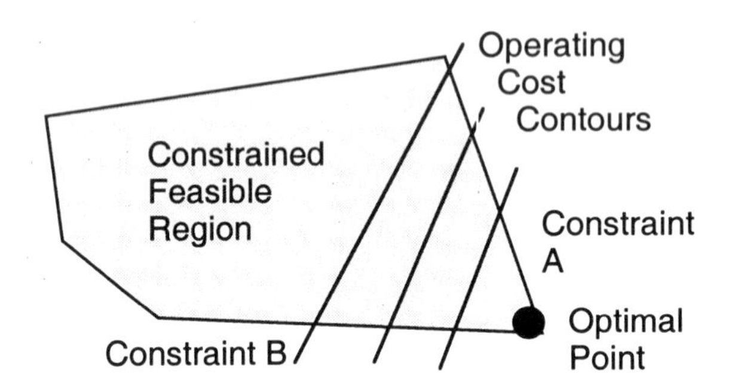

Figure 20.16: Constraint control.

The design procedure involves the following steps:

- Determine the degrees of freedom, the feasible operating region and the optimal operating point.
- For each active constraint (constraints A and B in Figure 20.16) select a simple, calculated or inferred measurement that indicates how close the process is to the constraint.
- Pair each *approach* measurement with an appropriate degree of freedom.
- Design controllers for each pair, generally slow acting integral or PI controllers. In some instances a second fast acting controller (or a gain scheduled controller) is required where violating the constraint at all can be dangerous, such as with compressor surge control.

In complex situations where the set of active constraints may change a more complex control system is necessary. This can generally be handled with extra measurements and high and/or low selectors to implement control system rearrangement.

Example 20.10 (Control of air blowers) *A simple wastewater example of constraint control is the control of multiple or variable stage air blowers. This is a one dimensional constraint control problem as illustrated in Figure 20.17. The single degree of freedom is the number of blowers operating. The operating cost is the power to the blowers and power minimisation involves the least pressure* *drop across the control valves regulating the DO in the aeration tanks. The active constraint is a minimum pressure drop (maximum valve opening) to enable the DO controllers to regulate properly and to absorb disturbances in the oxygen uptake rate.*

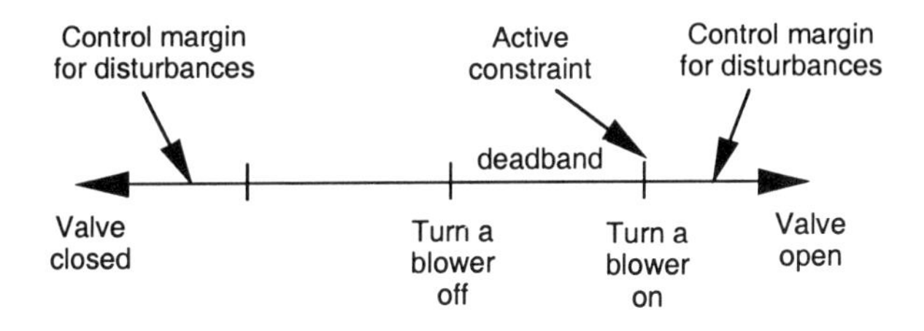

Figure 20.17: Blower constraint control.

The constraint measurement is the minimum of the average values of the valve position signals (there may be several valves regulating from the supply header). The signals need to be averaged over a sufficient time to achieve economy without excessive blower stops and starts due to the effects of disturbances. When the active constraint is violated another blower is turned on. A blower is turned off when the measurement is some deadband less than the constraint again to avoid excessive blower stops and starts.

Example 20.11 (Two-stage anaerobic digestion revisited) *This process was described in Example 20.3 together with the optimisation problem and a possible solution. The optimal operating point is shown in Figure 20.18 in the constraint corner of maximum recycle and second stage washout.*

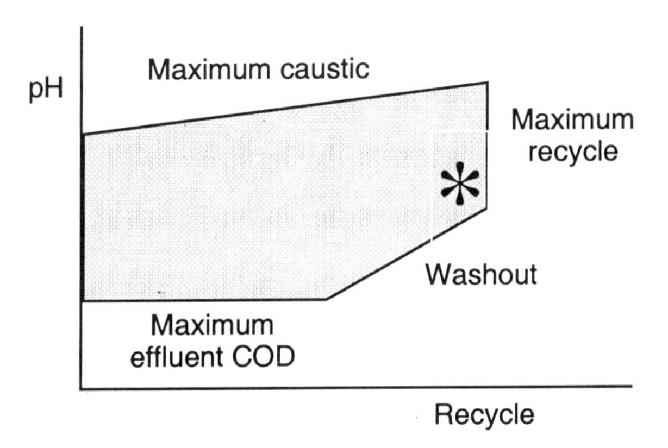

Figure 20.18: Constraint control for the anaerobic digestion process.

The constraint measurement for maximum recycle is a valve position signal which should be close to full open allowing a margin for use of the recycle to reject disturbances as in the previous example. Since the recycle is probably infrequently set by the operator, simple alarm messages can be used as a constraint controller to advise him that he is setting too low or too high a recycle. Measurement of the washout constraint is more difficult. Certainly a lower than normal methane flowrate at the same time as a higher than normal effluent COD is an indication. Another measurement that many believe is a precursor to washout is significant quantities of hydrogen in the methane gas. Either of these could be used with suitable averaging to adjust the setpoint of the pH controller.

20.3.3 Plant Wide Nutrient Removal Control during Highly Variable Loads

In a nutrient removal plant with highly variable load there is a strong need for a continuous adjustment, not only of the individual control loops such as dissolved oxygen control, but the complete plant operation in order to ensure a stable biological nitrogen and phosphorus removal. This example is published by Isaacs *et al.* (1999) and demonstrates how ammonia, nitrate and phosphate sensors are used to automatically make adjustments on a supervisory level. Due to high loads the design capacity was regularly exceeded before the control system was implemented.

The plant is an alternating plant (Section 19.5.2) having industrial wastewater input from a fishmeal factory. The industrial load is anywhere between 0 and 10 times the munipical load. Phosphate removal is performed by simultaneous precipitation using powdered iron sulphate in combination with biological P removal.

The supervisory control system is called STAR [1] (Superior Tuning and Reporting). It automatically calculates the setpoints for the individual control loops of the system. If the supervisory control system is unable to calculate a new setpoint then the setpoints are kept constant by the low level control system.

There are five supervisory control modules implemented in the system:

- *Phase control.* The phase control determines the switching between the Biodenitro phases as a function of ammonia and nitrate levels in the biological reactors. The phase control is coupled to the dissolved oxygen setpoint control.

- *Dissolved oxygen setpoint.* This control and the phase control serve to match the relative aerated nitrifying and the non-aerated denitrifying times to the current load of the plant. The dissolved oxygen setpoint during the aerated phase is adjusted according to the ammonia concentration level, as described in Section 17.7.1. The setpoint control is also used to save some carbon for use in the anoxic reaction and for the biological P removal.

[1] product of Krüger A. S., Søborg, Denmark

- *Pure oxygen addition.* During extreme loads the aeration is not sufficient and pure oxygen can be added, as discussed in Section 16.2.3. The control decision is based on measurements of ammonia, dissolved oxygen and estimates of the oxygen uptake rate.

- *Chemical dosage.* The iron dosing is a local control loop based on the phosphate and flowrate measurements, as described in Section 17.7.3.

- *Return sludge flowrate.* As discussed in Section 18.6.2 the return sludge flowrate is strongly coupled to its concentration. This control aims at stabilising the suspended solids concentration in the return sludge stream and is based on measurements of the suspended solids and the influent flowrate.

The system has been in full operation for some time and has been extremely successful at minimising the effect of the large load variations. The total effluent N has been decreased significantly. At the same time compressor energy has been saved by using up carbon in the denitrification and the biological P removal. Also the chemical dosage could be decreased up to 90% by keeping up the biological P removal.

20.3.4 Stormwater Control of an Activated Sludge System

During high hydraulic loads the clarifier is often the limiting factor (Section 16.1.1). During storm events it is then possible to increase the settling capacity simply by letting part of the aerator serve as a settler for a limited time. Thus an intermittent mixing is applied. This is the idea behind the Aeration Tank Settling (ATS) [2] process.

The treatment plant is operated in an alternating mode with two tanks in series. Raw wastewater is fed to the first tank and aeration is performed in order to obtain nitrification. In the second tank settling is carried out.

While the second tank serves as a settler it also can perform denitrification. Usually a post-denitrification process would need the addition of external carbon. However there is still carbon for the denitrification because the oxygen level in the first tank has been limited. The slowly biodegradable carbon becomes available just in time for the denitrification.

In principle the settling capacity of the system is adjusted according to the influent load. The approach is to control towards a constant sludge volume load to the clarifier. The required settling tank volume is expressed by:

$$Clarifiervolume = k \cdot Q \cdot SS_{to\ clarifier} \cdot SVI \tag{20.5}$$

where k is a constant, Q the flowrate into the settler, $SS_{to\ clarifier}$ equal to the $MLSS$ at the outlet of the aerator, and SVI the sludge volume index.

[2]developed by Krüger A. S., Søborg, Denmark

The settling volume is proportional to $MLSS_{outlet} \cdot Q$. By using the second tank as a settler the suspended solids concentration can be reduced considerably. This means that it reduces the hydraulic loading to the clarifier thus increasing the hydraulic capacity of the plant.

It is apparent that the control depends on several criteria taking both hydraulic and biological factors into consideration. The control in dry weather aims at a balance between nitrification, denitrification and P removal using minimum energy for aeration. At high hydraulic loads the control system enables maximum settling in the aeration tanks to increase the hydraulic capacity of the plant. This is an example of multi-objective optimisation.

Full scale tests at the Aalborg West Treatment plant in Denmark have proved the power of the control strategy. Still it is not confirmed how the operation will influence the sludge settling characteristics.

20.3.5 Controlling the Inflow at High Hydraulic Loads

Adding polymers

During a storm event there is a large impact of suspended solids to the plant. As a result the sludge blanket in the settler may rise too high and the surface load to the clarifier exceeds its capacity. Experiences in full scale operation in Sweden have shown that polymer addition can significantly reduce the suspended solids of the primary effluent. This will increase the nitrogen removal capacity of the plant on a long term basis. Both the average concentration and the variations of suspended solids could be decreased by adding polymer.

Monitoring the sewer system

On-line monitoring of wastewater transport systems has been less successful than wastewater treatment systems. There are several reasons for this. The investment cost is high since the system is geographically widespread. The instrumentation is installed in a rough environment and the sensors require adequate maintenance and calibration. Some of the locations may be difficult to reach, such as sewers under streets with heavy traffic which makes the maintenance cumbersome.

Signals from level transmitters need translating to flowrate which is not always simple.

Not all sewer systems offer capabilities for real-time control. If the only force influencing the flow of water is gravity then the controllability is not there.

By looking at upstream storages and pump stations it is possible to obtain interesting and meaningful information and control authority. With the combination of low cost instrumentation and information processing it has been possible to increase the potential for sewer flow control.

By making a dynamic model of a pump station the flowrate can be computed given the information on the water level in the wetwell, the electricity consumption of the pumps, and of the rain intensity in the nearby catchment.

A simple mass balance of the pump station can be formulated as:

$$\frac{dV}{dt} = Q_{wastewater} + Q_{rain} - Q_{pumps} - Q_{overflow} \tag{20.6}$$

where V is the water volume of the pump station, $Q_{wastewater}$ the predicted wastewater flowrate to the station, Q_{rain} the rain predicted from a run-off model, Q_{pumps} the pump flow rate calculated from the electrical power, and $Q_{overflow}$ the overflow to the recipient. The water volume is reconstructed from water level measurements.

By using Equation 20.6 it is possible to make reasonable predictions of the flowrates.

Predicting the plant influent flowrate

The classical objective of wastewater transport systems are to "get rid of the water" as efficiently as possible while the wastewater treatment requires that the water can be retained in the system. Therefore by-passing should always be considered the very last alternative. There is no apparent answer to these contradicting demands.

The classical approach has been to design large buffer volumes in the sewer system. The competing approach is to apply real-time control of the sewer flowrate.

One approach is described by Nyberg *et al.* (1996b). Physical data in combination with on-line measurements of flowrates at various locations were used to calibrate a model of the sewer system. For the treatment plant a dynamic model of the solids propagation in the plant was derived and calibrated to on-line suspended solids measurements. These sensors were located in the raw wastewater, in the primary and secondary effluent, in the mixed liquor and in the return sludge. Also the sludge blanket of the secondary settler was measured.

The investigations showed that the sludge blanket increased slowly below a certain influent flowrate, but rose quite rapidly once the influent flowrate exceeded a given value. The purpose of the real-time control was to avoid sludge loss due to overloaded clarifiers. This was achieved by controlling the influent flowrate.

In the Swedish Malmö-Klagshamn plant three strategies were applied to avoid overloading due to storm events, in-line storage in the sewer system, step-feed operation and an operating mode in which primary effluent by-passes the biological treatment step.

The effect of step-feed has been discussed in Section 18.6.3. It results in lower solids flux load to the settler but the price is a somewhat lower nitrogen removal efficiency. The combination of in-line storage and step feed proved to very efficient.

In the Malmö-Klagshamn case the implementation strategies were very simple once the numerical models had been derived and the strategies tested in simulation. The software package MOUSE was used for the sewer simulations (see Chapter 9).

20.4 Summary

A full plant operation can be described in several layers of control. In Chapters 17 to 19 we have described various local control loops, more or less advanced. To operate the full plants one needs to coordinate a lot of local controllers. Some of this has been the subject of this chapter. Furthermore, we have demonstrated that optimisation can be a powerful tool both for the design and for the operation of a plant. This becomes particularly useful when we have to take constraints into consideration.

Much more can be done to integrate the operation of wastewater treatment plants and sewer operations. There is increasing research and development in this area and we have just indicated some avenues of development.

This chapter concludes the Control part of the book. In the next chapter we will discuss some means of evaluation of the complete system operation.

20.5 Further Reading

- For more information on optimisation techniques see Adby & Dempster (1974) and Fletcher (1987).
- For more information on search strategies see Nelder & Mead (1965).
- For information on linear programming see Walsh (1985) and Gill *et al.* (1981).
- For information on sequential quadratic programming for nonlinear problems see Gill *et al.* (1981).
- For more information on multi-objective optimisation see Cohon (1978).
- The paper by Ayesa *et al.* (1998) demonstrates the use of optimisation in wastewater treatment.
- The book by Bryson & Ho (1969) is a good standard text on dynamic optimal control.
- A large scale optimiser, MISER, is described in Jennings *et al.* (1990) with application references by Wang & Cameron (1994) and Rossi & Figueroa (1997).
- Examples of the use of optimal control to improve wastewater processes can be found in Kabouris & Georgakakos (1990) and Kabouris & Georgakakos (1992).
- For more information on constraint control see Maarleveld & Rijnsdorp (1970) and Newell & Lee (1989).

- Constraint control applied to multiple or variable stage air blowers has been described by Nam *et al.* (1996).
- A systems approach to sustainable operation is described in Aspegren *et al.* (1997). How to formulate suitable objective functions for large operations and interconnected systems is discussed in Vanrolleghem *et al.* (1996a).
- A demonstration of an advanced system for sewer-plant interaction has been reported by Lumley *et al.* (1993). The optimal use of equalisation basins has been considered by Dold *et al.* (1984).
- The Automatic Tank Settling process has been described by Bundgaard *et al.* (1996).
- The stormwater management problem is described in Nyberg *et al.* (1993) and Nyberg *et al.* (1996b).
- The addition of polymers during storm events is described by Nyberg *et al.* (1996a).
- A grey box approach to pump station modelling and sewer flow prediction has been made by Carstensen *et al.* (1996).
- A numerical model for a large sewer system is described in Ji *et al.* (1996).

20.6 Exercises

1. A wastewater treatment process for BNR consists of an anaerobic tank, an anoxic tank, two aerobic tanks and a clarifier. Formulate an optimisation problem for selecting the DO setpoints for the two aeration tanks.
2. Consider a sequenced batch reactor (SBR) that has been retrofitted to an existing treatment plant for nitrogen removal. You can consult Chapters 4 and 14 and copy and use the MATLAB program `Example_4_1.m` if you wish.
   (a) Formulate an optimisation problem to select the reaction times for the aerobic (nitrification) and anoxic (denitrification) phases.
   (b) Write a MATLAB program to perform the optimisation using penalty functions and the simplex optimisation technique.
3. Consider a municipal BNR wastewater treatment plant with primary, secondary and tertiary treatment. The tertiary treatment includes anaerobic digestors for sludge treatment, chemical precipitation of residual phosphorus, and flocculation and sand filtration for residual suspended solids.
   (a) Formulate the problem for optimisation of the complete plant.

(b) Propose a multi-level optimisation strategy.

4. Use the MATLAB code `linesrch.m` shown in Section 20.2.1 to get some feeling for optimisation algorithms by maximising the following function in the range ($0 < x < 1$) from different starting points with different step sizes.
$J = e^{-500(x-0.2)^2} + 2e^{-100(x-0.4)^2} + 1.6e^{-200(x-0.7)^2}$

5. Use the MATLAB code `gradsrch.m` discussed in Section 20.2.1 to get some feeling for optimisation algorithms by minimising the following functions from different starting points with different step sizes.

   (a) $J = e^{x_1}\left(4x_1^2 + 2x_2^2 + 4x_1x_2 + 2x_2 + 1\right)$
   in the range ($-2 < x_1 < 2, -2 < x_2 < 2$)
   (see MATLAB code `exfunc1.m`).

   (b) $J = \left(x_1^2 - x_2\right)^2 + 100\left(1 - x_1^2\right)$.
   in the range ($-1.5 < x_1 < 1.5, -0.5 < x_2 < 1.5$)
   (see MATLAB code `exfunc2.m`).

6. Use the MATLAB code `simplex.m` discussed in Section 20.2.1 to get some feeling for optimisation algorithms by minimising the following functions from different starting points with different step sizes.

   (a) $J = e^{x_1}\left(4x_1^2 + 2x_2^2 + 4x_1x_2 + 2x_2 + 1\right)$.
   in the range ($-2 < x_1 < 2, -2 < x_2 < 2$)
   (see MATLAB code `exfunc1.m`).

   (b) $J = \left(x_1^2 - x_2\right)^2 + 100\left(1 - x_1^2\right)$.
   in the range ($-1.5 < x_1 < 1.5, -0.5 < x_2 < 1.5$)
   (see MATLAB code `exfunc2.m`).

   Compare the results with the previous Exercise.

Chapter 21

Benefit Studies

What is a benefit study? It is an examination of the operation of a process to identify and evaluate opportunities for improvements in process control and in operating strategies. It is what you should do to put a case to your management for funding improved performance of your plant.

The approach that we advocate in this chapter was developed during a joint academic and industrial study entitled "Advanced Process Control" conducted by the Warren Centre for Advanced Engineering at the University of Sydney. The study consisted of a number of project teams working on nominated industrial case studies over a period of a year. The case studies involved a variety of applications in different industries one of which was a wastewater treatment plant in western Sydney.

The term *advanced* is very much a relative term. We define it as something more sophisticated than you have at present. In some cases this means writing down operating instructions, in others it means feedback control loops, and in other cases it means multivariable optimising controllers. Likewise we view *process control* in a broad perspective. It may involve closed loop automatic control of the process or it may mean improved operating strategies for plant operators to implement by hand. Our goal is to give you some procedures and ideas that you can mould to your situation and apply to systematically identify and evaluate opportunities to improve your plant's performance.

21.1 Systematic Procedure

The systematic procedure identified from the Warren Centre study is summarised below:

- **Identification of Goals and Problems**. This is done by conducting a series of semi-structured interviews with representatives at each level of each group involved in the daily operation of the process. The emphasis in structuring the interview is to get the interviewees perceptions of their role, the goals, the problems and their ideas for improving the operation of the process.

- **Assembling a Ranked Opportunity List**. The goal here is to initially assemble an unordered list of opportunities for improving the process performance. These opportunities can come from the responses to the interviews, from the experience of the benefit study team and from brainstorming sessions with selected interviewees. Once a list is assembled it can then be ordered with the most promising opportunities at the top based on a qualitative judgement assessing the likely costs, the likely benefits, and the likelihood of realising the benefits.

- **Defining a Base Case**. Here the goal is to define the current performance and the economics of the process. This will form the basis for quantitative predictions of performance increases and the evaluation of the benefits and costs of selected opportunities.

- **Evaluation of Opportunities**. Here the goal is to quantify the ranked list of opportunities. Before starting it is wise to review the initial ranking in view of the information collected on the process in defining a base case. The process of quantifying may well change the order of the ranking. It involves predicting improved performance, defining the benefits and assessing the costs. It will usually involve at least a very preliminary design. Obviously we can only do this for a small number of opportunities which is why we rank them qualitatively in the first instance.

21.2 Interviews

Interviews would typically involve at least the following groups of people:

- plant operators,
- shift supervisors,
- supervising engineers,
- operations management,
- technical support engineers,
- technical support management,
- operations planners or schedulers,
- maintenance personnel,
- maintenance planning, and
- process accounting and economics.

Again the emphasis in structuring the interview is to get the interviewees perceptions of their role, the goals, the problems and their ideas for improving the operation of the process. At the same time the interviewee should be allowed freedom to follow a train of thought and to take the initiative. Some people make a list of key questions that they want answered and simply use them to initiate discussion or bring discussion back on track (see Example 21.1). Others use a much more structured system even resorting to questionnaires. If questionnaires are used they should be followed up by face-to-face discussions since many issues come up in a general discussion that people are hesitant to commit to paper.

Example 21.1 (Interview structuring) *The following questions were used by the Warren Centre project team involved with an activated sludge treatment plant:*

- *"Is more advanced control required in the sewage industry?*
- *How can energy usage be minimised? How can maintenance costs be minimised?*
- *Can manning reductions be made through automation?*
- *What are the implications on control strategy decisions of proposed future capital works to increase capacity?*
- *What new technology could affect plant operation?*
- *Are new methods or processes being used successfully elsewhere (e.g. overseas or at S.E.P.P. [1]) that have not been considered by the Sydney Water Board?*
- *What data is being collected and is it being used effectively?*
- *Is the current alarm system performing effectively?*
- *Are chemicals being wasted through inefficient dosing?*
- *Is automatic raw sludge withdrawal available?*
- *Can the process upsets resulting from heavy metals in trade wastes be minimised?"*

Some of these questions are general in nature while others are obviously very specific to the plant concerned. Some even involve design issues rather than just monitoring and control issues. If there are weaknesses in this list it is in the generality and the lack of questions on operational goals and objectives.

Some key responses from operator and management interviews were reported as follows:

- *"Their (operators) main tasks centred around manual tasks, troubleshooting, and supervising other personnel. Decision making is*

[1] South Eastern Purification Plant in Melbourne, Australia

left to other support staff, with hardly any attempts at maximising efficiency or improving the process to save energy or other costs. Although the flow to the plant fluctuates throughout the day, no attempts were made to operate the aeration equipment to accommodate such fluctuations and to save energy in the process. Operating practice is on the basis of large safety margins producing effluent quality far better in most respects than licence requirements."

- *"Operators felt that product specifications were being satisfied, with some product quality give-away when nitrification was being implemented. There is no incentive to operate near to constraints, as loading is not critical at present."*
- *"Management when interviewed undertook to support advanced process control techniques if they can be proven to be cost effective and can defer capital expenditure on future expansions of the plant. It was emphasised that cost constraints and stringent licence limits placed on effluent quality are the two main areas currently confronting management."*

Are these comments familiar? Management were concerned about costs and quality. Operators cared (knew?) nothing of costs and "felt" quality was good, but obviously were not sure. The two groups have quite different operating goals, and there is a definite lack of communication.

You will be amazed at the range of often conflicting perceptions that the different groups have as to the purpose, the goals and the problems of operating a process. Example 21.1 illustrates just this issue. Unfortunately it is quite a common occurrence, maybe even in your plant.

21.3 Possible Opportunities

In this section we will simply list examples of opportunities identified in benefit studies to give you ideas for your studies. In Table 21.1 we present a general list of the types of opportunities. In Table 21.2 we list opportunities specifically identified in the literature for wastewater treatment plants.

Example 21.2 (Opportunity list) *The initial list of opportunities developed by the Warren Centre project team following on from the interviews (see Example 21.1) is shown below:*

- *Blower control via dissolved oxygen, BOD, and inflow rate (*).*
- *Control of waste activated sludge flow from the aeration tanks.*
- *Control of return activated sludge flow to the aeration tanks.*
- *Clarifier sludge level control.*

- *Chemical dosing control.*
- *Control of filter operation cycle (*).*
- *Thickening of WAS prior to sludge digestion.*
- *Extent and form of data collection.*
- *Inflow pumping control (*).*
- *Monitoring and control of heavy metals in sewage.*
- *Telemetry and alarm system for unmanned plant.*
- *Postponement of expansion by running the plant above design capacity (*)."*

Those marked with an asterisk (*) were evaluated quantitatively.

Type	Description
Increased throughput	Increasing the throughput of the process often means that process expansion can be delayed. This can result in large savings in interest payments by deferring capital expenditure.
Yield improvements	This means either more product for a given amount of raw material or less raw material for a given amount of product. Which is a saving depends on process and market limitations.
Quality improvements	This can mean an upgrade in price for the higher quality product and/or an increase in market share. It also facilitates operating closer to constraints.
Utility reductions	These are generally direct cost savings. Occasionally it can be a deferred capital saving if expansion of generating capacity is the alternative.
Manpower savings	Reduction of manpower is seldom achievable. Some major instrumentation projects can make savings in the number of operating personnel by centralising operations.

Table 21.1: General types of opportunities.

Type	Description
Increased throughput	Influent flowrate control Quality improvements Tertiary filter control Sludge thickening control
Yield improvements	Heat pumps for energy recovery Methane production Nutrient recycle opportunities
Quality improvements	Influent flowrate control Influent pH control MLSS control Phase length control (intermittent plants) Return sludge control Control of nutrient recycles
Utility reductions	Improved DO control Constraint control on air flow valves Improved biological P removal Nitrate recycle control Chemical dosing control Filter backwash control
Manpower savings	Centralisation of control rooms Remote operation of plant Centralised laboratory service

Table 21.2: Opportunities in wastewater treatment plants.

21.4 Base Case Definition

The base case definition is the foundation for the quantitative evaluation and should be done with some care. It will involve gathering at least the following:

- process records (design and operating),
- plant operating instructions and records,
- time series graphs of selected process variables (disturbances, product qualities, utility consumptions, and key operating conditions),
- equipment running and maintenance records,
- raw material and product orders and deliveries (promised and actual),
- planning schedules (proposed and actual),
- costs of raw materials, services and manpower,

- prices of products and by-products (in particular marginal costs), and
- costs of plant expansion (the whole and parts thereof).

Part of this information can be obtained from plant performance audits, a service commonly provided by consultants. However these audits seldom probe deeply enough and are usually a *spot check* rather than gathering historical data over a period of time.

The components of the operating cost need to be identified and quantified. You will probably need to itemise the major components, in some cases by indirect calculations or by estimation.

21.5 Benefit Evaluation

Having identified and ranked likely opportunities and gathered basic operating and cost data for the base case we must evaluate the likely benefits from our top ranking opportunities. We will discuss five aspects, types of benefits, measures of performance, techniques used to predict the improved performance, investment costs and recurrent costs.

21.5.1 Tangibles and Intangibles

Tables 21.1 and 21.2 list tangible benefits where the economic return from improvements can be quantified. It is important in evaluating tangible benefits to use marginal dollar values. For example the true cost of air includes capital interest and depreciation costs for the air compressors. A reduction in air usage cannot decrease these cost components, it can only proportionally reduce daily running costs such as electricity and maintenance.

Intangible benefits can be very large but are very difficult to quantify (otherwise they would be tangible of course). Often operational improvements result from the improved understanding of the process and of management practises gained by doing a benefit study, even if none of the opportunities are implemented. However few people would dare to predict this quantitatively and even fewer would believe the predictions. The types of intangible benefits possible are listed below:

- process understanding,
- management systems understanding,
- goal and objective definitions,
- recognition of regular problems, and
- definition of operating cost components.

It is important to list intangible benefits as they may make a big difference if the return from the tangible benefits is marginal.

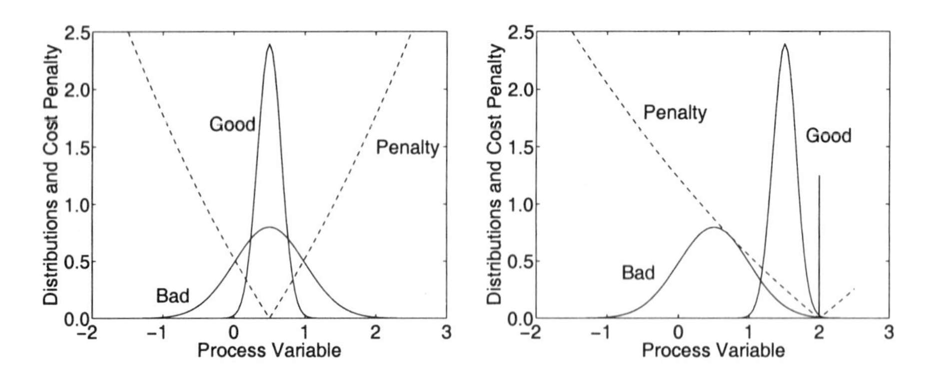

Figure 21.1: Control performance measured by Gaussian distributions.

21.5.2 Performance Measures

There are two basic types of control, regulatory (disturbance rejection) and stat driving. How do we measure the performance of these types of control?

The performance of regulatory control systems is most frequently quantifie by examining the frequency distribution plot of the controlled variable or of som related variable. This will generally be approximately Gaussian in form and ca therefore be characterised by a mean value and a standard deviation.

Improved regulation decreases the standard deviation, but how does that trans late to a financial return? Figure 21.1 illustrates the two cases, operation at a optimum (left plot) or operation at a constraint (right plot). For example ther are economic incentives to operate closer to effluent quality constraints.

Operation at an optimum is actually the less frequent situation. In this case by definition, the operating cost increases as you move away from the optimum Therefore the area under a cost times frequency curve decreases when the standar deviation is smaller (see left plot in Figure 21.2).

Operation at a constraint is the more common situation. Here we must specif that we will tolerate a certain percentage of the time outside the constraint, of th order of 0.5 to 5 per cent. A smaller standard deviation allows the mean value o the operating variable to be closer to the constraint (a smaller operating margin) The reason we are at the constraint is because it is cheaper to be there. Therefor the area under the cost times frequency curve will again be smaller (see right plo in Figure 21.2).

Figures 21.1 and 21.2 were produced by the MATLAB Toolbox routines `costopt.m` and `costcon.m`. The dotted line is a scaled cost penalty for operatin away from the optimum or the constraint respectively. The dashed lines represen old and new performance curves and the solid lines are the product of the penalt times the frequency. The areas under the solid curves represent the operating cos penalties for not operating exactly at the optimum or the constraint. The differ ence is the saving made by the performance improvement (reducing the standar

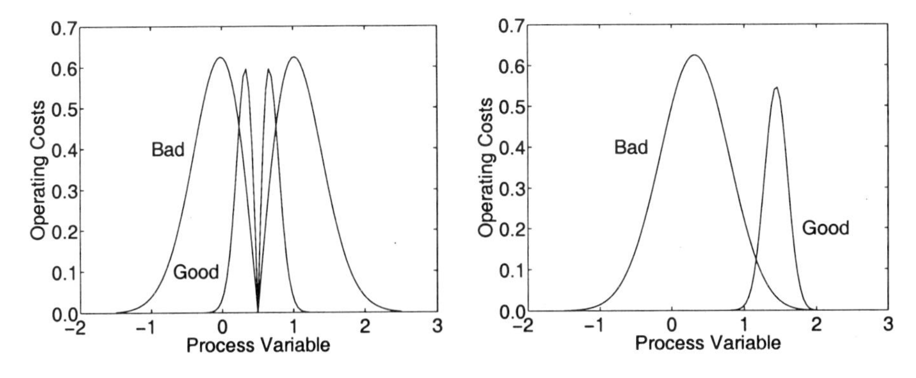

Figure 21.2: Controller performance cost curves.

deviation).

Example 21.3 (Regulatory control performance savings) *The current control performance for a feedback-controlled process variable is 100 ± 15 and the addition of gain scheduling (Chapter 18) is estimated to halve the standard deviation (Figure 21.1 left plot). If the cost penalty function is (2*PV2+1) dollars per day, the left plot in Figure 21.2 indicates a saving in operating cost of ($ 450.6 - $ 113.5) per day.*

The cost for the upgrade is $ 2500 installed with an estimated operating cost (primarily retuning) of three days per year at $ 1050 per day including overheads. The simple payback time would be:

$$\frac{2500}{450 - 113 - 3\frac{1050}{350}} = 7.6 days \tag{21.1}$$

which is a good buy in any language.

The economics of state driving control, moving from one operating point to another, is related to one or more of the following:

- the time,
- the utilities consumed, and
- the off-specification product.

One or more of these costs can often be reduced through improved sequencing of operations (Chapter 19) and by using optimal control strategies (see Chapter 20). The economic return can come from one or more of the following:

- reduced utility usage,
- reduction in off-specification product (reduced reprocessing or higher value),
- increased production (reduced change-over times), and
- more batches (reduced batch times).

21.5.3 Predicting Improvements

Having chosen the performance measure most appropriate for the opportunity being evaluated we can use our base case data to evaluate current performance. But how do we predict the size of the improvement that we can expect if we improve the control?

It is easy to say "use your experience and engineering judgement" but it is more difficult to sell the results to a sceptical management. It is better if we can use some sort of systematic procedure even if it comes down to experience and judgement in the end. There are several approaches that can be taken:

- previous implementations,
- manual implementation,
- best operator, and
- industry benchmarks.

Each of these will be discussed below.

Previous Implementations

Previous experience on similar plants within your own organisation is the ideal situation. This is why it is important to do post-mortem evaluations of projects even though there are difficulties and doubts involved in such exercises:

- Was it the control?
- Was it the little design changes also made?
- Was it just the improved process knowledge?
- Was it improved operator skills or motivation?
- Were the process conditions and disturbances the same?

You can always make some sort of distribution of the improvements amongst the various factors. Even with these problems such a figure will be more believable than a *guesstimate*.

Manual Implementation

Perhaps the next best approach is to set up a *manual* implementation. Using extra manpower and borrowed sensors or extra laboratory staff, and perhaps a bit of a bonus for all involved, perform a representative test run. You can then postulate that your automatic version will be able to achieve this performance all or most of the time.

Best Operator

Another approach is to evaluate the current performance of different operators or groups of operators. You can then postulate that improved automation or monitoring will enable all operators to achieve the performance of the best operator, in whole or in part. Of course if you don't have any really good operators, or if you have a small plant perhaps without full-time operators, you had best choose another basis.

Industry Benchmarks

These are probably better than a plain *guesstimate* but not by very much. You can generally find figures in your own industry journals and in professional surveys. There are also general process industry figures like the "six per cent of operating cost" quoted in the Warren Centre study.

Whichever basis you choose it is good to be honest and not inflate your figures. You could still be around when the post-mortem evaluation is performed.

21.5.4 Investment Costs

It is often necessary to perform a rough design in order to make a reasonable cost estimate. Be careful to state carefully the assumptions made and the reliability of figures used in your design. These figures must also be used to estimate an **in service** factor to apply to the benefits. Likely categories of costs are:

- sensors and other field instrumentation,
- sample preparation equipment,
- instrument and network cabling,
- control equipment,
- computing and communications hardware,
- computer software (purchased or developed),
- training materials or courses,
- preparation of documentation,
- design and engineering charges,
- project management costs,
- installation costs,
- control centre renovations (air-conditioning), and
- analyser houses or enclosures.

21.5.5 Recurrent Costs

Not to be forgotten are the daily running costs which will increase with more instrumentation, but hopefully not as much as the benefits will increase. The recurrent costs will include at least the following:

- sensor recalibrations and maintenance,
- analyser chemicals, filters, etc.,
- instrumentation maintenance,
- retuning of control applications,
- computer hardware and software maintenance contracts,
- computer software licence charges,
- increases in training costs, and
- increased power costs.

Running costs that are frequently overlooked are the additional manpower needed in the laboratory (sensor recalibrations and maintenance), the instrumentation shop (extra maintenance) and the process engineering group (retuning of control applications).

Too often is it assumed that once installed it will all run itself and achieve the same benefits forever. Experience shows just the opposite, a rapid deterioration in the benefits to none at all within one to three years unless regular monitoring and retuning occurs. Projects should not begin without a firm commitment to continuing support.

21.6 Summary

We have tried to present to you a systematic procedure and some hints for evaluating projects for improved monitoring and control. Such general procedures must always be adapted to the local organisation and often to the project itself, but some basis is always a good start. Experience has shown that systematic evaluations reduce the number of *disasters* and increase management receptiveness of new proposals.

21.7 Further Reading

- The Warren Centre reports, Marlin *et al.* (1987a) and Marlin *et al.* (1987b), have been the basis for most of the discussion and examples in this chapter.

- The report (Anon. (1997)) lists advanced control cost saving opportunities in the process industries in general.

- The paper of Vanrolleghem *et al.* (1996a) discusses the definition of design and operating goals in terms of the structure of a quantitative objective function.

- Information on the process improvement from utilizing process models and simulation is discussed in Speirs & Stephenson (1985) and van der Roest & Uijterlinde (1996).

- Savings from the addition of advanced control to Danish alternating plants are discussed in Isaacs *et al.* (1999).
- Monitoring the performance of control systems is discussed in Rhinehart (1995) and Harris *et al.* (1997).

21.8 Exercises

1. Your supervisor has asked you to talk with the operating team of a continuous BNR plant to identify the operating goals and problem areas.

   Construct a set of questions for semi-structured interviews with each of the operating team.

2. Using your treatment plant as an example construct a list of possible opportunities.

3. For each of the opportunities listed in the previous exercise:

   (a) Propose a methodology for quantitatively evaluating the benefits.

   (b) Propose a possible solution to obtain the benefit eveluated.

   (c) List the cost items involved in implementing the proposed solution.

4. Improved DO control in the aeration tank of an activated sludge plant allowed operation at 2 mg/l instead of 4.5 mg/l without increasing violations of the 1 mg/l constraint (assume the constraint remains at two standard deviations). Assume that the following equation can be used to calculate the cost penalty function and the reduced air flowrate given that $S_{O,sat}$ is 10 mg/l, q_a is 4 kl/s at S_O of 4.5 mg/l, OUR does not change and K_a is a constant.

   $OUR = K_a q_a (S_{O,sat} - S_O)$

   The annual cost of 4 kl/s of air is USD 127,000. Calculate the savings from the improved control using the MATLAB function `costcon.m`.

5. It is proposed to put the dosing of ferric chloride under automatic control using a phosphate analyser to replace the current system based on laboratory analyses once per shift. The current dosage is 35 g/m^3 wastewater with a standard deviation of 22 g/m^3. An intensive test run suggests that a dosage of 28 g/m^3 with a standard deviation of 5 g/m^3 is achievable.

   The annual cost of ferric chloride is currently USD 118,000. Calculate the savings from the improved control using the MATLAB function `costopt.m`.

Part D

Instrumentation

Chapter 22

Sensors for Primary Environmental Variables

22.1 Introduction

The sensors are the eyes and ears of the control system. We never really know what the process is doing, we only know what the sensors are telling us. They sense the values of the process variables (such as flowrates and temperatures) and transmit these values as signals to the controllers and/or display and reporting system.

Compared to computer technology the sensor technology lags far behind. In general the sensors are the weakest part of the chain in real time process control of wastewater treatment plants. We have seen that estimation and control theory can compensate for part of this. Sensors in wastewater treatment are often corrupted with significant noise and have to function in very harsh environments. As a result advanced signal processing and estimation are highly motivated so that maximum information can be extracted from the signal.

A sensor is an instrument that is capable of obtaining information directly from an object or process and that presents this information in such a way that one is able to deduce the variable to be measured. We may recognise some special features of an on-line sensor:

- the sensor should be located in the process,
- the sensor output should be considered continuous with respect to the process time scale, and
- the sensor should be operated without continuous human intervention even if periodic attention is often necessary.

The basic requirements of any sensor are described in Section 22.2. This is of course true for advanced sensors too. Since the sensor environment is usually quite hostile in a wastewater treatment plant, it is even more essential to prove the reliability of the sensors and the usefulness of the data. There is still a great challenge to convince design engineers, environmental managers and operators to use some of the mentioned instruments.

In Section 22.3 we will identify a number of physical sensors for temperature, gas and liquid flowrates, pumping and electrical power, level and pressure.

In Section 22.4 we will describe sensors for the measurement of physical or chemical properties of the wastewater such as turbidity, solids content, settleability, conductivity and oxidation-reduction potential (ORP or redox potential). In Chapter 23 we will also consider sensors for wastewater treatment properties where various kinds of analysis are required. In general these analysers perform automatic wet chemistry analysis.

22.2 Sensor Characteristics

Several factors have to be considered in the selection of sensors. Here we will mention the most important ones.

22.2.1 Selection

From the perspective of a control engineer the first design issue as far as sensors are concerned is the selection of the right one for the application from a bewildering range of possibilities. Factors to be considered are (refer to Table 22.1):

- **Type of sensor**. There are inevitably several basic types of sensor for measurements such as flowrate and temperature. Some selection tables and descriptions of common types are included in Section 22.3.

- **Range**. The range is the lowest measurement and the highest measurement that the sensor output signal registers, for example from 500^oC to 1000^oC. Choosing the desired sensor range can depend on the minimum and maximum values encountered during operation, on the desired sensitivity to deviations caused by disturbances and on the desired accuracy of measurement. Additional sensors in parallel with different ranges are occasionally required.

- **Linearity**. If the sensor output signal varies linearly over the measurement range subsequent display and processing is greatly simplified. For a *linear* sensor the deviation from linear should be less than the accuracy.

- **Accuracy**. Does the output signal always give the same value for the same measurement value? Of course this is unlikely, but how close is it? Typical values are less than one per cent of range.

Characteristic	Values
Type	DP Sensor
Range	10-100 kPa
Linearity	0.5 %
Accuracy	0.3 %
Drift	0.1 %
Speed of Response	5 seconds

Table 22.1: Typical sensor specifications.

- **Drift**. How quickly does the calibration change? Re-calibration of instruments is costly so you would like to keep it to a minimum. Modern electronic sensors are quite good in this regard.
- **Speed of response**. How quickly does the output signal respond to a change in the measurement value? Typically sensors will settle out in one or two seconds. This can be important where very fast acting control is necessary or for complex sensors which can respond much more slowly.
- **Cost**. All else being equal this is of course the deciding factor. Usually they are not of course, so it is a trade-off of cost against advantages in the above areas.

The following table gives a typical manufacturer's specifications statement. The most common shortcomings of these are no information on speed of response and no statement of the basis of the quoted percentages (Maximum error or mean error or standard deviation? Percentage of reading, calibrated range, or maximum range?).

22.2.2 Location of Sensors

The second design issue as far as sensors are concerned is where should they be located. There are several factors to be considered:

- **Dead time**. The time between the change in the process variable occurring and the time that change reaches the sensor should be as short as possible. For example if changes in intensive variables (temperature, concentration) occur in a vessel then the sensors should be in the vessel rather than 10 metres down a pipe where a flow velocity of say 1 m/s would give a dead time of 10 seconds. In essentially incompressible fluids (liquids and solids) variables such as flowrate and pressure can be measured anywhere down a pipe as the change is propagated essentially instantaneously.
- **Environment**. Most sensors are constructed for field mounting. That is they can be mounted basically anywhere. On-line composition analysers on

the other hand usually require a laboratory-type environment, for example in an air-conditioned or heated analyser house. In this case samples are usually piped large distances and the dead time must be minimised with fast loops bringing the sample quickly very close to the analyser before the actual sample is extracted.

- **Sample condition.** Many sensors require samples in a particular condition either to function at all or to give the specified accuracy. The flow pattern at an orifice plate must be well developed with straight lengths of pipe before and after the sensor. Analysers where the sample must be vaporised or condensed first require elaborate processing to ensure no changes in composition or excessive dead time. Shared analysers taking more than one measurement require elaborate valving and purging to eliminate cross-contamination. A detailed look at analyser sampling will occur in Section 23.2.
- **Maintainability.** For maintenance purposes the sensor should be easily accessible, beside a walkway rather than out of reach. Special piping or valving may be needed to allow removal without shutting down the process or to enable *in situ* recalibration.

Location should be specified on piping and instrumentation diagrams, preferably explicitly by notes if there is any ambiguity possible.

22.3 Sensors for Physical Variables

Flowrate measurements are of fundamental importance in any wastewater treatment process. Measuring flowrate is far from trivial and it is difficult to measure flowrates with great accuracy. There are three common principles for volumetric flowrate measurements (see also Table 22.2):

- differential pressure,
- turbine sensors, and
- magnetic flow sensors.

By making a constriction in a pipe and measuring the pressure difference Δp it is possible to obtain the fluid velocity. At the constriction the velocity will increase and a pressure drop appears. The flowrate is proportional to $\sqrt{\Delta p}$ where the proportionality constant depends on the constriction geometry. The pressure difference can be translated to motion which is monitored by a differential transformer. There are friction losses in the constriction so the pressure losses can be significant, but the hydrodynamic design of the constriction can be made more streamlined so that the pressure losses are minimised. This is done in the Venturi meter which is commonly used to measure flowrates in water pipelines.

A liquid flowrate can be measured with a turbine. Its rotation is transmitted by mechanical or magnetic means to a counter or some electronic equipment. A

Type	Mechanism	Applications
orifice plate	pressure drop over orifice type restriction	clean liquids and gases; nonlinear; low range
vortex	frequency of vortices in wake of bluff body	clean liquids and gases; linear; better range
electromagnetic	emf induced in magnetic field	slurries, dirty liquids, food; fluid must be conductive; expensive
turbine	speed of rotation of turbine or propeller	clean liquids and gases; very good repeatability and rangeability

Table 22.2: Typical sensors for flowrate.

turbine can only be used for clear liquids because the mechanical movement will be distorted by particles in the fluid.

Magnetic flowrate sensors are built on the induction principle which states that a voltage is induced in a conductor that is moved in a magnetic flux. In magnetic flowrate measurements the conductor is the flowing electrically conductive liquid. It can be shown that the voltage generated between two opposite points on the inner pipe wall is largely independent of the velocity profile. The magnetic field is generated by two coils which are energised by either the AC power line voltage or a pulsed DC voltage. The induced voltage is received by two insulated electrodes and is proportional to the liquid flow velocity. Since the flow cross sectional area is assumed constant the voltage is proportional to the flowrate.

Wastewater is commonly conveyed by open channel flow rather than in pressure conduits. There the **Parshall flume** is the most common device used to measure wastewater flowrates. A typical flume (Figure 22.1) consists of a converging and dropping open channel section. Flow moving freely through the unit can be calculated by measuring the upstream water level (H). Ultrasonics is a common method to measure the water level. There is a nonlinear relationship between the water level and the flowrate. For a Parshall flume under free-flowing conditions the equation for the flowrate q has the structure:

$$q = aBH^{bB^c} \tag{22.1}$$

where B is the width (See Figure 22.1), H the water level and a, b, and c are constant parameters. If the flowrate is expressed in m^3/s and the width and height in m then $a \approx 0.37$, $b \approx 1.57$ and $c \approx 0.026$.

The flume requires that the influent is in laminar flow. Another requirement is that the water flows out of the flume easily and follows the profile without getting

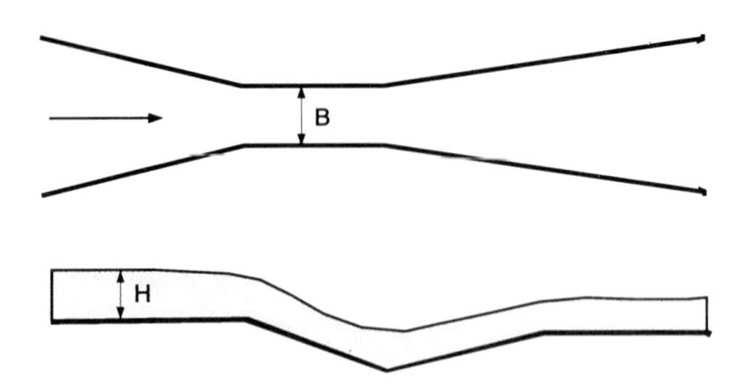

Figure 22.1: Parshall flume for measuring flowrate in an open channel.

clogged. A Parshall flume is self-cleaning in the sense that particles do not collect easily. The measurement accuracy is typically 3-5 %.

Weirs can also be used for measuring the flowrate of water in open channels. Equation 4.18 describes the flowrate in a weir. For example, the discharge over a 90 degree V-notch weir can be calculated by:

$$q = 1.4\, h^{5/2} \tag{22.2}$$

The height h is the vertical distance from the weir crest to the free water surface and is measured a short distance behind the weir in a stilling well.

Tables 22.2 to 22.5 list the most common sensors found in wastewater treatment plants for flowrate, level, pressure and temperature.

Type	Mechanism	Applications
displacement	weight change due to liquid displacement	clean liquids; levels or interfaces; installed in standpipe
differential pressure	hydrostatic head of liquid between taps	liquids; seals if dirty or corrosive
capacitance	changing dielectric constant between material and air	liquids or granular solids; insulate if material conductive
ultrasonic	echo from interface	liquids or solids; where contact is undesirable

Table 22.3: Typical sensors for level.

Type	Mechanism	Applications
Bourdon tube	motion or torque at closed end of curved or spiral tube	gauges; gases or clean liquids; seals if dirty or corrosive
diaphragm	force or displacement by capacitance, piezoelectric, strain gauges	sensors; gases or clean liquids; seals if dirty or corrosive

Table 22.4: Typical sensors for pressure.

Type	Mechanism	Applications
thermocouple	emf generated at junction of dissimilar metals	type T (-200°C to 350°C); type J (-200°C to 750°C) must be dry; type K (-200°C to 1100°C) oxidising sulphur-free environment; type S/R (0°C to 1450°C) oxidising environment
resistance (RTD)	change in resistance of metals with temperature, typically platinum	typically -260°C to 800°C
pyrometers	measures thermal radiation intensity focused optically	0°C to 4000°C; requires clear line of sight

Table 22.5: Typical sensors for temperature.

22.4 Primary Properties of Wastewater

In this section we will describe some fundamental measurements for wastewater treatment operation. They are all based on some kind of probe that can be used directly in the plant basins. Traditionally there have been manual procedures to define these variables but automatic measuring devices have been developed. The key variables are turbidity, suspended solids, settleability, conductivity and redox potential.

22.4.1 Turbidity

Insoluble particles of soil, organics, microorganisms and other materials impede the passage of light through water by scattering and absorbing the rays. This interference of light passage through water is referred to as turbidity.

Commercial turbidimeters (nephelometers) for determining low turbidities measure the intensity of light scattered at right angles to the incident light. *Nephelometric Turbidity Units* (NTU) are sometimes referred to as FTU since *Formazin* polymer is used as the reference turbidity standard suspension. There is an international standard for turbidity (ISO 7027).

Light is focused by a lens before passing horizontally through the sample. In an ordinary turbidimeter this measurement alone would be used to determine the turbidity. Some turbidimeters can compensate for colour in the water sample. Such an instrument also detects the transmitted light. At low or moderate turbidity levels the forward scatter signal is negligible compared to the transmitted signal and the output is simply a ratio of the 90 degree scattered light to transmitted light. This ratioing stabilises the instrument and negates the effect of colour. At high turbidity levels the instrument combines signals from the forward scattered and transmitted light detectors in a denominator of the ratio allowing linearity over the entire range.

Many sensors for mixed liquor suspended solids (MLSS) concentration in the aeration basin and the return sludge are based on the near infrared (NIR) measuring technique. Typical ranges for raw wastewater are 0 to 1000 mg/l and 0 to 200 mg/l for primary and secondary effluents.

22.4.2 Solids

Suspended solids, both organic and inert, are common tests for wastewater. Turbidity is related to the content of solids but there are laboratory tests for solids that can verify or calibrate turbidity measurements. Total solids, total residue on evaporation, is the term applied to material left in a dish after evaporation of a sample of wastewater and subsequent drying in an oven. The concentration is simply defined by the mass of the total residue divided by the volume of the sample. The weight loss on ignition is called “volatile solids”. Determination of volatile solids does not distinguish precisely between inorganic and organic matter because the loss on ignition is not confined to organic matter. It includes losses due to decomposition or volatilisation of some mineral salts. Better characterisation of organic matter can be made by such tests as total organic carbon (TOC), BOD, and COD (see Section 23.6).

The term *suspended solids* refers to the matter that does not pass through a filter. The matter passing through the filter is defined as the *dissolved solids.* Dissolved matter (nonfiltrable residue) is not determined directly but is calculated by subtracting suspended solids from the total residue on evaporation.

The type of filter, the pore size, area and thickness of the filter and the physical nature, particle size and amount of material deposited on the filter are the principal factors affecting separation of suspended from dissolved solids.

22.4.3 Settleability

A great deal of effort has been directed at modelling the physical phenomenon of sedimentation (Chapter 4). However the relation between settling behaviour and the biological state of activated sludge is hardly understood. This is why bulking sludge is such a challenge. For an effective cure of this problem the measurement of settling properties is a key issue. Currently there is a striking lack of sensors or analysers that can meet the need for settling information. The traditional way of quantifying sludge settleability is by measuring the sludge volume index (SVI).

Sludge Volume Index

SVI is traditionally measured in the laboratory using a standard 1 litre cylinder. Mixed liquid is drawn from the aerator and the sludge volume is measured by filling the graduated cylinder, allowing undisturbed settling for 30 minutes, and then reading the volume occupied by the settled solids in millilitres. The MLSS of the contents has to be measured as well. The sludge volume index is then calculated according to:

$$\mathrm{SVI} = \frac{V \cdot 1000}{\mathrm{MLSS}} \tag{22.3}$$

where V is the volume of the settled solids after 30 minutes (measured in ml/l). MLSS is measured in mg/l. With the conversion factor of 1000 the SVI is then measured in ml/g. Expressed in another way SVI is the volume in ml occupied by 1 g of a suspension after 30 minutes of settling. The value of an acceptable SVI may be lower than about 150.

Settlometers

Various approaches to designing on-line settlometers have been reported. The overall operation of a wastewater treatment plant depends crucially on the settling properties so a reliable and frequent measurement of sludge settling is highly motivated.

One approach is to install a measuring system that tracks the sludge blanket or concentration profiles in the full scale clarifier. From this it is theoretically possible to deduce sludge settling characteristics on-line in the form of a mathematical description. So far there are no applications reported.

Another methodology consists of using a down-scaled version of the device under study (in this case a final settler) in a measuring system and performing experiments in this model reactor. The sludge blanket level is monitored during a batch settling experiment as a function of time. The sensing system is based on light transmission measurements using a scanning system. From the settling curve obtained the settling properties are deduced using a dynamic model of the settling (Section 4.6). Still another approach is to use image analysis to reveal the relation between floc structure and settling properties.

Image processing

High resolution microscopes have been used to produce microorganism image data. On-line image processing has been developed where the microscope is submerged directly in the aeration tank. The instrument can obtain a very small amount of liquid sample. The cycle time is about 1.5 minutes. The instrument produces a high resolution image.

An image processing algorithm is used to recognize flocs and filamentous microorganisms. Using image processing a *bulking index* BI has been proposed to predict the occurrence of bulking phenomena in the activated sludge:

$$BI = \frac{R^2}{F/V} \tag{22.4}$$

where R is the floc size in μm, F the filamentous microorganism length (measured in m/mg MLSS), and V the floc volume. Typical values are in the following ranges: $BI \approx 0.05 - 0.2$, $R \approx 50 - 100\,\mu m$, $F \approx 10 - 30\ m/mgSS$. This should be compared with the range of a $SVI \approx 150 - 300\ ml/gSS$.

Using the BI has proven to be successful in predicting sludge bulking.

22.4.4 Conductivity

Conductivity is a measure of the ability of a solution to carry an electric current. This ability is dependent on the presence of ions, their total concentration, mobility, valence, relative concentrations and the temperature. Solutions of most inorganic acids, bases and salts are relatively good conductors. Conversely, molecules of organic compounds that do not dissociate in aqueous solution conduct a current very poorly if at all.

The physical measurement of conductivity is usually of resistance, measured in *ohms*. The resistance of a conductor is inversely proportional to its cross-sectional area and proportional to its length. The magnitude of the resistance measured in a liquid therefore depends on the characteristics of the conductivity cell used and is not meaningful without knowledge of these characteristics.

The reciprocal of resistance is *conductance* . It measures the ability to conduct a current and is expressed in reciprocal *ohms* or *mhos*. The term *conductivity* is preferred and is customarily reported in conductance per length, or millisiemens per meter (mS/m) where *siemens* is the SI unit of the reciprocal *ohm*. Typical values of the conductivity of potable water are 5 - 150 mS/m. The conductivity of domestic wastewater may be near that of the local water supply. Some industrial wastewater has conductivities above 1000 mS/m.

Conductivity can be a qualitative measure and rapid determination of large changes in inorganic content of water and wastewater. Conductivity sensors can give continuous measurements of conductivity if they are properly installed and maintained.

Conductivity is a function of temperature in electrolytes (unlike in metals) and increases with temperature at a rate of approximately 1.9 % per degree C. Each ion has a different temperature coefficient.

22.4.5 Redox potential

The oxidation-reduction potential (ORP), often called the redox potential, is an indication of the oxidation state of a specific monitored system. As we have seen in Section 19.5.4 the redox potential often displays a "knee" or a breakpoint in anoxic operation. Thus the redox provides information of biological processes in particular during anoxic and anaerobic conditions. Since the measurement is relatively unproblematic and quite accurate there is an obvious incentive to use the redox potential for control purposes.

Definition

Reactions in which both oxidation (loss of electrons) and reduction (gain of electrons) occur simultaneously are called oxidation-reduction reactions or simply *redox* reactions. The loss of electrons in an oxidation reaction has to be compensated by a corresponding gain of electrons in the reduction reaction. Since no electrons are created or destroyed in the process, in any redox reaction there can be no gain or loss of electrons.

In discussing redox reactions the phrases *oxidizing agent* and *reducing agent* are frequently used to designate the species responsible for oxidation and reduction. For example oxygen is an oxidizing agent since it brings about the oxidation of various compounds. The substance that reacts with oxygen is then the reducing agent, being responsible for the reduction of oxygen.

By means of an electrolytic cell it is possible to use electrical energy to bring about a non-spontaneous redox reaction. An important application of electrolytic cells is in the process of electroplating in which a thin layer of metal is deposited on an electrically conducting surface.

A *voltaic (galvanic) cell* is designed to achieve the opposite effect. A spontaneous redox reaction serves as a source of electrical energy. We are all familiar with certain types of voltaic cells such as the "dry cell" used in flashlights and the lead storage battery used in automobiles. The property of a voltaic cell which is of particular interest to us is its electromotive force, or potential, which is a measure of the driving force behind the reaction occurring within the cell. By properly interpreting cell voltages it is possible to obtain vital information as to the spontaneity and extent of redox reactions taking place in a reactor of the plant.

One can split a redox reaction into two half-reactions of reduction and oxidation. In the same fashion it is possible to divide a standard cell voltage into two parts, one corresponding to the reduction half-reaction and the other to the oxidation half-reaction. There is a potential corresponding to the reduction reaction called the *standard reduction potential.* Likewise for the oxidation there is a corresponding *standard oxidation potential.* The standard cell voltage is the difference between the potential of the oxidation half reaction and that of the reduction half reaction. This is what we call the **oxidation-reduction potential** (ORP).

It is possible to establish standard electrode potentials corresponding to a wide variety of oxidation and reduction half-reactions. One then arbitrarily assigns the value 0.00 V to the standard reduction potential of the H^+ ion which corresponds to the half-reaction

$$H^+ + e^- \Rightarrow \frac{1}{2}H_2 \tag{22.5}$$

The term "standard oxidation potential" is usually not used but is defined as the negative of the standard reduction potential.

A standard reduction potential can be designed for a large number of reduction half-reactions. A large negative potential signifies a half-reaction that is difficult to bring about, in other words some external energy has to be supplied for the reaction to take place. The more positive the potential the more spontaneous is the corresponding half-reaction.

We can now see that any reaction that can occur in a voltaic cell to produce a positive voltage must be spontaneous. To decide whether a given reaction is capable of taking place under a particular set of conditions all we have to do is to calculate the voltage associated with it. If the calculated voltage is positive there is an indication that the reaction is spontaneous. If we calculate a negative voltage the reaction cannot go by itself, the reverse reaction is energetically favoured. The standard voltage of any cell is the algebraic sum of the standard oxidation potential of the species being oxidized in the cell reaction and the standard reduction potential of the species being reduced.

Redox measurements

We have seen that the redox potential can be well defined for simple reactions. In a wastewater treatment plant there are many compounds present and there are several oxidation and reduction reactions that can take place. Therefore the redox potential measurement is an indicator of the sum of many phenomena and is not a universally unique number for all plants. Instead it has to be established empirically for each plant. In that way it can give significant information.

22.5 Summary

To measure is to know. In this chapter we have listed important sensors for physical properties of wastewater. Sensors are crucial for the operation of a wastewater treatment plant. However, it is important to define a number of properties of a sensor. Not only its accuracy but also its dynamics are of importance.

22.6 Further Reading

- There are several instrument handbooks available such as Considine (1974), Liptak (1982), de Silva (1989), Doebelin (1983), Kane (1987) and a wealth of manufacturer's literature available for the asking (especially if you wave an order form).
- Instrumentation for wastewater treatment systems is discussed in Manning & Dobs (1980) and Manross (1983).
- Flowrate sensors are described in detail in Moore (1986), level sensors in Gillum (1982), pressure sensors in Gillum (1984), and temperature sensors in Kerlin & Shepard (1982).
- Greenberg *et al.* (1992) is the standard reference for the examination of water and wastewater.
- A specialized workshop on sensors was arranged in 1995 by IAWQ (Lynggaard & Harremoes (1996)).
- There are several good overviews of commercially available sensors such as Manross (1983), Lynggaard-Jensen (1994) and Lynggaard-Jensen (1995).
- Experiences of suspended solids meters are reported by Nyberg *et al.* (1996a).
- Nyberg *et al.* (1993) describes a measuring system to track the sludge blanket.
- The ideas about deriving settling characteristics from sludge blanket measurements are described in Jeppsson & Diehl (1995).
- The idea of using an in-line detector for sludge properties is described in Vanrolleghem *et al.* (1996b).
- Image processing has been applied for sludge property analysis by Grijspeerdt & Verstraete (1996).
- Using ORP measurements for wastewater treatment applications is described in Charpentier *et al.* (1989).

22.7 Exercises

1. The height of flow above the crest of a 90 degree V-notch weir is 110 mm. What is the flowrate? (Answer: 0.0056 m^3/s)
2. Sensor dynamics are important in wastewater treatment systems. Consider the consequences of a slow sensor.

3. What is the difference between total, volatile, suspended, and dissolved solids?

4. What is the difference between solids content and turbidity?

5. Describe the possibilities and shortcomings of the SVI test.

6. List various possibilities where image analysis can be used for on-line monitoring in wastewater treatment systems.

7. What can be measured by conductivity?

8. Discuss the application of redox potential for control.

Chapter 23

Analysers

23.1 Introduction

It is obvious that advanced control of wastewater treatment systems will require sensors for continuous on-line information. In the past many treatment plants have been operated blindly. The knowledge of dynamics and computer control has to be matched by reliable sensors and analysers. In this chapter we will describe some important measuring principles for the analysis of wastewater composition. All of the sensors mentioned are commercially available even if their application sometimes is limited. Advanced instrumentation is still not widespread and requires specific knowledge in order to be reliable and give adequate information.

Most of the chemical and biological sensors should be used in-line with automatic sample conditioning. The handling of the samples is discussed in particular in the following section. It needs to be emphasised that proper installation is crucial. Many instruments have failed only for this reason. Also the importance of regular and frequent maintenance can not be overstated.

Various aspects of sampling systems will be discussed in Section 23.2. Nutrient analysers are complex instruments and some aspects of their design are discussed in Section 23.3. In the control chapters we have heavily relied on dissolved oxygen measurements. This is described in Section 23.4. Often it is crucial to know the acidity and alkalinity of the wastewater. Methods for pH and alkalinity measurements are discussed in Section 23.5. Organic content of wastewater is not trivial to define and many measurement principles exist where the most noted are BOD, COD and TOC, Section 23.6. Analysers for nitrogen and phosphorus are discussed in Section 23.7. Optical sensing of wastewater quality is becoming increasingly applied, Section 23.8. Respiration based control was discussed in the Control part. The basic principles of respirometers are described in Section 23.9. More refined methods to measure bacterial properties are mentioned in Section 23.10.

23.2 Sampling Systems

Pre-treatment of the samples is most often required to ensure satisfactory operation of the sensor equipment. Sensors requiring pre-treatment include nutrient analysers, that is ammonium, nitrate and phosphate analysers. The sampling unit is meant to remove threads and suspended solids that would otherwise block tubes, pumps and measuring cells or result in fouling of the electrode.

The desire for low chemicals consumption in on-line analysis leads to instruments with narrow pipe cross section and in the case of optical measuring processes also measuring cells that are too small. Therefore in order to prevent the analysers becoming clogged the wastewater samples must be pre-treated so that they are free of suspended solids. In the majority of plants a membrane process called **ultrafiltration** (UF) is employed for this purpose. Typical pore size is 20 μm. Tubular ultrafiltration modules are charged axially and the permeate is led off radially. A high flowrate prevents the formation of filter cakes and results in self-cleaning.

Typically a submerged pump leads the sample (of the order 5-10 m^3/h) to a measuring room in which an ultrafiltration system provides a filtered sample (0.5 - 30 l/h) according to the cross flow principle. The filtered sample is then led to the in-line sensor. Usually the reliability of such a system is considered high and the system seldom causes any problems. In pilot scale operations the problem of scale down comes in. The required flowrate for the ultrafiltration may be of the same order of magnitude as the total plant flow, so basically all the plant flow has to be pumped through the system.

The sample flowrate ranges from 1 ml/h to 2000 ml/h. The advantage of a small sample flowrate is partly that it entails lower consumption of reagents and partly that it allows the installation of smaller ultra-filters. A big sample quantity may be advantageous when coarsely filtered wastewater is measured (not ultra-filtered) or when the measurements are taken on a non-filtered sample for which a larger velocity is desirable to avoid blocking of tubes, pumps and valves.

An alternative sample handling system called **flow injection analysis** (FIA) has been used in bioprocesses and in water quality monitoring. It is increasingly used also for wastewater analysis. The advantage is a smaller reagent consumption. The sample is injected as a zone into a flowing carrier stream. As the sample zone is transported through the flow manifold it may undergo a variety of pre-treatment processes or chemical reactions with additional reagent streams before passing through a flow-through sensor which measures some analyte characteristic. Photometric, fluorescence and electrochemical detectors are typically employed for analyte detection in FIA.

Frequent cleaning (which can be automated) may be necessary for certain types of difficult wastewaters such as raw wastewater or certain industrial wastewaters.

New nutrient in-line sensors

New sensors for N and P measurements are being designed for direct submergence into open channels and process tanks. This avoids the whole pre-sampling loop. Consequently measurement and any filtration take place in one and the same unit inside the instrument. These sensors provide an interesting alternative in the future to conventional solutions. One drawback with a submerged sensor is that it can only measure at the point where it is located while an ultra-filtration system allows multiplexing from several measuring points.

Measuring Points

There are basically three measurement locations in the nutrient removal process:

- the plant inlet or the reactor inlet,
- the biological reactor, and
- the outlet of the plant.

By measuring the ammonia or phosphate concentration in the inlet the load may be estimated. However this has to be done with caution. The online measured concentrations are not always representative of the true concentrations since a significant part of the nitrogen and phosphorus is incorporated in compounds that are not hydrolysed until they reach the biological reactor.

If ammonia and phosphate are measured in the biological reactor the measurements can be used for feedback control. Some of these control actions have been discussed in Chapters 17 to 19.

By measuring at the outlet the effluent quality can be monitored. By combining the ammonia, nitrate and phosphate concentrations with on-line measurements of suspended solids the total effluent N and P concentrations can be estimated. This also requires laboratory analysis of the N and P in the suspended solids.

Measurements at the outlet are seldom useful for control since the measurements are too late. The control needs closer measurements to be effective.

23.3 Nutrient Sensor Design

There are a number of factors to consider when evaluating and using nutrient sensors. Certainly they are not like simple probes but adequately maintained and calibrated they can be used successfully. It may have to be emphasised that a sensor is properly maintained in the long run only if people are motivated to use it, so it has to be able to deliver meaningful information. Usually the information from such sensors is very meaningful, but one needs knowledge and training to be able to use the sensors and the information that they provide. One has to consider a number of factors:

- calibration,

- cleaning,
- response time,
- reagents,
- sample flowrate
- physical size,
- quality of components, and
- user-friendliness.

Both calibration and cleaning can be either manual or automatic. The time taken for the automatic cleaning and calibration may be from 1 to 60 minutes, but it may occur every 5 minutes or once a day. It is obvious that the time for cleaning and calibration has to be minimised since no information is obtained from the sensor during that time.

The response time for a nutrient sensor is crucial if it is going to be used for control purposes. This is particularly important in intermittent and alternating systems. There a response time of 5-15 minutes may be acceptable. Typically the response time varies between 1 and 30 minutes. The pre-treatment adds another 1-20 minutes so many instruments may be disqualified in sequential batch systems. However to use the instruments for recording historical data the response times are not crucial.

The reagents are a major part of the operating cost. The reagents may be bought ready-mixed or they may be mixed in the laboratory. The first alternative is probably more expensive but would guarantee a more uniform quality and would not require the time and special training of the laboratory personnel. The time intervals between replacements of reagents ranges typically from once a week to once every 12 weeks. There is always a risk of degradation of the reagent over a long time unless it is placed in a cool closed environment.

The physical size may be quite different. Some compact sensors may be wall mounted and require about 150 $\times$ 300 mm (width $\times$ height). Large sensors are mounted in big cabinets from floor to ceiling, typically 1 $\times$ 2 m, weighing more than 100 kg. Most sensors are about 0.5 $\times$ 1 m and are designed for mounting on a wall at working height.

The mechanical quality of the instrument is certainly important since the instrument has to operate around the clock. Previously instruments on the market were laboratory instruments upgraded to fit the purpose of a wastewater treatment plant and had frequent breakdowns. Today the component quality has been significantly improved.

The following accuracies can in general be expected when treating domestic wastewaters (these are standard deviations):

Ammonium:	0.3 mg/l	(range: 0-10 mg/l)
Nitrate:	0.5 mg/l	(range: 0-10 mg/l)
Phosphate:	0.2 mg/l	(range: 0-4 mg/l)

These values are only slightly higher than the average differences achieved in calibration investigations where a number of laboratories receive the same wastewater for analysis.

Signal conditioning must not be overlooked. The instrument signal is usually corrupted by noise and has to be properly filtered (see Sections 6.5 and 10.2.5). The noise can be both electronically generated and caused by air bubbles in the photometer for example.

23.4 Dissolved Oxygen Measurement

DO determination is a key measurement in wastewater treatment. For example we have seen that the DO concentration is the basis for the determination of respiration rates. It is crucial to keep the DO sufficiently high in aerators and sufficiently low in anoxic reactors.

23.4.1 Wet Chemistry - the Winkler or Iodometric Method

A standard chemical test for DO uses a BOD bottle for containing the water sample. The chemical reagents used in the test are manganese sulfate solution, alkali-iodide-azide reagent, concentrated sulfuric acid, starch indicator and standardised sodium thiosulfate titrant. The first step is to add a small amount of each of the first two reagents to the BOD bottle. If no oxygen is present the manganous ion (Mn^{++}) reacts only with the hydroxide ion to form a pure white precipitate of $Mn(OH)_2$ (indicated by the arrow pointing down on the right side of the reaction):

$$Mn^{++} + 2OH^- \rightarrow Mn(OH)_2 \downarrow \tag{23.1}$$

If oxygen is present some of the Mn^{++} is oxidised to Mn^{++++} and precipitates as a brown coloured oxide (MnO_2):

$$Mn^{++} + 2OH^- + \frac{1}{2}O_2 \rightarrow MnO_2 \downarrow + H_2O \tag{23.2}$$

After sufficient time has been allowed for all the oxygen to react, the chemical precipitates are allowed to settle leaving clear liquid in the upper portion. Concentrated sulfuric acid is added and the liquid turns yellow. The reaction that takes place with the addition of acid is:

$$MnO_2 + 2I^- + 4H^+ \rightarrow Mn^{++} + I_2 + 2H_2O \tag{23.3}$$

The manganic oxide is reduced to manganous manganese while an equivalent amount of iodide ion is converted to free iodine. The quantity of free iodine is equivalent to the dissolved oxygen in the original sample. The sample is now poured into a container for titration with thiosulfate solution. Thiosulfate in the titrant is oxidised to tetrathionate while the free iodine is converted back to iodide ion:

$$2S_2O_3^{--} + I_2 \rightarrow S_4O_6^{--} + 2I^- \tag{23.4}$$

Soluble starch is used as an end point indicator. It produces a blue colour in the presence of free iodine. The titration is continued to the first disappearance of blue colour.

Since there may be high concentrations of suspended solids and the biological activity of activated sludge flocs may have high oxygen utilisation rates, microbial activity must be stopped at the time of sample collection. It is common to use copper sulfate-sulfamic acid inhibitor solution to stop the biological activity and to flocculate the suspended solids.

23.4.2 Measuring DO by Membrane Electrodes

For on-line measurements the DO is measured without a chemical treatment of a sample. A DO probe is composed of two solid metal electrodes in contact with a salt solution that is separated from the water sample by a selective membrane. The recessed end of the probe containing the metal electrodes is filled with potassium chloride solution and is covered with a polyethylene or teflon membrane held in place by a rubber O-ring. The probe also has a sensor for measuring temperature. The unit can be submerged in an aeration tank for DO and temperature measurements. The membrane electrodes may be calibrated by reading against air as well as a water sample of known DO concentration determined by the iodometric method.

With membrane-covered electrode systems the problems of impurities are minimised because the sensing element is protected by an oxygen-permeable plastic membrane that serves as a diffusion barrier against impurities. Under steady-state conditions the current is directly proportional to the DO concentration. Fundamentally the current is directly proportional to the activity of molecular oxygen.

Membrane electrodes have been used for DO measurements in lakes and reservoirs, for control of industrial effluents and for continuous measuring of the DO in activated sludge systems. Being completely submersible membrane electrodes are suited for analysis *in situ*. Their portability and ease of operation and maintenance make them particularly convenient for field applications. They also provide an excellent method for DO analysis in highly coloured waters and in strong waste effluents.

Membrane electrodes exhibit a relatively high temperature coefficient largely due to changes in the membrane permeability. Temperature compensation can be made automatically by using thermistors in the electrode circuit.

23.4.3 Dissolved Oxygen Saturation Concentration

The solubility of oxygen in water is directly proportional to the pressure it exerts on the water. At a given pressure solubility of oxygen varies greatly with water temperature and to a lesser degree with salinity (chloride concentration). Some typical DO saturation values at a barometric pressure of 760 mm Mercury (approximately 100 kPa) are typically 9.2 mg/l at 20°C, 10.2 mg/l at 15°C and 11.3 mg/l at 10°C (assuming zero chloride concentration). It also decreases approximately linearly with decreasing barometric pressure.

23.5 Measuring Acidic or Basic Character

Measurement of pH is one of the most important and frequently used tests in water chemistry. Practically every phase of water supply and wastewater treatment such as acid-base neutralization, water softening, precipitation, coagulation, disinfection and corrosion control is pH dependent. pH is used in alkalinity and carbon dioxide measurements and many other acid-base equilibria. At a given temperature the *intensity* of the acidic or basic character of a solution is indicated by pH or hydrogen ion activity.

Alkalinity and acidity are the acid- and base-neutralizing capacities of water and usually are expressed as mg/l of $CaCO_3$. Buffer capacity is the amount of strong acid or base, usually expressed in moles per litre, needed to change the pH value of a 1 litre sample by 1 pH unit.

23.5.1 Hydrogen Ion Concentration (pH)

Water dissociates sparingly yielding a concentration of hydrogen ions equal to 10^{-7} mole per litre. Because water yields one hydroxyl (basic) ion for each hydrogen (acid) ion pure water is considered neutral:

$$H_2O \leftrightarrow H^+ + OH^- \tag{23.5}$$

The acidic nature of water is related to the concentration of hydrogen ions expressed as pH where:

$$\text{pH} = -\log\left[H^+\right] \tag{23.6}$$

Consequently neutral is defined as pH=7. In other words pH is the "intensity" factor of acidity. Measurement of pH is most frequently accomplished by using a meter that reads directly in pH units. A glass electrode in association with a calomel electrode dipped into the solution detects the hydrogen ions. Prepared standard solutions are used to calibrate the meter.

23.5.2 Alkalinity

Although only about 0.03 per cent of atmospheric air (though it is increasing every year), carbon dioxide plays a major role in water chemistry. It reacts readily with water forming bicarbonate and carbonate radicals. CO_2 is absorbed from the air or it can be produced by bacterial decomposition of organic matter in the water.

During nitrification alkalinity is consumed. It is produced during denitrification when nitrate is reduced to nitrogen gas.

Once in solution CO_2 reacts to form carbonic acid:

$$CO_2 + H_2O \leftrightarrow H_2CO_3 \tag{23.7}$$

When the pH of the water is greater than 4.5 carbonic acid ionises to form bicarbonate:

$$H_2CO_3 \leftrightarrow H^+ + HCO_3^- \tag{23.8}$$

Bicarbonate is transformed to the carbonate radical if the pH is above approximately 8.3:

$$HCO_3^- \leftrightarrow H^+ + CO_3^- \tag{23.9}$$

The bicarbonate-carbonate character of a water can be analysed by slowly adding a strong acid solution to a sample of water and reading resultant changes in pH. Such a titration is used to measure the alkalinity of a water. Acidity is measured by titrating with a strong basic solution. If two samples of pure water (pH 7) are titrated with sulfuric acid and sodium hydroxide solutions respectively, very small initial additions of either titrant result in significant changes in pH (Figure 23.1(a)). The addition or withdrawal of hydrogen ions is reflected immediately in changing pH readings.

Now consider what happens when acid is added to water containing a high initial concentration of carbonate ions, Figure 23.1(b). When acid is added the majority of the hydrogen ions from acid combine with the carbonate ions to form bicarbonates (following Reaction 23.9 from right to left). The excess hydrogen ions lower the pH gradually until at pH 8.3 all carbonate radicals have been converted to bicarbonates. Additional hydrogen ions reduce the bicarbonates to carbonic acid below pH 4.5 (following Reaction 23.8 from right to left).

Stirring of the sample at this time results in the release of carbon dioxide gas formed from the original carbonates (Reaction 23.7). The addition of a strong base results in the reverse titration behaviour, following the Reactions 23.8 and 23.9 from left to right.

Substances in solution such as the various ionic forms of carbon dioxide that offer resistance to changes in pH as acids or bases are added are referred to as **buffers**. To understand the buffering action is essential since many chemical and biological reactions in wastewater treatment are pH dependent and rely on pH control.

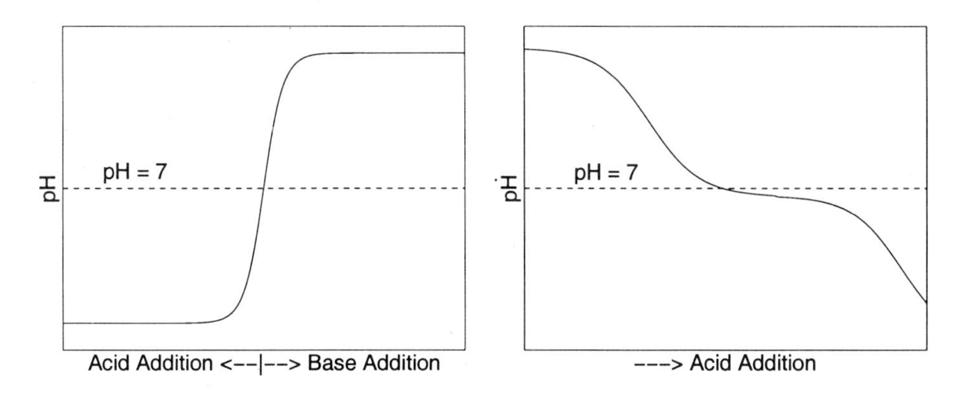

Figure 23.1: Titration curves (a) for pure water adding strong acid or base, (b) for carbonate in water adding strong acid.

Phosphorus removal by chemical precipitation is one example. The wastewater contains buffering compounds such as various phosphates ($H_2PO_4^-$, HPO_4^{--}, PO_4^{---}).

The alkalinity of a water is a measure of its capacity to neutralise acids, in other words to absorb hydrogen ions without a significant pH change. Alkalinity is measured by titrating a given sample with sulfuric acid. For highly alkaline samples the first step is titrating to a pH of 8.3. The second phase, or the first if the initial pH is less than 8.3, is titrating to an indicated pH of 4.5. The alkalinity is conventionally expressed in terms of mg $CaCO_3$ per litre and is calculated from:

$$\text{Alkalinity} = \frac{\text{titrant volume} \cdot \text{normality of acid}}{\text{sample volume}} \tag{23.10}$$

Normality is a method of expressing the strength of a chemical solution. A 1.00 N solution contains one gram of available hydrogen ions or its equivalent per litre of solution.

23.6 Organic Content

Biodegradable organic matter in municipal wastewater is classified into three major categories: carbohydrates, proteins and fats. Carbohydrates consist of sugar units containing the elements of carbon, hydrogen and oxygen. A single sugar ring is known as a monosaccharide; few of these occur naturally. Disaccharides are composed of two monosaccharide units. Examples are sucrose, common table sugar, and lactose, the most prevalent sugar in milk. Polysaccharides are long chains of sugar units and can be divided into two groups: readily biodegradable starches abundant in potatoes, rice, corn and other edible plants, and cellulose found in wood, cotton, paper and similar plant tissues. Cellulose compounds

Common name	Proper Name	Formula
Formic	Methanoic	$HCOOH$
Acetic	Ethanoic	CH_3COOH
Propionic	Propanoic	CH_3CH_2COOH
n-Butyric	Butanoic	$CH_3CH_2CH_2COOH$
Valeric	Pentanoic	C_4H_9COOH

Table 23.1: Carboxylic acids.

degrade biologically at a much slower rate than starches.

Proteins in the simple form are long strings of amino acids containing carbon, hydrogen, oxygen, nitrogen and phosphorus. They form an essential part of all living tissue and are necessary in the diet for all higher animals.

Fats refer to a variety of biochemical substances that have the common property of being soluble to varying degrees in organic solvents (like ether, ethanol, acetone and hexane) while being only sparingly soluble in water. Because of their limited solubility degradation by microorganisms occurs very slowly.

The majority of carbohydrates, fats and proteins in wastewater are in the form of large molecules that cannot penetrate the cell membrane of microorganisms. In order to metabolise high-molecular-mass substances bacteria must be capable of breaking down the large molecules into diffusible fractions for assimilation into the cell. The first step in bacterial decomposition of organic compounds is hydrolysis of carbohydrates into soluble sugars, proteins into amino acids, and fats to short fatty acids. Further aerobic biodegradation results in the formation of carbon dioxide (CO_2) and water. Anaerobic digestion, decomposition in the absence of oxygen, produces organic acids (see below), alcohols, and other liquid intermediates as well as gaseous entities of carbon dioxide (CO_2), methane (CH_4) and hydrogen sulfide (H_2S), identifiable by its rotten egg odour.

23.6.1 Organic Acids

All organic acids contain the carboxyl group written as $-COOH$. This is the highest state of oxidation that an organic radical can achieve. Further oxidation results in the formation of carbon dioxide and water. The simplest acids are listed in Table 23.1.

Acids with up to nine carbons are liquids but those with longer chains are greasy solids, hence the common name *fatty acid.* Organic acids are weak and ionise poorly. They all have sharp penetrating odours. Anaerobic decomposition of long-chain fatty acids results in the production of two- and three-carbon acids that are then converted to methane and carbon dioxide gas.

Volatile fatty acids (VFA) are important components in P removal as shown in Chapter 4. Measurement of organic acid concentration in anaerobically digesting sludge is used to monitor the digestion process. The acids, principally acetic, propionic, and butyric, are separated from water solution by column chromatography. The amount of total organic acids is measured by titration. The total organic acid concentration is expressed in mg/l as acetic acid.

23.6.2 Biochemical Oxygen Demand

Biochemical Oxygen Demand (BOD) is the classical parameter to define the "strength" of a municipal or organic industrial wastewater. BOD is by definition the quantity of oxygen utilised by a mixed population of microorganisms in the aerobic oxidation of the organic matter in a sample of wastewater at a temperature of 20°C. Measured amounts of a wastewater diluted with prepared water are placed in a bottle. The dilution water contains phosphate buffer (pH 7.2), magnesium sulfate, calcium chloride, ferric chloride and is saturated with dissolved oxygen. Seed microorganisms are supplied to oxidise the waste organics if sufficient microorganisms are not already present in the wastewater sample. The general biological reaction that takes place is:

$$\begin{aligned}\text{Organic matter } &\rightarrow^{DO}_{Bacteria} CO_2 + \text{Bacterial cells} \\ &\rightarrow^{DO}_{Protozoa} CO_2 + \text{Protozoal cells}\end{aligned} \tag{23.11}$$

The wastewater supplies the organic matter (biological food) and the dilution water furnishes the dissolved oxygen. The primary reaction is metabolism of the organic matter and uptake of DO by bacteria releasing carbon dioxide and producing a substantial increase in bacterial population. The secondary reaction represents the oxygen used by the protozoa, which consume bacteria in a predator-prey relationship. Depletion of oxygen in the test bottle is directly related to the amounts of degradable organic matter. The standard test has an incubation period of five days (BOD_5) or in some countries seven days (BOD_7) at 20°C.

Considering the time for the analysis the BOD test is certainly not suitable for operational purposes. Furthermore BOD is not a single point value but is time dependent. Figure 23.2 shows the BOD as a function of time. The carbonaceous oxygen demand progresses mainly as a first order batch reaction where the principal reactions are described by Equations 3.33 to 3.36 where the flow terms are set to zero. The BOD as a function of time can be described as:

$$BOD(t) = S_O(0) - S_O(t) \tag{23.12}$$

where S_O is the dissolved oxygen concentration at time 0 and at time t. The BOD is often approximated as a first order reaction:

$$BOD(t) = BOD_{\text{inf}} \left(1 - e^{-\alpha t}\right) \tag{23.13}$$

Actually the BOD_5 just shows a fraction of the ultimate BOD_{inf}, typically two-thirds in a municipal wastewater.

Nitrifying bacteria also have an oxygen demand in the BOD test where ammonia nitrogen is oxidised to nitrite by *Nitrosomonas* type bacteria and further

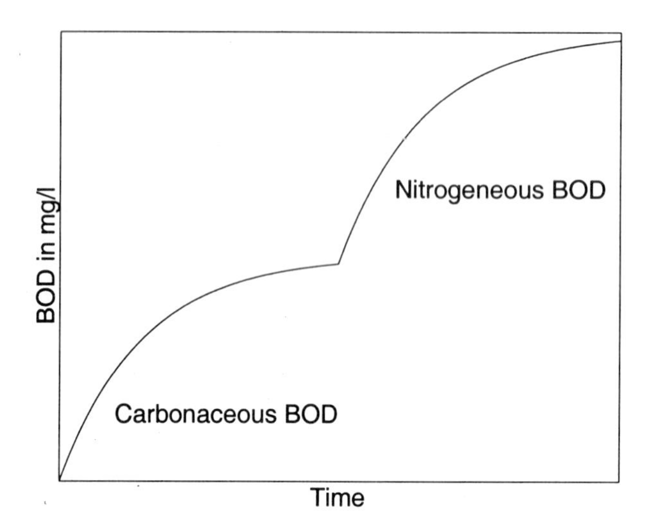

Figure 23.2: Qualitative shape of the carbonaceous and nitrogenous BOD curves.

oxidised to nitrate by the *Nitrobacter* bacteria as described in Section 4.1. Since the nitrifying bacteria grow much slower than the carbonaceous bacteria only a small fraction of the nitrogenous oxygen demand occurs after five or seven days. The standard method for preventing nitrification is the addition of trichloromethyl pyridine to the dilution water.

The BOD test is not a precise measurement and the reproducibility is quite poor. Tests on real wastewaters normally show standard deviations of 10-20 per cent.

23.6.3 Chemical Oxygen Demand

COD is widely used to characterise the organic strength of wastewaters. The test measures the amount of oxygen required for chemical oxidation of organic matter in the sample to carbon dioxide and water.

The laboratory test procedure is to add a known quantity of standard potassium dichromate solution, sulfuric acid reagent containing silver sulfate, and a measured volume of sample to a flask. This mixture is vaporised and condensed for 2 hours. Most types of organic matter are destroyed in this boiling mixture of chromic and sulfuric acid:

$$\text{Organics} + Cr_2O_7^{--} + H^+ \rightarrow_{Ag^+}^{\text{heat}} CO_2 + H_2O + 2Cr^{+++} \qquad (23.14)$$

After the mixture has been cooled and diluted with distilled water and the condenser has been washed down the dichromate remaining in the specimen is titrated with standard ferrous ammonium sulfate ($Fe(NH_4)_2(SO_4)_2$) using ferroin indicator. Ferrous iron reacts with dichromate ion with an end point colour change from blue-green to reddish brown:

$$6Fe^{++} + Cr_2O_7^{--} + 14H^+ \rightarrow 6Fe^{+++} + 2Cr^{+++} + 7H_2O \qquad (23.15)$$

A blank sample of distilled water is carried through the same COD testing procedure as the wastewater sample. The purpose of running a blank is to compensate for any error that may result because of the presence of extraneous organic matter in the reagents. COD is calculated according to Equation 23.16:

$$COD_{\text{mg/l}} = \frac{t_{\text{blank}} - t_{\text{sample}}}{V_{\text{sample}}} \cdot \text{norm}[Fe(NH_4)_2(SO_4)_2] \cdot 8000 \qquad (23.16)$$

where T is the titrant in ml, V is the volume in ml, and the 8000 factor appears since 1 litre contains 1000 ml and the equivalent weight of oxygen is 8.

There is no uniform relationship between the COD and BOD of wastewaters except that the COD value must be greater than the BOD. This is because chemical oxidation decomposes nonbiodegradable organic matter and the standard BOD test measures only the oxygen used in metabolising the organic matter for 5 or 7 days. The correlation of COD to BOD for a particular wastewater can be determined. Unfortunately such a relationship may be invalidated by the simple day-to-day variations in quality of the municipal wastewater.

23.6.4 Total Organic Carbon

We have seen that organic carbon in water and wastewater is composed of a variety of organic compounds in various oxidation states. Some of these carbon compounds can be oxidized further by biological (BOD) or chemical processes (COD). The TOC measures the organically bound carbon in a water or wastewater sample. Unlike BOD or COD it is independent of the oxidation state of the organic matter but does not provide the same kind of information. TOC does not measure other organically bound elements such as nitrogen, hydrogen and inorganics that can contribute to the oxygen demand measured by BOD and COD. Therefore TOC does not replace BOD or COD.

If a repeatable empirical relationship is established between TOC and BOD or COD then TOC can be used to estimate the accompanying BOD or COD.

To determine the quantity of organically bound carbon the organic molecules must be broken down to single carbon units and converted to a single molecular form that can be measured quantitatively. Commercially available TOC analysers oxidise the organic carbon to carbon dioxide by either heat and oxygen, ultraviolet radiation, chemical oxidants, or various combinations of these. The CO_2 can be measured directly by a nondispersive infrared analyser or it can be reduced to methane and measured by a flame ionisation detector in a gas chromatograph.

Free CO_2 in the sample may be titrated chemically. Inorganic carbon must be eliminated or compensated for since it is usually a very large portion of the total carbon in a water or wastewater sample. The determination of total carbon and total inorganic carbon with the estimation of total organic carbon by difference is a common procedure.

23.7 Nutrient Analysers

Since the purpose of nutrient removal plants is to remove nitrogen and phosphorus there is an obvious interest in measuring these substances. The focus has been placed on sensors for ammonium (NH_4), nitrate (NO_3) and phosphate (PO_4). The measurement of total N and P is interesting for influent and effluent monitoring purposes but to date there are only a few manufacturers of such sensors and the instruments are very expensive. The use of on-line N and P sensors (actually meaning ammonia, nitrate and phosphate) in wastewater treatment plants is widespread in the northern European countries. New sensor prototypes for the measurement of ammonium, nitrate and phosphate are being developed that will make on-line control using nutrient sensors much more realistic.

Most of the N and P sensors are based on some kind of automated chemistry. Most of the methods of the analysis are defined in Greenberg *et al.* (1992) but in some of the sensors the methods have been modified from the standard method used in laboratory analysis. The acid, base or the reducing agent may have been replaced and the reaction time may be different.

23.7.1 Ammonia Nitrogen

Common forms of nitrogen are organic, ammonia, nitrite, nitrate and gaseous nitrogen. The sum of ammonia and organic nitrogen is often referred to as total Kjeldahl nitrogen (TKN).

The laboratory test for ammonia nitrogen is a distillation process and is based on shifting the equilibrium between the ammonium ion and free ammonia. Gaseous ammonia is released along with the steam that is produced when the water is boiled. A buffer solution is added to increase the pH and to shift the equilibrium towards ammonia as shown by the formula

$$NH_4^+ \leftarrow_{\text{acidic}} \rightarrow^{\text{basic}} NH_3 \uparrow + H^+ \tag{23.17}$$

The mixture is distilled driving off steam and free ammonia. The steam containing ammonia is condensed and collected in a boric acid solution. The reaction with boric acid forms ammonium ions driving the reaction to the left while producing borate ions from the acid. The amount of ammonia in the sample is determined by the quantity of boric acid consumption in the collecting beaker. This may be measured by back titration of the solution with a standard acid to determine the amount of borate ion produced.

There are two main principles for ammonium sensors:

- the *gas electrode* sensor where the ammonium (NH_4) is converted to ammonia (NH_3) under alkaline conditions and an ammonia-selective electrode is used for detection. The measuring range is 0.1 to 1000 mg/l and the response time is less than 5 minutes.

- in a *colorimetric sensor* some colour is detected. Again ammonium is converted to ammonia and this results in a formation of a yellow compound which is detected by colorimetry. Alternatively there is a formation of blue indophenol also detected by colorimetry. With this method the measuring range is 0.1 to 100 mg/l and a typical response time is 5-20 minutes.

23.7.2 Nitrate and Nitrite

Total oxidized nitrogen is the sum of nitrate and nitrite nitrogen. Nitrate generally occurs in trace quantities in surface water but may attain high levels in some groundwater. In excessive amounts it contributes to the illness known as methemoglobinemia in infants. A limit of 10 mg/l nitrate nitrogen has been imposed on drinking water to prevent this disorder. Nitrate is found only in small amounts in fresh domestic wastewater but in the effluent of nitrifying biological treatment plants nitrate may be found in concentrations up to 30 mg/l. It is an essential nutrient for many photosynthetic autotrophs and in some cases has been identified as the growth-limiting nutrient.

Determination of nitrate (NO_3) is difficult because of the relatively complex procedures required. There is a high probability that interfering constituents will be present and there is a limited concentration range for the various techniques.

For nitrate there are three main principles used in common sensors:

- *electrode sensors*, one being directly a nitrate electrode and another based on an ammonia-selective electrode. In the latter, nitrate is first reduced to ammonia. The response time is less than 10 minutes.

- sensors using direct *photometry*. This is based on ultraviolet (UV) absorption and no reagents are used. The UV technique measures the absorbance of (NO_3) at 220 nm and is suitable for uncontamined waters (low in organic matter). The detection principle is then based on UV spectrophotometry. The measuring range is 0.1 to 100 mg/l and the response time is less than 5 minutes.

- sensors based on *colorimetry*. Nitrate is first reduced to nitrite. If cadmium reduction is used a red azo dye is formed and detected by colorimetry. If hydrazine reduction is used then a red azo dye is formed.

Nitrite (NO_2) is an intermediate oxidation state of nitrogen both in the oxidation of ammonia to nitrate and in the reduction of nitrate. Nitrite can be determined by photometric measurements in the range 5 to 50 $\mu g\ N/l$ if a 5 cm light path and a green colour filter are used. Higher concentrations can be determined by diluting a sample.

23.7.3 Organic Nitrogen

Organic nitrogen is defined functionally as organically bound nitrogen in the tri-negative oxidation state. It does not include all organic nitrogen components. Analytically organic nitrogen and ammonia can be determined together and have been referred to as Kjeldahl nitrogen. Organic nitrogen includes such natural materials as proteins and peptides, nucleic acids and urea, and numerous synthetic organic materials. Typical organic nitrogen concentrations vary from a few hundred μg per litre in some lakes to more than 20 mg/l in raw sewage.

The Kjeldahl method determines nitrogen in the tri-negative state. It fails to account for nitrogen in the form of nitrate and nitrite. If ammonia nitrogen is not removed in the initial phase of the procedure the result is Kjeldahl nitrogen. Should Kjeldahl nitrogen and ammonia nitrogen be determined individually then organic nitrogen can be obtained by difference.

The principle is as follows. In the presence of sulphuric acid (H_2SO_4), potassium sulfate (K_2SO_4) and mercuric sulfate ($HgSO_4$) catalyst the amino nitrogen of many organic materials is converted to ammonium sulfate [$(NH_4)_2SO_4$]. Free ammonia and ammonium-nitrogen also are converted to ammonium sulfate. During sample digestion a mercury ammonium complex is formed and then decomposed by sodium thiosulfate ($Na_2S_2O_3$). After decomposition the ammonia is distilled from an alkaline medium and absorbed in boric or sulfuric acid. The ammonia is determined colorimetrically or by titration with a standard mineral acid.

23.7.4 Phosphorus

Phosphorus occurs in natural waters and in wastewaters almost solely as phosphates. The common compounds of P are orthophosphates ($H_2PO_4^-$, HPO_4^{--}, PO_4^{---}), polyphosphate (polymers of phosphoric acid) and organic phosphorus. Polyphosphates are used in synthetic detergents. Orthophosphates applied to agricultural or residential cultivated land as fertilizers are carried into surface waters with storm runoff and to a lesser extent with melting snow. All polyphosphates gradually hydrolyse in water to the stable ortho form while decaying organic matter decomposes biologically to release phosphate. Orthophosphates are in turn taken up by the growing biomass. Organic phosphates are formed primarily by biological processes. They are contributed to sewage by body wastes and food residues and also may be formed from orthophosphates in biological treatment processes.

Phosphorus is essential to the growth of organisms and can be the nutrient that limits the primary productivity of a body of water. In instances where phosphate is a growth-limiting nutrient the discharge of raw or treated wastewater, agricultural drainage or certain industrial wastes to that water may stimulate the growth of photosynthetic aquatic micro- and macro- organisms in nuisance quantities.

Phosphate also occurs in bottom sediments and in biological sludges both as precipitated inorganic forms and incorporated into organic compounds.

Phosphate sensors are generally more complex than the ammonia or nitrate sensors since the reactions taking place are more complex. Some coloured substance is formed from the phosphate and the colour development is quantitatively measured using a spectrophotometer. There are two major colorimetric methods used:

- the vanadomolybdophosphoric acid method. Vanadium is used in combination with ammonium molybdate to produce a yellow colour with phosphate ion. The colour is detected with a photometer. The measuring range is 0.1 to 20 mg/l and the response time is less than 12 minutes.
- the molybdenum blue method. Using ascorbic acid a molybdenum blue is formed and can be detected with a photometer. The measuring range is 0.01 to 5 mg/l and the response time is less than 12 minutes.

Commercial phosphate analysers are based on the photometric principle and both the principles above are implemented.

Digestion using a strong acid converts all of the phosphorus in a sample including organic P to orthophosphate. Therefore in a total P test the sample is boiled in either a concentrated perchloric acid or sulfuric acid - nitric acid solution to digest the organic matter releasing bound P. The total P is measured by testing the digested sample for orthophosphate content.

The classification of P fractions in a sample includes the physical state of filterable (dissolved) and particulate as well as chemical types. A complete P analysis consists of conducting ortho, acid hydrolysable, and total P tests on measured portions of filtered and unfiltered samples. The particulate contents for each of the three P fractions are calculated by subtracting the filterable measurements from the respective total determinations on the whole sample.

23.8 Optical Sensing

A considerable amount of research has been devoted to the development of rapid techniques to replace the traditional BOD, COD, and TOC measurements for organic content. Absorption at particular wavelengths (such as 254 nm and 280 nm) has been found to correlate well with BOD, COD and TOC values. This research has led to commercial products. However the extensive use of such instrumentation has been severely frustrated by sensor fouling. Instruments based upon absorption require optical components to be in constant contact with the sample. In addition such instrumentation usually requires pre-sample filtration and frequent washing leading to increased maintenance. To overcome these difficulties there has been a shift towards the development of biosensors and flow injection, Section 23.2.

The light scattering properties of surface and wastewaters have been shown to correlate well with certain chemical and biochemical water quality parameters. Since light scattering processes can be initiated and detected remotely (only optical contact with the sample is required) sensor fouling problems associated with conventional contact (invasive) probes are avoided.

The interaction of light with molecules manifests itself by means of various scattering processes. The spectroscopic techniques are well established and the technique has been applied to wastewater for on-line noninvasive monitoring of wastewater quality. Good correlations between normalised fluorescence values (NF) and corresponding wastewater BOD values have been obtained without the need for sample pre-treatment. NF emission data have been used to successfully predict the long and short-term variations in the BOD for both industrial and domestic wastewater samples and a close to linear relationship has been found between the NF units and the BOD concentration (from almost zero to around 350 mg/l of BOD). It has been found that the NF intensities and the absorbance values at 254 nm are remarkably similar. This seems to indicate that both the fluorescence and absorption properties represent to some extent generically similar material although not necessarily identical specific chromophores. There are indications that the fluorescence data differentiates between biodegradable and non-biodegradable material but it is not completely clear how. A lot of research is going on in this area.

23.9 Respirometry

Respirometry is the measurement and interpretation of the respiration rate of activated sludge. The respiration rate is the amount of oxygen consumed by the microorganisms measured per unit volume and unit time. The respiration rate reflects two of the most important biochemical processes in a wastewater treatment plant, biomass growth and substrate consumption.

Respirometry has been the subject of many studies and a number of measurement techniques and instruments have been developed. There has been a great confusion about the principles of respiration and the various respirometers described. Some of this confusion has to do with the location of the instrument in the process and others with the interaction between the respirometer and the treatment plant. This situation created a need to make an extensive study of respirometric principles and their applications in wastewater treatment control. Within the IAWQ a task group was appointed in 1993 with the mission to write a Scientific and Technical Report on this subject. The task group were supposed to thoroughly examine existing literature and practice and write a state of the art report. The report was published in 1998.

Sometimes the term *BOD monitor* is used for respirometers. This is not to be confused with BOD_5 since the monitor measures oxygen consumption using a biomass adapted to the wastewater typically during a few minutes.

23.9.1 Measuring Principles

Substrate utilisation in an aerobic environment requires oxygen. A portion $(1-Y)$ of the consumed substrate is oxidised to provide the energy required to reorganise the remainder (Y) of the substrate molecules into new bacterial cell mass (see Chapter 3). The rate of oxygen consumption can be measured relatively easily by measuring physical variables like dissolved oxygen or partial pressure of oxygen. Oxygen is consumed both due to the removal of carbonaceous material by heterotrophic bacteria and the oxidation of ammonia nitrogen to nitrate nitrogen by autotrophic bacteria. Nitrification often accounts for approximately 40 % of the total oxygen demand.

Respiration rate has been identified as a very sensitive parameter to test the viability of an aerobic system. Using respiration rate measurements it is possible to estimate other parameters such as biomass growth rate and decay rate, nitrification rate and hydrolysis rate.

The simplest respirometer may be a manually operated bottle equipped with a DO sensor. In the respirometer different components, primarily biomass and substrate, are brought together. The oxygen consumption can be measured either in the gas phase or in the liquid phase. In other types of respirometers there may be a continuous flow of liquid or gas or both liquid and gas through the instrument.

The IAWQ task group has pointed out that there are only eight basic principles of respirometers according to these basic principles:

- measurement in liquid or in gas phase,
- flow regime of liquid (flowing or static), and
- flow regime of gas (flowing or static).

Respirometers based on the principle of measuring the DO concentration in the liquid are based on the DO mass balance in the container of the respirometer. This equation is identical with the DO mass balance in an aerator, Equation 3.35.

$$\frac{d}{dt}(V_L S_O) = q_{L,in} S_{O,in} - q_{L,out} S_{O,out} + r_O V_L + (K_L a)_L (S_{O,sat} - S_O) V_L \quad (23.18)$$

where the index L has been used to indicate that the volume, the flowrates and the oxygen transfer rate refer to the liquid part of the respirometer volume. In respirometers with a static flow regime the two flow terms are zero. In systems with flowing liquid the DO being transported by the advective flow terms can be measured. Usually the liquid volume is constant so the flowrates are equal. Measuring the various concentrations and knowing the parameters, the respiration rate r_O (negative) can be readily calculated.

In a simple respirometer the liquid is reaerated to bring the DO level to a high value and then left without any reaeration. Since there is no flowing liquid the DO level is governed by the simple expression:

$$\frac{d}{dt}S_O = r_O \tag{23.19}$$

The DO concentration is described by a straight line until the DO concentration is low enough that the respiration rate is limited by a lack of oxygen, as described in Chapter 3.

In many respirometers the liquid is continuously reaerated and then the oxygen transfer rate has to be represented as in Equation 23.18. This makes it necessary to determine the DO saturation concentration and the oxygen transfer rate on a regular basis.

Some respirometers are based on measurements of gaseous oxygen so the gas phase mass balance has to be considered. In addition to the DO mass balance in the liquid phase an oxygen mass balance for the gas phase has to be added:

$$\frac{d}{dt}(V_G C_G) = q_{G,in} C_{G,in} - q_{G,out} C_{G,out} - (K_L a)_L (S_{O,sat} - S_O) V_L \tag{23.20}$$

where V_G is the gas volume, C_G the oxygen concentration in the gas phase, q_G the gas flowrates into and out of the volume. The last term corresponds to the oxygen transfer from the gas phase to the liquid phase. Some respirometers avoid measuring the oxygen concentration and instead measure the pressure change. Since carbon dioxide is produced by the bioreaction this gas must be absorbed chemically to avoid incorrect measurements.

In Section 12.4.3 the whole aerator was used as a respirometer. The governing equation is the same (Equation 23.18). As long as the concentrations can be measured, estimated or are known then the respiration rate can be calculated.

23.10 Measuring Bacterial Properties

Research in biotechnology is crucial for advanced wastewater treatment control. Methods to directly measure key bacterial growth parameters have not been available until recently. It is now possible to measure bacterial *specific growth rates* and biomass by direct count. By measuring specific bacterial growth it is possible to distinguish between bacterial growth and metabolic activity. To put it simply the "activity" is the same if there are many organisms with low specific growth or if there are few organisms with a high specific growth.

Nutrient removal is a function of bacterial metabolism and/or growth. It is important to distinguish between these two concepts even if both of them are terms that describe bacterial physiology but at two different levels of complexity. Metabolism or metabolic activity is a term that includes reactions that are responsible for cell maintenance, growth, respiration and motion. Bacterial growth on the other hand is the ability of the cell to divide and to create new biomass. Being able to distinguish between bacterial growth and metabolism makes it possible to observe how the bacterial cell allocates its resources, in other words under what conditions the cells are likely to divide and create more biomass (sludge) or to allocate more energy to respiration (decrease the mass of sludge).

The so called *thymidine* method has been established to measure bacterial growth *in situ*, that is without changing the wastewater environment. Here bacterial growth means bacterial cell division. The method has been applied as early as 1960 in biotechnology. Growth is marked by the synthesis of new bacterial DNA and cell division. The bacterial cell increases in size until its biomass doubles then division occurs. New bacterial cells are radioactively tagged if their DNA is synthesized in the presence of radioactively labelled thymidine. The thymidine method can give a quantitative measure of the bacterial growth *in situ*. The growth rate can be calculated by converting the radioactivity incorporated into bacterial DNA in the number of new cells.

23.11 Summary

We have described the most common wastewater analysis procedures that can be used for on-line control of a treatment plant. Still some of these instruments are quite expensive and require a competent installation and regular maintenance. Used properly they can be the basis for successful control, which has been proven in many plants in particular in Europe.

23.12 Further Reading

- Details of standard water and wastewater analyses are found in Greenberg *et al.* (1992).

- Tables of DO saturation concentrations as a function of pressure, temperature and salinity can be found in most standard wastewater treatment textbooks such as Hammer (1986).

- A special sensor technology workshop was arranged by IAWQ in October 1995 (Lynggaard & Harremoes (1996)) where many experiences have been recorded.

- Operating experiences of nutrient analysers are reported in Teichgräber (1993), Thomsen & Kisbye (1996), Thomsen & Nielsen (1996), Wacheux *et al.* (1993), Wacheux *et al.* (1996), Londong & Wachtl (1996), Balslev *et al.* (1996) and Nielsen & Önnerth (1996).

- The application of flow injection analysers (FIA) is described in Pedersen *et al.* (1990) and Benson *et al.* (1996).

- A new class of sensors for the measurement of ammonium, nitrate and phosphate have been developed at Danfoss in Denmark in cooperation with the Danish Water Quality Institute. The principles are described in Lynggaard-Jensen *et al.* (1996). Due to their design they have the potential to make on-line control using nutrient sensors much more available than today.

- The state of the art for N and P sensors is also described by Schlegel & Baumann (1996). The accuracies of nutrient analysers presented here are reported by Thomsen & Kisbye (1996).

- Progress in optical sensing is reported in Iranpour *et al.* (1997), Reynolds & Ahmad (1997a) and Reynolds & Ahmad (1997b).

- A long term user and developer of respirometry, Bob Arthur, has given his perspective in Arthur (1994) and Arthur & Arthur (1997).

- Henri Spanjers has devoted a lot of research to respirometers. His PhD thesis contains detailed information about the design and use of respirometers (Spanjers (1993)). Various applications of respirometery are found in Spanjers *et al.* (1993) and Spanjers *et al.* (1994).

- The IAWQ scientific and technical report on respirometry (Spanjers *et al.* (1998)) summarises the state of the art in respirometry. Spanjers *et al.* (1996) was a status report of the work of the task group.

- Respirometry for toxic warning is reported by Temmink *et al.* (1993).

- Vanrolleghem & Verstraete (1992) and Vanrolleghem & Verstraete (1993) have worked with other types of respirometers for estimating kinetic parameters and for plant monitoring.

- The work on bacterial growth is reported by Pollard (1987) and Pollard & Greenfield (1997).

23.13 Exercises

1. Why is nutrient measurement at the outlet of the plant not very useful for control?

2. Where is the best location for an ammonia sensor for control purposes? For a nitrate sensor? For a phosphate sensor?

3. Relate the analyser response times to the dynamics of the activated sludge system. What would be an upper limit of the response time?

4. Explain qualitatively how the saturation value of dissolved oxygen depends on temperature and on barometric pressure.

5. Describe the three major categories of biodegradable organic compounds.
6. Explain why not all of the organic matter in wastewater is converted to carbon dioxide and water in biological treatment.
7. What is a fatty acid? Why is it important in a wastewater treatment system?
8. Explain the difference between short term BOD and BOD_5.
9. Why is the COD concentration higher than the BOD_5 or the BOD_7 concentration?
10. How does the TOC relate to COD or BOD?
11. What is included in Kjeldahl nitrogen and in total nitrogen?
12. Which fractions of nitrogen are not desirable in the effluent wastewater?
13. Why can phosphorus be harmful for the environment?
14. Explain the role of alkalinity in wastewater. Note its relationship to nitrification and denitirification.
15. Explain the basic principle of a respirometer. How is the oxygen measured?
16. Explain the main difference between metabolic activity and bacterial growth.

Chapter 24

Actuators and Controllers

The goal of this chapter is to complete the introduction of the hardware components which make up a process control system in its broad sense.

We will examine briefly actuators and controller equipment. We will also talk briefly of design guidelines for displays and reports. The other vital component, sensors, have been discussed in Chapters 22 and 23.

The detailed specification of control hardware is generally the responsibility of the instrument engineer. However it is useful for the process control professional to have enough knowledge to understand his colleague and to be able to offer advice on aspects which particularly relate to the process viewpoint. This interest should encompass matching the utility and reliability of the instrumentation system to process operating goals and ensuring that the type and rangeability of sensors and actuators match process control requirements.

24.1 Actuators

The control system interfaces with the plant through the actuators and sensors. There is no sense (if you excuse the term) making sophisticated decisions if they are not implemented. It stands to reason that these two classes of instrumentation are key to successful process control. Yet, they are most often taken for granted. So many "control problems" are caused by our poor selection of or our neglect of the basic sensors and actuators.

The predominant actuator in process plants remains the *control valve*, for adjusting the flowrate of gases, liquids, slurries and solids. Variable-speed drives (electric or hydraulic motors) are often used for solids conveyors and increasingly for pumps and compressors where they are displacing control valves. The cost of a variable-speed drive remains high but if the pressure loss across the control valve is expensive then it can be more economical in the longer term.

24.1.1 Pump and Compressor Drive Systems

Using a motor there are three main ways to control a flow or to transfer liquid, slurry, air, gas or other objects during a given period. The main methods are based on:

- constant speed motor,
- combinations of two or more sets of equipment providing step-by-step control of the drive, and
- variable speed motor.

Each main group can then be sub-divided into sub-groups according to the design and function of the drive system. For example a variable speed motor drive can be divided into three sub-groups:

- mechanical control,
- hydraulic control, and
- power electronic control.

Constant Speed

The simplest way to control a motor is to use on/off switches. By shutting off the motor when it is not needed unnecessary use of energy is eliminated in the same way as with a light switch. This is still quite common in treatment plants and has been shown for influent flowrates, Figures 2.1 and 15.5. From an *equipment* point of view there is a serious drawback due to the mechanical wear which is apparent when large flowrates and pressures are controlled. Frequent starts and stops can radically limit the life of the equipment due to the pressure surges. From a *process* point of view on/off control is mostly negative. We have seen (Chapter 15) that sudden flowrate changes will have a significant influence on the clarifier behaviour. Also wide swings of the dissolved oxygen control will deteriorate the biological behaviour (Chapter 17). Below we will show the consequences from an *energy* point of view.

Throttling is an old method of flow control and uses a damper or a throttle valve to restrict the flow for example in the piping. Independent of the valve position the pump operates continually at full speed. The power requirement decreases only marginally as the pump is continually forced to operate against the pressure and resistance caused by the throttling. This means that throttling which was earlier chosen because of its simplistic construction and low purchase price cannot in most cases be accepted today due to its low efficiency, extremely high energy loss and high total costs.

Guide vane control is a common way to control the air volume from a fan. The technique is used over a wide flow range. The angles of guide vanes built in the inlet to a "radial fan" can be adjusted so that the volume of air is made to rotate in the same direction as the impeller. An efficient way to reduce the power requirement in guide vane control is to equip the fan with a two-speed motor or with two motors for different operating ranges.

Compared to these control methods **variable pitch control** represents a more advanced type of control for both fans and pumps. The fan impeller and the pump impeller are provided with vanes, the angles of which can be adjusted according to the volumetric flow requirement. By means of the variable pitch blades it is possible to achieve a high efficiency over a wide flow range. The efficiencies of the methods described earlier cover only a limited part of the flow range.

Step-by-Step Control

To utilise energy more efficiently and to achieve a better pressure and flow capacity, fans as well as pumps can be connected in series or in parallel. It is then possible to use a constant-speed pump for base flow and a variable speed pump to cover the variable speed range. It is also possible to have two pumps that operate alternately with one running continuously in the daytime and the other in the night-time. The pumps then alternate as duty and standby pumps. Twin pumps are often used providing a stand-by and an option for different capacities.

Variable Speed Control

There are several ways to introduce variable speed control. The torque can be transferred between the motor and the load axes without any mechanical contact. In a *hydraulic coupling* the required speed is achieved through an oil slip between blades mounted on the motor shaft and the pump shaft in a rotating oil container. The speed on the pump shaft is varied by changing the oil level in the container. The motor runs at constant speed.

Eddy current couplings have been used for many decades to synchronise speeds between a constant-speed electric motor and a load. New and more sophisticated variations of the drive system have evolved. As in the hydraulic coupling the torque is transmitted without the motor shaft and the coupling shaft having any mechanical contact with each other. This is done by means of electromagnetic induction.

The most elegant and potentially the most versatile control method is by **power electronic control** using a **frequency converter**. Frequency converters have been developed since the 1960s and are today a mature and well proven technology. The development of power electronic circuits and control electronics has made frequency control affordable and reliable. When considering variable speed drives we mostly consider asynchronous (induction) motors since they are the dominating motor type in pump and compressor systems.

Today there are frequency converters available for a very wide range of power from less than one kW to several thousand kilowatts. A particularly used method for control is pulse width modulation (PWM) . The three phase current is first converted to a constant voltage. The voltage is divided into short pulses with a

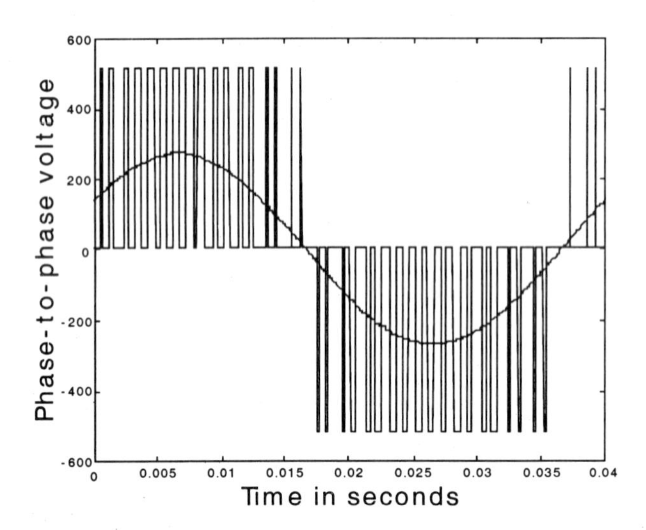

Figure 24.1: Development of the voltage during a cycle with PWM.

variable length and number so that the motor has the required levels of voltage and frequency in each instance. A high modulation frequency of the pulses creates an almost perfect sinusoidal curve. The better the modulation the better will be the efficiency of the motor and the drive system dynamics. Figure 24.1 shows an example of a PWM pattern. The switching of the voltage is performed by transistors or thyristors. In converters for power levels up to about 300 kW so called IGBTs (Insulated Gate Bipolar Transistors) are used. For higher power levels thyristors of a GTO (Gate Turn Off) type are often used.

The voltage is switched between three constant voltage levels. The sinusoidal curve shown is the actual curve form simulated from the given pulses.

A frequency converter can be used in both a new installation and in an old system. Since most pump or compressor systems already have an **asynchronous (induction) motor** they are in a sense prepared for frequency converter operation. In an advanced motor operation where a large operating range is going to be used the cooling of a standard motor may be insufficient at low speed and high torque. Then the motor may have to be supplied with extra cooling.

Energy Considerations

Both energy and money can be saved by adjusting the motor speed to the actual load requirement. This is particularly true for pumps and fans which continue to be controlled largely by energy consuming methods. The reason is that the load torque for turbulent flows increases with the square of the speed. Consequently the power need increases with the cube of the speed. In other words the power

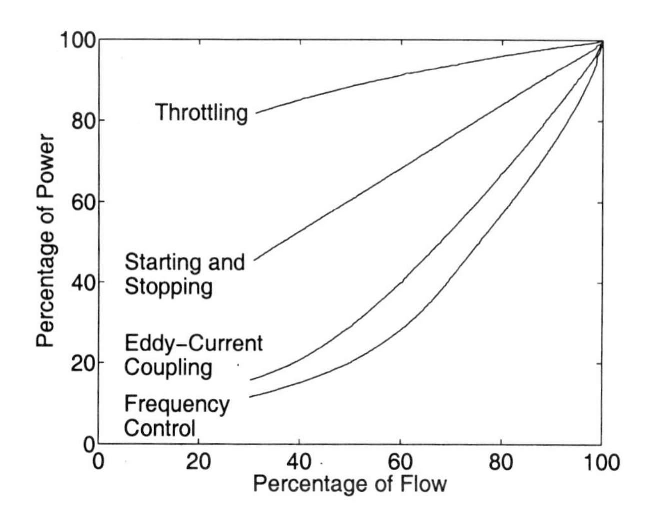

Figure 24.2: The difference between power need for throttling (upper curve) and for variable speed control (lower curve). Hydraulic and eddy current couplings are also shown. The example is an idealised centrifugal pump.

demand is increasing with the cube of the flowrate:

$$P \approx k \cdot Q^3 \tag{24.1}$$

Thus if the flowrate is decreased to 1/2 then the power need is only 1/8. At the same time the friction losses increase with the square of the flowrate. This is true in a system with zero static head. For a positive static head the possible saving will decrease. Figure 24.2 shows the principal difference in power need for different control methods in a system with a zero static head. In a throttled system (where we have neglected any pressure losses in the valve) there is a 12% power decrease for half the flowrate. For frequency control the corresponding power reduction is 80%. Hydraulic and eddy current couplings are also quite energy efficient.

Usually the investment cost of a motor represents a small fraction of the *service life* costs. It is not uncommon that the power costs are 90% of the life time cost. Therefore there is an obvious incentive to choose not only an adequate pump or fan but an efficient motor and drive system. The maintenance costs are usually much lower for variable speed systems than for constant speed systems. A speed that is adjusted to the need means a lower load torque and less mechanical wear both for the motor and for the load. Soft starts and stops can be achieved by the frequency converter's ramp functions. Thus dangerous and expensive pressure surges in the piping can be avoided at the same time as minimising mechanical stress of electric motors, pumps and valves. Furthermore it is possible to use the PWM converter so that critical frequencies and mechanical resonances in the drive system are avoided.

A PWM converter does not consume any reactive power from the power network. This means that it is usually not necessary to install capacitor banks for phase compensation of the power system. A frequency converter can be integrated into all kinds of integrated control systems since it can be controlled locally, from external controllers and through computerised communication systems.

Considering the process behaviour, energy consumption, electric installations, maintenance costs and control system structure there are strong reasons to use variable speed drives in wastewater treatment systems.

24.1.2 Control Valves

A **control valve** consists of a shaped plug which is mounted on a stem and which moves up and down within a usually circular seat. The stem is usually moved by air pressure on a diaphragm opposed by a spring. The spring either opens or closes the valve depending on its desired state in the event of air supply failure. Occasionally an electric or hydraulic actuator is used to move the stem. The plug and seat and the valve body design varies to accommodate the pressure drop across them, the type of fluid and the desired shape of the flowrate vs. stem position characteristic.

The valve body sizing is normally chosen to match the pipe size where the valve is located. The selection of valve body type and the sizing of the plug and seat combination requires consideration of the following factors:

- **Pressure drop**. Large pressure drops across control valves can make it difficult to move the valve stem. Special body designs divide the flow into opposite directions through two plug-seat combinations to cancel out the forces. Small pressure drops require different types of valves such as *butterfly valves*.

- **Maximum flowrate**. This should be the maximum design flowrate plus the maximum control action. Ideally the latter should be 30-50 % of the design flowrate although many non-control engineers hamstring control loops by cutting this margin to as little as 10 % (and then complain about the performance). The valve size is specified by the $C_{v,max}$ parameter calculated from the flowrate and pressure drop (see Equation 24.3).

- **Rangeability**. This is the ratio of the flowrates for stem positions of typically 15 and 85 % (100 % is full open). It is related primarily to the plug and seat design and the pressure drop vs. flowrate characteristic which is often related to the pump upstream of the valve (see below). Again the rangeability must account for the normal range of operating flowrates with an adequate control margin (preferably 30 to 50 %) both below and above that range. Different sized valves in parallel are occasionally required.

- **Sensitivity**. This relates to the rangeability and the amount of control action required to control to the desired accuracy. Occasionally a large valve

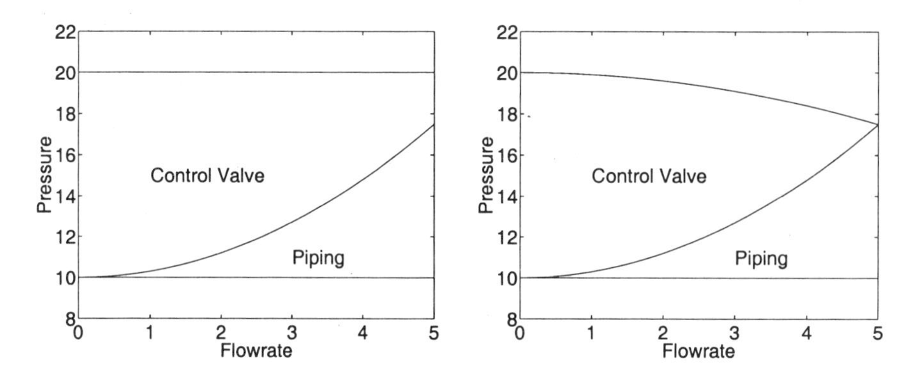

Figure 24.3: Pressure drop distributions for vessel and pump sources.

is required to set the nominal flowrate and a small valve in parallel is used to achieve the desired sensitivity.

- **Linearity**. For a control loop the objective is that the sensor output vs. controller output relation for the valve plus process plus sensor is linear. This means that often the valve characteristic and occasionally the sensor characteristic is chosen to compensate for nonlinearities of the process or sensor. If this linearity is not achieved poor control loop performance is likely without special controllers (see Section 17.5).

- **Hysteresis**. This is a common problem with control valves due to seal friction where the stem enters the body of the valve due to the fluid pressure drop across the valve. It is a common cause of small continuous oscillations in control loops. A valve positioner is the recommended solution. It is a special high-gain secondary control loop which measures stem position and applies strong control action to achieve the desired stem position requested by the primary controller (see Section 17.6).

Before discussing valve size or rangeability further it is important to understand the pressure drop vs. flowrate characteristic for control valves. There are two common cases. The first case is where the upstream and downstream pressures for the piping containing the valve are constant, which is shown in Figure 24.3 on the left. The second case is where the upstream pressure is the discharge pressure of a centrifugal pump which decreases as flowrate increases as shown in Figure 24.3 on the right.

The diagrams show two regions between the upstream and downstream pressures. The lower one is the pressure drop across the piping, pipe fittings and equipment not including the valve. The upper one is the pressure drop across the valve which can be seen to vary greatly with flowrate.

Pressure drops across piping and fittings and the internal pressure drops of pumps are often expressed in terms of a resistance R:

$$R = \frac{\Delta p}{\dot{m}^2} \tag{24.2}$$

where $\dot{m}$ is the mass flowrate and Δp is the pressure drop. The flowrate through the control valve can be calculated from the following equations.

$$\begin{aligned} \dot{m} &= C_{v,max} f(x)\sqrt{\Delta p} \qquad \text{(liquid)} \\ \dot{m} &= C_{v,max} f(x)\sqrt{p\,\Delta p} \qquad \text{(gas)} \end{aligned} \tag{24.3}$$

where $C_{v,max}$ is the valve size coefficient, $f(x)$ is the valve characteristic (see Equation 24.4), x is the valve stem position ($0 \leq x \leq 1$), p is the upstream pressure, and in this case Δp is the pressure drop across the valve.

The size coefficient $C_{v,max}$ is calculated by selecting the maximum flowrate, the corresponding pressure drop from the pressure drop vs. flowrate diagram, and setting $f(x)$ equal to one.

Valve rangeability can be calculated by calculating the two flowrates at x values of say 0.15 and 0.85. It is necessary to know the function $f(x)$ in this case which is the *inherent* valve characteristic. Typical commercially available characteristics are:

$$\begin{aligned} f(x) &= x \qquad \text{(linear)} \\ f(x) &= \sqrt{x} \qquad \text{(square root)} \\ f(x) &= A^{(x-1)} \qquad \text{(equal percentage)} \end{aligned} \tag{24.4}$$

where the design constant A is typically 20 to 50. These characteristics are shown in Figure 24.4. Notice that perfect equal percentage valves do not close fully. In practise they are often designed to approach linear characteristics at the very low openings and hence close fully.

The variation of the pressure drop across the valve shown in Figure 24.3 means that the *installed* valve characteristic (a plot of $\dot{m}/\dot{m}_{max}$ vs. x) is not the same as the inherent characteristic. This needs to be taken into account in valve trim selection. The MATLAB Toolbox program `cvchar.m` can be used to calculate and display inherent or installed valve characteristics.

Example 24.1 (Control valve design) *The task is to size and determine the rangeability of a control valve for the following application. The valve is located between a pump (maximum delivery pressure* 450 kPa*, internal resistance* 0.8 $kPa \cdot s^2/kg^2$*) and a vessel (operating pressure* 250 kPa*) with piping and equipment with a combined resistance of* 2.4 $kPa \cdot s^2/kg^2$*. The design operating flowrate is in the range* 3.5 *to* 6.2 kg/s*. The valve requires a linear installed characteristic. Output from the MATLAB Toolbox program* `cvchar.m` *shows:*

```
Maximum flowrate = 7.906
Control Margin = 27 %
DelP at maximum design flowrate = 169.2 ( 45 % )
Cvmax calculated = 1.167
Selected Cvmax ? 1.2
Maximum flowrate = 7.166
```

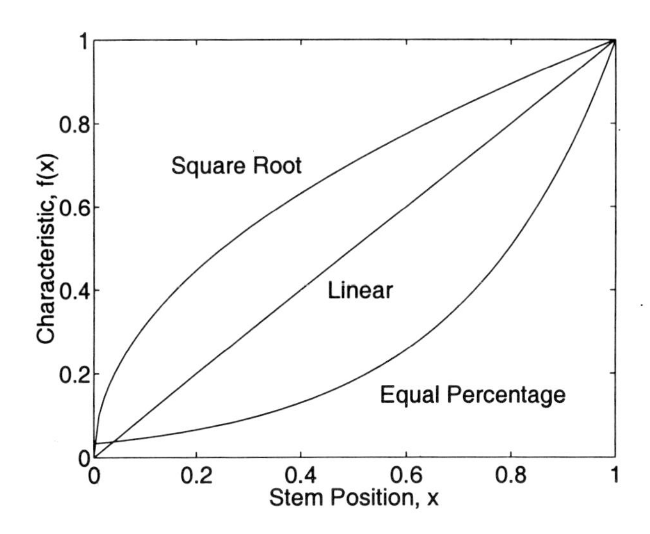

Figure 24.4: Inherent valve characteristics.

The installed and inherent characteristics for an equal percentage valve and a linear valve are shown in Figure 24.5. The installed characteristic on the left in Figure 24.5 is the closest to linear you can get with a normal commercial valve. The rangeability for valve openings of 10 and 90 % can be read from Figure 24.5 as 11 to 92 % of 7.166 corresponding to 0.8 to 6.6 which encompasses the desired operating range. That range of 3.5 to 6.2 is 49 to 87 % of maximum flow, which reads from the characteristic as a valve stem range of 57 to 84 %.

A valve contributes to nonlinearities in a control loop. This means that the controller gain may have to be different at low and at high flowrates. This is avoided by cascaded control which has been discussed in Section 17.6. The feed-

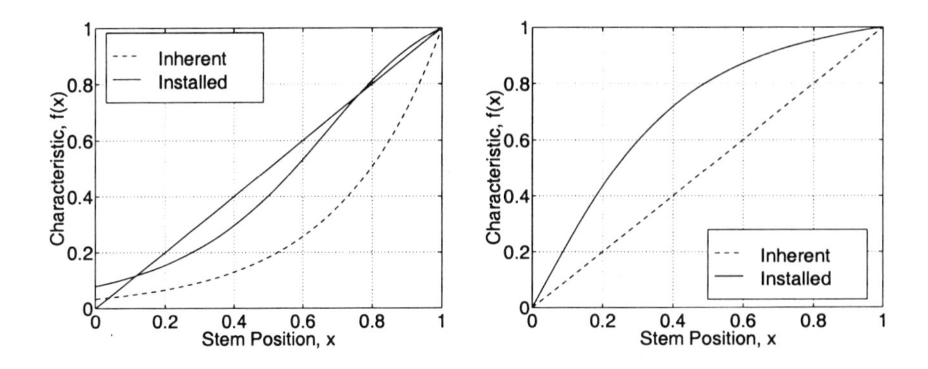

Figure 24.5: Installed characteristics from equal percentage and linear trims.

back around the valve makes it appear almost linear for the outer feedback loop.

24.2 Signal Transmission and Proccssing

Transmission today is almost exclusively electronic using a current in the range of 4 to 20 milliamps in a two-wire point-to-point cable. The *live zero* of 4 milliamps allows cabling problems to be detected. Old instrumentation systems or those in extremely hazardous environments use a pneumatic signal of 3 to 15 *psig* in copper tubing.

Digital transmission systems connecting many sensors to the one cable in a shared bus arrangement have not yet gained a wide acceptance. The recent popular acceptance of *intelligent sensors* and anticipated manufacturer acceptance of a *Fieldbus* standard, see Section 24.5.2, mean that we can expect to see all digital systems in the future.

Since most of the sensor signals are analog in nature we discuss some of the most important problems in transmitting and processing signals between the sensor and the computer.

24.2.1 Bandwidth and Noise

Bandwidth is an important parameter in a variety of contexts such as signal transmission, computer buses and feedback control where it means different things. In signal transmission and feedback control bandwidth is the range of frequencies amplified above a certain threshold. Here bandwidth is defined as the useful frequency range of a sensor or an actuator. This means that the sensor is sufficiently fast to faithfully measure the signal without sensor dynamics corrupting the signal. Although a wastewater treatment (WWT) process is quite slow the bandwidth of certain analysis instruments should be considered such as a COD monitor or a respirometer. Likewise an actuator has to have an adequate bandwidth in order to realise the desired control signal. The larger the bandwidth the faster the sensor or actuator response will be. Unfortunately the larger the bandwidth the more susceptible the device will be to unwanted high frequency disturbances.

Any measurement signal will be corrupted by noise and part of the signal transmission problem is to reduce the influence of the noise. The sources of noise have to be eliminated or minimised. Controllers can be designed to deal with noise in a systematic way as discussed in Sections 10.2.5 and 17.2.3.

24.2.2 Signal Conditioning

Analog signals generated by the measuring device often must be processed in some way before being sampled by the computer. A voltage signal has to be amplified in order to match the sensor voltage range to the computer interface range. Moreover the voltage level has to be shifted to align the minimum sensor output voltage with the minimum voltage for the input interface. This is called **signal conditioning**. The transmission of analog signals presents special problems due to electrical disturbances. A possible solution to noise is the conversion of the signal to a pulse rate or pulse duration proportional to the voltage level. This is useful when noise is influencing the same frequency band as the original signal. Pulse trains can be conducted either electrically or optically by using optical fibre technology.

The analog or pulse signals are converted to numerical (digital) form in the input interface. The computer processes the input data to generate a control signal or other output signal. Usually this digital value has to be converted to an analog form in the computer output interface. The signal level has to provide the right amplitude to drive an actuator.

24.2.3 Signal Grounding

Signal grounding is important in order to manage electrical noise problems and to ensure intrinsic safety. Extreme care is needed where low level signals are used such as those from thermocouples or some analysers. Poor grounding is the most frequent contributor to signal transmission problems.

24.2.4 Electrical Disturbances

Disturbances are generated in many ways, of which the most important are those caused via coupling by:

- resistance (via the conductor),
- capacitance,
- induction (magnetic), and
- radiation.

Resistive (galvanic) coupling via a conductor is independent of the frequency of the disturbance. However in a capacitive or inductive coupling the degree of coupling depends on the frequency. At higher frequencies more energy is transmitted. This means in practice that fast circuits (with fast voltage or current changes) are more serious disturbance sources than slow circuits. Usually there is a combination of different types of electrical couplings. Typically difficult problems arise as soon as signal conductors are located close to power cables.

A lot of research and development is going on to develop a better noise environment for electronic equipment. The goal is to achieve **electromagnetic compatibility (EMC)** within and between electronic and communication systems. A device has to be insensitive to external disturbances and should not generate such disturbances that other equipment will pick up.

Resistive Couplings

Noise currents flow through conductors. Any conductor joining a sensor with its signal conditioning circuitry is a potential collector of electrical noise. For example resistive coupling is obtained when a common power unit delivers voltage to different electronic systems and they in turn are connected to a common ground (see next section). Other common noise sources may be poorly shielded motors and frequency converters with semiconductor switches. One way to avoid the problem is to supply different sensitive electronic devices with separate power supplies. Another way is to galvanically separate different power units and devices.

Capacitive Couplings

Usually there is a capacitance (leaking capacitance) between two conductors or between a noise source and a conductor. Capacitive coupling is characterised by the fact that a variable voltage induces a current i in the conductor that is proportional to the voltage time derivative, $i = C(dv/dt)$ where C is the capacitance. Capacitive coupling should be minimised and it becomes smaller if the conductors (the noise source and the receiver) are kept far from each other.

A good way to decrease capacitive coupling is to supply the measurement conductor with an electrostatic shield to break the disturbance route. The shield has to be grounded so that its potential is zero. Where the measurement conductor to the sensor or to the computer terminates the shield does not surround the complete conductor. As some small parts at the ends are not protected there is a little capacitive coupling. Therefore it is important to make this connection as short as possible.

Inductive (Magnetic) Coupling

An electrical conductor generates a magnetic field around itself with a magnetic flux density dependent on the amplitude of the current. Therefore magnetic coupling is a great nuisance close to power cables. A variable current generates a variable magnetic field and in accordance with the induction theorem the varying magnetic field generates a voltage. This induced current increases with the area that encloses the magnetic flux.

There are several ways to eliminate inductive couplings. The area of the circuit that encloses the magnetic flux can be decreased by twisting the cables. Also conductors can be kept close to each other so that the area between them is minimal. Furthermore the small area *changes sign* at each turn so that the net magnetic flux becomes very small.

The measurement conductor should be located as far away as possible from the disturbance. In particular sensitive electronics need to be placed as far as possible away from transformers and inductors. Cables should be placed so that probable disturbance fields propagate along the cable. Good rules to follow are to avoid power cables and signal wires in the same cable conduit, to maintain some minimum distance between them, and to cross them at right angles.

The magnetic field can be dampened by shielding. A copper or aluminium shield has very high conductivity and due to eddy currents in the shield the magnetic flux is reduced. Shielding can also be realised with high permeability material such as iron. A magnetic shield is often clumsy since it has to be thick in order to dampen the magnetic flux. Therefore shielding is used mostly in devices that produce large magnetic fluxes.

Some Rules of Thumb

We summarise some of the basic rules to eliminate or dampen the influence of electrical noise on the measurement environment. Evidently one should first try to reduce the noise source. Other important factors are:

- galvanic couplings,
- the distance between the noise source and the object, and
- the noise frequency content.

Capacitive couplings can be reduced by:

- using shielded cables, and
- minimising the length of the unshielded wires at the termination panel.

Magnetic couplings are dampened if:

- the cables are twisted, so that the area of the circuit that encloses the magnetic flux is decreased and the orientation is altered,
- individually twisted pair wires are used, one for each sensor,
- power cables and signal wires are not placed together,
- low voltage and high voltage cables are crossed at right angles, and
- low voltage wires are held at some distance (at least half a metre) from interference sources and power cables.

Sampling of Analog Signals

Analogue signals cannot be read continuously into the computer and instead are fetched only at intermittent intervals. Thus a signal is represented in the computer only by a sequence of discrete values. The operation of reading a signal only at determined instances in time is called **sampling**.

The sampling interval has to be sufficiently short so that the continuous signal variations are truthfully described by the discrete time signal. If the sampling time is too long the computer will get a wrong picture of the original signal. This may seem a trivial problem in a wastewater treatment system where a computer may read a signal once every second. However the electrical circuits usually add some high frequency noise to the signal. Even if the sampling rate is sufficiently high to follow the signal related to the process dynamics the additional high-frequency noise will greatly distort the sampled signal.

Example 24.2 (Distortion from AC power cable) *High frequency signals are often overlapping low frequency signals. An example is 50 Hz alternating current that is picked up by signal wires from power cables (Figure 24.6). We demonstrate the* **alias** *problem by a simulated example. A pure sinewave signal of 2 Hz is sampled with $f = 60$ Hz. The disturbed signal is the 2 Hz signal with the 50 Hz sinewave overlapped. The sampling points are shown by the stems. The resulting signal that can be reconstructed from the sampling points (the low frequency signal in (c) and (d)) contains another false frequency that is not present in the initial signal, a 10 Hz signal which is actually the difference 60-50 Hz.*

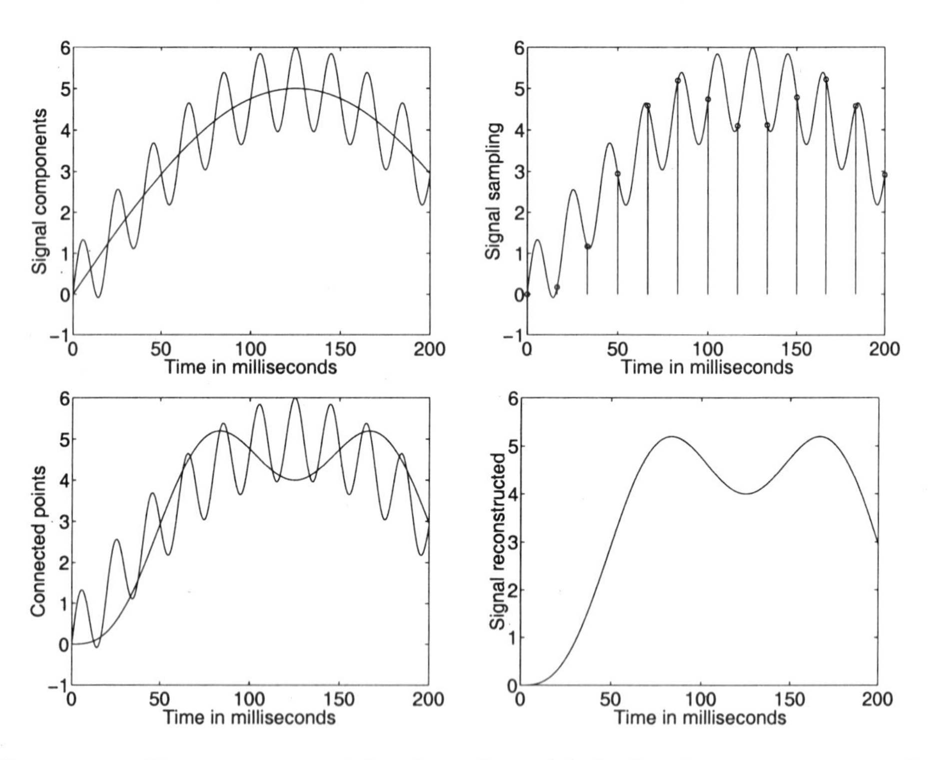

Figure 24.6: Demonstration of the alias effect: (a) the low frequency sinewave (2 Hz) signal to be sampled and the overlapping 50 Hz sinewave, (b) sampling of the high frequency signal with 60 Hz sampling, (c) the signal reconstructed from the sampling points, (d) the reconstructed signal isolated.

The false frequency in the example is called an **alias frequency** *and is simply a result of the sampling rate. In fact it is necessary to sample the signal at least* *twice as fast as the highest frequency component in the signal. In the example this means at least 100 Hz.*

Another example was shown in Example 6.13. It shows that the sampling rate depends not only on the "theoretical" settler dynamics but on other disturbances. Figure 6.13 demonstrates that the sludge scraper adds much higher frequencies than expected. The choice of sampling rate has to take this into consideration.

In order to avoid excessive sampling rates the analog signals have to be filtered. The frequency f_N called the **Nyquist frequency** is defined to be half the sampling frequency. All signal components above the Nyquist frequency have to be removed before sampling. This is done by analog filters called **anti-alias filters**. Note that analog filtering should be used to reduce electric noise only *after* proper shielding and differential amplification has reduced the noise as much as possible.

An anti-alias filter is a low pass filter and is designed to pass frequencies below a specified corner frequency and dampen higher frequencies.

After analog filtering and A/D conversion further filtering possibilities are offered through digital filtering. Digital filtering provides a great deal of flexibility since the filter characteristics can easily be changed by tuning a few parameters in the computer. Digital filters are discussed in Section 10.2.5.

24.3 Basic Control Equipment

The functional view of a control system as a collection of more or less independent control loops made up of the *process*, a *sensor*, a *controller* and an *actuator* typically does not correspond to the actual hardware that makes up the control system and is only part of the total instrumentation system. Figure 24.7 defines what we mean by an instrumentation system again from a functional viewpoint. There are three levels of feedback obvious in this diagram, the automatic controllers, the plant operators and the plant management. As the dashed lines indicate, only the first of these is hardware, the others being humanware with all its frailties in perception, decision making and action (or usually lack of?). It is the job of the engineer to design good hardware (sensors, controllers, actuators, displays and reports) and it is the job of the recruiter and trainer to supply good humanware.

The displays and reports are often overlooked by the engineer. However they are very important in ensuring that the humanware gets the right message and hence has the best chance of making the right decision and enhancing rather than degrading the performance of the system as a whole.

In selecting a hardware instrumentation system what are the important criteria? Typically they include:

- **Suitability**. The equipment must of course do the job required.
- **Availability**. The instrumentation should not cause undue interruptions to production. The importance of this depends upon the process. It is a

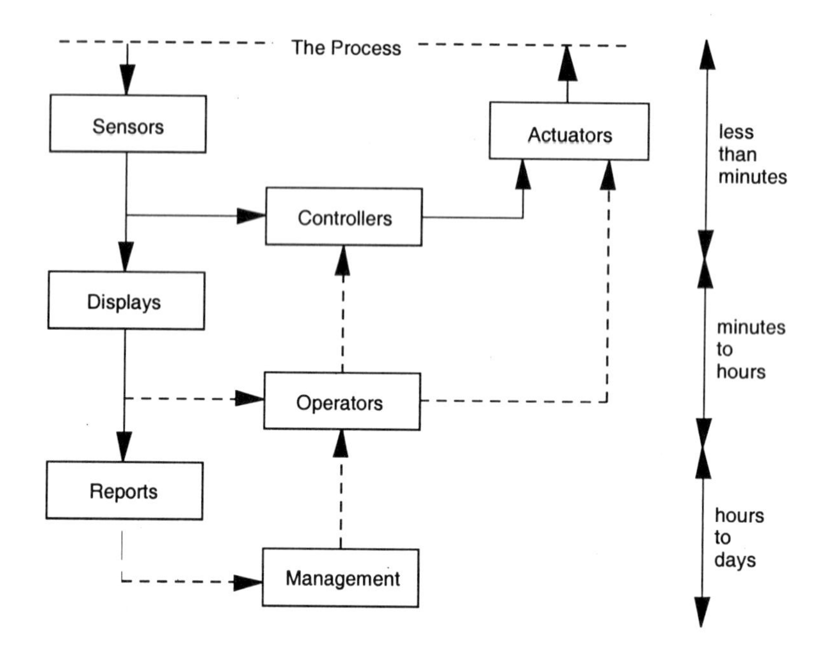

Figure 24.7: An instrumentation system.

key issue for large continuous processing plants but less of an issue for small batch processing plants. This determines the chosen architecture and the degree of redundancy.

- **Reliability**. Even if the availability criterion is met we don't want frequent problems.
- **Ease of use**. This a key issue in achieving the best possible interface between the hardware and the humanware. What are the facilities for simple operator interaction, developing good displays and reporting systems, and interfacing to plant information systems?
- **Ease of maintenance**. Is the equipment accessible? Can faulty parts be replaced on-line? What sort of diagnostics are available? What is the backup in parts and expertise from the manufacturer?
- **Cost**. Again this is the deciding factor all else being equal. Again it is a trade-off of cost against advantages in the above areas since usually things are not equal.

We will examine the architecture of the different classes of commercial products and how that affects the availability and the ease of use. The classes we will discuss are:

- single loop controllers,
- distributed control systems,
- programmable (logic) controllers, and
- personal computer and workstation systems.

They are in approximate historical order with single loop controllers ruling from before World War II until the sixties, distributed control systems (DCS) from the seventies to the present, programmable logic controllers (PLC) from the eighties to the present, and PC or workstation systems from the nineties. The discussion will omit the systems based on mini to mainframe computers, the so-called direct digital control (DDC) systems, which were developed in the sixties and are now largely extinct.

24.3.1 Single Loop Controllers

Sensors and actuators are nearly always separate pieces of hardware (some exceptions in the case of sensors will be discussed later). Controllers can also be completely separate hardware each with its own displays and adjustments for individual loops. You will see them in old process plants, very small systems (on local control panels) or critical high-availability systems.

They are the ultimate from the high availability viewpoint as only one loop at a time is likely to be affected by a hardware failure, assuming separate and secure power supplies. The problem is that the display and adjustment functions as well as the control function are fully distributed. This resulted in the often very large control panels which made the interface with the operator difficult. It also required more operators with the inherent communication and coordination problems.

24.3.2 Distributed Control Systems

The development of the microprocessor led to the rise of the distributed control system or DCS for short. Despite its name the DCS is a compromise between single loop controllers and the now extinct direct digital control (DDC) system. The latter implemented large numbers of control loops on a single computer which centralised the display and reporting function. However they were an availability nightmare especially since computers in the seventies and eighties were not nearly as reliable as they are now. The DCS basically implements the control loops in small groups, each group having its own microprocessor which is often duplicated. The microprocessors are then connected via a data highway to which one or more centralised display stations are connected. The data highway is also frequently duplicated.

This arrangement gives us the best of both worlds. We can have the control functions as separated as we like (a cost vs. availability trade-off) and the display, adjustment and reporting functions centralised for ease of operation. The DCS is most suited to medium (tens of loops) to large (hundreds of loops) continuous processing plants although more recent systems can also be applied to large batch processing plants.

The major problem with the DCS is that there are several major manufacturers (such as Honeywell, Bailey, Yokogawa, Foxboro, ABB, Siemens) in a very competitive market with essentially proprietary systems. These systems will not interconnect and in many cases connect with general purpose computer (information) systems only at considerable expense. The latter problem is at least being addressed in more recent systems but interconnection at the operator level remains virtually impossible.

24.3.3 Programmable (Logic) Controllers

Before the advent of the microprocessor sequences of operations (batch plants and equipment startup and shutdown) were implemented with complex systems of relays as described in Section19.3. These were expensive to build and maintain and changes were a major exercise.

The programmable logic controller (PLC) was developed to address this need and has now virtually eliminated relay systems in processing and manufacturing plants. The PLC consists of a microprocessor, digital inputs for detecting switch positions, digital outputs for activating solenoids or switches, and a simple interface for programming the sequences into the microprocessor. Early interfaces used the ladder logic terminology (intelligible only to instrument personnel) but more recent interfaces use Boolean logic or even higher level representations (GRAFCET). See Chapter 19 for more information in this area.

Some more recent PLCs have also incorporated analog signal interfaces and even the capability to implement feedback and feedforward control loops. One or more of these can be interfaced to a personal computer as an operator console. These PLCs are often used for batch processing plants and small (low tens of loops) continuous processing plants.

24.3.4 Personal Computer and Workstation Systems

More recently a large number of control packages have become available on personal computers (PCs) and UNIX workstations. The connection to plant signals is implemented by interface cards which plug into the computer, by interface cards which plug into separate boxes connected to the computer, or by a PLC connected to the computer. In some cases a dedicated microprocessor or a PLC is used to implement some or all of the control functions leaving the PC for display and higher level control functions.

These systems are used for small batch or continuous processing applications (low tens of loops). They can suffer from reliability problems particularly those which utilise personal computers and operating systems not designed for real-time applications (DOS and Windows for example). The response time in a WWT plant application should not be a problem and an operating system like Windows NT is considered sufficiently fast. Using the techniques of 1999 a typical response time for a PC for real-time control is about 3-500 ms. Since the operating system works with interrupts it also gives a lot of flexibility. Each real-time program can be processed in its own time scale so the real-time demands would not be crucial for a process control application like WWT. On the other hand the reliability is still not up to the requirements in many applications. Who has heard of viruses in PLCs? Unfortunately they are a reality in PCs. However PCs are a relatively low-cost solution at least in the short term.

Recently there has been an obvious development where PLCs are completely replaced by personal computers. The development has been called "soft PLCs". General Motors (USA) has decided to choose personal computers instead of PLCs as much as possible. The reasons are obvious. There are already personal computers available in the plant for operator communication, data storage, and off-line calculations. There are also many objections to the "soft PLC" solution. Since the PC is a standard product it may be much less efficient for some control tasks. Probably open systems like PCs will probably create difficulties when components of different brands are to be connected. On the other hand the economy is a strong driving force. The 1996 market for personal computers is about USD 100 billion and for PLCs only about USD 4 billion. This opens up a huge market for general control software. There are already some control packages available on the market.

Usually the real limitation in a PC based computer control system is the bus capacity. In order to connect the computer to the physical processes you can use an interface card to which each signal has to be connected or you can use some bus to connect the signals. Then there is a remote processor close to the sensor where the primary signal processing takes place. Such fieldbus structures are further discussed in Section 24.5.2. One interpretation of the current development is that the small PLCs will still be available at the sensor locations performing a very specialised task.

24.4 Displays and Reports

While the interface between the control system and the process is important to get right, the really important interface which is also a much more difficult task is the interface between the instrumentation system and the humans operating the plant.

Human perception is both amazingly adaptable and amazingly fickle. People will very often see what they want to see even if quite the opposite is actually the case (politicians for example?). The perception of the operating personnel from the operator to the manager of the state of the plant is no more reliable. At the same time it must be said that the interface designs by instrument companies, consultants and local operating staff generally leave much to be desired.

We will very briefly touch on the design of operator screens and of reports.

We will refer to some literature also for further reading on these important topics. But first there are some general guidelines that should be followed no matter what sort of human interface is involved.

- Always employ the KISS principle in any interface design (in this case we mean "Keep It Short and Simple").
- Keep the signal to noise ratio as high as possible as humans are easily distracted.
- Only tell people what they should know when they should know it but give them a way to look for things they might want.
- Organise information hierarchically with customisable (per person) "hotlists" for fast access to key information.

Information more specific to the type of display is covered below.

24.4.1 Operator Displays

The first requirement for designing good operator displays is to perform a thorough analysis of the process to be controlled, the user of the display (the skills and knowledge level of the operator) and the tasks the operator is to perform. This will enable you to write a specification of the information requirements of the operator.

There are then two aspects of display design, the information content (what to display) and the visual form (how to display it). The information content will be large, certainly large enough for tens to even hundreds of display screens. The information will include:

- the standard measurements from the sensors on the process,
- calculated information based on the measurements,
- actuator signals and status indicators,
- alarms indicating abnormal conditions or equipment failures,
- equipment design data such as pump curves and valve characteristics,
- maintenance information, status and instructions, and
- operating goals and economics.

The huge amount of information required raises the immediate issue of organisation and access. The information should be grouped by operator task (some information may appear in several different groups) arranged hierarchically with overviews then summaries then details and also cross-linked for fast access to related tasks or problems. More modern features such as large display screens and multiple windowing should be exploited. Each operator station is likely to require two screens (more are likely to confuse rather than help).

The visual form or how the information is presented is at least as important as the information itself. There are numerous examples of information that is there not being seen, because of poor presentation. The following guidelines should be considered:

- The *layout* of like screens should be consistent. There should be a small number, no more than five or ten display classes with templates used to enforce consistency of appearance and consistent location of navigating and help buttons.

- Screens should be laid out by *function.* Perform an operator task analysis and then design a screen or two for each task with all the operator needs for that task.

- The use of *attributes* such as colour, shape, sounds, etc. should be used sparingly and consistently. Christmas tree displays confuse rather than help the user. Beware of colour blindness which is rather common.

- *Clarity* is crucial. Everything on every display should be tested to ensure no possibility of ambiguity or interpretation.

- *Simplicity* is a key word. It is better to use three simple clear displays than one cluttered and confusing display.

- *Redundancy* of information is key to the operator being able to validate abnormal behaviour. He must continually answer the question "is it the instrument or the process?".

- *Transparency* is important. The operator needs to see the state of the process, not the state of the control system or the display.

- *Pattern recognition* is much more powerful and faster for discrimination than text and cognitive processes (thinking). It is better for the screen design engineers to think it through once slowly and give the operators patterns to recognise and detect quickly.

- *Predictive capability* is valuable in detecting changes before they become problems. Displays of rates-of-change and directions-of-change are often more powerful than current absolute values. Computer implemented process models with continual look-ahead capability are useful if they are mostly right.

- *Goals* and measures of attainment are important to enable the operator to optimise operation. Don't expect him to maximise yield or minimise operating costs if he doesn't know what they are.

The economic viability of processes frequently lies in how well they are operated and operator displays play an important role in this, as do factors such as operator training, control system tuning, plant and instrument maintenance.

24.4.2 Reports

How often do you marvel when you visit process plants and see the production report being printed off, all fifty five pages of densely packed numbers. What would any management person do with that amount of data. It has been said of the process industries (perhaps the world) that we are "data rich, but information poor". The computer is a fabulous information handling tool and yet we churn out numbers by the thousand, and leave it to the poor unfortunate human recipients to find what they want. The result is predictable. They soon stop looking and manage "by the seat of their pants" just as they did before the *information age* (perhaps more aptly named the *data age*).

Reports must be targeted to the recipient and the general guidelines must be followed. Employ the KISS principle, keep the signal to noise ratio high, and only tell people what they should know when they need it. If your software won't let you do it then buy some software that will. The cost of plant incidents caused because information was not seen is immeasurably greater than the cost of good report generating software, even if you have to buy ten packages before you find the one you like.

What sort of information might the management at a wastewater treatment plant need to know? We will try to give you a checklist:

- effluent quality information: just those in the license preferably on SQC charts plus the running regulatory averages.
- an operating cost summary: a *bar chart* comparing it with the previous ten reports and a *pie chart* giving a summary of the breakdown.
- influent flowrates and compositions: just the key ones presented as a time series plot.
- a summary list of equipment breakdowns: preferably as a *Pareto* chart.
- a summary list of safety incident reports.
- a filtered list of employee suggestions.

That is about five or six pages. Then give them a terminal and a really friendly way of getting more details if they really want them presented any way they choose. Teach them and their secretary how to use it. Read about the charts mentioned above in Chapter 11. Also, read Section 10.5. If you are a manager write and tell us what we missed.

24.5 System Communications

We have examined the instrumentation system components but we must also discuss how they talk to each other. The demand for communication in any process or manufacturing plant is steadily growing. Any user today demands flexible and open communications preferably following some recognised standard. Here we will just mention some of the concepts that are essential for any sizeable instrumentation system.

24.5.1 Architecture

Figure 24.7 gave a functional view of an instrumentation system, and in Section 24.3 we discussed the different types of controllers in common use: single-loop controllers, the DCS, the PLC and general-purpose computers. How do we connect these into an *integrated system*? There are a number of factors we must consider when choosing a system architecture:

- **Accessibility**. Ideally all appropriate devices should be able to access and share data and programs. It is a matter of drawing up an information flow specification for the different types of data, where it originates and where and to whom it should go.
- **Reliability**. In principle point-to-point links are more reliable than networks in the sense that the mode of failure means fewer communications are lost for a given breakdown. This can be alleviated to a certain extent by duplicate networks run over different routes. In practise the reliability of point-to-point links (and duplicate networks) is frequently compromised in any case by shared cables/routes and shared power supplies.
- **Flexibility**. Networks with an open architecture offer the most from the point of view of interconnecting a variety of devices from different manufacturers. Future expansion will occur and will inevitably involve different types of equipment from different suppliers.
- **Equipment constraints**. Certain equipment such as sensors, actuators, single-loop controllers and the DCS will be restricted in the possible choices open to you for interconnection. The PLC and general-purpose computers generally have a wider range of communication options.
- **Cost**. Shared networks instead of point-to-point links save much money in terms of capital cost, installation and maintenance. On the other hand the cost of a network increases steeply with capacity so that careful traffic planning is needed to optimise current and future costs.

Figure 24.8 shows a typical architecture for an industrial instrumentation system. The different parts will be discussed in more detail below.

To overcome the difficulties of having to deal with a large number of incompatible standards the International Organisation for Standardisation (ISO) has defined the **open systems interconnection (OSI)** scheme. OSI itself is not a standard but offers a framework to identify and separate the different conceptual parts of

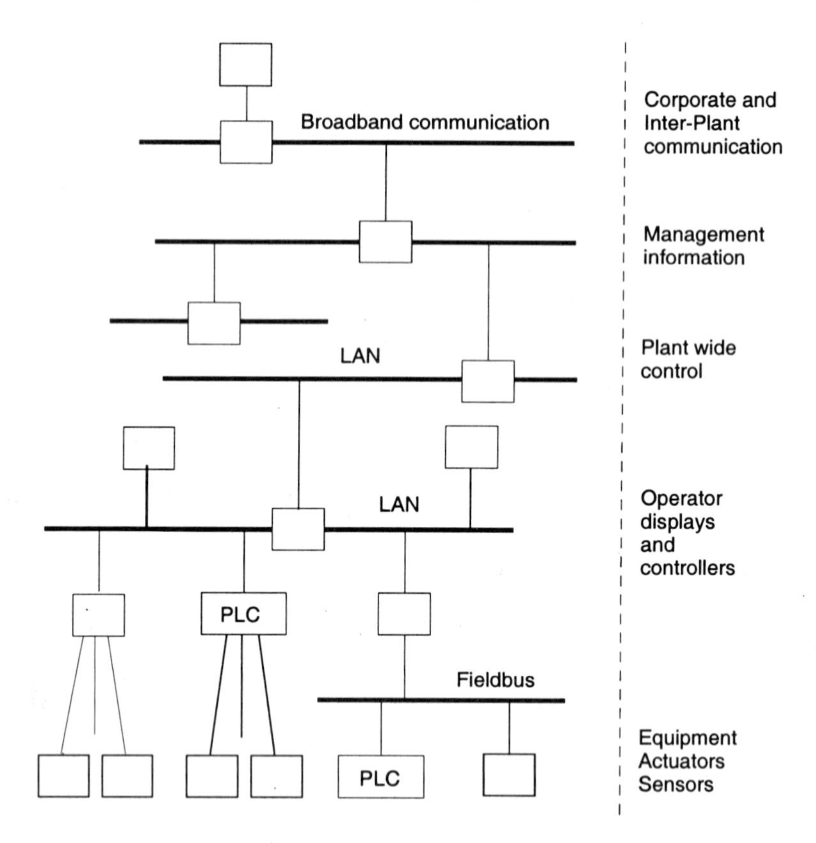

Figure 24.8: Typical plant communications architecture.

the communication process. In practice OSI does not indicate what voltage levels, which transfer speeds or which protocols need to be used to achieve compatibility between systems. It says that there has to be compatibility for voltage levels, speed and protocols as well as for a large number of other factors. The practical goal of OSI is optimal network interconnection in which data can be transferred between different locations without having to waste resources for conversion purposes, with the related delays and errors.

24.5.2 Fieldbus - Communication at the Sensor Level

There is a trend to replace conventional cables from sensors with a single digital connection. Thus a single digital loop can replace a large number of 4-20 mA conductors. This has been implemented not only in manufacturing plants but also in aircraft and automobiles and will become more common in process plants. It is obvious that each sensor needs an interface to the bus and standardisation is necessary. This standard is known as **Fieldbus**. There is no single Fieldbus standard yet but different solutions have been presented by the industry and by research institutions. In the course of time what has been proposed and is operating in the field will crystallise around one or a few technologies that will then become part of a more general Fieldbus standard.

When all communicating units are located in a close workcell and are connected to the same physical bus there is no need for multiple end-to-end transfer checks as if the data were routed along international networks. For the connection of computers in the restricted environment of a factory, the data exchange definition of OSI layers 1 (physical layer) and 2 (data link layer) and an application protocol at the OSI level 7 are more than enough.

The possibilities opened by fieldbuses are notable. A large share of the intelligence required for process control is moving out to the field. The maintenance of sensors becomes much easier because operations like test and calibration can be remotely controlled and require less direct intervention by maintenance personnel. And as we have already pointed out the quality of the collected data influences directly the quality of process control.

The International Electrotechnical Commission (IEC) is working on an international fieldbus standard. The standard should ensure interconnectivity of different devices connected to the same physical medium. National projects have already started in different countries to define how the future standard will look. A final agreement has not been reached yet but nobody wants to wait until a general standard is introduced. Some companies have already defined their products and are marketing them, and projects have been carried out in some countries to define national fieldbus standards. In the end all experiences and proposals may come together into a single and widely accepted standard but it might turn out that the different already existing proposals will live their own parallel lives.

Some examples of fieldbuses are FIP from France and PROFIBUS from Germany as well as the industrial Bitbus developed by Intel. There is a need for low cost complements. Therefore buses like SDS (Smart Distributed System) based on the CAN (Controller area network) bus (developed by Honeywell), ASI (actuator-sensor-interface) and Opus have been developed. They are more restricted than FIP or Profibus and are meant to be a low end alternative (but still compatible upwards) with the more advanced fieldbuses. Many semiconductor manufacturers like Motorola, Intel and NEC make circuits for communication or single chip computers with built-in CAN interfaces. Sensor and controller manufacturers form user groups for the different fieldbus concepts. In 1997 there were far more than 100 different brands of fieldbuses available in the market.

24.5.3 Local Area Networks (LANs)

In order to communicate between different PLC and DCS systems and computers within a plant there is a clear trend to using Ethernet as the medium. Ethernet is a widely used local area network (LAN) for both industrial and office applications. Jointly developed by the companies Xerox, Intel and Digital Equipment,

Ethernet was introduced in the market in 1980. Ethernet follows the IEEE 802.3 specifications.

Ethernet has a bus topology with branch connections. At the physical level Ethernet consists of a screened coaxial cable to which peripherals are connected with "taps" or individual twisted-pair cables to "hubs". Ethernet does not have a network controlling unit and all devices decide independently when to access the medium. Consequently because the line is entirely passive there is no single-failure point on the network. Ethernet supports communication at different speeds as the connected units do not need to decode messages not explicitly directed to them. Maximum data transfer rate is 10 Mbit/s for standard Ethernet or 100 Mbit/s for fast Ethernet.

Ethernet's concept is flexible and open. There is little capital bound in the medium and the medium itself does not have active parts like servers or network control computers which could break down or act as a bottleneck and tie up communication capacity. Some companies offer complete Ethernet-based communication packages which may also implement higher layer services in the OSI hierarchy.

Within DCS systems proprietary synchronous or token ring LANs are common. The justification is based on the need for *guaranteed* data delivery times for real-time data. Such systems usually have a bus master device which often requires duplication for improved reliability. These proprietary networks require bridging devices to connect them to open networks such as Ethernet which can be quite expensive.

24.5.4 Plant Wide Networks

Two comprehensive concepts for information exchange in industrial processes are MAP (Manufacturing Automation Protocol) and TOP (Technical and Office Protocol). They are largely compatible with each other and are oriented to different aspects of industrial processes (production and administration). Both MAP and TOP are resource-intensive products and support the interconnection of a large number of devices in medium-size to large plants.

The **Manufacturing Automation Protocol (MAP)** is not a standard, an interface or a kind of electric cable, but a comprehensive concept to realise interconnectivity between different equipment at plant floor level and to higher level planning and control systems. But the realisation of the conceptually simple principle to have different units communicate together using common protocols has taken about thirty years and is far from being completed. The principal goals of open communication are **interoperability** (all information should be understandable by the addressed units without need for conversion programs) and **interchangeability** (a device replaced with another of a different model or manufacturer should be able to operate without changes in the rest of the connected system).

MAP follows the OSI layering scheme. For every one of the OSI layers there is a defined standard as part of the MAP scheme. The standards at levels 1 to

6 are used also in other applications than MAP. The MAP specific part is the **Manufacturing Message Specification (MMS)**. MMS is a kind of language (a collection of abstract commands) for the remote monitoring and control of industrial equipment. MMS defines the content of monitoring and control messages as well as the actions which should follow, the expected reactions, acknowledging procedures, etc.

A MAP application must have a physical connection which follows the LAN Token Bus standard with Logical Link Control according to IEEE 802.2. It must code data following ASN.1 (ISO 8824) and the Basic Encoding Rules of ISO 8825 and has to exchange MMS Messages (ISO 9506). Any other combination even if it is technically feasible is not consistent with the MAP scheme. For instance a solution where Ethernet is used instead of Token Bus for the data link and physical connection is not a MAP application. However in recent MAP applications Ethernet has been accepted for the connections provided that the network has sufficient transfer capacity so that the real-time requirements can be satisfied.

In process control or factory automation there are generally speaking three operational levels as indicated in Figure 24.8, general management, process control or production line control and field control. MAP supports the central levels of communication. It coordinates the operations of multiple cells on a production line and of several lines at plant level. MAP is not used for communication and control down to the sensor level. MAP is a very "heavy" product because of all the involved layers with the related protocols and does not match the need for simple, fast and cheap technology that is required at the lowest levels. Here the Fieldbus is used. However MAP remains the key concept for the practical realisation of Computer-Integrated Manufacturing (CIM) applications and is likely to have an increasing influence in process control systems.

24.6 Summary

We have given a brief summary of basic process instrumentation, actuators, sensors and controllers, and then a discussion of displays and reports and communications. Rather than a comprehensive list of equipment we have tried to give examples and then give you selection criteria and design guidelines. We hope you find them useful.

24.7 Further Reading

- Standard texts on power electronics and electric drive systems are Mohan *et al.* (1995) and Leonhard (1997).

- For more information on control valves see Hutchison (1976).

- Pessen (1989) describes standard components in automation.

- For more information on instrumentation systems see Considine (1974), Kane (1987) and de Silva (1989).
- For more information on operator displays see Hollnagel *et al.* (1988), Foley *et al.* (1990) and Wickens (1992).
- For signal transmission and general instrumentation problems we also recommend Doebelin (1983).
- Sampling of signals is described in control books like Åström & Wittenmark (1990) and Olsson & Piani (1992).
- There are literally hundreds of different brands of PLC systems. The best way to learn more about them is to consult journals and periodicals on automation. Also the manufacturers have a lot of information to offer. Olsson (1996) presents an overview of PLCs.
- An integrated view of computers for automation is presented in Olsson & Piani (1992) including signal processing and transmission.
- For more information on networking and communications see Kaminski Jr. (1986) for MAP/TOP, Wood (1988) for Fieldbus and Göddertz (1990) for PROFIBUS.
- The standard IEC1138 (1993) defines fieldbus specifications.
- The books by Tanenbaum (1996) and King (1990) provide a lot of information on communications.
- For updated information on fieldbus standardisation you should look at the web address: *www.fieldbus.org*

24.8 Exercises

1. A valve is needed to control the flowrate of a liquid additive to one of the zones in the aeration tank. You need to add 5 litres/minute, you have a pump with a delivery pressure of 100 kPa gauge and an internal resistance of $720 kPa.s^2/kg^2$, you have an orifice plate flowmeter, and the meter and piping will have a pressure drop of 30 kPa at the design flowrate. The control margin should be from 25 to 140 % of the design flowrate. Size a control valve for the application and choose an appropriate characteristic. You can use the MATLAB Toolbox program `cvchar.m`.
2. Consider the compressors for air supply at your plant. What is the average air flowrate compared to the design flowrate? How much energy is throttled away?

3. In a modern process industry it is common to distribute computers. This has many implications such as reliability, cable costs, disturbances. What are the consequences for electric disturbances?

4. Why should signal cables not run parallel to power cables?

5. A signal contains noise from a 50 Hz AC source. The process demands that the signal is sampled only once every 5 seconds. There are two ways to get rid of the 50 Hz noise in the signal. What can be done?

6. Many control tasks are today solved in PLCs. Today there is a tendency to replace PLCs with PCs. What are the advantages of using PLCs in an industrial environment?

7. What are the major advantages of replacing analog signals with fieldbuses? Are there any drawbacks?

8. Select an instrument on your plant, or one you would like on your plant, and see how much of the information listed in the sensor checklists above you can determine. Would you have selected another instrument or located it elsewhere? If you are doing the exercise for a new instrument make a selection and choose a location.

9. Select a small unit from your plant (prefermenter, aeration tank, clarifiers) and design the ideal operator display(s). Analyse operational goals, operator tasks, etc. and use the guidelines above. Compare this with what you have.

10. Repeat the last exercise designing the ideal management report. Compare it with what you get or what you give, depending on where you sit.

Part E

The Future

Chapter 25

The Future

Assuming you have been intelligent and concerned enough about wastewater treatment to have purchased our book, this chapter will give you something to think about.

It will also give you a laugh when you read it again in five years time. But then hindsight is easy and it can sometimes even be useful but speculation about the future is what takes us forward. We are going to speculate for you and even suggest where things should go. Let us know what you think and whether you agree or disagree with our analysis.

Firstly we will discuss the likely future trends of the water and wastewater industry. Then we will do a traditional SWOT analysis (strengths, weaknesses, opportunities, threats) from strategic planning methodology. This will identify what we think are the important issues in the application of a systems approach to wastewater treatment. From this we will suggest some policies organisations can consider to implement their systems approach.

Finally we will paint a picture of urban water systems in twenty five years time and consider what is required to achieve this in terms of R&D and education.

25.1 Society Needs and Industry Trends

Water and wastewater treatment were originally developed to help solve sanitary problems particularly in the large cities. The hygienic standard remains a key driving force and today we are used to a high standard with drinkable water in our clean water system. We believe that in the future the focal points in water and wastewater treatment will be the recipients, the users, and the resource utilization in terms of water, nutrients and energy.

During the last twenty years removal of nutrients from wastewater has been a key issue. Some of the techniques which have been developed to mitigate this problem are criticised for being wasteful with valuable resources. In particular the fate of nitrogen which is lost to the atmosphere in the present system is discussed in this context. Urine contains most of the nitrogen and phosphorus and could possibly be directly utilised in agriculture. As a consequence it is sometimes suggested that techniques for the separation of urine will be required in the future. Furthermore phosphorus is a limited resource so any recycle of phosphorus will be increasingly important. It may be used directly as a fertiliser, in the manufacturing of new detergents, or it may be fractionated into new components.

According to the Brundtland report "Sustainable is development that meets the needs of the present without compromising the ability of future generations to meet their own needs". The general criterion for sustainable development can easily be accepted by most people as it is very vague and unspecified. The problem with this type of general definition is of course that it can be interpreted in many different ways and that it does not give any guidance on how to implement sustainable development in the real world. Sustainable development also requires consensus on certain basic ethical and moral aspects and that basic human needs must be acknowledged.

It is argued that an integrated (holistic) performance index has to be derived for the design and operation of water and wastewater treatment systems. In Chapter 20 we discussed some criteria that would have to be considered in any future system. They include hygienic aspects, environmental impact, resource utilization and user aspects. We looked at resource handling in Chapter 14. A trade-off should be made between the quality of the process outputs and the effort required to achieve this. In the performance index we included not only economic terms but also some sustainability terms. One approach is to translate the sustainability terms into a corresponding land area needed to produce the materials, energy and manpower for treatment in a sustainable way. In all these approaches it is crucial to define where the system limits are. For example to compare decentralised and the more traditional centralised solutions for the urban water system the whole infrastructure of the city has to be considered. A metric to judge the sustainability of different options will facilitate a fruitful dialogue between those involved such as politicians, ecologists, engineers and economists. Only a truly interdisciplinary approach can help to solve the challenges ahead.

This means that wastewater from urban areas in future should be used to improve conditions in existing water bodies. Furthermore in areas with scarce water resources the water needs to be more recycled in the future.

25.2 SWOT Analysis

This involves listing key strengths (S), weaknesses (W), opportunities (O) and threats (T). In this case we do this from an operational perspective with emphasis on instrumentation, control and automation (ICA).

25.2.1 Strengths

- *Environmental demands* ensure that the industry is not under threat and that there is a lively atmosphere for development driven by:
  - ever cleaner effluent demands,
  - a demand for greater consistency (no bypassing during plant failures or high loads), and
  - a looming demand for sustainability (partial or complete recycling).
- *Efficiency demands* are increasing both in terms of capital expenditure and operating costs. This requires:
  - better use of plant volumes,
  - savings in energy (air) and chemicals, and
  - minimal or unattended human supervision.
- *Traditional overdesign* enables current plants (excluding those that are currently overloaded) to:
  - absorb disturbances,
  - operate with low levels of relatively unskilled supervision, and
  - absorb some increases in load.

25.2.2 Weaknesses

- *Industry organisation* which has meant little integration between water supply and distribution, water users, wastewater collection (sewers), wastewater treatment and receiving waters. Even where some of these are within one organisation they have been traditionally compartmentalised with little co-ordination.
- *Design tradition* often does not take ICA into account and certainly not early enough in the design process. This has led to:
  - limited control authority (ability to manipulate the process),
  - little demand for instruments developed for wastewater, and
  - an inability to recognise the advantages of ICA.
- *Process complexity* is increasing rapidly with the demands for nutrient removal and recycling.
- *Sensor technology* now lags behind the needs due to the design tradition mentioned above and the tendency of legislation to specify laboratory methods rather than data accuracies.
- *Skill levels* in systems engineering are relatively low in the industry including:

- management which frequently lacks process and ICA knowledge which leads to a failure to recognise opportunities,
- designers and consultants usually have no skills in dynamics and control and often limited microbiological knowledge,
- engineers are seldom employed for technical support as is common in other process industries, and
- operators frequently lack process knowledge and the skills to achieve efficiency and reliability.

25.2.3 Opportunities

- *New and cheaper sensors* particularly for nutrient measurement are becoming available at an increasing rate with reliable ultrafiltration units for sample preparation.
- *Cheap computing power* is available today far surpassing our current requirements allowing us to contemplate far more sophisticated applications.
- *Universal communications* enable us to gather data from and disseminate information to distant sites enabling a greater degree of integration.
- A *systems approach* to operation is now established in many process industries and there is now sufficient knowledge of wastewater processes to utilise sophisticated techniques.

25.2.4 Threats

- *Instrumentation installation and maintenance* must be done correctly over the lifetime of the hardware and software. Processes and influent characteristics change necessitating periodic updating and tuning of sophisticated systems to maintain the financial return on investment.
- *Skilled manpower* must be available and their skills must be continuously updated to enable ICA systems to be effectively utilised.
- *Management attitudes* towards quality, optimal operation and ICA must be changed especially in the middle ranks.
- *Privatisation* of urban water systems and especially of parts thereof threatens the free exchange of information and the effective integration of the complete system.
- *Political decisions* to change systems without a global examination of the issues and tradeoffs can lead to entrenched inefficiencies.

25.3 Management Policies

What should be done to meet society needs, benefit from the opportunities, build on the strengths, overcome the weaknesses and avoid the threats? Some of the latter are a little out of our league although a little lobbying would probably not go astray. The following is a set of six management policies for an enlightened wastewater treatment operating company.

1. *Recruit a systems group* to educate, plan, monitor and maintain ICA systems within the company. Following sections discuss the duties of a systems group and define the ideal ICA systems for wastewater treatment operations.

2. *Implement a quality ethic* judiciously from the top down. We say "judiciously" because many meticulous bureaucratic quality systems are counter-productive, stifling innovation and local ownership.

3. *Educate on the systems methodology* from the top down with increasing detail sufficient for informed decision making.

4. *Implement advanced ICA systems* on the basis of benefits with user involvement and ownership and with regular systematic performance assessment and maintenance.

5. *Work for integration* of policy, decision making and ICA systems with the interfacing parts of the urban water system (sewers, water supply and receiving waters).

6. *Work for sustainability* by defining criteria and policies with all parties involved in your urban water system.

Several of the policies say "top down". This is for the very good reason that successful implementation of change, particularly if it involves a change of attitudes, will only succeed by example from the top. You might also call it "management committment".

25.4 Systems Group Duties

What is a systems group? What does a systems group do? We heard your questions and here are the answers. Firstly we will define the duties of a systems group in wastewater treatment operations:

- train people at all levels from the board to the operators,
- plan, monitor and evaluate quality systems,
- perform benefit studies for potential ICA systems,
- plan and specify new ICA systems, and
- monitor and evaluate ICA systems.

Secondly we will look at the people required to perform these duties, one or more depending on the group size but at least two process systems people for continuity (people do come and go):

- training persons,
- process systems engineers,
- computer and communications engineers,
- instrumentation engineers, and
- microbiologists.

There can of course be many variations but we believe this defines a pragmatic compromise between paradise and the current situation in most wastewater organisations.

25.5 ICA Systems

What constitutes an ICA system? Contrary to popular belief it is not one large black box with one or two flashing lights in a dark room devoid of human habitation (at least not yet). There are four functional components to the ideal ICA system:

- *A Quality Team* of people who feel a deep sense of ownership of the system and the treatment plant and who are committed to the continuous improvement ethic.

- *An intelligent monitoring system* which:

  - gathers data from the instrumentation system,
  - processes and displays the data for the Quality Team,
  - detects abnormal situations and alerts the Quality Team,
  - assists the Quality Team with diagnosis and advice,
  - simulates the likely effects of plant adjustments, and
  - evaluates and records treatment plant performance.

- *An advanced control system* will be necessary to meet efficiency goals. As treatment plant volumes are more fully utilised it will be important to reject process disturbances using model-based process control both locally within treatment plants and by coordination with connected systems such as the sewers.

- *An instrumentation system* which implements data gathering, performs low-level control and automation, and supplies communications within the plant and with remote plants and information systems.

The first component is humanware, the second and third is software, and the fourth is firmware and hardware - for those of you who talk in "wares".

25.6 Urban Water Systems in 2025

Speaking of "wares", where will the wastewater industry be in about 25 years? Here are our thoughts:

Holistic. Our view is that the wastewater water industry in 2025 will become just a part of a "water industry", an industry concerned with all aspects of the resource we call "water".

Full cost recovery. Users of water will pay what it costs. No hidden subsidies from our taxes. Only then will we all appreciate water as much as those of us who struggle to survive in arid lands.

Zero dry weather discharge. In dry weather the water system will be almost a "closed system". That is complete recycling of treated wastewater, no explicit discharges to receiving waters and only requiring fresh water makeup to account for leaks in the distribution system and garden use (which will be much smaller than in 1999 now there is full cost recovery). Indeed many of our water storages of today will simply collect wet weather treated discharges for future use, wet weather untreated bypassing will not exist.

Normally unmanned plants. The Quality Team will visit maybe once a week spending a half day at the plant, probably when the maintenance team visits. They will spend the time reviewing past operation and deciding on those all important continuous improvements. Monitoring and control will be automatic with adjustments made from the car or from home (assuming we still have cars). Perhaps the laptop and modem already used at a few plants will even be replaced by the home "Internet TV", or maybe a second or third set if you have children, or maybe every room will have a wall screen by 2025.

Periodic preventive maintenance. The maintenance team will also visit once a week for preventive maintenance and repairs. Rarely will they visit at other times. Our systems will become reliable enough with sufficient redundancy to continue to operate despite the inevitable unexpected breakdowns.

Centralised Quality Teams. Our Quality Team will now look after ten or maybe even twenty plants.

Centralised systems group. The systems group will probably be even more centralised, probably just one group for the complete urban water system or even responsible for several complete systems.

One manager. I guess we still need someone to answer the E-mail (complaints will be rare though), to make sure our teams get their pay and to take around the school groups. The rest should be done by the information system, even all that stuff the government will want.

Optimistic? Crazy? Progress will only occur if some of us dare to dream.

25.7 Research and Development

While there are real tangible and intangible gains from embracing a systems engineering approach to wastewater treatment, there are also many aspects where our knowledge and technology will limit what can be achieved. We have listed some of these below.

Detection. We saw in Chapter 14 that we needed effective control for disturbance rejection in order to maximise the use of volumes. In order to get effective control we need early detection of disturbances and probably integration with sewers and pumping stations. In Part B we also saw that detection is a major component in the early diagnosis of process problems. Applied R&D into *sensors*, *soft sensors* and *signal processing* is required.

Diagnosis. While it is easy to enunciate the strategies for diagnosis, as we did in Chapters 10 and 13, finding and implementing effective algorithms requires a considerable effort. Pure and applied R&D into *diagnosis algorithms* is required.

Soft sensors. We have seen that sensors are the eyes of ICA and that even the newer analysers are relatively complex and hence costly to install and maintain (Chapter 23). *Model based estimators* which reliably predict costly measurements from related less costly measurements are needed (Chapter 12).

Biomass measurements. Perhaps the most urgent sensor need of all are measurements of the quantity and effectiveness of the major functional classes of organisms such as nitrifiers, denitrifiers, and phosphorus accumulating organisms. Cooperation between microbiologists and instrumentation developers and systems engineers will be essential to achieve this.

Models. Once past the main bioreactors the sophistication of existing models falls off rapidly. The use of simple empirical models which frequently extrapolate poorly is dangerous for design and optimisation where we are deliberately pushing the limits. Better settling and digester models and models for clarification, prefermenters, fixed-film processes and sludge handling processes are required.

Simulators. You could say that there are more simulators than needed available for wastewater treatment. However those available have some serious limitations (Chapter 9). Lack of *transparency*, knowing exactly your model equations and limitations, is a serious flaw of most simulators. You simply don't know if you can trust the answers and you have no way of finding out (short of building the plant). Some of the simulators do not allow *user models* which limits their usefulness. More complex plants where the returns from simulation are the greatest frequently have equipment not available in the simulator model libraries.

Model management. We have seen throughout the book how many different models of different complexities are needed for sophisticated ICA. Managing model changes, updates and calibrations has become a serious issue throughout the process industries. There are no effective software tools or tested strategies available.

Model calibration. Our more effective models are not easily identifiable and there are no established protocols (Chapter 7).

Process knowledge. There remain serious gaps in our understanding of the processes that we use (Chapter 4). Among those we need to know more about are:

- biomass population development,
- flocculation, settling and clarification,
- phosphorus removal,
- simultaneous nitrification and denitrification,
- hydrolysis and fermentation, and
- fixed-film life cycles.

Process design. There are a number of issues facing consultants in designing treatment plants of the future. These include the following:

- The tradeoff between excess volumes and more control authority in manipulated variables is an issue that must be faced in the design of treatment plants. This will require dynamic simulation during design.
- On-line flexibility and optimisation of the process flowsheet will become a necessity and will likely include more use of batch processing to achieve affordable flexibility. This is already the trend in other process industries.
- Process intensification through the use of fixed-film processes is already promoted by a few suppliers and is likely to become a mainstream option.
- Processes for water recycling and for the extraction of byproducts and impurities (such as heavy metals) will become more common.

Most of these directions require considerable R&D. Current wastewater design consultants also face big changes in paradigm or they may be overtaken by the larger and more general design and construction companies of the mainstream process industries.

It is clear that so much remains to be done yet the next few decades are likely to be times of diminishing funding for R&D. The industry may have some catching up to do at present, but unless they *fund research now*, they will face a brick wall in further reducing costs much sooner than they appreciate.

25.8 Systems Education

It should be clear by now that an increasingly sophisticated wastewater industry will begin competing with the mainstream process industries for process engineers and particularly process systems engineers. The latter are in particularly short supply and intense competition for people will develop not only between wastewater and mainstream process industries but also between treatment plants, consulting companies, R&D organisations and educational institutions. How might the current and looming skills shortage be alleviated? Some options are:

- university courses for new graduates (Systems concepts in environmental engineering courses? Wastewater courses in chemical engineering courses?),
- distance education for current graduates (by universities? by industry groups (VAV in Sweden)? by professional institutions (IWA)? by commercial companies?), and
- dare we say "books like this"?

Internet dissemination for both formal distance education and of educational material such as books and manuals will become commonplace. In fact we hope that a number of courses based on this book will become available over the Internet, for a fee of course.

25.9 Summary

There were a few purposes for this chapter:

- to provide our ideas on where wastewater treatment is headed, towards a holistic and integrated "water industry", towards sustainability, and towards a systems-oriented future,
- to provide you with some thought-provoking ideas on how we might progress towards this future and on what is needed to get there, and
- to be a sort of closing summary of our book.

From now we hope that *you* will be included in the "our" of "our book" by giving us your comments and ideas. You can contact us at any time:

Gustaf Olsson
Email: Gustaf.Olsson@iea.lth.se
Lund, Sweden
Bob Newell
Email: newell@gil.com.au
Queensland, Australia

Please do!

25.10 Further Reading

- For a good survey of the critical variables for the success of an ICA project see Winter & Manning (1993). Successful enterprise wide ICA examples can be found in Sweeney & Manning (1993) and Parker *et al.* (1993).
- Swedish experiences with unmanned operation can be found in Lumley *et al.* (1993).
- To get further views on the issue of sustainability in water systems see Henze *et al.* (1997).

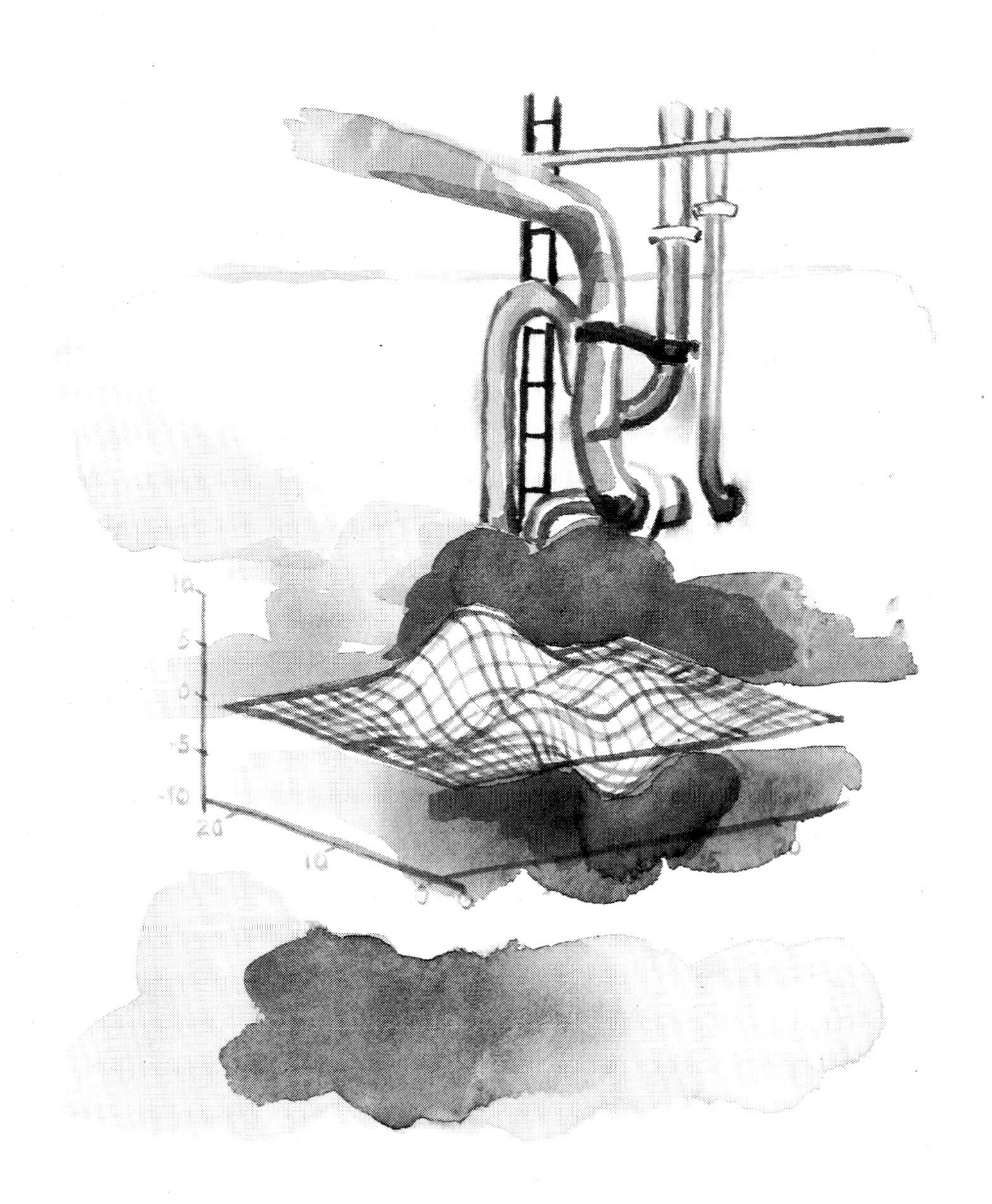

Part F

Appendices

Appendix A

MATLAB Support Software

A.1 Software Summary

Engineering today whether in the design offices or in the research labs or on the plant makes extensive use of computer software. It was impossible to write a book on modelling, simulation and control without involving software. The primary purpose of the software supplied with this book is to enhance your understanding of the concepts presented. Much of the software could also be used in your work with wastewater treatment and it has been written with that in mind.

Why MATLAB? Well for one, it is a great calculation and data visualisation tool. It also has an enormous following in both education and in the industry. There is a huge pool of public-domain programs and functions accessible on the Internet. It is available for many computing platforms with both full commercial and student versions available. Finally it is easily understood whatever basic programming language you were taught.

We have deliberately written the software so it will work on both full and student versions and without the need to have any of the commercial toolboxes. Where we borrowed other software we did so from public domain sources.

Of course there are bugs and maybe just plain mistakes in the software, it would be dishonest to pretend otherwise. All care has been taken but no guarantees are given. You may use it, hack it and adapt it as you will at your own risk. If you publish work where you use it give us an acknowledgement to help spread the information throughout the industry.

The software is available on the Web (where else?) at either:
http://www.iea.lth.se/
or:
http://daisy.cheque.uq.edu.au/
There are three major categories:

- the Exercise, Example and Figures collection of programs and simulations,

- the RTD simulator for residence time studies, and
- the WWT and Graphing Toolbox collections of general-purpose tools.

More details on the second two categories can be found in the following sections. You may find it convenient to add the Toolbox directories to your MATLABPATH.

A.2 The RTD Simulator

A.2.1 Introduction

The purpose of this simulator is to help illustrate the principles of fitting residence time distributions of real vessels with combinations of *ideal vessels*. It also enables you to fit tailored combinations to real plant data.

The simulator will calculate simple tracer responses to an impulse, pulse or step in tracer concentration at the inlet. Most flowsheets can be configured from a combination of instances of five types of object:

- a tracer-injection,
- a well-mixed-tank,
- a plug-flow-tank (with adjustable dispersion),
- a mixing-tee, and
- a sink.

Each object can direct up to four flows to other objects allowing for flowsheets with parallel objects, bypasses and recycles.

The output concentration of each object is plotted and the plot can be tailored by adjusting a scaling attribute for each object (a zero scaling removes the object from the plot).

The flowsheet definition can be loaded from and/or saved to a file with an `.rtd` extension. The simulation results can be saved to a file with a `.dat` extension. They are saved in ASCII format for compatibility with other software.

A compare function allows plotting against a file of external data with a `.dat` extension (time then tracer concentration data in two columns).

A.2.2 Object Definition File Format

One row of 12 numbers for each object. The first object is always a *tracer injection* object which cannot be deleted. The object number is the row number. The columns represent:

Row	Attribute
1	object type 1=tracer, 2=well-mixed, 3=plug-flow, 4=mixing-tee, 5=sink
2	first object-type dependent attribute (see below)
3	second object-type dependent attribute (see below)
4	outlet 1 destination object number (0=unused)
5	outlet 1 flowrate (Megalitres per day)
6 & 7	ditto for outlet 2
8 & 9	ditto for outlet 3
10 & 11	ditto for outlet 4
12	plot scaling factor (0=not plotted)

The two object-type dependent attributes are defined as follows:

Type	Attribute 1	Attribute 2
1	injection type 0=impulse 1=pulse 2=step (height)	tracer amount
2	volume (Megalitres)	(not used)
3	volume (Megalitres)	dispersion factor (see below)
4 & 5	(not used)	(not used)

The dispersion factor controls the number of well-mixed compartments used to simulate the plug-flow vessel. The dispersion factor can take a value between zero (50 compartments) and one (10 compartments) and a linear interpolation is used to determine the number of compartments for intermediate values.

There must be a mass balance for object types 2 to 4 before calculations will be made, that is the sum of all inflows must equal the sum of all outflows.

The file is ASCII format with a `.rtd` extension automatically added to the user specified name.

A.2.3 Results File Format

The results are saved as a matrix with 300 rows and N+1 columns where N is the number of objects.

- Column one is time in days (currently 0 to 3 days in steps of 0.01 days).
- Columns 2 to N+1 are outlet tracer concentrations for objects 1,2,..,N respectively.

The file is ASCII format with a `.dat` extension automatically added to the user specified name.

A.2.4 External Data File Format

External data can be compared to simulated tracer responses using the compare function.

The external data file is expected to contain a full matrix with N+1 columns of M rows where column one is the time in days (between 1 and 3) followed by columns for N concentration curves. M is the number of time points.

The points are plotted as asterisks (*) and if M is greater than about 30 the plot is not visually pleasing.

The file should be in ASCII format with a `.dat` extension which will be automatically added to the user specified name.

A.3 The WWT Toolbox

This is a collection of general-purpose MATLAB functions used throughout the book. In general they are not specific to wastewater treatment and could be just as useful to a process engineer in an oil refinery or a copper refinery.

The toolbox contains the following functions:

annnet	Evaluate ANN neural network
betacf	Used with "tdist[inv]" and "fdist[inv]"
betai	Used with "tdist[inv]" and "fdist[inv]"
costcon	Operating Cost against a Constraint
costopt	Operating Cost about an Optimum
cvchar	Control valve characteristics plus
datarec	Linear Data Reconciliation
detrend	Remove trend from a time series
dfdp	Default gradient calculator for "leasqr"
dostr	DO self-tuning regulator simulation
ellclust	Ellipsoidal clustering
euler	Euler integration with similiar call to "ode45"
fcmalcut	Fuzzy C-means membership grid
fcmeans	Fuzzy C-means clustering
fdist	Probability for given F-value
fdistaux	Auxiliary function for "fdist"
fdistinv	F-value for given probability
fillin	Replace NaN's in a time series
filt	Filter a time series
fitdist	Simple fit of influent time series
fitdistf	Auxiliary function for "fitdist"
fitmodel	Simple fit of time series input-output model
fitnoise	Simple fit of time series disturbance model
fitrbfn	Fit a radial basis function network

fitstep	Simple fit of time constants to step response
fitstepf	Auxiliary function for "fitstep"
fitwavnt	Fit a wavelet function network
frspec	Frequency Spectrum of Signal
fsolve	Solve set of nonlinear algebraic equations
fsolve2	Auxiliary function for "fsolve"
gammaln	Auxiliary function for "tdist" and "fdist"
gradsrch	Simple steepest descent search
imctune	IMC tuning for PID controllers
kmeans	K-means clustering
leasqr	General nonlinear least squares regression
levelfit	Fit one wavenet level, called by "fitwavnt"
linesrch	Simple axes-in-turn search
linreg	2D Linear Regression
missing	Fix Missing Points or Uneven Spacing
ne*	Twelve auxiliary functions for "fsolve"
nlreg	2D Nonlinear Regression
nntrain	Train an ANN neural network
noise	Generate coloured noise from a model
normtest	Tests if data set normally distributed
outlier	Detection and Removal of Outliers
pctiles	Plot and print percentile points of time series
plotprbs	Plot a PRBS Signal
prbs	Pseudo-Random Binary Sequence Generator
rbfnet	Evaluate a radial basis function network
resdist	Plot Residuals Distribution
rls	Recursive least-squares algorithm
search	Simple linear search routine
season	Correlogram for Seasonality Detection
shcorr	Shifted correlation between two time series
shcorrfn	Auxiliary function for "shcorr"
simplex	Simplex search routine
srchstat	Auxiliary function for search functions
tdist	Probability for given t-value
tdistaux	Auxiliary function for "tdist"
tdistinv	t-value for given probability
trend	Detection and estimation of trends
usopid	Simulate PID control of underdamped 2nd order
values	Extracts all values which are not NaN's
wavefunc	Generate dyadic wavelet (called by wavelet functions)
wavenet	Evaluate a wavelet network
wwpca	NIPALS principal components algorithm

Type "help" followed by the function name for information on how to use it. If you find it useful, install it as a subdirectory in the MATLAB toolbox directory and add it to the MATLAB path (consult you MATLAB manuals for details).

A.4 The Graphing Toolbox

This is basically the Styled Text Toolbox from the MATLAB Web Site with some modifications for MATLAB Version 5 and a couple of additional functions.

Appendix B

Simple Carbon Removal Model

We will consider a simple carbon removal process with a perfect clarifier shown in Figure B.1.

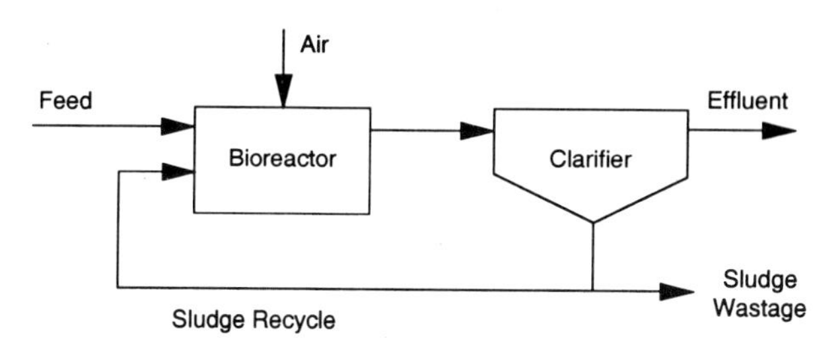

Figure B.1: Simple carbon removal process.

The variables and their standard values are defined in Table B.1.
The model will be developed with the following assumptions:

- No microorganisms in the effluent.
- No dynamics in the clarifier.
- Same substrate in the effluent, sludge wastage and sludge recycle.
- No oxygen in the sludge wastage and sludge recycle.
- A single constant volume perfectly mixed aeration tank.
- Exponential relationship between air flowrate and oxygen mass transfer parameter.

Symbol	Description	Value	Units
q_F	influent flowrate	50	Ml/d
S_{SF}	influent substrate concentration	150	mg/l
S_{OF}	influent oxygen concentration	1.0	mg/l
X_{HF}	influent heterotrophic concentration	5	mg/l
q_A	air flowrate	6.3	m^3/min
q_R	recycle flowrate	50	Ml/d
q_W	wastage flowrate	1.615	Ml/d
V	aeration tank volume	12.5	Ml
S_S	aeration tank substrate concentration	4.4	mg/l
S_O	aeration tank oxygen concentration	3.0	mg/l
X_H	aeration tank heterotrophic concentration	850	mg/l
$\hat{\mu}_H$	maximum heterotrophic growth rate	3.733	1/d
K_S	substrate saturation coefficient	20	mg/l
K_{OH}	oxygen saturation coefficient	0.2	mg/l
Y_H	yield coefficient	0.6	-
b_H	heterotrophic decay rate	0.4	1/d
f_P	fraction inerts on decay	0.1	-
a	K_La value at infinite air flowrate	166	1/d
b	K_La exponent coefficient	16	m^3/min
$S_{O,sat}$	saturated oxygen concentration	10	mg/l

Table B.1: Model variables and their values.

A perfect settler without dynamics implies that no organisms are lost. The organism mass balance of the settler is then:

$$(q_F + q_R)\, X_H = (q_R + q_W)\, X_{HR} \tag{B.1}$$

where X_{HR} is the underflow organism concentration. So we have:

$$X_{HR} = \left(\frac{q_F + q_R}{q_R + q_W}\right) X_H \tag{B.2}$$

The organism mass balance for the aerator can be written,

$$\frac{d}{dt}(V\, X_H) = q_F X_{HF} + q_R X_{HR} - (q_F + q_R) X_H + r_H V \tag{B.3}$$

The mass balances for carbon substrate, dissolved oxygen and heterotrophic organisms take the following form (based on Equations B.2 and B.3 and Equations 3.33 to 3.37 from Chapter 3):

$$\begin{aligned}\frac{dS_S}{dt} &= \frac{q_F}{V}(S_{SF} - S_S) \\ &\quad - \frac{\hat{\mu}_H}{Y_H}\left(\frac{S_S}{K_S+S_S}\right)\left(\frac{S_O}{K_{OH}+S_O}\right) X_H + (1 - f_P) b_H X_H\end{aligned} \tag{B.4}$$

$$\frac{dS_O}{dt} = \frac{q_F}{V} S_{OF} - \frac{q_F+q_R}{V} S_O + \frac{Y_H-1}{Y_H} \hat{\mu_H} \left(\frac{S_S}{K_S+S_S}\right) \left(\frac{S_O}{K_{OH}+S_O}\right) X_H$$
$$+a\left(1 - e^{-\frac{q_A}{b}}\right)(S_{O,sat} - S_O) \tag{B.5}$$

$$\frac{dX_H}{dt} = \frac{q_F}{V} X_{HF} - \frac{q_W}{V} \left(\frac{q_F+q_R}{q_W+q_R}\right) X_H$$
$$-\frac{\hat{\mu_H}}{Y_H} \left(\frac{S_S}{K_S+S_S}\right) \left(\frac{S_O}{K_{OH}+S_O}\right) X_H + (1 - f_P) b_H X_H \tag{B.6}$$

A typical variable classification might be as follows:

- disturbances: q_F, S_{SF}, S_{OF}, X_{HF}.
- manipulated variables: q_R, q_W, q_A.
- states: S_S, S_O, X_H.
- outputs: S_S, S_O.
- known parameters: V, $S_{O,sat}$
- unknown parameters: $\hat{\mu}_H$, K_S, K_{OH}, Y_H, b_H, f_P, a, b.

The plots in Figure B.2 show responses to an increase in influent substrate. The different time scales of nutrient and biomass responses are obvious. The responses were generated by the MATLAB program `Example_B_1.m`.

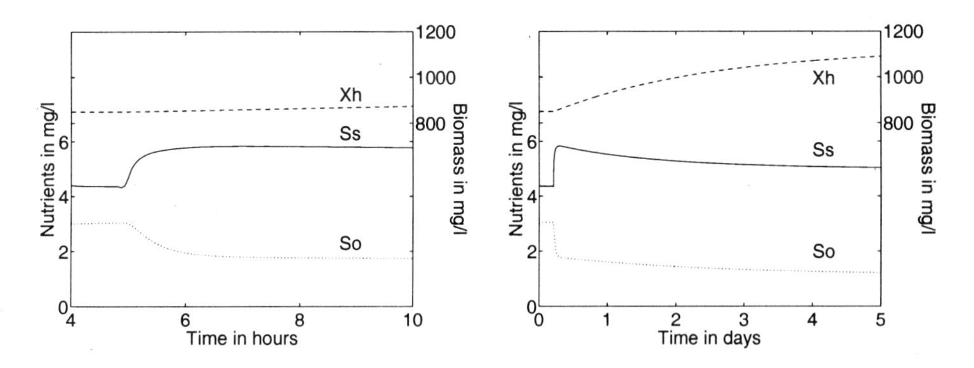

Figure B.2: Responses to an increase in influent substrate.

The MATLAB program `Example_B_2.m` was used to calculate the dynamic and steady state sensitivities of the model to the unknown parameters and the results are shown below.

Dynamic sensitivities for states S_S, S_O, X_H to parameters $\hat{\mu}_H$, K_S, K_{OH}, Y_H, b_H, f_P, a, b are:

1.0000	-0.8940	-0.3576	0.5364
2.0000	0.7328	0.2931	-0.4397
3.0000	0.0559	0.0224	-0.0335
4.0000	0.8940	0.8940	0
5.0000	0.3060	0	-0.3400
6.0000	-0.0340	0	0
7.0000	0	0.3782	0
8.0000	0	-0.3086	0

Steady state sensitivities for states S_S, S_O, X_H to parameters $\hat{\mu}_H$, K_S, K_{OH}, Y_H, b_H, f_P, a, b are:

1.0000	0.1211	0.0067	0.0034
2.0000	0.0995	0.0055	0.0028
3.0000	0.0075	0.0004	0.0002
4.0000	0.0076	0.1885	0.1445
5.0000	0.0782	0.0331	0.0446
6.0000	0.0011	0.0107	0.0055
7.0000	0.0149	0.1983	0.0004
8.0000	0.0121	0.1618	0.0003

These have been ranked in Chapter 3 in Tables 3.6 and 3.7 by degree of sensitivity.

Appendix C

Introduction to MATLAB

MATLAB is a versatile interactive or programmed computing environment ideally suited to engineering calculations, data analysis and data visualisation. Many engineers around the world are now taught MATLAB as their only or primary computing language.

The special power of MATLAB is in its Toolboxes, collections of specialist functions for performing a wide variety of calculations. The toolboxes include Statistics, Signal Processing, Control, Neural Networks and Chemometrics to name but a few. There are also many powerful user-contributed programs, functions and toolboxes.

The following is a mini-tutorial which we believe is sufficient to enable you to use and enjoy the Exercise and Example Programs and the WWT Toolbox supplied with this book.

Information on MATLAB and its Toolboxes is available on the World Wide Web at the address: http://www.mathworks.com/

C.1 Starting and Stopping MATLAB

- The MATLAB command window can be opened by clicking on the appropriate icon or typing the command `matlab`. Which works best depends upon the computer and operating system you are using, the version of MATLAB installed and your own preference.
- To abort a MATLAB program which is taking longer than you wish, press `control-C`, that is hold down the *ctrl* key and press the letter *C* key.
- To remove all the open plot windows, type `close all` in the command window, and to remove all data from your workspace, type `clear all` in the command window.
- To terminate MATLAB type `quit` in the command window.

C.2 MATLAB Data

Data is stored in a *workspace* in the form of real double-precision numbers and is addressed by a *variable name.* A variable name can be from one to about sixteen letters or numbers and can include the underline character, but it must start with a letter and letters are case sensitive, that is variable "a" is different to variable "A".

A variable can represent a *scalar* (a single number), a *vector* (a row or column of numbers), or a *matrix* (N rows of M numbers) as shown below:

$$a\_scalar = 3.45 \tag{C.1}$$

$$a\_vector = \begin{pmatrix} 1 & 0 & -8 \end{pmatrix} \quad \text{or} \quad \begin{pmatrix} 1e5 \\ -32 \end{pmatrix} \tag{C.2}$$

$$a\_matrix = \begin{pmatrix} 1 & -6 & 3.4 \\ 6 & 8.2 & -12 \end{pmatrix} \tag{C.3}$$

The number "1e5" means "1" times "10" to the power "5". Variables are not explicitly declared but take their data type from the context in which they are used.

A list of all variables stored in the workspace can be obtained with the command who or whos. A variable and its data can be removed from your workspace by using the command clear <variable>. You can clear all variables with the command clear all.

The variables and data in your workspace can be stored in a disk file with the command save <name>. The disk file will have the name "¡name¿.mat". The workspace can be restored from the disk file with the command load <name>.

C.3 MATLAB Files

In order to access disk files, you must tell MATLAB which *directory* contains the files you wish to use. To do this use the command chdir <path> or cd <path> for short. If <path> is omitted, the command will print the path name of the current directory.

MATLAB programs or functions are contained in so-called *m-files* which have the suffix or extension ".m". These files can be examined or modified with the system editor, usually accessible through a *File-Open* option in the command window menu bar, or sometimes by using the command edit <name>.m. They can then be executed from the command window by simply typing their <name>. Functions may also require parameters. Most programs and functions are written with helpful comments at the beginning which are displayed if you type the command help <name>. You can of course create your own m-files in one of your own directories (use chdir to choose the directory first).

MATLAB *data files* can be in MATLAB's own binary format, and will have the extension ".mat", or they can be plain ASCII text files, with any chosen extension. Again they must be in the current directory or somewhere in the MATLAB path so that they can be found. The `save` and `load` commands are those most commonly used with data files.

MATLAB *toolboxes* are simply collections of programs and functions under a particular theme. They are generally stored in their own subdirectory in the `/matlab/toolbox` directory (or its equivalent on your computer). Each toolbox subdirectory must also be in the MATLAB path so that the contents can be found. The `path` command can be used to manipulate this path, though it is usually set up by the software installing the toolboxes.

C.4 Entering Data

Data can be entered simply with an assignment statement within a program or function or interactively by typing it as a command. Example assignment statements are:

```
a = 3.2
b = [ 2 1 3 4 ];
c = [ 5 1 -6 ; 3 2 -1 ];
```

The first command will echo on the screen:

```
a =
        3.2
```

A semicolon at the end of the line, seen in the second and third assignments, suppresses this echo. This is general for all MATLAB *statements* and *commands.*

Vectors and arrays are entered row-wise, with a semicolon to indicate the end of a row. The second assignment creates a row vector, and the third assignment creates a matrix with two rows of three columns. The function `size` gives the row and column count for a variable, for example `size(c)` will echo:

```
ans =
     2    3
```

The command `[ n, m ] = size(c);` will set `n` to the row count and `m` to the column count.

Strings of characters must be placed within *single quotes* and are simply stored in a column vector, for example:

```
>> a = 'hello';
>> size(a)
ans =
     1    5
```

The ">>" is the normal interactive prompt displayed by MATLAB in the command window. The variable **ans** is the default result of a command when an assignment is not entered.

Within a *program* or *function*, data can be interactively requested from the user with the `input` statement. Data is typed just as it was on the right-hand sides of the assignments above. For example:

```
>> a=input('enter a column vector > ')
enter a column vector > [1;2;3]
a =
     1
     2
     3
```

The string argument in the `input` statement is echoed as the interactive prompt.

C.5 Displaying Data

As implied in the examples in the previous section, data can be simply displayed by typing the variable name:

```
>> a
a =
     1
     2
     3
```

Alternatively, the function disp can be used:

```
>> disp(a)
     1
     2
     3
```

The results of an interactive session, both what you typed and what MATLAB answered, can be stored in a disk file for later reference or for printing using the `diary` command.

```
>> diary info.txt
>> a
a =
     1
     2
     3
>> diary
```

The first diary command initiated the recording into file "info.txt", and the second diary command with no argument terminated the recording.

MATLAB is particularly convenient for graphical data visualisation. Suppose you have a plain ASCII text file called "data.txt" containing two columns of numbers. These can be displayed and plotted as follows:

```
>> load data.txt
>> data
data =
     5    32
     6    37
     7    45
     8    57
     9    62
    10    81
>> plot(data(:,1),data(:,2))
```

The following plot will appear in a plot window (Figure C.1).

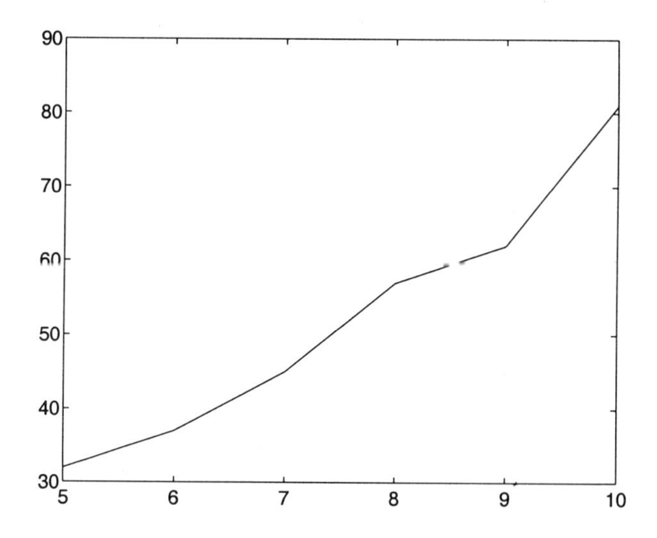

Figure C.1: Example plot.

The plot command supports other arguments for plotting multiple curves and controlling the appearance of the curves (see help plot). Commands such as title, xlabel, ylabel, grid, legend, and text can be used to add the frills, and the command axis can be used to control the scales.

Plot windows can be closed in the normal way for individual windows or by close all in the command window. *Beware of the* Quit MATLAB *option in individual window menus with the Windows version.* It does just that. The *Close* option is what you use to close just the plot window.

There are many other plotting commands including a number for three dimensional surface and contour plots (see `help` on `plotxy`, `plotxyz`, `graphics`, and `color`).

C.6 Exercises

1. Enter the following matrix, take its inverse (see `help inv`), and show that a matrix multiplied by its inverse is the identity matrix (ones on the diagonal, otherwise zeros).

$$\begin{pmatrix} 1 & 5 & -3 \\ 4 & 1 & 7 \\ -2 & 4 & -6 \end{pmatrix} \tag{C.4}$$

2. Plot the following data so that it looks like Figure C.2.

Time	5	12	16	21	25	33	36	41	44	50
Fall	11	21	33	38	51	62	68	78	92	105

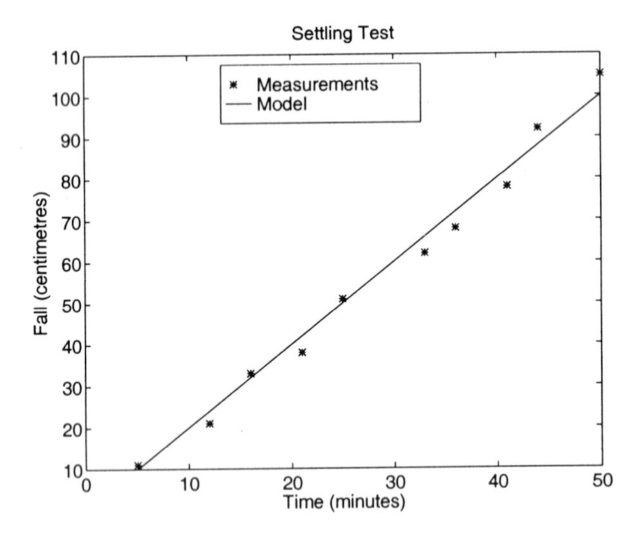

Figure C.2: Plot for exercise.

Appendix D

Introduction to Matrices

This appendix contains a brief introduction to matrices, vectors and scalars sufficient to be able to read and understand the book. It is intended for those of you who, like us, are a long time from college but who, unlike us, have not had the opportunity to retain and practise their knowledge in this area.

Why matrices? Matrices are simply a convenient way of representing and manipulating data involving many equations and/or many variables or values. They are particularly applicable to sets of linear equations.

D.1 Matrices, Vectors and Scalars

A matrix (plural: matrices) simply looks like a two-dimensional table of variables or values.

$$\begin{pmatrix} x & 5 & -1 \\ 3.1 & -5 & y \end{pmatrix} \tag{D.1}$$

Individual elements of the matrix are referred to by row and column numbers.

$$\begin{pmatrix} a_{11} & a_{12} & a_{13} \\ a_{21} & a_{22} & a_{23} \end{pmatrix} \tag{D.2}$$

A vector is a matrix with one row (a row vector) or one column (a column vector).

$$\begin{pmatrix} a_{11} & a_{12} & a_{13} \end{pmatrix} \quad \begin{pmatrix} a_{11} \\ a_{21} \\ a_{31} \end{pmatrix} \tag{D.3}$$

Then a scalar is simply a matrix with one row and one column.

$$a_{11} \tag{D.4}$$

In other words, the numbers that we are all familiar with on an every day basis are scalars.

D.2 Addition and Subtraction

Adding and subtracting matrices are some of the more straight-forward operations. Firstly, matrices must be the same size (the same numbers of rows and columns) before we can add or subtract them. Then we simply add or subtract each corresponding element.

$$A_{n \times m} = B_{n \times m} + C_{n \times m} \tag{D.5}$$

$$a_{ij} = b_{ij} + c_{ij} \tag{D.6}$$

where the $n \times m$ indicates n rows by m columns, and the ij indicates that this applies to each corresponding element. By convention, matrix variables are generally denoted in upper-case and vectors or scalars in lower-case.

D.3 Multiplication and Transpose

Multiplying a matrix by a scalar is very simple. We simply multiply each element of the matrix by the scalar.

$$a \times \begin{pmatrix} b_{11} & b_{12} \\ b_{21} & b_{22} \\ b_{31} & b_{32} \end{pmatrix} = \begin{pmatrix} a \times b_{11} & a \times b_{12} \\ a \times b_{21} & a \times b_{22} \\ a \times b_{31} & a \times b_{32} \end{pmatrix} \tag{D.7}$$

Matrix multiplication is a little more complex. Each element in the product is the corresponding row in the first matrix "times" the corresponding column in the second matrix. In this case "times" means the sum of the product of the respective elements.

$$B \times C = \begin{pmatrix} b_{11} & b_{12} \\ b_{21} & b_{22} \\ b_{31} & b_{32} \end{pmatrix} \times \begin{pmatrix} c_{11} & c_{12} \\ c_{21} & c_{22} \end{pmatrix} \tag{D.8}$$

$$= \begin{pmatrix} b_{11}c_{11} + b_{12}c_{21} & b_{11}c_{12} + b_{12}c_{22} \\ b_{21}c_{11} + b_{22}c_{21} & b_{21}c_{12} + b_{22}c_{22} \\ b_{31}c_{11} + b_{32}c_{21} & b_{31}c_{12} + b_{32}c_{22} \end{pmatrix} \tag{D.9}$$

This means that the number of columns in the first matrix must equal the number of rows in the second matrix, and the size of the product is the number of rows in the first matrix by the number of columns in the second matrix.

$$A_{n \times m} \times B_{m \times p} = C_{n \times p} \tag{D.10}$$

It also means that in general matrix multiplication is not commutative, unlike with scalar multiplication or multiplication by a scalar.

$$A \times B \neq B \times A \tag{D.11}$$

The other operation that is often associated with multiplication is the transpose, denoted by the subscript T.

$$A = \begin{pmatrix} a_{11} & a_{12} & a_{13} \\ a_{21} & a_{22} & a_{23} \end{pmatrix} \quad A' = \begin{pmatrix} a_{11} & a_{21} \\ a_{12} & a_{22} \\ a_{13} & a_{23} \end{pmatrix} \tag{D.12}$$

A common representation of the sum-of-squares of a set of numbers arranged in a column vector is:

$$SOS = a^T \times a = a_1 \times a_1 + a_2 \times a_2 + a_3 \times a_3 + \ldots \tag{D.13}$$

D.4 Division and Inverse

Dividing a matrix by a scalar is also very simple. We simply divide each element of the matrix by the scalar.

$$\begin{pmatrix} b_{11} & b_{12} \\ b_{21} & b_{22} \\ b_{31} & b_{32} \end{pmatrix} \div c = \begin{pmatrix} b_{11} \div c & b_{12} \div c \\ b_{21} \div c & b_{22} \div c \\ b_{31} \div c & b_{32} \div c \end{pmatrix} \tag{D.14}$$

We can also think of it as multiplying by the inverse of the scalar.

In one sense matrix division is the simplest operation - there is no division with matrices.

However, it raises the related operation of the matrix inverse. The matrix inverse is difficult to calculate, which is where computer packages like MATLAB (Appendix C) are useful. The matrix inverse is defined by a relationship that looks very much like division.

$$A \times A^{-1} = A^{-1} \times A = I \tag{D.15}$$

where I is the so-called unit matrix.

$$I_{3 \times 3} = \begin{pmatrix} 1 & 0 & 0 \\ 0 & 1 & 0 \\ 0 & 0 & 1 \end{pmatrix} \tag{D.16}$$

A unit matrix is "square", the same number of rows as columns, and "diagonal", only the elements a_{ii} non-zero.

The inverse A^{-1} is only defined for square matrices and the matrix A must also be "non-singular", in one way equivalent to a scalar being non-zero. In both cases (singular matrix and zero scalar) the inverse is invalid.

D.5 Solution of Linear Equations

A set of linear equations,

$$a_{11} \times x_1 + a_{12} \times x_2 + a_{13} \times x_3 = b_1 \quad \text{(D.17)}$$
$$a_{21} \times x_1 + a_{22} \times x_2 + a_{23} \times x_3 = b_2 \quad \text{(D.18)}$$
$$a_{31} \times x_1 + a_{32} \times x_2 + a_{33} \times x_3 = b_3 \quad \text{(D.19)}$$

can be written in matrix notation as:

$$A \times x = b \quad \text{(D.20)}$$

If there is a unique solution, the same number of equations (rows in A) as variables (columns in A), then we can pre-multiply each side by the inverse of A to obtain the solution.

$$x = A^{-1} \times b \quad \text{(D.21)}$$

The inverse will only exist if A is non-singular which is also equivalent to saying that all the equations must be independent for a unique solution to exist.

If there are more equations than variables, then we look for the so-called least-squares solution, commonly called linear regression. This involves multiplying each side by the transpose of A and then the inverse of $A^T \times A$ to obtain the solution.

$$x = (A^T \times A)^{-1} \times A^T \times b \quad \text{(D.22)}$$

Of course, if there are more variables than equations, not even matrices can help us - which is probably an appropriate point to finish.

D.6 Further Reading

- Introductory material on matrices can be found in the following texts: Barnett (1979), Golub & Van Loan (1983) and Coulson (1965).
- Computer packages for matrix calculations include MATLAB and Matrix-X and codes can be found in Press *et al.* (1986).

Appendix E

Introduction to Differential Equations

This appendix contains a brief introduction to differential equations sufficient to be able to read and understand the book. It is intended for those of you who, like us, are a long time from college but who, unlike us, have not had the opportunity to retain and practise their knowledge in this area.

Why differential equations? Differential equations are necessary for the mathematical modelling of dynamic systems which is the primary interest in our case. They are also used in the modelling distributed parameter systems, for example trickling filters and sludge blankets, but will not be covered in detail here.

E.1 Ordinary Differential Equations

The ordinary differential equations of primary interest to us involve not only process state variables, x, but also their rate of change with respect to time, dx/dt. This inclusion in the equations of a rate or differential term makes them a "differential equation".

In dynamic modelling it is customary to write our differential equations in the explicit "state equation" form.

$$\frac{dx}{dt} = f(x, u, p) \tag{E.1}$$

The term "explicit" refers to the fact that the differential term can be separated out on the left-hand side with the other variables appearing on the right-hand side. Of course in general the states x, inputs u, and parameters p are vectors so that Equation E.1 represents all the differential equations making up our dynamic model.

As discussed in Chapters 3 and 4, these differential equations are the mass, energy and occasionally momentum conservation balances.

E.2 Solution of Ordinary Differential Equations

If our differential equations are linear in the states and inputs, or if we linearise our nonlinear equations (see Chapter 3), then we can write Equation E.1 in the linear form:

$$\frac{dx}{dt} = Ax + Bu \tag{E.2}$$

where the elements of the matrices A and B are functions of the parameters p.

In this special case there is an analytical solution:

$$x = e^{At}x_0 + \int_0^t e^{A(t-\tau)}Bu(\tau)d\tau \tag{E.3}$$

where x is the state at time t and x_0 is the state at time zero. This does introduce the thorny issue of evaluating the matrix exponentials. Fortunately MATLAB has a function which does this. You will also find some references at the end of this Appendix.

It is more often the case that our differential equations are nonlinear in the states and inputs. Only in very special cases do we find an analytical solution in this case. Generally we are forced into a numerical solution, which will be discussed in Appendix F and software for this purpose was discussed in Chapter 9 on simulators.

E.3 Partial Differential Equations

In certain instances our model equations will involve differential terms in more than one independent variable, typically a spatial variable such as the length of an aeration tank or the depth in a clarifier. In this case we get a partial differential equation typically of the form:

$$\frac{\partial x}{\partial t} = -\frac{1}{\tau}\frac{\partial x}{\partial z} \tag{E.4}$$

In general such equations are not possible to solve analytically. Instead some technique is used to discretise the spatial dimension and decompose the problem into a much larger set of ordinary differential equations which are more amenable to solution.

Such techniques include finite differences, finite elements, the method of lines and orthogonal collocation.

Using the simplest application of finite differences the following approximation can be applied.

$$\frac{\partial x}{\partial z} \approx \frac{x_{k+1} - x_k}{\Delta z} \tag{E.5}$$

This means that the n equations represented by Equation E.4 will become

$$\frac{n z_{tot}}{\Delta z} \tag{E.6}$$

which could easily lead to an order of magnitude or more ordinary differential equations to solve. However this is possible.

E.4 Further Reading

- Introductory material on differential equations can be found in the following texts: Lambert (1973) Hoffman (1992).
- Information on computer codes for solving ordinary differential equations can be found in Chapter 9 and in Press *et al.* (1986).
- Material on evaluating matrix exponentials can be found in introductory texts on modern control theory such as Brogan (1974) Shinners (1992).
- References on the solution of partial differential equations include Smith (1978).

Appendix F

Introduction to Numerical Methods

This appendix contains a brief introduction to numerical methods sufficient to be able to read and understand the book. It is intended for those of you who, like us, are a long time from college but who, unlike us, have not had the opportunity to retain and practise their knowledge in this area.

Why numerical methods ? We are primarily interested in numerical methods that solve our models or solve optimisation problems. The complexity of real problems seldom allows us the luxury of the analytical solutions we learnt of in our mathematics courses.

F.1 Solving Linear Algebraic Equations

We saw in Appendix D that the solution of a set of linear algebraic equations involved the matrix inverse.

$$x = A^{-1} \times b \tag{F.1}$$

These problems can be solved by direct methods such as Gaussian Elimination with or without matrix LU decomposition or by iterative methods such as the Gauss-Seidel method.

The details of these methods are outside the scope of this book. Computer programs such as MATLAB (Appendix C) implement most of these methods and books of computer codes are readily available.

F.2 Solving Nonlinear Algebraic Equations

The solution of sets of nonlinear algebraic equations $f(x) = 0$ is typically a difficult problem with difficulties with convergence if the initial guess is not close to the final solution.

Various Newton methods based on a truncated Taylor series are common. Methods discussed above are used to solve for δx at x^n

$$J\delta x = -f(x^n) \tag{F.2}$$

where J is the Jacobian

$$J = \left\{ \frac{\partial f_i}{\partial x_j} \right\}_{x^n} \tag{F.3}$$

and then an iteration is made

$$x^{n+1} = x^n + \delta x \tag{F.4}$$

and the process is repeated until δx becomes sufficiently small.

F.3 Solving Ordinary Differential Equations

In dynamic simulation we are interested in the relatively simple integration of

$$\frac{dx}{dt} = f(x, u, p) \tag{F.5}$$

to give x with time given the initial value x_0 at time zero.

There are a multitude of integration techniques. Mostly we hear of the Euler method because it is a good conceptual place to start (see Chapter 9) and Runge-Kutta techniques because most computer codes use some variation of these.

The Euler method uses a first-order approximation to the derivative

$$\frac{dx}{dt} \approx \frac{x_{n+1} - x_n}{\Delta t} \tag{F.6}$$

leading to the familiar Euler integration function

$$x_{n+1} = x_n + \Delta t f(x_n) \tag{F.7}$$

The typical fourth-order Runge-Kutta method improves the accuracy with more evaluations of the derivative

$$k_1 = \Delta t f(x_n) \tag{F.8}$$

$$k_2 = \Delta t f(x_n + \frac{k_1}{2}) \tag{F.9}$$

$$k_3 = \Delta t f(x_n + \frac{k_2}{2}) \tag{F.10}$$

$$k_4 = \Delta t f(x_n + k_3) \tag{F.11}$$

combined in the integration function

$$x_{n+1} = x_n + \frac{\Delta t}{6}(k_1 + 2k_2 + 2k_3 + k_4) \tag{F.12}$$

The primary issues with integration are accuracy, stability and stiffness. These are discussed in Chapter 9 on Simulators.

F.4 Solving Differential Algebraic Equations

Most dynamic mathematical models involve some combination of nonlinear differential and algebraic equations. By themselves they are trouble, but together they are double-trouble.

As discussed in Chapter 9, most simulators try to avoid the problem by substitution and ordering of the equations to eliminate or at least minimise the need to iterate the algebraic variables. If they need to iterate their efficiency drops rapidly.

The other approach taken by a few simulators is to use sophisticated DAE (differential algebraic equation) solvers which combine the tasks of integration and solution of the algebraic equations. Again the most common of these algorithms are based on Runge-Kutta techniques, usually adapting the step length to meet some accuracy criterion. The DIRK (diagonally implicit Runge Kutta) codes are typical in this class but are far too complex to discuss here.

F.5 Solving the Optimisation Problem

Optimisation remains as much an art as a science. There are a multitude of techniques which generally fall into the classes:

- simple inefficient techniques which are generally robust (they will solve most problems),
- more complex and very efficient techniques which have been tuned to solve particular classes of problem, and
- very complex and sophisticated techniques which are generally both reasonably robust and reasonably efficient.

The "art" is to recognise the class of problem and pick the optimisation technique from the second class of techniques that solves the problem most efficiently.

On the other hand, many advocate the "brute force" approach of throwing computing power at the problem by selecting a favourite method from either the first or second class of techniques and use it to solve everything.

Methods in the first class (simple and robust) are discussed in some detail in Chapter 20.

The most commonly used method in the second class (tuned to the problem) is LP (linear programming) and MILP (mixed integer linear programming), typically used in very large production scheduling problems.

Methods in the third class (sophisticated yet reasonably robust and efficient) include SQP (sequential quadratic programming) which is used by many simulators as an add-on optimiser.

F.6 Further Reading

- Introductory material on numerical methods can be found in the following texts: Burden & Faires (1985), Coute (1965) and Reklaitis *et al.* (1983).
- Information on computer codes for solving ordinary differential equations and differential algebraic equations can be found in Chapter 9.
- Computer codes for common numerical methods can be found in Press *et al.* (1986) and Carnahan *et al.* (1969).

Appendix G

Glossary

A/D converter - analog-digital converter. A device which transforms analog signals into digital representation (binary number).

accuracy - the extent to which a measured value approaches the true value.

actuator - a transducer which reacts to a control signal and performs the desired action.

adaptive control - control such that the properties of the controller equipment automatically adjust to the input signals of the system.

aliasing - the introduction of error caused when a signal is sampled at a rate which does not allow the proper analysis of high frequencies.

alkalinity - a measure of the capacity of water to neutralise acids, in other words, to absorb hydrogen ions without significant pH change. Conventionally expressed in terms of mg $CaCO_3$ per litre.

analog - a continuous signal that represents proportionally a physical variable over some range, for example, the analog of a temperature is a representative voltage.

amplifier - a device capable of increasing the magnitude or power level of a time-varying input without distorting the wave shape of the quantity.

analog signal - signals which have information contained in their incremental variations in magnitude.

anaerobic - conditions in a biological treatments system characterised by the absence of oxygen in any of its forms.

anoxic - nitrate serves as terminal electron acceptor.

ANSI - American National Standards Institute.

ASCII - American Standard Code for Information Interchange, a code which provides information compatibility between digital devices.

ASI - fieldbus acronym for actuator-sensor-interface.

autoregressive (AR) model - a mathematical equation describing the dependency of a signal, $y(t)$, on previously observed values of the signal, $y(t) + a_1 y(t-1) + ... + a_n y(t-n) = e(t)$. AR models are able to describe serial or autocorrelation in a time series. Serial correlation indicates that values of a measurement taken at one time are correlated or related to values of the same measurement taken at a different time.

autoregressive moving average (ARMA) model - a mathematical equation consisting of an autoregressive and a moving average component. ARMA models are frequently used to describe the behaviour of time series. See also autoregressive model, and moving average model.

ARMAX - an ARMA model with an external input.

backward chaining - a search technique used in production systems (if-then rule) that begins with the action clause of a rule and works backward through a chain of rules in an attempt to find a verifiable set on condition clauses. See also expert systems and forward chaining.

BI - bulking index.

bias - the systematic under- or over-estimation of the true measurement value.

binary code - a code for which there are only two acceptable values, normally one and zero.

bit - (BInary digiT) a single digit in a binary number (1 or 0).

black box model - a type of mathematical model which relates observed outputs to observed inputs without attempting to explain the mechanisms which govern the process.

BNR - biological nutrient removal.

BOD - biochemical oxygen demand.

Boolean variables - variables for which only two values are allowable: true or false.

box plot - a plot made up of five values: the maximum, the upper quartile, the median, the lower quartile and the minimum. The points are joined, and the three inner values are boxed to form an impression of the distributional properties of the data.

byte - is made up of 8 bits, and represents a single alphanumeric character in computer storage.

CAN - fieldbus acronym for controller area network.

cascade control - system where one controller supplies the set point for another controller.

causal filter - a type of digital filter which only uses past historic data to calculate a present filtered value; can be used for on-line filtering of measurement values.

CEN - the European Committee for Standardisation.

CIM - Computer Integrated Manufacturing.

CISC - Complex Instruction Set Computer. There are a large number of relatively specialised instructions available. The Intel 80X86 series of microprocessors is an example. See also RISC.

closed loop control - system where the controlled variable is measured and the result of the measurement used to manipulate one of the process variables. Also called feedback control.

COD - chemical oxygen demand.

constant - a fixed value.

cumulative probability distribution function - a description of how the ranked values in a data set are associated with their frequency of occurrence.

current leakage - the current that flows between the input common terminal and the output common terminal (across the isolation barrier) with a specified voltage applied to it.

DAE - differential algebraic equation.

database - a bank of information contained by a computer usually in the form of one or more tables.

data file - a computer collection of records containing data.

DCS - distributed control system where the control functions are distributed over a number of microprocessors linked by a data highway. The system may or may not allow significant geographic distribution of functions.

DDC - direct digital control originally meaning that the actuators were manipulated directly by a computer. Later it was used where there was a single computer performing all the control functions. See also DCS.

DDE - dynamic data exchange links pieces of data in different computer programs in a Microsoft Windows environment. For example it can implement dynamic updating in one program when the piece of data is changed in another.

dead band - a range of values through which a measured value can be changed without triggering a response.

dead time - time interval from when a change in an input variable is made until any change is observed in an output variable.

denitrification - the conversion of nitrate-nitrogen to gaseous nitrogen through anoxic heterotrophic cell growth.

dependent variable - if y is a function of x, then y is the dependent variable, that is a value of x results in a value of y.

DES - discrete event system.

digital - data expressed in the form of digits.

digital filter - a mathematical procedure for separating data into frequency components and evaluating the properties in each frequency interval. A highpass filter passes the high frequencies (noise) whereas a lowpass filter removes noise so that a trend can be detected.

digital signal - signals of uniform amplitude which are represented by only two possible values, high and low.

directory path - a hierarchical link between directories in a computer.

distribution - a description of a set of data, with the object of identifying the density of the occurance of different values in the data.

DMC - dynamic matrix control.

DN - denitrification.

DO - dissolved oxygen.

drift - a gradual deviation from a set adjustment, such as frequency or balance current, or from a direction.

DSP - digital signal processor is a very fast computer with specialist circuitry for common signal processing functions. It is generally used as embedded processors within an instrument.

dynamic, dynamics - conditions which change with time and are transient or unstable.

dynamic models - an equation or set of equations which predict the time-varying performance of a process. Dynamic models are usually differential equations.

dynamic simulation - see simulation.

electrical impedance - the resistance of the circuit material to a current, described as the ratio of the potential difference to the current applied.

EMC - electromagnetic compatibility.

Ethernet - a widely used local area network (LAN).

expert system - a computer program containing knowledge about a problem area and which emulates the reasoning processes of human experts in that particular area. A typical expert system consists of the knowledge base, the inference engine, and the user interface.

error - the extent from which a measured value deviates from the true value.

FA - factor analysis (see further PCA).

FCM - fuzzy C-means algorithm.

FIA - flow injection analysis.

F/M - food to microorganism ratio; the substrate load applied to the process per unit of biomass in the aeration tank per unit of time.

filter - an electronic device or mathematical equation which transmits signals in a certain frequency range while other signals outside the range are attenuated.

filtering - the separation of desired signals from undesired signals (noise).

floating point number - a notation used in computer systems where two or more binary words are used to express a non-integer number with a certain number of significant digits, for example, 0.08 is a floating point number.

forward chaining - a search technique used in production systems (if-then rule) that begins with the condition clause of a rule and works forward through a chain of rules in an attempt to activate implied action clauses. See also expert system and backward chaining.

Fourier analysis - analysis of when and how well a function is represented by its Fourier series or transform and of the convergence or divergence of the series or transform.

FTU - measure of turbidity, see NTU.

fuzzy controller - a controller based on rules about how the control variable is selected based on linguistically quantised values of the measured variables. Uses fuzzy logic to describe the quantised values.

fuzzy logic - a logical theory where the truth values 'true' and 'false' have been replaced by more approximate values like 'not very true', 'not likely' or 'very unlikely'.

gain - the ratio of the output signal to the associated input signal of a device.

GMC - Generic Model Control.

grounding - electrical connection to a grounding conductor or earth.

GTO - Gate turn off transistor.

high pass filter - a high pass filter dampens low frequency variations and highlights high frequency variations.

histogram - a graphical representation using rectangles whose widths span intervals within the range of the values and whose heights represent the number of values falling into each interval. Also known as a bar graph.

hysteresis - non-linearity in a measurement device, generally in the form of a lagging of a current or voltage behind the corresponding measured value.

IAWPRC - International Association on Water Pollution Research and Control (now IAWQ).

IAWQ - International Association on Water Quality.

ICA - instrumentation, control and automation.

identification - the process of distinguishing which one of a number of linear equations best describes a system and, subsequently, estimating the values of the parameters in the equations.

IEC - The International Electrotechnical Commission.

IEE - Institution of Electrical Engineers, UK.

IEEE - Institution of Electrical and Electronic Engineers, USA.

IFAC - International Federation of Automatic Control.

IGBT - Insulated Gate Bipolar Transistor.

impedance - see electrical impedance.

impedance match - a state where the external impedance is equal to the internal impedance; this results in maximum transfer of energy, minimum reflection, and minimum distortion for AC signals.

independent variable - if y is a function of x, then the independent variable is x.

inductive coupling - coupling of two circuits by means of the mutual inductance.

integrator windup - phenomenon in a PI or a PID controller. If the control variable reaches a constraint and the setpoint cannot be achieved, then the integral or the error is theoretically unbounded. Has to be overcome in a real implementation.

interface - a boundary between systems or functional units with different characteristics.

ISO - International Organisation for Standardisation.

isolation amplifier - a device which provides ohmic isolation (breaks ohmic continuity of an electric signal) between the input and the output of the device.

IWA - International Water Association (merger of the IAWQ and IWSA).

IWSA - International Water Supply Association.

KBS - knowledge based system.

KISS - Keep It Short and Simple.

K_La - overall oxygen mass transfer rate coefficient.

knowledge based system - see expert system.

LAN - local area network.

least squares method - a method of fitting a curve to given points such that the sum of the squares of the deviations between the curve and the points is minimised.

linearity - the relationship that exists between two quantities when a change in one of them produces a directly proportional change in the other.

log-normal distribution - a mathematical model which is used to describe many environmental data sets. A histogram of the logarithms of the values results in a symmetrical bell-shaped plot. The values of the logarithms plot as a straight line on normal probability paper.

low pass filter - a low pass filter dampens high frequency variations and highlights low frequency variations.

MAP - Manufacturing Automation Protocol.

mean - a measure of central tendency where the sum of the values is divided by the number of values, thus the mean is typical of the values.

median - a value that, when arranged with a particular series in numerical order, has an equal number of values below as above it. Also a measure of central tendency.

MLR - multiple linear regression.

MLSS - mixed liquor suspended solids.

MMS - Manufacturing Message Specification.

model - equation or set of equations used to describe a process. Variables in the equation(s) represent the inputs, outputs and internal states of the process. The equations with associated notation and terminology provide a means of conceptualising and communicating important aspects of a process.

moving average (MA) model - a mathematical equation describing the behaviour of a signal, $y(t)$, as the sum of a white noise sequence, $y(t) = e(t) + b_1e(t-1) + ... + b_ne(t-n)$. MA models can be used to simulate a large number of stochastic processes.

multiple regression - a technique for finding a best fit expression for a dependent variable which is a function of several independent variables.

multiplexer - a switch arrangement which allows several input channels to be serviced by one amplifier and A/D converter.

NF - normalised fluorescence.

NIR - near infrared measuring technique.

nitrification - the conversion of ammonia-nitrogen to nitrite and nitrate-nitrogen through autotrophic cell growth.

Nitrobacter - bacteria capable of oxidising ammonia to nitrate.

Nitrosomonas - bacteria capable of oxidising ammonia to nitrite.

NOC - normal operating conditions.

noise - disturbances in the current or voltage from an on-line sensor that interfere with the true signal and are unwanted.

non-causal filter - a type of digital filter which uses both past and future values to calculate a present filtered value; this type of filter is only used off-line.

NTU - Nephelometric Turbidity Units, measuring turbidity. Often expressed as FTU since Formazin polymer is used as the reference turbidity standard suspension.

Nyquist rate - half the sampling rate.

on-line estimation - estimation of model parameters during the operation of the system.

On-Off control - a type of control where only two positions are available: on, when the measured value is below the setpoint, and off, when the measured value is above the setpoint.

open loop control - system where information about the controlled variable is not used to manipulate any of the system inputs to compensate for variations in the process variables. Also called sequencing.

ORP - oxidation-reduction potential or redox.

OSI - Open System Interconnection.

oxygen transfer coefficient - see K_La.

oxygen uptake rate (OUR) - rate at which oxygen is utilised by microorganisms, typically expressed in mg/L/hr.

PAO - phosphorus accumulating organisms.

parameter - a constant or variable such that different values result in different cases of the phenomenon observed.

Parshall flume - a device for measuring the flow of liquids in open channels by measuring the upper and lower heads at a specified distance from an obstructing sill.

PC - personal computer.

PC - principal components (in PCA).

PCA - principal component analysis.

percentile - a value in a range of values such that a certain percentage of the values fall below the percentile.

PHA - poly-hydroxyl-alkanoates.

PID controller - three mode controller combining the actions of proportional (gain), integral (reset) and derivative (rate) elements into a single unit.

PLC - programmable (logical) controller.

PLS - partial least squares (or projection to latent structures).

PRBS - pseudo random binary sequence.

precision - a measure of how close measurement values are to one another when collected under the same conditions.

process stability - a stable process is one in which the output is always bounded for any bounded input. Although most processes are inherently stable, automatic controllers, under some circumstances, may change a stable process into an unstable one.

proportional band - defined as 100/K in a PID controller, where K is the gain.

prototype - full-scale and functional but a trial form of a design.

psi - pounds per square inch.

PWM - pulse width modulation.

random - if a process is governed by chance and not by some underlying cause-and-effect mechanism, the outcome is said to be random. For example, lotteries and bingos, provided that they are run fairly, produce random results.

RAS - recycle activated sludge.

RBCOD - readily biodegradable chemical oxygen demand.

real numbers - set of numbers including zero, all positive and negative integers and fractions.

recursive estimation - the use of a recursive algorithm to estimate the values of model parameters from field data. Such algorithms enable parameter estimates to be continually updated; updated estimates at time t are calculated based on data collected at time t and the values of the parameter estimates for the previous time period, $t-1$.

redox - oxidation-reduction potential or ORP.

regression - a technique for finding a best-fit set of parameters for an expression relating some dependent variable to one or more independent variables.

residence time - the time taken for fluid to pass through a vessel. See residence time distribution, RTD.

residence time distribution - an attribute of a vessel which describes the spread of residence times of fluid passing through the vessel. See residence time, RTD.

residual - the difference between a value derived from theory and a value resulting from experimental procedures.

RISC - Reduced Instruction Set Computer which has a small number of very fast and very simple instructions. More complex functions must be performed by a number of these instructions. See also CISC.

RTD - a MATLAB computer program written to illustrate the residence time distribution of different arrangements of "ideal" vessels and to match it to experimental data. See residence time, residence time distribution.

sample-and-hold circuit - "grabs" the present value of a signal just before the beginning of an A/D conversion and holds it constant, despite a changing input, until the A/D conversion is complete.

sampling interval - the interval of time between successive samplings.

SBR - sequential batch reactor.

SCOUR, specific oxygen uptake rate - rate of oxygen utilisation by the microorganisms per mass of microorganisms.

secondary storage - storage of data in an accessible source external to the main computer.

sequencing - see open loop control.

serial interface - an interconnection between devices or systems where the data may pass only in bit-serial form (see interface).

setpoint - the desired value for a control system, that is, a temperature, flowrate, pressure or level at which a process should operate.

shunt resistor - a resistor used in an ammeter which increases the ammeter's range.

siemens - the SI unit of the reciprocal *ohm*, used to measure conductivity. Conductivity is usually expressed in millisiemens per metre (mS/m).

signal conditioning - pre-processing of input signals which can include amplification, isolation, voltage division, surge suppression, current-to-voltage conversion and filtering.

signal-to-noise ratio - the ratio of the amplitude of the true signal to the amplitude of the noise.

simulation - the use of a computer program to predict the performance of a process under different conditions. Simulations are often employed when an actual system is difficult or costly to test.

simulator - a computer program or piece of equipment that shows the effects of various applied changes to a process. See simulation.

sludge age - solids retention time, typically expressed in days.

SND - simultaneous nitrification and denitrification.

SPC - statistical process control. A term generally used when the measurements are continuous in nature.

SQC - statistical quality control. A term generally used when the measurements are discrete in nature.

SRP - standard reduction potential

SRT - solids retention time (sludge age).

SS - suspended solids.

stability - a system is stable when any bounded input signal results in a bounded output signal.

steady-state - a system state in which the conditions at each point do not change with time, that is, after initial transients or fluctuations have disappeared.

step feed - moving the position of the influent to the aeration basin towards the settler, thereby reducing the solids to the settler.

stochastic - concerning random variables.

SVD - singular value decomposition.

SVI - sludge volume index

time series - a sequence of values collected at equal intervals in time.

TKN - Total Kjeldahl Nitrogen, the sum of ammonia and organic nitrogen.

TOC - Total Organic Carbon.

TOP - Technical and Office Protocol.

trend - the general tendency of a set of values as related to time or another set of data.

trickling filter - a biofilm process in which the applied wastewater trickles down over an unsaturated bed of rocks or plastic packing. A clarifier normally follows the trickling filter.

TS - total solids.

TSA, time-series analysis - a set of statistical techniques which are used to describe and predict a time series. A special problem of most time series, accounted for in time-series analysis, is that successive measurements in the time series are not completely independent.

UF - ultrafiltration.

variance - square of the standard deviation. A measure of the variation of a variable about the mean value.

VFA - volatile fatty acids.

VSS - volatile suspended solids.

WAS - waste activated sludge.

WWT - wastewater treatment.

WWTP - wastewater treatment plant.

References

ACSL. *ACSL/GM dynamic simulation language.* MGA Software, 919-B Willowbrook Drive, Huntsville, AL 35802, USA. phone +1 800 647 2275, +1 256 883 5516, software@mga.com.

Adby, P. R. & Dempster, M. A. H. 1974. *Introduction to optimization methods.* London: Chapman and Hall.

Adeyemi, S. O., Wu, S. M. & Berthouex, P. M. 1979. Modeling and control of a phosphorus removal process by multivariate time series method. *Wat. Res.*, **13**, 105–112.

Albertson, O. E. 1991. Bulking sludge control - progress, practice and problems. *Wat. Sci. Tech.*, **23**(4–6), Part 2, 835–846.

Alty, J. L. & Coombs, M. J. 1984. *Expert systems. concepts and examples.* Manchester, England: NCC Publications.

Anderson, T. W. 1984. *Introduction to multivariate statistical analysis.* second edn. Wiley.

Andersson, B., Nyberg, U. & Aspegren, H. 1995. Methanol and ethanol as carbon sources for denitrification. *In: Nordic seminar: Nitrogen removal from municipal wastewater.*

Andersson, M. 1994. *Object-oriented modelling and simulation of hybrid systems.* Ph.D. thesis, Dept. of Automatic Control, Lund Inst. of Technology, Lund, Sweden. reference TFRT-1043.

Andrews, J. F. 1969. Dynamic model of the anaerobic digestion process. *J. Sanit. Engng. (Am. Soc. of Civil Engrs. Sanit. Engng. Div.)*, **1**, 95–116.

Andrews, J. F. 1974. Dynamic models and control strategies for wastewater treatment processes. *Wat. Res.*, **8**, 261–289.

Andrews, J. F. (ed). 1992. *Dynamics and control of the activated sludge process.* Technomic Publishing Company Inc., Lancaster, Pennsylvania, USA.

Andrews, J. F. & Graef, S. P. 1971. Dynamic modeling and simulation of the anaerobic digestion process. *Adv. Chem. Ser. 105.*

Angelidaki, I., Ellegaar, L. & Ahring, B. K. 1993. A mathematical model for dynamic simulation of anaerobic digestion of complex substrates: Focusing on ammonia inhibition. *Biotechnology and Bioengineering*, **42**, 159–166.

Anon. 1996. Real-world applications of standards. *Quality Progress*, **29**(6), 29–49.

Anon. 1997. Controlling those costs. *The Chemical Engineer*, 6th February, 26–32.

AQUASIM. *Aquatic systems dynamic simulator.* Aquasystem, Habsburgsw. 30, CH-8600 Winterthur, Switzerland.

Arthur, R. M. 1994. Thirty years of respirometry. *In: Purdue Industrial Waste Conf.* Chelsea, Michigan: Lewis.

Arthur, R. M. & Arthur, B. 1997. Respirometry: Not just another "gimmick". *Operations Forum, Water Environment Federation*, **14**(11), 19–23.

Årzèn, K. E. 1989. Knowledge-based control systems - aspects on the unification of conventional control systems and knowledge-based systems. *Pages 2233–2238 of: Proc. American Control Conference.*

Årzèn, K. E. 1990. Knowledge-based control systems. *Pages 1986–1991 of: Proc. American Control Conference.*

Aspegren, H., Andersson, B., Nyberg, U. & la Cour Jansen, J. 1992. Model and sensor based optimisation of nitrogen removal at Klagshamn wastewater treatment plant. *Wat. Sci. Tech.*, **26**(5-6), 1315–1323.

Aspegren, H., Hellström, B.-G. & Olsson, G. 1997. The urban water system - a future Swedish perspective. *Wat. Sci. Tech.*, **35**(5), 33–43.

Åström, K. J. & Hägglund, T. 1988. *Automatic tuning of PID controllers.* Research Triangle Park, NC, USA: Instrument Society of America.

Åström, K. J. & Wittenmark, B. 1990. *Computer controlled systems - theory and design.* Second edn. Prentice-Hall.

Åström, K. J. & Wittenmark, B. 1995. *Adaptive control.* Second edn. Reading, MA, USA: Addison-Wesley.

Åström, K. J., Anton, J. J. & Årzèn, K. E. 1986. Expert control. *Automatica*, **22**, 277–286.

Åström, K. J., Elmqvist, H. & Mattsson, S. E. 1998. Evolution of continuous-time modeling and simulation. *In: The 12th European Simulation Multiconference, ESM '98.*

Ayesa, E., Florez, J., Garcìa-Heras, J. L. & Larrea, L. 1991. State and coefficients estimation for the activated sludge process using a modified Kalman filter algorithm. *Wat. Sci. Tech.*, **24**(6), 235–247.

Ayesa, E., Florez, J., Larrea, L. & Garcia-Heras, J.L. 1993. Evaluation of sensitivity and observability of the state vector for system identification and experimental design. *Wat. Sci. Tech.*, **28**(11-12), 209–218.

Ayesa, E., Goya, B., Larrea, A., Larrea, L. & Rivas, A. 1998. Selection of operational strategies in activated sludge processes based on optimisation algorithms. *Wat. Sci. Tech.*, **37**(12), 327–334.

Bakshi, B. R. & Stephanopoulos, G. 1993. Wave-Net: a multi-resolution hierarchical neural network with localized learning. *AIChE Journal*, **39**(1), 57–81.

Balslev, P., Lynggaard-Jensen, A. & Nickelsen, C. 1996. Nutrient sensor based real-time on-line process control of a wastewater treatment plant using recirculation. *Wat. Sci. Tech.*, **33**(1), 183–192.

Barker, P. S. & Dold, P. L. 1997. General model for biological nutrient removal activated sludge systems: Model application. *Water Environment Research*, **69**(5), 985–991.

Barnett, M.W. & Gall, B. 1996. A robust rule-based system for on-line diagnosis of nitrification problems in activated sludge treatment. *A.I.Ch.E. Symp. Series*, **92**, 343–346.

Barnett, S. 1979. *Matrix methods for engineers and scientists.* UK: McGraw-Hill.

Barton, D. A. & McKeown, J. J. 1986. Evaluation of an aerator control strategy utilising time varying mathematical model simulations. *Wat. Sci. Tech.*, **18**(6), 189–201.

Bastin, G. & Doachain, D. 1990. *On-line estimation and adaptive control of bioreactors.* Amsterdam: Elsevier.

Batstone, D., Keller, J., Newell, R.B. & Newland, M. 1997. Model development and full-scale validation for anaerobic treatment of protein and fat based wastewater. *Wat. Sci. Tech.*, **36**(6–7), 423–431.

Bechmann, H., Nielsen, M. K., Madsen, H. & Poulsen, N. K. 1998. Control of sewer systems and wastewater treatment plants using pollutant concentration profiles. *Wat. Sci. Tech.*, **37**(12), 87–93.

Beck, M. B. 1977. Modelling and control in practice. *Progress in Water Technology*, **9**(5/6), 557–564.

Beck, M. B. 1981. Operational estimation and prediction of nitrification dynamics in the activated sludge process. *Wat. Res.*, **15**(12), 1313–1330.

Beck, M. B. 1984. Modelling and control studies of the activated sludge process and Norwich Sewage Works. *Transactions of the Inst. of Measurement and Control*, **6**(3), 117–131.

Beck, M. B. 1986. Identification, estimation and control of biological waste-water treatment processes. *IEE Proc.*, **133**, 254–264.

Beck, M. B. 1989. System identification and control. *Chap. 9 of:* Patry, G. G. & Chapman, D. (eds), *Dynamic Modelling and Expert Systems in Wastewater Engineering.* Chelsea, MI, USA: Lewis Publishers Inc.

Beck, M. B., Latten, A. & Tong, R. M. 1978. Modelling and operational control of the activated sludge process in wastewater treatment. *In: Intl. Inst. for Applied Systems Analysis (IIASA).* Professional Paper PP-78-10.

Benson, R. L., Truong, Y. B., McKelvie, I. D., Hart, B. T., Bryant, G. & Hilkmann, W. 1996. Monitoring of dissolved reactive phosphorus in wastewaters by flow injection analysis. part 2. on-line monitoring system. *Wat. Res.*, **30**, 1965–1971.

Bergh, S. G. 1996. *Diagnosis problems in wastewater settling.* Licentiate thesis, Dept. of Industrial Electrical Engineering and Automation, Lund Inst. of Technology, Lund, Sweden. CODEN:LUTEDX/TEIE-1011, ISBN 91-88934-01-2.

Bergh, S. G. & Olsson, G. 1996. Knowledge based diagnosis of solids-liquid separation problems. *Wat. Sci. Tech.*, **33**(2), 219–226.

Berthouex, P. M. & Hunter, W. G. 1981. Simple statistics for interpreting environmental data. *J. WPCF*, **53**(2), 167–175.

Berthouex, P. M., Lai, M. & Darjatmoko, A. 1987. A statistics-based information and expert system for plant control and improvement. *Pages 146–150 of:* Carrol, W. E. (ed), *Proceedings 5th National Conf. on Microcomputers in Civil Engineering.*

Bezdek, J. C. 1981. *Pattern recognition with fuzzy objective function.* New York: Plenum Press.

Bidstrup, S. M. & Grady Jr., C. P. L. 1987. *A user's manual for SSSP simulation of single-sludge processes for carbon oxidation, nitrification and denitrification.* Environmental Systems Engineering, Clemson Univ., Clemson, SC, USA.

Bidstrup, S. M. & Grady Jr., C. P. L. 1988. SSSP - simulation of single sludge processes. *J. WPCF*, **60**(3), 351–361.

BioWin. *Wastewater treatment simulator.* EnviroSim Associates Ltd., 482 Anthony Drive, Oakville, Ontario L6J 2K5, Canada or Reid Crowther International Ltd., 300, 4170 Still Creek Drive, Burnaby, B.C. V5C 6C6, Canada. phone +1 905 648 9814, fax +1 905 648 4410, info@envirosim.com.

Birtwistle, G. M., Dahl, O. J., Myhrhaug, B. & Nygaard, K. 1973. *Simula begin.* Auerbach Publishers Inc.

Bocken, S. M., Braae, M. & Dold, P. L. 1989. Dissolved oxygen control and oxygen utilisation rate estimation - extension of the Holmberg/Olsson method. *Wat. Sci. Tech.*, **21**(10–11), Part 4, 1197–1208.

Box, G. E. P. & Jenkins, G. M. 1976. *Time series analysis: Forecasting and control.* San Francisco: Holden Day.

Box, G. E. P., Hunter, W. G. & Hunter, J. S. 1978. *Statistics for experimenters.* Wiley.

Briggs, R. (ed). 1997. *7th IAWQ workshop on instrumentation, control and automation of water and wastewater treatment and transport systems.* Wat. Sci. Tech.

Brogan, W. L. 1974. *Modern control theory.* New York: Quantum Publishers Inc.

Bruce, A., Donoho, D. & Gao, H.-Y. 1996. Wavelet analysis. *IEEE Spectrum*, October, 26–35.

Bryson, A. E. & Ho, Y. C. 1969. *Applied optimal control: optimization, estimation, and control.* New York, USA: Hemisphere Publ. Co.

Buhr, H. O., Andrews, J. F. & Keinath, T. M. (eds). 1975. *Research needs for automation of wastewater treatment systems.* USA: Clemson University and U. S. Environmental Protection Agency.

Bundgaard, E. 1988. Nitrogen and phosphorus removal by the Bio-Denitro and Bio-Denipho processes. *In: Proc. of the Int. Workshop on Wastewater Treatment Technology.*

Bundgaard, E., Nielsen, M. K. & Henze, M. 1996. Process development by full-scale on-line tests and documentation. *Wat. Sci. Tech.*, **33**(1), 281–287.

Burden, R.L. & Faires, J.D. 1985. *Numerical analysis.* Third edn. Boston, USA: Prindle, Webe & Schmidt.

Burk, A.F. 1992. Strengthen process hazard reviews. *Chem. Engng. Prog.*, **88**, 90–94.

Busby, J. B. & Andrews, J. F. 1975. Dynamic modeling and control strategies for the activated sludge process. *J. WPCF*, **47**, 1055–1080.

Butcher, J. C. 1987. *The numerical analysis of ordinary differential equations.* Wiley.

Cameron, F. & Seborg, D.E. 1983. A self-tuning controller with a PID structure. *Int. J. Control*, **38**(2), 401–417.

Cameron, I. T. & Hangos, K. 1999. *Process modelling and model analysis.* London: Academic Press.

Carlsson, B. 1993. On-line estimation of the respiration rate in an activated sludge process. *Wat. Sci. Tech.*, **28**(11–12), 427–434.

Carlsson, B. & Wigren, T. 1993 (July 19-23). On-line identification of the dissolved oxygen dynamic in an activated sludge process. *In: 12th IFAC World Congress.*

Carlsson, B., Lindberg, C. F., Hasselblad, S. & Xu, S. 1994. On-line estimation of the respiration rate and the oxygen transfer rate at Kungsängen wastewater plant in Uppsala. *Wat. Sci. Tech.*, **30**(4), 255–263.

Carnahan, B., Luther, H.A. & Wilkes, J.O. 1969. *Applied numerical methods.* USA: Wiley.

Carstensen, J. 1994. *Identification of wastewater processes.* Ph.D. thesis, IMSOR, Danish Univ. of Technology, Denmark.

Carstensen, J., Harremoës, P. & Madsen, H. 1994. Statistical identification of monod-kinetic parameters from on-line measurements. *In:* Henze, M. & Harremoës, P. (eds), *IAWQ Specialised Seminar: Modelling and Control of Activated Sludge Processes.*

Carstensen, J., Nielsen, M. K. & Harremoes, P. 1996. Predictive control of sewer systems by means of grey-box models. *Wat. Sci. Tech.*, **34**(3-4), 189–194.

Carstensen, J., Nielsen, M. K. & Strandbæk, H. 1997. Prediction of hydraulic load for urban storm control of a municipal WWT plant. *In: IAWQ 7th Int. Workshop on Instrumentation, Control and Automation of Water and Wastewater Treatment and Transport Systems.*

Cech, J. S., Chudoba, J. & Grau, P. 1985. Determination of kinetic constants of activated sludge microorganisms. *Wat. Sci. Tech.*, **17**(2–3), 259–272.

Cellier, F. E. 1993. Integrated continuous-system modelling and simulation environments. *In:* Linkens, D. A. (ed), *CAD for Control Systems.* New York: Marcel Dekker Inc.

Chan, W. T. & Koe, L. C. C. 1991. A knowledge based framework for the diagnosis of sludge bulking in the activated sludge process. *Wat. Sci. Tech.*, **23**, 847–855.

Chang, W.C., Ouyang, C.F., Chiang, W.L. & Hou, C.W. 1998. Sludge pre-recycle control of dynamic enhanced biological phosphorus removal system: an application of on-line fuzzy controller. *Wat. Res.*, **32**(3), 727–736.

Charpentier, J., Godart, H., Martin, G. & Mogno, Y. 1989. Oxidation-reduction potential (ORP) as a way to optimize aeration and C, N and P removal: Experimental basis and various full-scale examples. *Wat. Sci. Tech.*, **21**, 1209–1223.

Chatfield, C. 1984. *The analysis of time series.* New York: Chapman and Hall.

Chen, J. & Beck, B. 1993. Modelling, control and on-line estimation of activated sludge bulking. *Wat. Sci. Tech.*, **28**(11-12), 249–256.

Chynoweth, D. P., Svoronos, S. A., Lyberatos, G., Harmon, J. L., Pullammanappallil, P., Owens, J. M. & Peck, M. J. 1994 (23-27 January). Real-time expert system control of anaerobic digestion. *In: Seventh International Symposium on Anaerobic Digestion.*

Cohon, J. L. 1978. *Multiobjective programming and planning.* New York: Academic Press.

Considine, D. 1974. *Process instrument and controls handbook.* Second edn. McGraw-Hill Inc.

Cook, S. & Marsili-Libelli, S. 1981. Estimation and control problems in activated sludge processes. *Wat. Sci. Tech.*, **13**(11–12), 737–742.

Cooper, B.E. 1969. *Statistics for experimentalists.* Pergamon Press.

Costello, D. J. 1989. *Modelling, optimisation and control of high-rate anaerobic reactors.* Ph.D. thesis, University of Queensland, St Lucia, Australia.

Costello, D. J., Greenfield, P. F. & Lee, P. L. 1991a. Dynamic modelling of a single-stage high-rate anaerobic reactor - I model derivation. *Wat. Res.*, **25**, 847–858.

Costello, D. J., Greenfield, P. F. & Lee, P. L. 1991b. Dynamic modelling of a single-stage high-rate anaerobic reactor - II model verification. *Wat. Res.*, **25**, 859–871.

Cote, M., Grandjean, B. P. A., Lessard, P. & Thibault, J. 1995. Dynamic modelling of the activated sludge process: Improving prediction using neural networks. *Wat. Res.*, **29**(4), 995–1004.

Coulson, A.E. 1965. *An introduction to matrices.* UK: Longmans.

Coute, S.D. 1965. *Elementary numerical analysis.* USA: McGraw-Hill.

Crosby, P. B. 1989. *Let's talk quality.* McGraw-Hill.

Cryer, J. D. 1986. *Time series analysis.* Boston: Duxbury Press.

Curds, C. R. 1973a. A theoretical study of factors influencing the microbial population dynamics of the activated sludge process - I the effects of diurnal variations of sewage and carnivorous ciliated protozoa. *Wat. Res.*, **7**, 1269–1284.

Curds, C. R. 1973b. A theoretical study of factors influencing the microbial population dynamics of the activated sludge process - II a computer-simulation study to compare two methods of plant operation. *Wat. Res.*, **7**, 1439–1452.

Cutler, C.R. & Ramaker, B.L. 1980. Dynamic matrix control - a computer control algorithm. *Pages WP5-B of: Proceedings of Joint Auto. Control Conf.*

David, R. & Alla, H. 1992. *Petri nets and grafcet.* London, United Kingdom: Prentice Hall Intl.

Davis, J. F., Stephanopoulos, G. & Venkatasubramanian, V. (eds). 1996. *Conf. on intelligent systems in process engineering.* AIChE Symp. Series 312, vol. 92.

Davis, J. J., Kermode, R. I. & Brett, R. W. J. 1973. Generic feed forward control of activated sludge. *Proc. of the Am. Soc. of Civil Eng., J. of the Env. Eng. Div.*, **99**(EE3), 301–314.

de Silva, C. W. 1989. *Control sensors and actuators.* Prentice-Hall.

DeMarco, T. 1978. *Structured analysis and system specification.* New York: Yourdon Press.

Deming, W. E. 1982. *Quality, productivity and competitive position.* Cambridge, Mass., USA: Massachusetts Inst. of Technology.

Diehl, S. & Jeppsson, U. 1998. A model of the settler coupled to the biological reactor. *Wat. Res.*, **32**(2), 331–342.

Diehl, S., Sparr, G. & Olsson, G. 1990. Analytical and numerical description of the settling process in the activated sludge operation. *Pages 471–478 of:* Briggs, R. (ed), *Advances in Water Pollution Control.* Pergamon Press.

Doebelin, E. O. 1983. *Measurement systems.* McGraw-Hill.

Dold, P. L. 1990. A general activated sludge model incorporating biological excess phosphorus removal. *In: CSCE Annual Conf.*

Dold, P. L. 1992. *Activated sludge system model incorporating biological nutrient (N & P) removal.* Tech. rept. Dept. of Civil Eng. & Eng. Mech., McMaster University, Hamilton, Ontario, Canada.

Dold, P. L. & Marais, G. v. R. 1986. Evaluation of the general activated sludge model proposed by the IAWPRC task group. *Wat. Sci. Tech.*, **18**, 63–89.

Dold, P. L., Ekama, G. A. & Marais, G. v. R. 1980. A general model for the activated sludge process. *Prog. Water Technology*, **12**, 47–77.

Dold, P.L., Buhr, H.O. & Marias, G.v.R. 1984. An equalisation control strategy for activated sludge process control. *Wat. Sci. Tech.*, **17**(2-3), 221–234.

Drainkov, D., Hellendoorn, H. & Reinfrank, M. 1993. *An introduction to fuzzy control.* Berlin: Springer Verlag.

Draper, N. R. & Smith, H. 1981. *Applied regression analysis.* New York: Wiley.

Dupont, R. & Henze, M. 1992. Modelling of the secondary clarifier combined with the Activated Sludge Model No. 1. *Wat. Sci. Tech.*, **25**(6), 285–300.

Dymosim. *Dymamic modeling package.* Dynasim AB, Ideon Research Park, Lund, Sweden. phone +46 46 , fax +46 46, http://www.dynasim.se.

EASY-5. *General-purpose simulator.* The Boeing Company, Seattle, WA, USA. phone +1 905 522 0012, fax +1 905 522 0031, http://www.boeing.com/assocproducts/easy5.

Eckenfelder Jnr., W.W. 1980. *Principles of water quality management.* Boston, Mass.: CBI Publishing Company, Inc.

EFOR. *Wastewater treatment simulator.* EFOR Aps., c/o Krüger Systems, Gladsaxevei 363, DK-2860 Söborg, Denmark. phone +45 39 57 24 18, fax +45 39 69 08 06, efor@efor.dk.

Ekama, G. A. & Marais, G. v. R. 1979. Dynamic behaviour of the activated sludge process. *J. WPCF*, **51**, 534–556.

Elmqvist, H. 1978. *A structured model language for large continuous systems.* Ph.D. thesis, Dept. of Automatic Control, Lund Inst. of Technology, Lund, Sweden. Reference TFRT-1015.

Elmqvist, H. 1995. *Dymola - user's manual.* Dynasim AB, Ideon Research Park, Lund, Sweden.

Elmqvist, H., Mattsson, S. E. & Otter, M. 1998. Modelica - the new object-oriented modeling language. *In: The 12th European Simulation Multiconference, ESM '98.*

Enbutsu, I., Baba, K., Hara, N., Waseda, K. & Nogita, S. 1993. Integration of multi AI paradigms for intelligent operation support systems – fuzzy rule extraction from a neural network. *Wat. Sci. Tech.*, **28**(11–12), 333–340.

EXPERT. *Expert rules engine.* Neuron Data Inc., 1310 Villa Street, Mountain View, CA 94041, USA. http://www.neurondata.com.

Eykhoff, P. 1974. *System identification, parameter and state estimation.* John Wiley.

Fabian, M. & Lennartson, B. 1995. Applying supervisory control theory to discrete event systems modeled by object oriented principles. *In: INRIA/IEEE Conference on Emerging Technologie and Factory Automation EFTA'95.*

Feigenbaum, A. V. 1961. *Total quality control: Engineering and management.* McGraw-Hill.

Fisher, T. G. 1990. *Batch control systems.* Instrument Society of America.

Flanagan, M. J. 1979. Upgrading the activated sludge process through automatic control. *AIChE Symposium Series*, **190**(75), 232–242.

Flanagan, M. J., Bracken, B. D. & Roesler, J. F. 1977. Automatic dissolved oxygen control. *Proc. Am. Soc. Civil Eng., J. of the Env. Eng. Div.*, **103**(EE4), 707–722.

Fletcher, R. 1987. *Practical methods of optimization.* Wiley.

Foley, J. D., van Dam, A., Feiner, S. K. & Hughes, J. F. 1990. *Computer graphics: Principles and practice.* Second edn. Reading, Mass, USA: Addison-Wesley.

Font, R. & Ruiz, F. 1993. Simulation of batch and continuous thickeners. *Chem. Engng. Sc.*, **48**(11), 2039–2047.

Franklin, G. F., Powell, J. D. & Emami-Naeimi, A. 1986. *Feedback control of dynamic systems.* Reading, MA, USA: Addison-Wesley.

Fröse, G. & Köhler, St. 1995. Practical experiences and newer developments with process control via Redox-potential. *In: First IAWQ Specialised Conference for Sensors in Wastewater Technology.*

G2. *Real-time expert system.* Gensym Corporation, 125 Cambridge Park Drive, Cambridge, MA, USA. http://www.gensym.com.

Gabb, D. M. D., Still, D. A., Ekama, G. A., Jenkins, D. & Marais, G. v. R. 1991. The selector effect on filamentous bulking in long sludge age activated sludge systems. *Wat. Sci. Tech.*, **23**(4–6), 867–877.

Gall, B. R. A. & Patry, G. P. 1989. Knowledge-based system for the diagnosis of an activated sludge plant. *Chap. 7 of:* Patry, G. G. & Chapman, D. (eds), *Dynamic Modelling and Expert Systems in Wastewater Engineering.* Chelsea, MI, USA: Lewis Publishers Inc.

Garrett, M. T. 1958. Hydraulic control of activated sludge growth rate. *Swge Indust. Wastes*, **30**, 253–261.

Garvin, D. A. 1984. What does 'product quality' really mean? *Sloan Management Review*, **26**(1), 25–43.

Gear, C. W. 1971. *Numerical initial value problems in ordinary differential equations.* Prentice-Hall.

Gilbert, R. O. 1987. *Statistical methods for environmental pollution monitoring.* New York: Van Nostrand Reinhold Co.

Gill, P. E., Murray, W. & Wright, M. H. 1981. *Practical optimization.* Academic Press.

Gillblad, T. & Olsson, G. 1977. Computer control of a medium sized activated sludge plant. *Progress in Water Technology*, **9**(5/6), 427–433.

Gillum, D. R. 1982. *Industrial level measurement.* Instrument Society of America.

Gillum, D. R. 1984. *Industrial pressure measurement.* Instrument Society of America.

Ginsberg, D. W. & Whiten, W. J. 1991. Cluster analysis for mineral processing applications. *Trans. Instn. Min. Metall. (Sect. C: Mineral process. Extr. Metall.)*, **100**, C139–C146.

Göddertz, J. 1990. *PROFIBUS.* Klockner-Möller GmbH, Postfach 1880, W-5300 Bonn 1, Germany. Reference VER 27.759 GB.

Godfrey, K. (ed). 1993. *Perturbation signals for system identification.* London: Prentice Hall.

Golub, G.H. & Van Loan, C.F. 1983. *Matrix computations.* Oxford, UK: North Oxford Academic.

GPS-X. *Wastewater treatment dynamic simulator.* Hydromantis Inc., 1685 Main Street West, Suite 302, Hamilton, Ontario L8S 1G5, Canada. phone +1 905 522 0012, fax +1 905 522 0031, info@hydromantis.com, http://www.hydromantis.com.

Grady, C.P.L. & Lim, H.C. 1980. *Biological wastewater treatment. theory and applications.* New York: Marcel Dekker.

Graef, S. P. 1972. *Dynamics and control strategies for the anaerobic digester.* Ph.D. thesis, Clemson University, South Carolina, USA.

Graef, S. P. & Andrews, J. F. 1974. Mathematical modelling and control of anaerobic digestion. *AIChE Symposium Series*, **122**, 101–131.

Greenberg, A.E., Clesceri, L.S. & Eaton, A.D. 1992. *Standard methods for the examination of water and wastewater.* eighteenth edn. Washington DC, USA: American Public Health Association.

Griffiths, P. 1994. Modifications to the IAWPRC task group general activated sludge model. *Wat. Res.*, **28**(3), 657–664.

Grijspeerdt, K. & Verstraete, W. 1996. A sensor for the secondary clarifier based on image analysis. *Wat. Sci. Tech.*, **33**(1), 61–70.

Grijspeerdt, K., Vanrolleghem, P. & Verstraete, W. 1995. Selection of one dimensional sedimentation models for on-line use. *Wat. Sci. Tech.*, **31**(2), 193–204.

Grinyer, M. & Goldsmith, H. 1995. *Benchmarking for competitive advantage - facilitators guide.* UK: BBC for Business.

Gujer, W. & Henze, M. 1990. Activated sludge modelling and simulation. *Wat. Sci. Tech.*, **23**, 1011–1023.

Gujer, W. & Kappeler, J. 1992. Modelling population dynamics in activated sludge systems. *Wat. Sci. Tech.*, **25**(6), 93–104.

Gujer, W., Henze, M., Mino, T., Matsuo, T., Wentzel, M. C. & Marais, G. v. R. 1994 (22-24 August). Basic concepts of the activated sludge model no 2: Biological phosphorus removal processes. *In: IAWQ Specialized Seminar, Modelling and Control of Activated Sludge Processes.*

Gupta, A., Flora, J. R. V., Gupta, M., Sayles, G. D. & Suidan, M. T. 1994. Methanogenesis and sulfate reduction in chemostats - I kinetic studies and experiments. *Wat. Res.*, **28**(4), 781–793.

Gustafsson, K. 1994. Traps and pitfalls in simulation. *In: SIMS (Scandinavian Simulation Society) Simulation Conference.*

Gustafsson, L. G., Lumley, D. J., Lindeborg, C. & Haraldsson, J. 1993. Integrating a catchment simulator into wastewater treatment plant operation. *Wat. Sci. Tech.*, **28**(11–12), 45–54.

Hairer, E. & Wanner, G. 1991. *Solving ordinary differential equations II - stiff and differential-algebraic problems.* Springer Series in Computational Mathematics, vol. 14. Springer-Verlag.

Hairer, E., Nørsett, S. P. & Wanner, G. 1987. *Solving ordinary differential equations I - nonstiff problems.* Springer Series in Computational Mathematics, vol. 8. Springer-Verlag.

Hallin, S., Lindberg, C.-F., Pell, M., Plaza, E. & Carlsson, B. 1996. Microbial adaptation process performance and a suggested control strategy in a predenitrifying system with ethanol dosage. *Wat. Sci. Tech.*, **34**(1–2), 91–99.

Halling-Sørensen, B. & Jørgensen, S. E. 1993. *The removal of nitrogen compounds from wastewater.* Amsterdam, The Netherlands: Elsevier Science Publishers B. V.

Hammer, M. 1986. *Water and wastewater technology.* Prentice-Hall.

Harris, T.J., Seppala, C.T. & Desborough, L.D. 1997. A review of performance assessment and process monitoring techniques for univariate and multivariate control systems. *In: ADCHEM 97.*

Härtel, L. & Pöpel, H. J. 1992. A dynamic secondary clarifier model including processes of sludge thickening. *Wat. Sci. Tech.*, **25**(6), 267–284.

Hartigan, J. 1975. *Clustering algorithms.* New York: Wiley.

Hasselblad, S. & Hallin, S. 1996. Intermittent dosage of ethanol in a pre-denitrifying activated sludge process. *Wat. Sci. Tech.*, **34**(1–2), 387–389.

Hasselblad, S. & Xu, S. 1996. On-line estimations of settling capacity in secondary clarifier. *Wat. Sci. Tech.*, **34**(3–4), 323–328.

Hayes-Roth, F., Waterman, D.A. & Lenat, D.B. 1983. *Building expert systems.* Mass., USA: Addison-Wesley.

Haykin, S.S. (ed). 1994. *Neural networks, a comprehensive foundation.* Macmillan International.

Heinzle, E., Dunn, I. J. & Ryhiner, G. B. 1993. Modeling and control for anaerobic wastewater treatment. *Advances in Biochemical Engineering Biotechnology*, **48**, 80–112.

Hellström, B. G. & Bosander, J. 1990. *Styrstrategier för optimering av extern kolkälla vid fördenitrifikation (control strategies for optimization of external carbon in pre-denitrification) (in Swedish).* Tech. rept. SYVAB, Himmerfjärdsverket, S-14792 Grödinge, Sweden.

Hellström, B. G., Vopatek, P. & Österman, A. 1984. Ferrous sulphate - dissolution tanks and a computer application for controlling the dosage at the Himmerfjärden sewage treatment plant. *Vatten*, 40–45.

Henze, M. 1991. Capabilities of biological nitrogen removal processes from wastewater. *Wat. Sci. Tech.*, **23**, 669–679.

Henze, M. 1992. Characterisation of wastewater for modelling of activated sludge processes. *Wat. Sci. Tech.*, **25**(6), 1–15.

Henze, M., Grady Jr., C. P. L., Gujer, W., Marais, G. v. R. & Matsuo, T. 1987a. *Activated sludge model no 1.* IAWPRC Scientific and Technical Reports. London, UK: IAWPRC (IAWQ).

Henze, M., Grady Jr., C. P. L., Gujer, W., Marais, G. v. R. & Matsuo, T. 1987b. A general model for single-sludge wastewater treatment systems. *Wat. Res.*, **21**(5), 505–515.

Henze, M., Gujer, W., Mino, T., Matsuo, T., Wentzel, M. C. & Marais, G. v. R. 1994 (22-24 August). Wastewater and biomass characterization for modelling of biological phosphorus removal in the activated sludge model no 2. *In: IAWQ Specialized Seminar, Modelling and Control of Activated Sludge Processes.*

Henze, M., Harremoes, P., la Cour Jansen, J. & Arvin, E. 1995. *Wastewater treatment, biological and chemical processes.* Germany: Springer Verlag.

Henze, M., Somlyody, L., Schilling, W. & Tyson, J. (eds). 1997. *Sustainable sanitation.* Vol. 35. Wat. Sci. Tech.

Hermanowisz, S. W. 1987. Dynamic changes in populations of the activated sludge community: Effects of dissolved oxygen variations. *Wat. Sci. Tech.*, **19**(5–6), 889–895.

Hickey, R. F. & Goodwin, S. 1989. Anaerobic processes. *J. WPCF*, **61**(6), 814–821.

Hiraoka, M. & Tsumura, K. 1989. System identification and control of the activated sludge process by use of a statistical model. *Wat. Sci. Tech.*, **21**(10–11), 1161–1172.

Hobson, P. N. & Wheatly, A. D. 1992. *Anaerobic digestion: Modern theory and practice.* London: Elsevier.

Hoffman, J.D. 1992. *Numerical methods for engineers and scientists.* USA: McGraw-Hill.

Hofland, A., Morris, A. & Montague, G. 1992. Radial basis function networks applied to process control. *Pages 480–484 of: Proc. American Control Conference.*

Hollnagel, E., Mancini, G. & Woods, D. D. (eds). 1988. *Cognitive engineering in complex dynamic worlds.* London: Academic Press.

Holmberg, A. 1981. Micro-processor based estimation of oxygen utilisation in the activated sludge wastewater treatment process. *International Journal of Systems Science*, **12**(6), 703–718.

Holmberg, A. 1982. On the practical identifiability of microbial growth models incorporating Michaelis-Menten type nonlinearities. *Math. Bioscience*, **62**, 23–43.

Holmberg, A. & Ranta, J. 1982. Procedures for parameter and state estimation in microbial growth process models. *Automatica*, **18**, 181–193.

Holmberg, U. 1986. Adaptive dissolved oxygen control and on-line estimation of oxygen transfer and respiration rates. *In: AIChE Annual Meeting.*

Holmberg, U. 1987. *Adaptive dissolved oxygen control and on-line estimation of oxygen transfer and respiration rates.* M.Phil. thesis, Dept. of Automatic Control, Lund Inst. of Technology, Lund, Sweden. Reference TFRT-3189.

Holmberg, U. & Olsson, G. 1985. Simultaneous online estimation of the oxygen transfer rate and respiration rate. *Pages 185–189 of:* Johnson, A. (ed), *Modelling and Control of Biotechnological Processes.* Oxford: Pergamon Press.

Holmberg, U., Olsson, G. & Andersson, B. 1989. Simultaneous DO control and respiration estimation. *Wat. Sci. Tech.*, **21**(10–11), 1185–1195.

Hoskins, J.C. & Himmelblau, D.M. 1988. Artificial neural network models of knowledge representation in chemical engineering. *Comput. Chem. Engng.*, **12**(9–10), 881–890.

Hudson, M. 1991. Dynamic simulation as a design tool. *The Chemical Engineer (UK)*, 24–27.

Hutchison, J. W. (ed). 1976. *ISA handbook of control valves.* Pittsburgh, PA.: Instrument Society of America.

IAWPR. 1977. International workshop on instrumentation and control for water and wastewater treatment and transport systems. *Progress in Water Technology*, **9**(5–6), 616.

IAWPR. 1981. Practical experiences of control and automation in wastewater treatment and water resources management. *Wat. Sci. Tech.*, **13**(8–12), 845.

IAWPRC. 1985. 4th IAWPRC workshop on instrumentation and control of water and wastewater treatment and transport systems. *In:* Drake, R. A. R. (ed), *Advances in Water Pollution Control.* Pergamon Press.

IAWPRC. 1990. 5th IAWPRC workshop on instrumentation, control and automation of water and wastewater treatment and transport systems. *In:* Briggs, R. (ed), *Advances in Water Pollution Control.* Pergamon Press.

ICCA. 1998. *Responsible care status report 1998.* Tech. rept. http://www.icca-chem.org/rcreport98. International Council of Chemical Associations, London, UK.

IEC1131. 1993. *IEC standard 1131: Programmable controllers Part 1: General information; Part 2: Equipment requirements and tests; Part 3: Programming languages.* International Electrotechnical Commission, Geneva.

IEC1138. 1993. *IEC standard 1158: Fieldbus standard for use in industrial control systems - Part 2: Physical layer specification and service definition.* International Electrotechnical Commission, Geneva.

ILOG. *Rules inference engine.* ILOG Inc., 1901 Landings Drive, Mountain View, CA 94043, USA. http://www.ilog.com.

Iranpour, R., Straub, B. & Jugo, T. 1997. Real-time BOD monitoring for wastewater process control. *J. of Environmental Engineering*, **123**, 155–159.

Iri, M., Aoki, K., O'Shima, E. & Matsuyama, H. 1979. An algorithm for diagnosis of system failures in the chemical process. *Comput. Chem. Engng.*, **3**, 489–493.

Irvine, R. L. & Ketchum Jr., L. H. 1989. Sequencing batch reactors for biological wastewater treatment. *CRC Critical Reviews in Environmental Control*, **18**(4), 255–294.

Isaacs, S. H., Søeberg, H. & Kümmel, M. 1992. On the monitoring and control of a biological nutrient removal process. *Pages 2199–2208 of: Proc. of the 6th Forum of Applied Biotechnology*, vol. 58. Brugges, Belgium: Med. Fac. Landbouww. Univ. Gent.

Isaacs, S. H., Henze, M., Søeberg, H. & Kümmel, M. 1994. External carbon source addition as a means to control an activated sludge nutrient removal process. *Wat. Res.*, **28**, 511–520.

Isaacs, S. H., Nielsen, M. & Hansen, N.P. 1999. STAR solution to variable loads. *Wat. Qual. Int.*, January-February, 42–44.

Ischikawa, M., Simizu, K. & Iwahori, K. 1993. Diagnosis expert system of activated sludge process using biota observed by microscopic examination. *Wat. Sci. Tech.*, **28**(11–12), 231–238.

Ishikawa, K. 1989. How to apply company-wide quality control in foreign countries. *Quality Progress*, **22**, 70–74.

Jacob, A. 1992. Responsible Care - is it working? *Chemical Engineer (London)*, **514**, 25.

Jank, B. (ed). 1993. *6th IAWQ workshop on instrumentation, control and automation of water and wastewater treatment and transport systems.* Vol. 28 (11-12). Wat. Sci. Tech.

Jantzen, J. 1994. *Fuzzy control.* Tech. rept. Electric Power Engineering Dept., Technical University of Denmark.

Jenkins, D., Richard, M. G. & Daigger, G. T. 1993. *Manual on the causes and control of activated sludge bulking and foaming.* Second edn. Lewis Publishers.

Jennings, L.S., Fischer, M.E., Teo, K.L. & Goh, G.J. 1990. *Miser 3: Optimal control software - theory and user manual.* Australia: EMCOSS Pty Ltd.

Jeppsson, U. 1993. *On the verifiability of the activated sludge system dynamics.* M.Phil. thesis, Dept. of Ind. Elec. Eng. and Automation, Lund Inst. of Technology, Lund, Sweden. Reference TEIE-1004.

Jeppsson, U. 1994. *A comparison of simulation software for wastewater treatment processes - a COST 682 program perspective.* Tech. rept. Dept. of Industrial Electrical Engineering and Automation, Lund Inst. of Technology, Lund, Sweden.

Jeppsson, U. 1996. *Modelling aspects of wastewater treatment processes.* Ph.D. thesis, Dept. of Industrial Electrical Engineering and Automation, Lund Inst. of Technology, Lund, Sweden. ISBN 91-88934-00-4, Reference CODEN:LUTEDX/TEIE-1010.

Jeppsson, U. & Diehl, S. 1995. Validation of a robust dynamic model of continuous sedimentation. *In: Proc. Workshop Modelling, Monitoring and Control of Wastewater Treatment Plants.*

Jeppsson, U. & Olsson, G. 1993. Reduced order models for on-line parameter identification of the activated sludge process. *Wat. Sci. Tech.*, **28**(11–12), 173–184.

Ji, S., Vitasovic, Z. & Zhou, S. 1996. A fast hydraulic numerical model for large sewer collection systems. *Wat. Sci. Tech.*, **34**(3–4), 17–24.

Johansson, P. 1994. *SIPHOR - a kinetic model for simulation of biological phosphate removal.* Ph.D. thesis, Dept. of Water and Environment Engineering, Lund Institute of Technology, Lund, Sweden.

Johansson, R. 1993. *System modeling and identification.* Prentice Hall Intl. Editions.

Johnson, R. & Wichern, D. 1992. *Applied multivariate statistical analysis.* Second edn. Hall.

Joint Committee of ASCE and WEF. 1998. *Design of municipal wastewater treatment plants.* American Society of Civil Engineers. ISBN 0-7844-0342-2, 2500 page book & CD-ROM set.

Jones, R. M. 1989. State estimation in wastewater engineering: Application to an anaerobic process. *Environmental Monitoring and Assessment*, **12**, 271–282.

Jones, R. M., MacGregor, J. F., Murphy, K. L. & Hall, E. R. 1992. Towards a useful dynamic model of the anaerobic wastewater treatment process: A practical illustration of process identification. *Wat. Sci. Tech.*, **25**(7), 61–71.

Jönsson, K., Johansson, P., Christensson, M., Lee, N., Lie, E. & Welander, T. 1996. Operational factors affecting enhanced biological phosphorus removal at the waste water treatment plant in Helsingborg, Sweden. *Wat. Sci. Tech.*, **34**(1–2), 67–74.

Juran, J. M. & Gryna, F. M. 1980. *Quality planning and analysis.* Second edn. New York, NY, USA: McGraw-Hill.

Kabouris, J. C. & Georgakakos, A. P. 1990. Optimal control of the activated sludge process. *Wat. Res.*, **24**(10), 1197–1208.

Kabouris, J.C. & Georgakakos, A.P. 1992. Optimal control of the activated sludge process of sludge storage. *Wat. Res.*, **26**(4), 507–517.

Kabouris, J.C. & Georgakakos, A.P. 1996a. Parameter and state estimation of the activated sludge process - I. model development. *Wat. Res.*, **30**(12), 2853–2865.

Kabouris, J.C. & Georgakakos, A.P. 1996b. Parameter and state estimation of the activated sludge process: on-line algorithm. *Wat. Res.*, **30**(12), 3115–3129.

Kaminski Jr., M. A. 1986. Protocols for communicating in the factory. *IEEE Spectrum*, April, 56–62.

Kane, L. (ed). 1987. *Handbook of advanced process control systems and instrumentation.* Houson, USA: Gulf Publishing Co.

Kappeler, J. & Gujer, W. 1992. Estimation of kinetic parameters of heterotrophic biomass under aerobic conditions and characterisation of wastewater for activated sludge modelling. *Wat. Sci. Tech.*, **25**(6), 125–139.

Kappeler, J. & Gujer, W. 1994a. Development of a mathematical model for "aerobic bulking". *Wat. Res.*, **28**(2), 303–310.

Kappeler, J. & Gujer, W. 1994b. Scumming due to *actinomycetes*: towards a better understanding by modelling. *Wat. Res.*, **28**(4), 763–779.

Kappeler, J. & Gujer, W. 1994c. Verification and applications of a mathematical model for "aerobic bulking". *Wat. Res.*, **28**(2), 311–322.

Kavuri, S.N. & Ventkatasubramanian, V. 1993. Representing bounded fault classes using neural networks with ellipsoidal activation functions. *Computers and Chem. Engng.*, **17**(2), 139–163.

Kayser, R. 1990. Process control and expert systems for advanced wastewater treatment plants. *Pages 203–210 of:* Briggs, R. (ed), *Advances in Water Pollution Control.* Pergamon Press.

Keller, J. 1992. *Anaerobic digestion model Part 1: Transfer to NIMBUS.* Tech. rept. Dept. of Chemical Engineering, The University of Queensland.

Keller, J., Lee, P. L., Newland, M. & Greenfield, P. F. 1994. Application of a structural, dynamic model on pilot scale experiments and large scale design. *In: Seventh International Symposium on Anaerobic Digestion AD-94.*

Kendall, M. G. & Ord, J. K. 1990. *Time series.* New York: Oxford University Press.

Kerlin, T. W. & Shepard, R. L. 1982. *Industrial temperature measurement.* Instrument Society of America.

Kiendl, H. & Rüger, J. 1995. Stability analysis of fuzzy control systems using facet functions. *Fuzzy Sets and Systems*, **70**, 275–285.

King, P. J.B. 1990. *Computer and communication systems performance modelling.* Englewood Cliffs, NJ (USA): Prentice Hall Intl. Series in Computer Science.

Klapwijk, A., Spanjers, H. & Temmink, H. 1993. Control of activated sludge plants based on measurement of respiration rates. *Wat. Sci. Tech.*, **28**(11–12), 369–376.

Klapwijk, A., Brouwer, H., Vrolijk, E. & Kujawa, K. 1998. Control of intermittantly aerated nitrogen removal plants by detection endpoints of nitrification and denitrification using respirometry only. *Wat. Res.*, **32**(5), 1700–1703.

Klawonn, F., Kruse, R. & Michels, K. 1997. *Fuzzy-regler (in German).* München, Germany: Carl Hanser Verlag.

Kletz, T.A. 1986. *HAZOP and HAZAN - notes on the identification and assessment of hazards.* Second edn. Rugby, UK: Inst. Chem. Eng.

Klir, G. J. & Folger, T. A. 1988. *Fuzzy sets, uncertainty, and information.* Englewood Cliffs, NJ, USA: Prentice-Hall.

Ko, K. Y. J., McInnis, B. C. & Goodwin, G. C. 1982. Adaptive control and identification of the dissolved oxygen process. *Automatica*, **18**, 727–730.

Kosko, B. 1990. *Neural networks and fuzzy systems.* Prentice-Hall.

Krebs, P. 1991. The hydraulics of final settling tanks. *Wat. Sci. Tech.*, **23**(4–6), 1037–1046.

Krebs, P. 1995. Success and shortcomings of clarifier modelling. *Wat. Sci. Tech.*, **31**(2), 181–191.

Kresta, J., MacGregor, J. F. & Marlin, T. E. 1989. Multivariate statistical monitoring of process operating performance. *In: AIChE Annual Meeting.*

Kristensen, G.H., Jørgensen, P.E. & Henze, M. 1992. Characterisation of functional microorganism groups and substrate in activated sludge and wastewater by AUR, NUR and OUR. *Wat. Res.*, **25**(6), 43–57.

Kruse, R., Gebhardt, J. & Klawonn, F. 1994. *Foundations of fuzzy systems.* New York, USA: Wiley.

Ladiges, G. & Kayser, R. 1993. On-line and off-line expert systems for the operation of wastewater treatment plants. *Wat. Sci. Tech.*, **28**(11–12), 315–323.

Lambert, T.D. 1973. *Computational methods in ordinary differential equations.* UK: Wiley.

Larrea, L., García-Heras, J. L., Ayesa, E. & Florez, J. 1991 (August 21-23). Designing experiments to determine the coefficients of activated sludge models by identification algorithms. *In: IAWQ Seminar on Interactions of Wastewater, Biomass and Reactor Configurations in Biological Treatment Plants.*

Laukkanen, R. & Pursiainen, J. 1991. Rule-based expert systems in the control of wastewater treatment systems. *Wat. Sci. Tech.*, **24**(6), 299–306.

Lear, J.B. 1995. Implementing safe, operable control systems. *Pages 163–167 of: Proc. Control 95.* Barton, Australia: Inst. Eng. (Aust).

Lee, P. L. (ed). 1993. *Nonlinear process control: Applications of generic model control.* London: Springer-Verlag.

Lee, P. L. & Sullivan, G. R. 1988. Generic model control. *Comput. Chem. Engng.*, **12**(6), 573–580.

Lee, P. L., Newell, R. B. & Cameron, I.T. 1998. *Process control and management.* London: Chapman and Hall.

Lee, T. T., Wang, F. Y. & Newell, R. B. 1999. On the modelling and simulation of a BNR activated sludge process based on distributed parameter approach. *Wat. Sci. Tech.*, **39**(6), 79–88.

Lennartson, B., Egardt, B. & Tittus, M. 1994. Hybrid systems in process control. *In: Proc. 33rd CDC.*

Lennartsson, B., Tittus, M., Egardt, B. & Pettersson, S. 1996. Hybrid systems in process control. *IEEE Control Systems*, **16**(5), 45–56.

Leonard, J.A. & Kramer, M.A. 1991. Radial basis function networks for classifying process faults. *IEEE Control Systems*, **11**(3), 31–38.

Leonhard, W. 1997. *Control of electrical drives.* 2nd edn. Springer.

Levenspiel, O. 1972. *Chemical reaction engineering.* Wiley.

Lewis, R.W. 1997. *Programming industrial control systems using IEC 1131-3.* London, United Kingdom: The Institution of Electrical Engineers.

Lind, M. 1987. *Multilevel flow modelling - basic concepts.* Tech. rept. The Servo Laboratory, Technical University of Denmark, Copenhagen.

Lind, M., Harder, E., Jensen, H. & Agger, S. 1987. *Systembeskrivelse og präsentation i proceskontrol, SIP project technical report (in Danish).* Tech. rept. Institute of Automatic Control Systems, Technical University of Denmark, Kobenhavn.

Lindberg, C. F. 1997. *Control and estimation strategies applied to the activated sludge process.* Ph.D. thesis, Dept. of Materials Science, Systems and Control Group, Uppsala Univ., Sweden.

Lindberg, C.-F. & Carlsson, B. 1996a. Adaptive control of external carbon flow rate in an activated sludge process. *Wat. Sci. Tech.*, **34**(3–4), 173–180.

Lindberg, C.-F. & Carlsson, B. 1996b. Estimation of the respiration rate and oxygen transfer function utilizing a slow DO sensor. *Wat. Sci. Tech.*, **33**(1), 325–333.

Lindberg, C. F. & Carlsson, B. 1996c. Nonlinear and set-point control of the dissolved oxygen dynamics in an activated sludge process. *Wat. Sci. Tech.*, **34**(3–4), 135–142.

Linde, M. 1993. *An approach to methanol dosage control in post-denitrification processes.* Tech. rept. Dept. of Industrial Electrical Engineering and Automation, Lund Inst. of Technology, Lund, Sweden.

Liptak, B. 1982. *Instrument engineer's handbook.* PA 19089, USA: Chilton Book Co. ISBN: 0-8019-6971-9.

Ljung, L. 1987. *System identification: Theory for the user.* Prentice-Hall.

Ljung, L. & Söderström, T. 1983. *Theory and practice of recursive identification.* Cambridge, MA, USA: MIT Press.

Londong, J. & Wachtl, P. 1996. Six years of practical experience with the operation of on-line analyzers. *Wat. Sci. Tech.*, **33**(1), 159–164.

Luenberger, D.G. 1971. An introduction to observers. *IEEE Trans. Autom. Control*, **AC-16**, 596–603.

Lumley, D. J., Sahlberg, K.-Å. & Haraldsson, J. 1993. Use of part-time unmanned operation at a large wastewater treatment plant. *Wat. Sci. Tech.*, **28**(11–12), 29–36.

Luyben, W. L. 1990. *Process modelling, simulation and control for chemical engineers.* Second edn. McGraw-Hill.

Lynggaard, A. & Harremoes, P. (eds). 1996. *IAWQ workshop on sensors in waste water technology.* Vol. 33 (1). Wat. Sci. Tech.

Lynggaard, A. & Nielsen, M. K. 1993. Superior Tuning and Reporting (STAR) - a new concept for on-line process control of wastewater treatment plants. *In: 6th IAWPRC conference on Instrumentation and Control.*

Lynggaard-Jensen, A. 1994. Sensors for water quality and their potential. *Pages 162–181 of: Advances in Water Quality Monitoring.* Technical Reports in Hydrology and Water Resources, vol. 42. World Meteorological Organization (WMO).

Lynggaard-Jensen, A. 1995. Status for on-line sensors and automated operation of wastewater treatment plants. *In: Proc. Nordic Seminar Nitrogen Removal from Municipal Wastewater.*

Lynggaard-Jensen, A., Eisum, N. H., Rasmussen, I., H., Svankjaer-Jacobsen & Stenstrøm, T. 1996. Description and test of a new generation of nutrient sensors. *Wat. Sci. Tech.*, **33**(1), 25–35.

Maarleveld, A. & Rijnsdorp, J.E. 1970. Constraint control on distillation columns. *Automatica*, **6**, 51–55.

MacFarlane, A. G. J. 1993. Information, knowledge and control. *Chap. 1 of:* Trentelman, H. L. & Willems, J. C. (eds), *Essays on Control: Perspectives in the Theory and its Applications.* Boston, Basel, Berlin: Birkhäuser.

MacGregor, J. F., Marlin, T. E., Kresta, J. & Skagerberg, B. 1991. Multivariate statistical methods in process analysis and control. *Pages 79–99 of: Proc. 4th Int. Conf. on Chemical Process Control.*

Manning, A. W. & Dobs, D. M. 1980. *Design handbook for automation of activated sludge wastewater treatment plants.* Municipal Environmental Research Laboratory, Cincinnati, OH 45268, USA. Reference USEPA 600/8-80-028.

Manross, R. C. 1983. *Wastewater treatment plant instrumentation handbook.* Water Engineering Research Laboratory, Office of Research and Development, US Environmental Protection Agency, Cincinnati, Ohio 45268, USA. Reference Contract 68-03-3130.

Marais, G. v. R. & Ekama, G. A. 1976. The activated sludge process: Part 1 - steady state behaviour. *Water South Africa*, **2**, 163–200.

Marlin, T. E., Barton, G.W., Brisk, M.L. & Perkins, J.D. 1987a. *Advanced process control project report.* Tech. rept. ISBN 0 86758 264 2. The Warren Centre, University of Sydney, Sydney, Australia.

Marlin, T. E., Perkins, J.D., Barton, G.W. & Brisk, M.L. 1987b. *Advanced process control technical papers*. Tech. rept. ISBN 0 86758 265 0. The Warren Centre, University of Sydney, Sydney, Australia.

Marsili-Libelli, S. 1989. Modelling, identification and control of the activated sludge process. *Advances in Biochemical Engineering/Biotechnology*, **38**, 89–148.

Marsili-Libelli, S. 1990. Adaptive estimation of bioactivities in the activated sludge process. *IEE Proceedings*, **137**(6), 349–356.

Marsili-Libelli, S. 1991. Fuzzy clustering of ecological data. *Chap. 15, pages 173–184 of:* Feoli, E. & Orlóci, L. (eds), *Computer Assisted Vegetation Analysis*. Kluwer.

Marsili-Libelli, S. 1992. Deterministic and fuzzy control of the sedimentation process. *Med. Fac. Landbouww. Univ. Gent*, **57**, 2229–2238.

Marsili-Libelli, S. 1997. Estimation of respirometric activities in bioprocesses. *J. of Biotechnology*, **52**, 181–192.

Marsili-Libelli, S. & Giovannini, F. 1997. On-line estimation of the nitrification process. *Wat. Res.*, **31**(1), 179–185.

Marsili-Libelli, S. & Müller, A. 1996. Adaptive fuzzy pattern recognition in the anaerobic digestion process. *Pattern Recognition Letters*, **17**, 651–659.

Marsili-Libelli, S., Fois, G. & Morneschi, D. 1978. Modeling and control of an activated sludge process. *Pages 805–812 of:* Vansteenkiste (ed), *Modeling, Identification and Control in Environmental Systems*, vol. 1. North-Holland Publishing Company.

Martens, H. & Naes, T. 1989. *Multivariate calibration*. Wiley.

MATLAB. *MATLAB language, Simulink simulator, and Real-Time Workshop*. The MathWorks, Inc., 24 Prime Park Way, Natick, MA 01760-2500, USA. phone +1 508 653 1415, fax +1 508 653 6284, info@mathworks.com, http://www.mathworks.com.

MATRIX$_X$. *MATRIX$_X$ language, SystemBuild modeller, and RealSim simulator*. Integrated Systems, 3260 Jay Street, Santa Clara, CA 95054, USA. phone +1 408 980 1500, fax +1 408 980 0400, http://www.isi.com.

Maurath, P.R., Mellichamp, D.A. & Seborg, D.E. 1985. Predictive controller design by principal components design. *Pages 1059–1065 of: Proc. ACC*.

Maurer, M. & Gujer, W. 1998. Dynamic modelling of enhanced biological phosphorus and nitrogen removal in activated sludge systems. *Wat. Sci. Tech.*, **38**(1), 203–210.

McCarty, P. L. & Mosey, F. E. 1991. Modelling of anaerobic digestion processes (a discussion of concepts). *Wat. Sci. Tech.*, **24**(8), 17–33.

McFarlane, I. 1995. *Automatic control of food manufacturing process.* 2nd edn. London: Blackie.

McMillan, G. & Weiner, S. 1994. *How to become an instrument engineer.* Instrument Society of America.

Metcalf & Eddy Inc. 1972. *Wastewater engineering: Collection, treatment, disposal.* McGraw-Hill.

MINITAB. *Statistics package.* Minitab Inc., 3081 Enterprise Drive, State College, PA 16801-3008, USA. phone +1 800 448 3555, fax +1 814 238 4383, http://www.minitab.com.

Mino, T., van Loosdrecht, M.C.M. & Heijnen, J.J. 1998. Microbiology and biochemistry of the enhanced biological phosphate removal process. *Wat. Res.*, **32**(11), 3193–3207.

Mittal, G.S. (ed). 1997. *Computerized control systems in the food industry.* New York: Marcel Dekker.

Mohan, N., Undeland, T.M. & Robbins, W.P. 1995. *Power electronics.* New York: John Wiley & Sons.

Moore, A. 1986. Selecting a flowmeter. *The Chemical Engineer*, April, 39–45.

Morari, M. 1993. Some control problems in the process industries. *Chap. 3 of:* Trentelman, H. L. & Willems, J. C. (eds), *Essays on Control: Perspectives in the Theory and its Applications.* Boston, Basel, Berlin: Birkhäuser.

Mosey, F. E. 1983. Mathematical modelling of the anaerobic digestion process: Regulatory mechanisms for the formation of short-chain volatile acids from glucose. *Wat. Sci. Tech.*, **15**, 209–232.

Motard, R. L. & Joseph, B. 1994. *Wavelet applications in chemical engineering.* Mass., USA: Kluwer Academic.

Náhlîk, J. & Burianec, Z. 1988. On-line parameter and state estimation of continuous cultivation by extended Kalman filter. *Appl. Microb. and Biotech.*, **28**, 128–134.

Naidoo, V., Urbain, V. & Buckley, C.A. 1998. Characterisation of wastewater and activated sludge from European municipal wastewater treatment plants using the NUR test. *Wat. Sci. Tech.*, **38**(1), 303–310.

Naidu, S. R., Zafiriou, E. & McAvoy, T. J. 1990. Use of neural networks for sensor failure detection in a control system. *IEE Control Systems*, **10**(3), 49–55.

Nakazawa, H. & Tanaka, K. 1991. Kinetic model of sequencing batch activated sludge process for municipal wastewater treatment. *Environmental Science and Technology*, **23**, 1097.

Nam, W., Nam, J.M. & Lee, K.S. 1996. On-line integrated control system for an industrial activated sludge process. *Wat. Env. Res.*, **68**(1), 70–75.

Negoită, C. V. 1985. *Expert systems and fuzzy systems.* Menlo Park, Ca. USA, London, Amsterdam, Sydney: The Benjamin/Cummings Publ. Comp.

Nelder, J.A. & Mead, R. 1965. A simplex method for function minimisation. *The Computer Journal*, **7**, 308.

Newell, R. B. & Cameron, I. T. 1991. *NIMBUS dynamic process simulator user's manual.* Computer Aided Process Engng. Centre, The University of Queensland, Queensland 4072, Australia. http://daisy.cheque.uq.edu.au/Nimbus.

Newell, R.B. & Lee, P.L. 1989. *Applied process control - a case study.* Prentice-Hall.

Newell, R.B., Bailey, J., Islam, A., Hopkins, L., Steffens, M.A. & Lant, P.A. 1997. Characterising bioreactor mixing with RTD tests. *In: IAWQ 7th Int. Workshop on Instrumentation, Control and Automation of Water and Wastewater Treatment and Transport Systems.*

Nie, M. & Xu, S. 1991. Technical and economic analysis of stabilisation ponds. *Wat. Sci. Tech.*, **24**(5), 55–62.

Nielsen, M. K. & Onnerth, T. B. 1995. Improvement of a recirculating plant by introducing STAR control. *Wat. Sci. Tech.*, **31**(2), 171–180.

Nielsen, M. K. & Önnerth, T. B. 1996. Strategies for handling of on-line information for optimising nutrient removal. *Wat. Sci. Tech.*, **33**(1), 211–222.

Nielsen, M. K., Persson, O. & Kümmel, M. 1981. Computer control of nitrifying and denitrifying activated sludge process. *Wat. Sci. Tech.*, **13**(9), 285–291.

Nielsen, M. K., Carstensen, J. & Harremoes, P. 1996. Combined control of sewer and treatment plant during rainstorm. *Wat. Sci. Tech.*, **34**(3–4), 181–188.

NIMBUS. *Dynamic process simulator.* The University of Queensland, Queensland 4072, Australia. http://daisy.cheque.uq.edu.au/Nimbus.

Nimmo, I. 1994. Extend HAZOP to computer control systems. *Chem. Engng. Prog.*, **90**, 32–44.

Novotny, V., Jones, H., Feng, X. & Capodaglio, A. 1991. Time series analysis models of activated sludge plants. *Wat. Sci. Tech.*, **23**(4–6), 1107–1116.

Nyberg, U., Aspegren, H. & Andersson, B. 1993. Integration of on-line instruments in the practical operation of the Klagshamn wastewater treatment plant. *Pages 2019–2028 of: Proc. Workshop Modelling, Monitoring and Control of Wastewater Treatment Plants*, vol. 58. Med. Fac. Landbouww., Univ. Gent. also in Vatten, 49 (4) 235-244.

Nyberg, U., Andersson, B. & Aspegren, H. 1996a. Experiences with on-line measurements at a wastewater treatment plant for extended nitrogen removal. *Wat. Sci. Tech.*, **33**(1), 175–182.

Nyberg, U., Andersson, B. & Aspegren, H. 1996b. Real time control for minimizing effluent concentrations during storm water events. *Wat. Sci. Tech.*, **34**(3–4), 127–134.

Ødegaard, H. 1995. An evaluation of cost efficiency and sustainability of different wastewater treatment processes. *Vatten*, 291–299.

Oles, J. & Wilderer, P. A. 1991. Computer aided design of sequencing batch reactors based on the IAWPRC activated sludge model. *Wat. Sci. Tech.*, **23**, 1087–1095.

Olsson, G. 1977. State of the art in sewage treatment plant control. *AIChE Symposium Series*, **72**, 52–76.

Olsson, G. 1985a. *Analys av mätdata från eskilstuna reningsverk (in Swedish)*. Tech. rept. Swedish Environmental Protection Board (SNV) and Swedish Water and Waste Water Works Association (VAV), Lund, Sweden. Sewage Works Evaluation Project, Document 45.

Olsson, G. 1985b. Control strategies for the activated sludge process. *Chap. 65, pages 1107–1119 of:* Moo-Young, M. (ed), *Comprehensive Biotechnology*. Pergamon Press.

Olsson, G. 1989. Practical experiences of identification and modelling from experiments. *Chap. 10 of:* Patry, G. G. & Chapman, D. (eds), *Dynamic Modelling and Expert Systems in Wastewater Engineering*. Chelsea, MI, USA: Lewis Publishers Inc.

Olsson, G. 1993. Advancing ICA technology by eliminating the constraints. *Wat. Sci. Tech.*, **28**(11–12), 1–7.

Olsson, G. 1996. Programmable controllers. *Chap. 18 of:* Levine, William (ed), *Control Handbook*. CRC Press and IEEE Press.

Olsson, G. & Andrews, J. F. 1978. The dissolved oxygen profile - a valuable tool for control of the activated sludge process. *Wat. Res.*, **12**, 985–1004.

Olsson, G. & Chapman, D. 1985. Modelling the dynamics of clarifier behaviour in activated sludge systems. *Pages 405–412 of:* Drake, R. A. R. (ed), *Advances in Water Pollution Control.* Pergamon Press.

Olsson, G. & Hansson, O. 1976. Modelling and identification of an activated sludge process. *In: Proc. IFAC Symp. on Identification and System Parameter Estimation.*

Olsson, G. & Piani, G. 1992. *Computer systems for automation and control.* Prentice-Hall.

Olsson, G. & Stephenson, J. 1985. The propagation of hydraulic disturbances and flow rate reconstruction in activated sludge plants. *Environmental Technology Letters*, **6**, 536–545.

Olsson, G., Holmberg, U. & Wikström, A. 1985a. A model library for dynamic simulation of activated sludge systems. *Pages 721–728 of:* Drake, R. A. R. (ed), *Advances in Water Pollution Control, IAWPRC.* Houston, Texas, USA: Pergamon Press.

Olsson, G., Rundqwist, L., Eriksson, L. & Hall, L. 1985b. Self-tuning control of the dissolved oxygen concentration in activated sludge systems. *Pages 473–480 of:* Drake, R. A. R. (ed), *Advances in Water Pollution Control, IAWPRC.* Houston, Texas, USA: Pergamon Press.

Olsson, G., Stephenson, J. & Chapman, D. 1986. Computer detection of the impact of hydraulic shocks on plant performance. *J. WPCF*, **58**, 954–959.

Olsson, G., Aspegren, H. & Nielsen, M. K. 1997. Operation and control of wastewater treatment - a Scandinavian perspective over 20 years. *In: IAWQ 7th Int. Workshop on Instrumentation, Control and Automation of Water and Wastewater Treatment and Transport Systems.*

Önnerth, T. B., Nielsen, M. K. & Stamer, C. 1996. Advanced computer control based on real and software sensors. *Wat. Sci. Tech.*, **33**(1), 237–246.

Ossenbruggen, P. J., Spanjers, H., Aspegren, H. & Klapwijk, A. 1991. Designing experiments for model identification of the nitrification process. *Wat. Sci. Tech.*, **24**(6), 9–16.

Ozog, H. & Bendixen, L.M. 1987. Hazard identification and quantification. *Chem. Engng. Prog.*, **83**, 55–64.

Parker, M. J., Casey, R. J., Reynolds, L. K., de Vries, R. R., Brueck, T. M. & Williams, C. N. 1993. Functional approach leads to successful ICA across 124 water and 389 wastewater treatment works. *Wat. Sci. Tech.*, **28**(11–12), 37–43.

Patry, G. G. & Olsson, G. 1987. Dynamic modeling and expert systems in wastewater engineering. *Wastewater Technology Centre Newsletter*, **August**.

Patry, G. G. & Takács, I. 1990. Modular/multi-purpose modelling system for the simulation and control of wastewater treatment plants: an innovative approach. *Pages 385–392 of:* Briggs, R. (ed), *Advances in Water Pollution Control.* Pergamon Press.

Pavlostathis, S. G. & Giraldo-Gomez, E. 1991. Kinetics of anaerobic treatment: A critical review. *Crit. Rev. Environ. Control*, **21**(5–6), 411–490.

Pearson, K. 1901. On lines and planes of closest fit to systems of points in space. *Philosophical Magazine*, **2**, 559–572.

Pedersen, K. M., Kümmel, M. & Søeberg, H. 1990. Monitoring and control of biological removal of phosphorus and nitrogen by flow injection analysers in a municipal pilot-scale waste-water treatment plant. *Anal. Chim. Acta*, **238**, 191–199.

Pernitsky, D.J., Finch, G.R. & Huck, P.M. 1995. Disinfection kinetics of heterotrophic plate count bacteria in biologically treated potable water. *Wat. Res.*, **29**(5), 1235–1241.

Pessen, D. W. 1989. *Industrial automation: Circuit design and components.* New York: John Wiley & Sons.

Plission-Saune, S., Capdeville, B., Mauret, M., Deguin, A. & Baptise, P. 1996. Real time control of nitrogen removal using three ORP bending-points: Signification, control strategy and results. *Wat. Sci. Tech.*, **33**(1), 275–280.

Pollard, P. 1987. Dialysis: a simple method of separating labelled bacterial DNA and tritiated thymidine from aquatic sediments. *J. Microbiol. Methods*, **8**, 91–101.

Pollard, P. & Greenfield, P. F. 1997. Measuring *in situ* bacterial specific growth rates and population dynamics in wastewater. *Wat. Res.*, **31**(5), 1074–1082.

Pons, M. N., Potier, O., Roche, N., Colin, F. & Prost, C. 1993. Simulation of municipal wastewater treatment plants by activated sludge. *Computers and Chem. Engng.*, **17**(5), 227–232.

Press, W. H., Flannery, B. P., Teukolsky, S. A. & Vetterling, W. T. 1986. *Numerical recipes.* Cambridge University Press.

Preul, H.C. & Wagner, R.E. 1987. Waste stabilisation pond prediction model. *Wat. Sci. Tech.*, **19**(12), 205–211.

Pullammanappallil, P., Svoronos, S. A., Lyberatos, G. & Chynoweth, D. P. 1992. An expert system for the control of anaerobic digesters. *Pages 1911–1916 of: Proc. ACC.*

Qasim, S. R. 1999. *Wastewater treatment plants: Planning, design and operation.* Second edn. Technomic Publishing.

Ramsay, I.R., Pullammanappallil, P., Newell, R.B., Lee, P.L. & Keller, J. 1994 (September 25-28). Basis for dynamic modelling of anaerobic degradation of protein-based wastewater. *Pages 235–242 of: Proc. 22nd Australasian Chem. Engng. Conf.*

Randall, C.W., Barnard, J.L. & Stensel, H.D. (eds). 1992. *Design and retrofit of wastewater treatment plants for biological nutrient removal.* Lancaster, Pa.: Technomic Publishing Coy.

Rasmussen, J. & Lind, M. 1981. *Coping with complexity.* Tech. rept. Risø National Laboratory, DK-4000 Roskilde, Denmark.

Reichert, P. 1994. Aquasim - a tool for simulation and data analysis of aquatic systems. *Pages 21–30 of: Water Quality International 94, IAWQ 17th Biennial Int. Conf.*, vol. 2.

Reklaitis, G.V., Ravindran, A. & Ragsdell, K.M. 1983. *Engineering optimisation - methods and applications.* USA: Wiley.

Reynolds, D. M. & Ahmad, S. R. 1997a. The absorbance and fluorescence properties of wastewater. *Int. Environmental Technology*, **7**(3), 17–19.

Reynolds, D. M. & Ahmad, S. R. 1997b. Rapid and direct determination of waste water BOD values using a fluorescence technique. *Wat. Res.*, **31**(8), 2012–2018.

Rhinehart, R. R. 1995. A watchdog for controller performance monitoring. *In: Proceedings of ACC 95.* Paper TM10-6.

Richalet, J., Rault, A., Testud, J.L. & Papon, J. 1978. Model predictive heuristic control: applications to industrial processes. *Automatica*, **14**, 413–428.

Ristuccia, R. M. & Searle, J. 1985. A comparison of mycin and inferno as reasoning mechanisms. *Aust. Comp. J.*, **17**(1), 14.

Rivera, D. E., Morari, M. & Skogestad, S. 1986. Internal model control 4. PID controller design. *Ind. Eng. Chem. Process Des. Dev.*, **25**(1), 252–265.

Rodrigues, J. 1994. *Directory of simulation software.* Vol. 5. The Society for Computer Simulation. ISBN 1-56555-064-1.

Rosén, C. 1998. *Monitoring wastewater treatment systems.* Licentiate thesis, Dept. of Ind. Elec. Eng. and Automation, Lund Inst. of Technology, Lund, Sweden. Reference ISBN 91-88934-10-1.

Rosén, C. & Olsson, G. 1997. Disturbance detection in wastewater treatment systems. *In: IAWQ 7th Int. Workshop on Instrumentation, Control and Automation of Water and Wastewater Treatment and Transport Systems.*

Rosenof, H. P. & Ghosh, A. 1987. *Batch process automation.* van Nostrand-Reinhold.

Rossi, A. P. & Figueroa, J. L. 1997. Economic performance of optimal linear control in process industries: A case study. *Latin American Applied Research*, **27**(4), 235–243.

Rundqwist, L. 1988. *Self-tuning control of the dissolved oxygen concentration in the Käppala plant.* Tech. rept. Dept. of Automatic Control, Lund Inst. of Technology, Lund, Sweden. Reference TFRT-7383.

Ryhiner, G. B., Heinzle, E. & Dunn, I. J. 1993. Modeling and simulation of anaerobic wastewater treatment and its application to control design: Case whey. *Biotech. Prog.*, **9**(3), 332–343.

Sackellares, R. W., Barkley, W. A. & Hill, R. D. 1987. Development of a dynamic aerated lagoon model. *J. WPCF*, **59**(10), 877.

Sahm, H. 1984. Anaerobic wastewater treatment. *Adv. in Biochem. Biotechnol.*, **29**, 83–115.

Sasaki, K., Yamamoto, Y., Tsumura, K., Hatsumata, S. & Tatewaki, M. 1993. Simultaneous removal of nitrogen and phosphorus in intermittently aerated 2-tank activated sludge process using DO and ORP bending point control. *Wat. Sci. Tech.*, **28**(11–12), 513–521.

Sbarbaro-Hofer, D., Neumerkel, D. & Hunt, K. 1992. Neural control of a steel rolling mill. *Pages 122–127 of: IEEE Int. Symp. on Intelligent Control.*

Schenck, H. 1968. *Theories of engineering experimentation.* New York: McGraw-Hill.

Schink, B. 1988. Principles and limits of anaerobic degradation: Environmental and technological aspects. *In:* Zehnder (ed), *Biology of Anaerobic Microorganisms.* Wiley.

Schlegel, S. & Baumann, P. 1996. Requirements with respect to on-line analyzers for N and P. *Wat. Sci. Tech.*, **33**(1), 139–146.

Seborg, D. E., Edgar, T. F. & Mellichamp, D. A. 1989. *Process dynamics and control.* Wiley.

Self, K. 1990. Designing with fuzzy logic. *IEEE Spectrum*, **27**(11), 42–105.

Shinners, S.M. 1992. *Modern control system theory and design*. New York: Wiley-Interscience.

Shiozaki, J., Matsuyama, H., O'Shima, E. & Iri, M. 1979. An improved algorithm for diagnosis of system failures in the chemical process. *Comput. Chem. Engng.*, **9**(3), 285–293.

Shortcliffe, E. H. 1978. *Medical consultation systems: Designing for doctors*. Tech. rept. Report HPP-78-29. Stanford University, Stanford, California.

Simba. *Wastewater collection and treatment simulator*. Otterpohl Wasserkonzepte GbR, Kanalstrasse 52, D-23552 Lübeck, Germany. see also Institut für Automation und Kommunikation (IFAK), Germany, phone +49 39203 81044, ali@ifak.fhg.de.

SIMCA. *SIMCA-P and SIMCA 4000 data analysis packages*. Erisoft AB, Box 920, S-97128 Luleå, Sweden, or Umetri AB, Box 7960 S-90719 Umeå, Sweden. Erisoft: phone +46 920 42700, fax +46 920 97545, e-mail: mvainfo@lu.erisoft.se. Umetri: phone +46 90 154840, fax +46 90 197685, e-mail: info@umetri.se, www: http://www.umetri.se.

Simnon. *Simnon - user's guide for UNIX systems*. SSPA Maritime Consulting AB, P.O. Box 24001, S-400 22 Gothenburg, Sweden. phone +46 31 772 9000, fax +46 31 771 9124, simnon@sspa.se.

Smith, G.D. 1978. *Numerical solution of PDEs: finite difference methods*. UK: Oxford.

Söderström, T. & Stoica, P. 1989. *System identification*. Prentice-Hall.

Sollfrank, U. & Guyer, W. 1990. simultaneous determination of oxygen uptake rate and oxygen transfer coefficient in activated sludge systems by an on-line method. *Wat. Res.*, **24**(6), 725–732.

Sørensen, J., Thornberg, D. E. & Nielsen, M. K. 1994. Optimisation of a nitrogen removing biological wastewater treatment plant using on-line measurements. *WEF Research Journal*, **66**(3), 236–242.

Sorour, M. T., Olsson, G. & Somlyody, L. 1993. Potential use of step feed control using the biomass in the settler. *Wat. Sci. Tech.*, **28**(11–12), 239–248.

Spanjers, H. 1993. *Respirometry in activated sludge*. Ph.D. thesis, Wageningen Agricultural University, Wageningen, The Netherlands.

Spanjers, H., Olsson, G. & Klapwijk, A. 1993. Determining influent short-term biochemical oxygen demand by combined respirometry and estimation. *Wat. Sci. Tech.*, **28**(11–12), 401–414.

Spanjers, H., Olsson, G. & Klapwijk, A. 1994. Determining short-term biochemical oxygen demand and respiration rate in an aeration tank by using respirometry and estimation. *Wat. Res.*, **28**, 1571–1583.

Spanjers, H., Vanrolleghem, P., Olsson, G. & Dold, P. 1996. Respirometry in control of activated sludge processes. *Wat. Sci. Tech.*, **34**(3–4), 117–126.

Spanjers, H., Vanrolleghem, P., Olsson, G. & Dold, P. 1998. *Respirometry in control of activated sludge processes.* IAWQ Scientific and Technical Reports. London, UK: IAWQ.

SPEEDUP. *Dynamic process simulator.* Aspen Technology Inc., Ten Canal Park, Cambridge, Mass. 02141-2201, USA. phone +1 617 949 1000, fax +1 617 949 1030, info@aspentech.com, http://www.aspentech.com.

Speirs, G. W. & Stephenson, J. P. 1985. Operational audit assists plant selection for demonstration of energy savings and improved control. *Pages 457–464 of: Proc. of IAWPRC Conference on Instrumentation and Control of Water and Wastewater Treatment and Transport Systems.* New York: Pergamon Press.

SSSP. *Simple activated sludge simulator.* Prof. C. P. Les Grady, Clemson University, Clemson, SC 29634-0919, USA.

Stearns, S. D. & David, R. A. 1988. *System identification.* Prentice-Hall.

Stephanopoulos, G. 1984. *Chemical process control - an introduction to theory and practice.* Prentice-Hall.

STOAT. *Dynamic modelling of sewage treatment works and process modelling software.* WRc plc, Software Services, Frankland Road, Blagrove, Swindon, Wiltshire SN5 8YF, UK. phone +44 1793 511711, fax +44 1793 511712, stoat@wrcplc.co.uk.

Sullivan, L. P. 1986. The seven stages in company-wide quality control. *Quality Progress*, **19**, 77–83.

Sweeney, M. & Manning, A. 1993. Realizing the benefits of enterprise-wide computing. *Wat. Sci. Tech.*, **28**(11–12), 21–27.

Switzenbaum, M. S., Plante, T. R. & Woodworth, B. K. 1992. Filamentous bulking in Massachusetts: Extent of the problem and case studies. *Wat. Sci. Tech.*, **25**(4–5), 265–271.

Taiwo, O. 1993. Comparison of four methods of on-line identification and controller tuning. *IEE Proceedings-D*, **140**(5), 323–327.

Takács, I., Patry, G. G. & Nolasco, D. 1991. A dynamic model of the clarification/thickening process. *Wat. Res.*, **25**(10), 1263–1271.

Tanenbaum, A. S. 1996. *Computer networks.* Third edn. Englewood Cliffs, NJ, USA: Prentice-Hall.

Teichgräber, B. 1993. Control strategies for a highly loaded biological ammonia elimination process. *Wat. Sci. Tech.*, **28**(11–12), 531–538.

Temmink, H., Vanrolleghem, P., Klapwijk, A. & Verstraete, W. 1993. Biological early warning systems for toxicity based on activated sludge respirometry. *Wat. Sci. Tech.*, **28**(11–12), 415–426.

Thomsen, H. A. & Kisbye, K. 1996. N and P on-line meters: Requirements, maintenance and stability. *Wat. Sci. Tech.*, **33**(1), 147–157.

Thomsen, H. A. & Nielsen, M. K. 1996 (April 8-10). Practical experience with on-line measurements of NH4, NO3, PO4, Redox, MLSS and SS in advanced activated sludge plants. *Pages 378–388 of: Proc. of the HYDROTOP Conf., Colloquium "The City and the Water"*, vol. 2.

Thornberg, D. E. 1988. Nitrogen removal by computer control of a simple activated sludge plant. *Water Supply*, **6**, 361–369.

Thornberg, D. E., Nielsen, M. K. & Andersen, K. L. 1993. Nutrient removal, on-line measurements and control strategies. *Wat. Sci. Tech.*, **28**(11–12), 549–560.

Tong, R. M., Beck, M. B. & Latten, A. 1980. Fuzzy control of the activated sludge wastewater treatment process. *Automatica*, **16**, 695–701.

Tsai, Y. P., Ouyang, C. F., Wu, M. Y. & Chiang, W. L. 1993. Fuzzy control of a dynamic activated sludge process for the forecast and control of effluent suspended solids concentration. *Wat. Sci. Tech.*, **28**(11–12), 355–367.

Tsai, Y.P., Ouyang, C.F., Chiang, W.L. & Wu, M.Y. 1994. Construction of an on-line fuzzy controller for the dynamic activated sludge process. *Wat. Res.*, **28**(4), 913–921.

United States Environmental Protection Agency. 1989. *Fine pore aeration systems.* United States Environmental Protection Agency. Reference EPA/625/1-89/023.

US Congress Subcommittee on Energy and the Environment. 1979. *Accident at the Three Mile Island nuclear power plant.* Hearings May 9th, 10th, 11th and 15th, 1979. Washington DC.

van der Roest, H. & Uijterlinde, C. 1996. Model answer. *Wat. Qual. Int.*, March-April, 13–15.

van Dongen, G. & Geuens, L. 1998. Multivariate time series analysis for design and operation of a biological wastewater treatment plant. *Wat. Res.*, **32**(3), 691–700.

Van-Niekerk, A. M., Jenkins, D. & Richards, M. G. 1988. A mathematical model of the carbon-limited growth of filamentous and floc-forming organisms in low F/M sludge. *J. WPCF*, **60**(1), 100–106.

Vanrolleghem, P. & Jeppsson, U. 1994. Simulators for modelling of WWTP. *Chap. 8 of:* Dochain, D., Vanrolleghem, P. & Henze, M. (eds), *COST 682 Environment, Report 1992-95.* Rue de la Loi 200, B-1049 Brussels: European Commission. ISSN 1018-5593.

Vanrolleghem, P. & Verstraete, W. 1993. On-line monitoring equipment for wastewater treatment processes: State of the art. *In: Proc. TI-KVIV Studiedag Optimalisatie van Waterzuiveringsinstallaties door Proceskontrole en Sturing.*

Vanrolleghem, P., Vermeersch, L. K., Dochain, D. & Vansteenkiste, G. C. 1993. Modelling of a nonlinear distributed parameter bioreactor, optimisation of a nutrient removal process. *Pages 563–567 of:* Pavé, A. (ed), *Modelling and Simulation.* San Diego, CA, USA: SCS.

Vanrolleghem, P., Jeppsson, U., Carstensen, J., Carlsson, B. & Olsson, G. 1996a. Integration of wastewater treatment plant design and operation - a systematic approach using cost functions. *Wat. Sci. Tech.*, **34**(3–4), 159–172.

Vanrolleghem, P., van der Schueren, D., Krikilion, G., Grijspeerdt, K., Willems, P. & Verstraete, W. 1996b. On-line quantification of settling properties with in-sensor-experiments in an automated settlometer. *Wat. Sci. Tech.*, **33**(1), 37–51.

Vanrolleghem, P. A. & Verstraete, W. 1992. Simultaneous biokinetic characterization of heterotrophic and nitrifying populations of activated sludge with an on-line respirographic biosensor. *Wat. Sci. Tech.*, **28**(11–12), 415–426.

Vanrolleghem, P. A., Van Impe, J. F., Vandewalle, J. & Verstraete, W. 1991 (November 6-8). Model based monitoring and control of activated sludge wastewater treatment processes, Part I: On-line estimation of crucial biological variables with the RODTOX. *In: Proc. of the European Simulation Symp.*

Vitasovic, Z. 1986. *An integrated control strategy for the activated sludge process.* Ph.D. thesis, Rice University, Houston, Texas, USA.

Vitasovic, Z. C., Zhou, S., McCorquodale, J. A. & Lingren, K. 1997. Secondary clarifier analysis using data from the clarifier research technical committee protocol. *Water Environment Research*, **69**(5), 999–1007.

Vuillot, M. & Boutin, C. 1987. Waste stabilisation ponds in Europe: a state of the art review. *Wat. Sci. Tech.*, **19**(12), 1–6.

Wacheux, H., Da Silva, S. & Lesavre, J. 1993. Inventory and assessment of automatic nitrate analyzers for urban sewage works. *Wat. Sci. Tech.*, **28**(11–12), 489–498.

Wacheux, H., Million, J.-L., Guillo, C. & Alves, E. 1996. NH4 automatic analysers for wastewater treatment plant: Evaluation test at laboratory and field level. *Wat. Sci. Tech.*, **33**(1), 193–201.

Walpole, R. E. & Myers, R. H. 1989. *Probability and statistics for engineers and scientists.* New York: Macmillan.

Walsh, G. R. 1985. *An introduction to linear programming.* Second edn. Wiley.

Wang, F.Y. & Cameron, I.T. 1994. Control studies on a model evaporation process - constrained state driving with conventional and higher relative degree systems. *J. Process Control*, **4**(2), 59–75.

Wareham, D. G., Hall, K. J. & Mavinic, D. S. 1993. Real-time control of wastewater treatment systems using oxidation-reduction potential (ORP). *Wat. Sci. Tech.*, **28**(11–12), 273–282.

Watanabe, S., Baba, K., Yoda, M., Wu, W., Enbutsu, I., Hiraoka, M. & Tsumura, K. 1993. Intelligent operation support system for activated sludge process. *Wat. Sci. Tech.*, **28**(11–12), 325–332.

Waterman, D.A. 1986. *A guide to expert systems.* Reading, Mass.: Addison-Wesley.

Wells, C. H. 1979. Computer control of fully nitrifying activated sludge processes. *Instrumentation Technology*, **4**, 32–36.

WEST. *Waste-water treatment plant engine for simulation and training.* Hemmis n.v., Koning Leopold III laan 2, 8500 Kortrijk, Belgium. phone +32 56 372637, fax +32 56 372324, info@hemmis.be.

Wickens, C. 1992. *Engineering psychology and human performance.* USA: Harper-Collins.

Willis, M.J., Di Massimo, C., Montague, G. A., Tham, M. T. & Morris, A. J. 1991. Artificial neural networks in process engineering. *IEE Proceedings D*, **138**(3), 256–266.

Willsky, A. S. 1976. A survey of design methods for failure detection systems. *Automatica*, **12**, 601–611.

Winter, M. & Manning, A. 1993. Critical variables in system success: Top managements' commitment is key. *Wat. Sci. Tech.*, **28**(11–12), 15–19.

Witteborg, A., van der Last, A., Hamming, R. & Hemmers, I. 1996. Respirometry for determination of the influent S_S-concentration. *Wat. Sci. Tech.*, **33**(1), 311–323.

Wold, H. 1982. Soft modelling, the basic design and some extensions. *In:* Joreskog, K. & Wold, H. (eds), *Systems under Indirect Observation.* Amsterdam: North Holland.

Wold, S., Ruhe, A., Wold, H. & Dunn, W. 1984. The colinearity problem in linear regression: The partial least squares approach to generalized inverses. *SIAM J. Sci. Stat. Comput.*, **5**(3), 735–743.

Wold, S., Esbensen, K. & Geladi, P. 1987. Principal component analysis. *Pages 37–52 of: Chemometrics and Intelligent Laboratory Systems*, vol. 2. The Netherlands: Elsevier Science Publishers.

Wood, G. G. 1988. International standards emerging for Fieldbus. *Control Engineering*, October, 22–25.

Wood, M.G., Greenfield, P.F., Howes, T., Johns, M.R. & Keller, J. 1995. Computational fluid dynamic modelling of wastewater ponds to improve design. *Wat. Sci. Tech.*, **31**(12), 111–118.

Wood, M.G., Howes, T., Keller, J. & Johns, M.R. 1998. Two dimensional computational fluid dynamic models for waste stabilisation ponds. *Wat. Res.*, **32**(3), 958–963.

Yust, L. J., Stephenson, J. P. & Murphy, K. L. 1981. Dynamic step feed control for organic carbon removal in a suspended growth system. *Wat. Sci. Tech.*, **13**(11–12), 729–736.

Yust, L. J., Stephenson, J. P. & Murphy, K. L. 1984. Control of the specific oxygen utilisation rate for the step-feed activated-sludge process. *Trans. of the Inst. of Measurement and Control*, **6**(3), 165–172.

Zadeh, L. A. 1984. Making computers think like people. *IEEE Spectrum*, August, 26–32.

Zayac, J. L., Dolan, R. J. & Pearl, J. C. 1996. A quality initiative for a sanitary district. *Operations Forum*, **13**(1).

Zhao, H., Isaacs, S. H., Søeberg, H. & Kümmel, M. 1994a. Modeling and identification of an alternating activated sludge process. *Pages 795–801 of: Proc. of PSE 94.*

Zhao, H., Isaacs, S. H., Søeberg, H. & Kümmel, M. 1994b. A novel control strategy for improved nitrogen removal in an alternating activated sludge process - Part I: Process analysis; Part II: Control development. *Wat. Res.*, **28**, 521–542.

Zimmermann, H.-J. 1991. *Fuzzy set theory and its applications.* Boston: Kluwer.

Index